AMERICA'S CHILDREN
AND THE ENVIRONMENT

THIRD EDITION

EPA 240-R-13-001
JANUARY 2013

I am pleased to present the U.S. Environmental Protection Agency's *America's Children and the Environment, Third Edition*. This report marks the important progress we have made as a nation to reduce environmental risks to children's health.

The report contains good news for children and families including significant improvements in the quality of the air we breathe, substantial decreases in childhood blood lead levels, and a steady reduction in children's exposure to secondhand smoke. We are encouraged by these findings. We also know that there is still much work to be done, including further research on the causes of increases in asthma rates, the potential impacts of early life exposures to chemicals, and disease disparities in minority children and children in low-income families. *America's Children and the Environment* will help focus our efforts in addressing these challenges and others.

Protecting children's health is central to the EPA's mission, and the agency has taken great strides to improve the environment for children where they live, learn, and play, including:

- **Finalizing the Mercury and Air-Toxics Standards Rule** to limit mercury and other air toxics emissions from electric generating utilities. These new standards address the largest remaining domestic source of mercury emissions to the environment—a well known neurotoxin in children. The controls put in place by these standards will also avoid 130,000 asthma attacks every year—which disproportionately impact children especially in underserved communities.
- **Updating the National Ambient Air Quality Standard for fine-particle pollution (PM$_{2.5}$)** to improve public health protection. Exposure to PM$_{2.5}$ can aggravate asthma and lead to other respiratory symptoms in children.
- **Establishment of new National Ambient Air-Quality Standards for nitrogen dioxide (NO$_2$) and sulfur dioxide (SO$_2$)**, and a network of monitors to limit near roadway exposures to NO$_2$. These new standards will limit respiratory-related emergency room visits and hospital admissions and will improve public health protection, especially for children, the elderly, and people with asthma.
- **Supporting cutting-edge research through the Centers for Children's Environmental Health and Disease Prevention Research**, in partnership with the National Institute of Environmental Health Sciences, to enhance scientific understanding of the relationships between environmental contaminants and children's health.
- **Launching new voluntary guidelines** that promote environmentally safe siting of schools and the establishment of school environmental health programs by states.
- **Working with other federal agencies to develop and implement the Coordinated Federal Action Plan to Reduce Racial and Ethnic Asthma Disparities** to reduce the disproportionate impact of asthma on minority and low-income children.

As we move forward, the EPA is committed to continuing the success of our children's health efforts. The national indicators presented in this comprehensive report are important for informing future research related to children's health. We will continue to partner with other federal agencies to develop increasingly reliable information that will help us to further improve children's health.

I want to thank the many individuals who contributed to this report for their hard work and efforts. By monitoring trends, identifying successes, and shedding light on areas of concern, we can continue to improve the health of our children and all Americans.

Lisa P. Jackson

Administrator

Table of Contents

Authors

Daniel Axelrad, Office of Policy

Kristen Adams, Office of Policy (Association of Schools of Public Health Fellow)

Farah Chowdhury, Office of Policy

Louis D'Amico, Office of Children's Health Protection (American Association for the Advancement of Science, Science and Technology Policy Fellow)

Erika Douglass, Office of Policy (Association of Schools of Public Health Fellow)

Gwendolyn Hudson, Office of Children's Health Protection (Association of Schools of Public Health Fellow)

Erica Koustas, Office of Policy (Oak Ridge Institute for Science and Education Fellow)

Juleen Lam, Office of Policy (Oak Ridge Institute for Science and Education Fellow)

Alyson Lorenz, Office of Children's Health Protection (Association of Schools of Public Health Fellow)

Gregory Miller, Office of Children's Health Protection

Kathleen Newhouse, Office of Children's Health Protection (on detail from Office of Research and Development)

Onyemaechi Nweke, Office of Environmental Justice

Doreen Cantor Paster, Office of Policy (on detail from Office of Chemical Safety and Pollution Prevention)

Julie Sturza, Office of Policy

Lindsay Underhill, Office of Policy

Kari Weber, Office of Policy

Contributors

Robin Schafer, Office of Solid Waste and Emergency Response

Jennifer Parker, Centers for Disease Control and Prevention, National Center for Health Statistics

Contractor Support

Jonathan Cohen, ICF International

Brad Hurley, ICF International

Internal Reviewers

EPA Reviewers

John Bowser, Office of Chemical Safety and Pollution Prevention

Rich Cook, Office of Air and Radiation

Glinda Cooper, Office of Research and Development

Elizabeth Corr, Office of Water

Rebecca Dzubow, Office of Research and Development

Alison Freeman, Office of Air and Radiation

Jane Gallagher, Office of Research and Development

Anne Grambsch, Office of Research and Development

Michael Hadrick, Office of Air and Radiation

Beth Hassett-Sipple, Office of Air and Radiation

David Hrdy, Office of Chemical Safety and Pollution Prevention

Jyotsna Jagai, Office of Research and Development

Michael Kolian, Office of Air and Radiation

Toni Krasnic, Office of Chemical Safety and Pollution Prevention

Jade Lee-Freeman, Office of Water

Danelle Lobdell, Office of Research of Development

Phil Lorang, Office of Air and Radiation

Matthew Lorber, Office of Research and Development

David Miller, Office of Chemical Safety and Pollution Prevention

Patricia Murphy, Office of Research and Development

Maureen O'Neill, Region 2

Ted Palma, Office of Air and Radiation

Andrea Pfahles-Hutchens, Office of Chemical Safety and Pollution Prevention

Robin Schafer, Office of Solid Waste and Emergency Response

Andrew Simons, Office of General Counsel

Maryann Suero, Region 5

Nicolle Tulve, Office of Research and Development

Prentiss Ward, Region 3

Melanie Young, Office of Water

Reviewers from Other Agencies

Linda Abbott, United States Department of Agriculture

Lara Akinbami, Centers for Disease Control and Prevention, National Center for Health Statistics

Peter Ashley, Department of Housing and Urban Development

John T. Bernert, Centers for Disease Control and Prevention, National Center for Environmental Health

Benjamin Blount, Centers for Disease Control and Prevention, National Center for Environmental Health

Laurie A. Brajkovich, California Department of Pesticide Regulation

Debra Brody, Centers for Disease Control and Prevention, National Center for Health Statistics

Antonia Calafat, Centers for Disease Control and Prevention, National Center for Environmental Health

Basil Ibewiro, California Department of Pesticide Regulation

Lisa Marengo, Texas Department of State Health Services

Pauline Mendola, Centers for Disease Control and Prevention, National Center for Health Statistics

Cynthia Ogden, Centers for Disease Control and Prevention, National Center for Health Statistics

John Osterloh, Centers for Disease Control and Prevention, National Center for Environmental Health

Jennifer Parker, Centers for Disease Control and Prevention, National Center for Health Statistics

Patricia Pastor, Centers for Disease Control and Prevention, National Center for Health Statistics

Kenneth Schoendorf, Centers for Disease Control and Prevention, National Center for Health Statistics

Children's Health Protection Advisory Committee, *America's Children and the Environment* Task Group

Laura Anderko, Georgetown University

Lynda Knobeloch, Wisconsin Department of Health Services

Amy Kyle, University of California, Berkeley

Robert Leidich, BP Products North America Inc.

Melanie Marty, California Environmental Protection Agency

Elise Miller, Collaborative on Health and the Environment

Jerome Paulson, George Washington University

Jennifer Roberts, Exponent, Inc.

External Peer Reviewers

Jennifer Adibi, University of California, San Francisco

Dana Barr, Emory University

Paloma Beamer, University of Arizona

Cynthia Bearer, University of Maryland

Alan Becker, Florida Agricultural and Mechanical University

Michelle Bell, Yale University

Deborah Bennett, University of California, Davis

Jason Booza, Wayne State University

Carla Campbell, Drexel University

Gang Chen, Penn State University

Timothy Church, University of Minnesota

Julianne Collins, Greenwood Genetic Center

Lucio Costa, University of Washington

Susan Duty, Simmons College/Harvard University

Diane Heck, New York Medical College

Susan Jobling, Brunel University

Kurunthachalam Kannan, New York State Department of Health

Catherine Karr, University of Washington

Judy LaKind, University of Maryland

Peter Langlois, Texas Department of State Health Services

Bruce Lanphear, Simon Fraser University

An Li, University of Illinois at Chicago

Morton Lippmann, New York University

Larry Lowry, University of Texas, Tyler

Alex Lu, Harvard University

Kristen Malecki, University of Wisconsin, Madison

Kathleen McCarty, Yale University

John Meeker, University of Michigan

Dawn Misra, Wayne State University

Vlasta Molak, Gaia Foundation/Gaia Unlimited

Ardythe Morrow, Cincinnati Children's Hospital Medical Center

Erica Phipps, Canadian Partnership of Children's Health and Environment

Gregory Pratt, Minnesota Pollution Control Agency

Beth Resnick, Johns Hopkins University

James Roberts, Medical University of South Carolina

Leslie Rubin, Emory University

Barry Ryan, Emory University

Arnold Schecter, University of Texas, Dallas Regional Campus

Kellogg Schwab, Johns Hopkins University

Perry Sheffield, Mount Sinai School of Medicine

Martha Stanbury, Michigan Department of Community Health

Jennifer Straughen, Wayne State University

Michael Wilson, University of California, Berkeley

Catherine Yeckel, Yale University

About this Report

What is *America's Children and the Environment*?

America's Children and the Environment (ACE) is EPA's report presenting data on children's environmental health. ACE brings together information from a variety of sources to provide national indicators in the following areas: Environments and Contaminants, Biomonitoring, and Health. Environments and Contaminants indicators describe conditions in the environment, such as levels of air pollution. Biomonitoring indicators include contaminants measured in the bodies of children and women of child-bearing age, such as children's blood lead levels. Health indicators report the rates at which selected health outcomes occur among U.S. children, such as the annual percentage of children who currently have asthma. Accompanying each indicator is text discussing the relevance of the issue to children's environmental health and describing the data used in preparing the indicator. Wherever possible, the indicators are based on data sources that are updated in a consistent manner, so that indicator values may be compared over time.

This report is the third edition of ACE (referred to as ACE3); previous editions of ACE were published in 2000 and 2003. EPA has provided updated indicator values on its website on a regular basis beginning in 2006, and will provide online updates for the indicators published in this edition (see www.epa.gov/ace).

What are the purposes of *America's Children and the Environment*?

America's Children and the Environment has three principal objectives:

- First, it compiles data from a variety of sources to present concrete, quantifiable indicators for key factors relevant to the environment and children's health in the United States.
- Second, it can inform discussions among policymakers and the public about how to improve data on children's health and the environment.
- Third, it includes indicators that can be used by policymakers and the public to track trends in children's environmental health, and ultimately to help identify and evaluate ways to minimize environmental impacts on children.

This report is motivated by EPA's belief that it is valuable to be aware of, and to share with the public, information on trends in children's environmental health. The purpose of ACE is to compile information, and make it available to a broad audience, that can help identify areas that warrant additional attention, potential issues of concern, and persistent problems. Some of the indicators can also support efforts to evaluate whether past environmental policies and actions have been effective. EPA hopes that the development and presentation of these indicators will motivate continuing research, additional data collection, and, when appropriate, necessary interventions.

The information in ACE3 is not intended to serve as a definitive basis for planning specific policies or projects. EPA and other federal agencies rely on a wide range of technical information to inform their activities on children's environmental health. Emerging and ongoing research will help shape these efforts for years to come. The presentation of findings from the scientific literature in ACE3 is not intended to constitute an authoritative summary or conclusion on the weight of scientific evidence.

What are children's environmental health indicators?

For ACE3, an indicator is defined as a quantitative depiction of an aspect of children's environmental health that summarizes the underlying data in a relevant, understandable, and technically appropriate manner. The data may represent measurements of environmental conditions, of chemicals measured in the bodies of children and women of child-bearing age, or of the frequency of certain childhood diseases and health outcomes. Federal data on children's environmental health issues come from a variety of agencies and are often very detailed and complex; ACE brings this information together into one report and summarizes the data in graphics that convey the key information. The ACE indicators generally focus on presenting data at the national scale in order to meet the three principal objectives described above.

Many indicators in this report provide a time series of data (i.e., a "trend" graph), to evaluate whether conditions have changed over time. Other indicators provide a "snapshot" that focuses on data from a single time period. These indicators may depict differences in conditions for different population groups (defined by race/ethnicity or income), or they may provide data for different children's health hazards for a single time period.

The World Health Organization defines environmental health as "all the physical, chemical, and biological factors external to a person, and all the related factors impacting behaviors. It encompasses the assessment and control of those environmental factors that can potentially affect health."[1] In concordance with this definition, use of the term "children's environmental health" in ACE3 refers to external physical, chemical, and biological factors that are known to affect children's health or may potentially affect children's health. The evidence of a relationship between environmental exposure and children's health continues to evolve for many of the indicators presented in this report. Inclusion of an indicator in this report does not necessarily imply a known relationship between environmental exposure and children's health effect. EPA aims to develop increasingly informative indicators of children's environmental health as more data become available to reduce these uncertainties.

The ACE3 indicators are intended to be easy to understand and to cover a broad range of topics. More extensive analyses are available for most of the datasets featured by ACE3 indicators; links to the associated studies and reports will be provided on the ACE website.

Although the ACE3 indicators focus on national statistics, environmental exposures and health may vary significantly across communities. Patterns of environmental exposure may vary due to the nature and extent of pollutants found in each community. Patterns of health may vary across communities due to demographic and socioeconomic characteristics. Links to online resources with community-level information will be provided on the ACE website.

Why did EPA focus on indicators for children?

Under Executive Order 13045, EPA and other federal agencies are directed to "make it a high priority to identify and assess environmental health risks and safety risks that may disproportionately affect children."[2] Environmental contaminants can affect children quite differently than adults, both because children may be more highly exposed to contaminants, and because they are often more vulnerable to the toxic effects of contaminants.

Children generally eat more food, drink more water, and breathe more air relative to their size than adults do, and consequently may be exposed to relatively higher amounts of environmental chemicals. Children's normal activities, such as putting their hands in their mouths or playing on the ground, can result in exposures to chemicals that adults do not face. In addition, some environmental contaminants may affect children disproportionately because their bodies are not fully developed and their growing organs can be more easily harmed.

How is *America's Children and the Environment* organized?

After this introduction, ACE3 features four main sections: Environments and Contaminants, Biomonitoring, Health, and Supplementary Topics. Each section presents information on a series of children's environmental health topics, and at least one indicator is provided for each topic.

The Environments and Contaminants section presents information on chemicals and pollutants in environmental media to which children are commonly exposed (through air, drinking water, and food), along with other important aspects of children's environments. Topics addressed in the Environments and Contaminants section include criteria air pollutants, hazardous air pollutants, indoor environments, drinking water contaminants, chemicals in food, contaminated lands, and climate change.

The Biomonitoring section presents information on selected chemicals measured in the blood and urine of children and women of child-bearing age. Biomonitoring indicators for women ages 16 to 49 years are included based on concern for potential adverse health effects in children born to women who have been exposed to certain chemicals. Topics addressed in the Biomonitoring section include lead, mercury, cotinine (a marker for exposure to environmental tobacco smoke), perfluorochemicals (PFCs), polychlorinated biphenyls (PCBs), polybrominated diphenyl ethers (PBDEs), phthalates, bisphenol A (BPA), and perchlorate.

The Health section presents information on diseases, conditions, and outcomes that may be influenced by environmental exposures. Many factors contribute to children's health, including genetic inheritance, nutrition, and exercise, among others. The adverse health consequences of some environmental exposures may occur through interactions with other risk factors, and it is often difficult to determine the extent to which the environment (or any other factor) contributes to children's health outcomes of concern. Topics addressed in the Health section include respiratory diseases, childhood cancer, neurodevelopmental disorders, obesity, and adverse birth outcomes.

The Supplementary Topics section presents topics for which adequate national data are not available, but for which more targeted data collection efforts could be used to provide measures illustrating additional children's environmental health issues of interest. Data sets used for these measures are representative of particular locations (such as a single state) and/or were surveys conducted a single time rather than on a continuing or periodic basis. Since these data sets are lacking in certain key elements desirable for ACE3 indicators, data presentations for the Supplementary Topics are referred to as "measures" rather than "indicators." Topics addressed in this section include birth defects in Texas and contaminants in schools and child care facilities.

How were the topics and indicators in the third edition of *America's Children and the Environment* selected?

In choosing indicators for ACE3, EPA considered a variety of factors, including public interest, magnitude of prevalence and/or trend in prevalence, extent of exposure, severity of health outcome, past EPA actions to address the issue, and research findings indicating or suggesting that an environmental exposure may contribute to a children's health outcome. ACE3 includes topics for which scientific evidence is sufficient to conclude there is a causal relationship between exposure and health effects, as well as topics for which there is less extensive scientific evidence. Inclusion of a topic in ACE3, therefore, does not imply that a cause-effect determination has been made.[i]

ACE3 includes updates and revisions to topics and indicators included in the 2003 ACE report, as well as new topics and indicators developed for this edition. Although ACE3 addresses a substantially expanded set of children's environmental health topics compared with the 2003 edition of this report, it is not intended to be inclusive of all children's environmental health issues.

The selection of topics involved generating a list of children's environmental health issues of potential interest, evaluating availability of suitable databases relevant to those topics, and considering indicators that might be derived from those databases. EPA obtained input from members of EPA's Children's Health Protection Advisory Committee (CHPAC) on each stage of this process, including input on the ultimate selection of topics and indicators presented in ACE3.[3,4] Independent external peer reviewers provided their opinions to EPA regarding the suitability of the indicators and other information provided for each topic. EPA revised the report based on the peer review comments, and comments received from the public.

---

[i] See "What is known about the role of the environment in contributing to adverse children's health outcomes?" below for more information.

Selection of a topic for inclusion in ACE depended, in part, on whether data appropriate for indicator development were available. Available databases were considered in the context of the following:

- Relevance to the topic of interest.
- Degree to which scientifically sound data collection methodologies and quality assurance procedures were used.
- Availability of data documentation.
- Availability of raw data (individual measurements or survey responses).
- Degree to which the database can be used to characterize national patterns.
- Ongoing (continuous or periodic) data collection, with relatively recent data available.
- Comparability of target population, sample selection, and data collection methods across time.
- Ability to stratify data by race/ethnicity, income, and location (region, state, county, or other geographic unit).

The suitability of each database was determined through an overall weighing of these considerations. Some databases ranked comparatively better than others with respect to each of these considerations. For example, some databases contain the results of nationally representative surveys that cannot be stratified geographically but are excellent in other respects; inability to extract statistics for regions, states, or other geographic divisions does not preclude the use of these databases in ACE3. Similarly, some monitoring data sets are not explicitly designed to be nationally representative; however, they may still be informative as long as their limitations are understood.

ACE3 presents one or more indicators to illustrate status and/or trends for each selected topic with a suitable database. In some cases, a topic is represented with multiple indicators that portray different aspects of the underlying data or make use of different types of data.

Considerations that EPA used in developing specific indicators from each selected database include some of the same factors used in selecting the database, as well as others, including:

- Utility of the indicator for portraying some aspect of children's environmental health.
- Sensitivity to changes in the condition of interest.
- Robustness (unaffected by changes in factors not relevant to the condition of interest).
- Degree to which the indicator offers an appropriate summary of the underlying data.
- Ability to be presented as population-based statistic (for example, the indicator takes the form of "percentage of children affected," or as defined points in the population. distribution of values, such as medians), particularly a national population-based statistic.
- Clarity.

Indicators that do not satisfy all considerations may still be considered suitable; for example, some indicators may lack data for presenting a trend over a number of years, but present useful information for some relatively recent time period (a single year or set of years). To help guide reader evaluations, text boxes are provided that summarize the characteristics of the data used for each indicator.

What are the sources of data for the indicators in *America's Children and the Environment?*

Federal agencies provided the data for most of the indicators. Data for the Environments and Contaminants indicators are generally from data systems maintained by EPA and by state environmental agencies. Data on indoor lead hazards are from surveys conducted by the U.S. Department of Housing and Urban Development. Pesticide residue data are from the Pesticide Data Program of the U.S. Department of Agriculture. Health and biomonitoring data are from the National Center for Health Statistics in the Centers for Disease Control and Prevention. Cancer data are from the National Cancer Institute. Child population data from the Census Bureau were used for calculations in several of the Environments and Contaminants indicators.

Data for the Supplementary Topics measures are from more targeted data collection efforts that illustrate some aspect of a children's environmental health issue of interest in the absence of a more comprehensive data source. Childcare facility measures are derived from a national study, and a study performed in North Carolina and Ohio. For schools, a measure on indoor pesticide application is derived from data reported by California schools and collected by the California Department of Pesticide Regulation. The data on birth defects are from the Texas birth defects monitoring program. Data from individual states are not intended to describe national conditions or conditions in other states.

What years are included in the *America's Children and the Environment* indicators?

ACE3 aims to include indicators that present trends over at least the past 10 years; however, for some indicators, data are not available for this length of time. When sufficient data are not available to show changes over time, indicators present the most current data available, frequently focusing the presentation on demographic comparisons of race/ethnicity and income. Some topics include both a trend indicator and a separate indicator with demographic comparisons using current data.

All ACE3 indicators incorporate the most current data that were available at the time of analysis. For some indicators, additional data were released prior to the publication of ACE3. Newer data will be incorporated for the indicator updates provided online at www.epa.gov/ace.

What groups of children are included in the *America's Children and the Environment* indicators?

Census Bureau data indicate that there were 74.2 million children ages 17 years and younger in the United States in 2010. The age range used for each indicator depends on data availability and the nature of the topic being addressed. Each indicator clearly identifies the age range in the title of the figure.

ACE3 presents (where possible) indicators for groups of children of different races and ethnicities and for children living in households with various levels of income. In some cases, these breakouts by race/ethnicity and family income are shown in the graphs, while in other cases they are included in the data tables.

The specific race/ethnicity categories used for each indicator depend on the underlying data source, and are further discussed in the introduction to each section of the report.

Many of the indicators also provide separate indicator values for children living in homes with family income below the poverty level (shown in graphs and tables as < Poverty) and those in homes at or above the poverty level (≥ Poverty). "Poverty level" is defined by the federal government and is based on income thresholds that vary by family size and composition. In 2010, for example, the poverty threshold was $22,113 for a household with two adults and two related children.[5] Based on this federal definition, 22% of children were living below poverty level in 2010, an increase from 18% of children in 2007.[6]

Why does *America's Children and the Environment* compare indicator values by race/ethnicity and income?

Under Executive Order 12898,[7] EPA (along with other federal agencies) is directed to "make achieving environmental justice part of its mission by identifying and addressing, as appropriate, disproportionately high and adverse human health or environmental effects of its programs, policies, and activities on minority populations and low-income populations." Comparing indicator values across these demographic groups helps identify differences in the distributions of exposures and health outcomes, which are factors in investigating the potential for disproportionate impacts.

Comparing indicator values by demographic groups is also in keeping with the goals of Healthy People 2020, the federal government's program of objectives for improving the health of all Americans. Among the overarching goals of Healthy People 2020 are to "achieve health equity, eliminate disparities, and improve the health of all groups" and to "create social and physical environments that promote good health for all."[8] Healthy People 2020 defines a health disparity as "a particular type of health difference that is closely linked with social, economic, and/or environmental disadvantage. Health disparities adversely affect groups of people who have systematically experienced greater obstacles to health based on their racial or ethnic group; religion; socioeconomic status; gender; age; mental health; cognitive, sensory, or physical disability; sexual orientation or gender identity; geographic location; or other characteristics historically linked to discrimination or exclusion."[9] Presentation of ACE indicator values by race/ethnicity and income groups provides information useful for investigating possible health disparities and possible environmental contributors to health disparities. Additionally, EPA's regulations implementing Title VI of the Civil Rights Act state, in part, "No person shall be excluded from participation in, be denied the benefits of, or be subjected to discrimination under any program or activity receiving EPA assistance on the basis of race, color, [or] national origin" (40 C.F.R. 7.30). Where comparison data are available on a state-specific basis, it may help EPA and its assistance recipients (for example, state environmental agencies) assess whether discriminatory impacts are occurring.

What information is presented for each topic and indicator?

Presentation of each topic includes a discussion of the scope of the issue and a brief snapshot of the relevant scientific literature regarding associations between exposures and health effects. If an authoritative source has published conclusions regarding the strength of evidence relevant to the topic, such as a determination of a cause-effect relationship between exposure and outcome, these findings are summarized. In the absence of such a source, the discussion describes selected literature and highlights significant sources of uncertainty, but this review should not be considered either an evaluation of the available literature or a statement regarding the strength of the evidence.

This is followed by an explanation of the indicator chosen to represent the topic, including a discussion of the data source, a description of the data provided in the indicator, and information to aid in interpreting the indicator, including data limitations.

Following this background text, one or more indicators are provided. Each indicator is presented in a figure. A text box is provided to help readers understand the characteristics of the data displayed. Bullet points that highlight key data points from the indicator are included.

Appendices to the report provide data tables for each indicator. Detailed explanations of the methods for calculating each indicator are provided in the online materials available at www.epa.gov/ace.

What is known about the role of environmental exposures in contributing to adverse children's health outcomes?

Some environmental exposures have a well-established cause-effect relationship with children's health, such as effects of lead exposure on childhood IQ and effects of certain air pollutants on respiratory outcomes. For some other environmental exposures, there is evidence suggestive of a relationship to children's health outcomes but not enough evidence to conclude the existence of a cause-effect relationship; and for many other environmental exposures there is very little information on potential health consequences of the exposure levels typically experienced by children in the United States. Furthermore, for many of the children's health effects discussed in this report, our scientific knowledge regarding causes is somewhat limited.

A major focus of environmental health research is to expand our understanding of the possible role of environmental contaminant exposures, as well as other environmental risk factors, in childhood diseases and disorders. Research is increasingly pointing to interactions of genetic factors and environmental factors as critical to the process for most diseases.

Even when a clear relationship between exposure to a particular hazardous environmental contaminant or factor has been documented, some children will have worse outcomes and others will be unaffected or have outcomes that are less severe. Exposure characteristics—such as the length of exposure, the magnitude of the exposure, the route of exposure and the developmental stage when a child is exposed—explain much of the variation in outcome. However, genetic variability in the population can mean that individuals vary greatly in how their body metabolizes a chemical and in their susceptibility to diseases that may result from an environmental exposure. In addition, variability in concurrent or prior exposures to other environmental contaminants and to non-chemical stressors can also lead to substantial variability of outcomes within an exposed population.[10]

A child's genetic inheritance can often play an important role in disease. However, as scientific methods for examining the role of genetics in disease have advanced, it has become clear that much of human chronic disease cannot be explained by genetic factors alone, and that environmental factors (broadly defined to include nutrition, exercise, exposures to environmental contaminants, and other factors) and their interactions with genetic factors also play an important role in chronic disease.[11-13]

The effects of an environmental exposure on children's health often depend on the developmental stage at which the exposure occurs. Different organ systems in a child's body go through critical developmental stages at different times, from conception through the entire period of fetal development, in infancy and early childhood, and continuing through adolescence. Some chemical exposures can result in adverse effects if they occur during a particular critical window, and may have different effects or no effect at all when occurring at a different stage of development.[14] For this reason, even some environmental exposures to adults can be important for children's health, as research has found that the prenatal period is the most sensitive developmental stage for adverse effects of some chemicals.[15] In some cases, the effects of a harmful exposure may not become evident until many years later; exposure during early developmental stages may even contribute to the onset of chronic diseases in adulthood.[15]

What types of scientific studies provide evidence about the potential relationships between environmental exposures and children's health outcomes?

Developing conclusive evidence that environmental factors cause or contribute to the incidence of childhood health effects is difficult. Many health outcomes are hypothesized to be multi-factorial, with contributions from genetics, underlying health conditions, and lifestyle, as well as the social and physical environment. Scientific evidence linking the environment to health outcomes consists primarily of laboratory assays, experimental studies in animals, and epidemiological studies in humans. Each of these methods has limitations, but together they can provide complementary evidence in assessing how exposures can influence the development of health outcomes.

A major advantage of animal studies is that they are controlled experiments in which exposures are imposed upon the study subjects and all other variables are held constant. In many cases, animals can provide good models of human physiological systems and thus indicate how humans might respond to exposures. However, it is not always straightforward to interpret findings of animal studies and their meaning or importance for human health. Furthermore, animal studies are often conducted using exposure levels much greater than those typically experienced by humans, and some uncertainty exists as to whether the same effects would be seen at lower exposure levels.

In contrast, observational epidemiological studies are advantageous because they evaluate the relationship between environmental conditions and health outcomes in exposed human populations. Since this type of study is not a controlled experiment, there may be factors related to both the exposure and the health effect in the study population that can create false associations, or mask true associations, between the exposure and the health effect. Observational epidemiological studies provide the strongest evidence when they have been replicated in multiple populations to minimize the likelihood that an association between exposure and health outcomes occurred due to something other than a true causal relationship.

Some epidemiological studies are conducted in samples of the U.S. general population, or in other countries with similar exposure levels, and thus reflect exposures that occur on a routine basis. Sometimes studies in the United States or in other countries may be focused on communities that experience higher exposure levels than the rest of the country; these studies would be considered to have greater-than-average exposures but are still within the range of exposures occurring in the United States. In other cases, exposures in epidemiological studies are conducted in populations with substantially higher exposure, such as workers exposed to chemicals on the job, populations in other countries that have higher levels of pollution, residents of communities where disasters or accidental poisoning incidents have occurred, or populations in the United States or other industrialized countries before environmental protection efforts to reduce exposure occurred. In such cases, some uncertainty exists as to whether the same effects would be seen at lower exposure levels observed today.

An important additional consideration is the extent to which toxicological or epidemiological studies are available for environmental contaminants and chemicals in commerce. For many environmental exposures of interest, the epidemiological research is very limited and there are significant gaps in the available animal testing data.

How are the findings of scientific studies regarding children's health and the environment represented in *America's Children and the Environment*?

The level of knowledge regarding the relationship between environmental exposures and health outcomes varies widely among the topics presented in this report. Some associations between contaminants and health outcomes are supported by a large body of consistent evidence from rigorously designed and conducted studies. In other cases, research findings may suggest reason for potential concern but may not be sufficient to draw conclusions regarding the nature or strength of the relationship between an environmental contaminant exposure and a children's health outcome.

Where available, ACE3 relies on authoritative reviews of the scientific literature and reports their conclusions regarding the strength of the evidence for a causal role of specific environmental factors in the development of childhood diseases and disorders. Examples of authoritative sources are the National Research Council, the Institute of Medicine, and the National Toxicology Program.[ii] When such reviews are unavailable, selected findings from the epidemiological literature that address the potential role of environmental factors in contributing to an effect are summarized. Literature on animal studies is discussed in certain cases when epidemiological data are lacking. These reviews of scientific information are intended to summarize the concerns that have led to inclusion of the topic in this report. The literature summaries are not intended as reviews of the literature determining the strength of the evidence, which is an undertaking beyond the scope of this report.

How are children's environmental health indicators different from epidemiological research?

The presentation of children's environmental health indicators in ACE3 is intended to highlight issues of interest, describe indicator values over time, and describe indicator values for different demographic groups. However, the indicators themselves are not intended as a basis for reaching conclusions that an environmental factor is or is not related to a particular children's health outcome. Comparison of trends in Health indicators to trends in Environments and Contaminants or Biomonitoring indicators may suggest hypotheses for further research, but their presentation here cannot be used to conclude that a causal

---

[ii] The National Research Council and the Institute of Medicine are private, nonprofit institutions that provide expert advice on science and health matters. The National Toxicology Program, part of the U.S. Department of Health and Human Services, provides evaluations of substances of public health concern.

relationship may or may not exist. Indicators cannot account for the multiple factors that should inform these judgments.

Epidemiological studies can be designed that consider both individual-level or community-level exposures (or surrogates for exposures) and outcomes within the same population, along with other factors such as the timing of exposure relative to the timing of outcome, related variables that could influence the health outcome, and appropriate statistical models of a hypothesized relationship.

ACE3 indicators do not incorporate these factors, and thus are not intended as a basis for conclusions about associations between exposures and outcomes. Rather, the value of indicators is in their ability to reveal trends (or absence of trends) and variations (or lack of variation) within the population, which can then be used to identify areas for closer review. Since they are based on ongoing data collection programs, the indicators can be updated regularly and can be used to alert policy makers and the public when unexpected patterns emerge from new data, or to provide an indication of whether recently adopted exposure reduction interventions and actions are having an impact.

What is the difference between this report and an EPA risk assessment?

Human health risk assessment is the process used to estimate the nature and probability of adverse health effects in populations who may be exposed to chemicals. A risk assessment typically focuses in depth on a particular environmental contaminant to identify potential adverse health outcomes, likely exposure pathways, the estimated magnitude of exposure, and the likelihood of health outcomes occurring at different levels of exposure experienced by a population.

The indicators in this report do not constitute a risk assessment. The indicators present observed data on status and trends in environmental conditions, biomonitoring, and health outcomes; they do not attempt to provide the information relating exposures and outcomes provided in a risk assessment. The scope of a risk assessment involves a much more detailed examination of the health effects literature and of exposure data, including estimation of the relationship between particular levels of exposure and potential outcomes. The indicators in this report should not be construed to indicate the level of risk to children's health from exposures to environmental contaminants.

More information about risk assessment may be found at EPA's Risk Assessment Portal at http://www.epa.gov/risk_assessment/.

Key Findings

These Key Findings summarize the observations obtained from each of the indicators presented in this report. Statistically significant trends or differences are identified by the terms "increase," "decrease," "higher," or "lower." Please see the body of the report for background helpful in understanding and interpreting each of these findings, including definitions, descriptions of data sets, and summaries of relevant scientific findings. The years for which data are available varies across the indicators.

The evidence of a relationship between environmental exposure and children's health continues to evolve for many of the indicators presented in this report. Inclusion of an indicator in this report does not necessarily imply a known relationship between environmental exposure and children's health effect.

Environments and Contaminants

Criteria Air Pollutants

- From 1999 to 2009, the proportion of children living in counties with measured pollutant concentrations above the levels of one or more national ambient air quality standards decreased from 75% to 59%. This includes both concentrations above the level of any current short-term standard at least once during the year as well as average concentrations above the level of any current long-term standards.

- In 2009, 6% of children lived in counties with measured ozone concentrations above the level of the 8-hour ozone standard on more than 25 days. An additional 3% of children lived in counties with measured concentrations above the level of the ozone standard between 11 and 25 days, and 12% of children lived in counties where concentrations were above the level of the standard between 4 and 10 days.

- In 2009, 1% of children lived in counties with measured fine particle ($PM_{2.5}$) concentrations above the level of the 24-hour $PM_{2.5}$ standard on more than 25 days. An additional 2% of children lived in counties with measured concentrations above the level of this standard between 11 and 25 days, and 1% of children lived in counties with measured concentrations above the level of the 24-hour $PM_{2.5}$ standard between 8 and 10 days.

- Based on categories from EPA's Air Quality Index, the percentage of children's days that were designated as having "unhealthy" air quality decreased from 9% in 1999 to 3% in 2009. The percentage of children's days with "good" air quality increased from 41% in 1999 to 57% in 2009. The percentage of children's days with "moderate" air quality was approximately constant at 21–23% from 1999 to 2007, and then decreased to 16% in 2009.

Hazardous Air Pollutants

- In 2005, nearly all children (99.9%) lived in census tracts in which hazardous air pollutant (HAP) concentrations combined to exceed the 1-in-100,000 cancer risk benchmark. Seven percent of children lived in census tracts in which HAPs combined to exceed the 1-in-10,000 cancer risk benchmark. Fifty-six percent of children lived in census tracts in which at least one HAP exceeded the benchmark for health effects other than cancer.

Indoor Environments

- In 2010, 6% of children ages 0 to 6 years lived in homes where someone smoked regularly, compared with 27% in 1994.

- In 2005–2006, 15% of children ages 0 to 5 years lived in homes with either an interior lead dust hazard or an interior deteriorated lead-based paint hazard, compared with 22% in 1998–1999.

Drinking Water Contaminants

- The estimated percentage of children served by community drinking water systems that did not meet all applicable health-based standards was 19% in 1993 and about 5% in 2001. Since 2002, this percentage has fluctuated between 7% and 13%, with the most recent estimate being 7% in 2009.

- Between 1993 and 2009, the estimated percentage of children served by community water systems that had at least one monitoring and reporting violation fluctuated between about 11% and 23%, and was 13% in 2009.

Chemicals in Food

- In 1999, 81% of sampled apples had detectable organophosphate pesticide residues, and in 2009, 35% had detectable residues. In 2000, 10% of sampled carrots had detectable organophosphate pesticide residues, and in 2007, 5% had detectable residues. In 2000, 21% of sampled grapes had detectable organophosphate pesticide residues, and in 2009, 8% had detectable residues. In 1998, 37% of sampled tomatoes had detectable organophosphate pesticide residues, and in 2008, 9% had detectable residues.

Contaminated Lands

- As of 2009, approximately 6% of all children in the United States lived within one mile of a Corrective Action or Superfund site that may not have had all human health protective measures in place.

- Approximately 21% of all children living within one mile of a Corrective Action or Superfund site that may not have had all human health protective measures in place were Black, while 15% of children in the United States as a whole are Black.

Biomonitoring

Lead

- The median concentration of lead in the blood of children between the ages of 1 and 5 years dropped from 15 micrograms per deciliter (µg/dL) in 1976–1980 to 1.2 µg/dL in 2009–2010, a decrease of 92%. At the 95th percentile, blood lead levels dropped from 29 µg/dL in 1976–1980 to 3.4 µg/dL in 2009–2010, a decrease of 88%.

■ The median blood lead level in Black non-Hispanic children ages 1 to 5 years in 2007–2010 was 1.6 µg/dL, higher than the level of 1.2 µg/dL in White non-Hispanic children, Mexican-American children, and children of "All Other Races/Ethnicities."

Mercury

■ The median concentration of total mercury in the blood of women ages 16 to 49 years has shown little change between 1999–2000 and 2009–2010, and was 0.8 micrograms per liter (µg/L) in 2009–2010.

■ Among women in the 95th percentile of exposure, the concentration of total mercury in blood decreased from 7.4 µg/L in 1999–2000 to 3.7 µg/L in 2001–2002. From 2001–2002 to 2009–2010, the 95th percentile of total blood mercury remained between 3.7 and 4.5 µg/L.

Cotinine

■ The median level of cotinine (a marker of exposure to environmental tobacco smoke) measured in blood serum of nonsmoking children ages 3 to 17 years dropped from 0.25 nanograms per milliliter (ng/mL) in 1988–1991 to 0.03 ng/mL in 2009–2010, a decrease of 88%. Cotinine values at the 95th percentile decreased by 34% from 1988–1991 to 2009–2010.

■ The median level of cotinine measured in blood serum of nonsmoking women ages 16 to 49 years dropped from 3.2 ng/mL in 1988–1991 to 2.1 ng/mL in 2009–2010, a decrease of 86%. Cotinine values at the 95th percentile decreased by 35% from 1988–1991 to 2009–2010.

Perfluorochemicals (PFCs)

■ Between 1999–2000 and 2007–2008, median blood serum levels of perfluorooctane sulfonic acid (PFOS) and perfluorooctanoic acid (PFOA) in women ages 16 to 49 years showed a decreasing trend; median levels of perfluorononanoic acid (PFNA) showed an increasing trend; and median levels of perfluorohexane sulfonic acid (PFHxS) remained relatively constant over time.

Polychlorinated Biphenyls (PCBs)

■ In 2001–2004, the median level of polychlorinated biphenyls (PCBs), summing together four selected PCBs, in blood serum of women ages 16 to 49 years was 30 nanograms per gram (ng/g) lipid. Data are not yet available for comparing these PCB levels over time.

Polybrominated Diphenyl Ethers (PBDEs)

■ The median concentration of polybrominated diphenyl ethers (PBDEs) in blood serum of women ages 16 to 49 years was 44 ng/g lipid in 2003–2004. Data are not yet available for comparing these PBDE levels over time.

Phthalates

■ From 2001–2002 to 2007–2008, the median level of di-2-ethylhexyl phthalate (DEHP) metabolites in urine of women ages 16 to 49 years varied between 41 µg/L and 51 µg/L, and was 51 µg/L in 2007–2008. From 1999–2000 to 2007–2008, the median level of dibutyl phthalate (DBP) metabolites in women ages 16 to 49 years varied between 27 µg/L and 36 µg/L, and was 36 µg/L in 2007–2008. From 1999–2000 to 2007–2008, the median level of butyl benzyl phthalate (BBzP) metabolites in women ages 16 to 49 years varied between 10 µg/L and 14 µg/L, and was 12 µg/L in 2007–2008.

■ From 2001–2002 to 2007–2008, the median level of DEHP metabolites in urine of children ages 6 to 17 years varied between 45 µg/L and 62 µg/L, and was 45 µg/L in 2007–2008. From 1999–2000 to 2007–2008, the median level of DBP metabolites in children ages 6 to 17 years varied between 36 µg/L and 42 µg/L, and was 41 µg/L in 2007–2008. The median level of BBzP metabolite in children ages 6 to 17 years decreased from 25 µg/L in 1999–2000 to 16 µg/L in 2007–2008.

Bisphenol A (BPA)

■ From 2003–2004 to 2009–2010, the median concentration of bisphenol A (BPA) in urine among women ages 16 to 49 years varied between 2 µg/L and 3 µg/L. From 2003–2004 to 2009–2010, the concentration of BPA in urine at the 95th percentile varied between 10 µg/L and 16 µg/L, and was 10 µg/L in 2009–2010.

■ Among children ages 6 to 17 years the median concentration of BPA in urine decreased from 4 µg/L in 2003–2004 to 2 µg/L in 2009–2010. The concentrations of BPA in urine at the 95th percentile decreased from 16 µg/L in 2003–2004 to 10 µg/L 2009–2010.

Perchlorate

■ From 2001–2002 to 2007–2008, the median level of perchlorate in urine among women ages 16 to 49 years was 3 µg/L with little variation over time. Over the same period, the 95th percentile varied between 13 and 17 µg/L.

Health

Respiratory Diseases

■ The proportion of children reported to currently have asthma has increased from 8.7% in 2001 to 9.4% in 2010.

■ In 2007–2010, the percentages of Black non-Hispanic children and children of "All Other Races" reported to currently have asthma, 16.0% and 12.4% respectively, were greater than for White non-Hispanic children (8.2%), Hispanic children (7.9%), and Asian non-Hispanic children (6.8%).

■ The rate of emergency room visits for asthma decreased from 114 visits per 10,000 children in 1996 to 103 visits per 10,000 children in 2008. Between 1996 and 2008, hospitalizations for asthma and for all other respiratory causes decreased from 90 hospitalizations per 10,000 children to 56 hospitalizations per 10,000 children.

Childhood Cancer

- The age-adjusted annual incidence of cancer increased from 1992–2009. The incidence ranged from 153 to 161 cases per million children between 1992 and 1994 and from 172 to 175 cases per million children between 2007 and 2009.

- Childhood cancer mortality has decreased from 33 deaths per million children in 1992 to 24 deaths per million children in 2009.

- Leukemia was the most common cancer diagnosis for children from 2004–2006, representing 28% of total cancer cases. Incidence of acute lymphoblastic (lymphocytic) leukemia increased from 30 cases per million in 1992–1994 to 35 cases per million in 2004–2006. The rate of acute myeloid (myelogenous) leukemia was 7 cases per million in 1992–1994 and 9 cases per million in 2004–2006.

Neurodevelopmental Disorders

- From 1997 to 2010, the proportion of children ages 5 to 17 years reported to have ever been diagnosed with attention-deficit/hyperactivity disorder (ADHD) increased from 6.3% to 9.5%.

- In 2010, 8.6% of children ages 5 to 17 years had ever been diagnosed with a learning disability. There was little change in this percentage between 1997 and 2010.

- The percentage of children ages 5 to 17 years reported to have ever been diagnosed with autism increased from 0.1% in 1997 to 1.0% in 2010.

- In 2010, 0.7% of children ages 5 to 17 years were reported to have ever been diagnosed with intellectual disability (mental retardation). There was little change in this percentage between 1997 and 2010.

Obesity

- Between 1976–1980 and 2007–2008, the percentage of children identified as obese showed an increasing trend. In 1976–1980, 5% of children ages 2 to 17 years were obese. This percentage reached a high of 17% in 2007–2008. Between 1999–2000 and 2007–2008, the percentage of children identified as obese remained between 15% and 17%.

- In 2005–2008, a higher percentage of Mexican-American and Black non-Hispanic children were obese at 22% and 20%, respectively, compared with 14% of White non-Hispanic children and 14% of children of "All Other Races/Ethnicities."

Adverse Birth Outcomes

- Between 1993 and 2008, the rate of preterm birth showed an increasing trend, ranging from 11.0% in 1993 to its highest value of 12.8% in 2006.

- Between 1993 and 2008, the rate of term low birth weight for all races/ethnicities stayed relatively constant, ranging between 2.5% and 2.8%.

Supplementary Topics

The Supplementary Topics section presents topics for which adequate national data are not available, but for which more targeted data collection efforts could be used to provide measures illustrating additional children's environmental health issues of interest. Data sets used for these measures are representative of particular locations (such as a single state) and/or were surveys conducted a single time rather than on a continuing or periodic basis. Since these data sets are lacking in certain key elements desirable for ACE3 indicators, data presentations for the Supplementary Topics are referred to as "measures" rather than "indicators."

Birth Defects

- The rates for all categories of birth defects in Texas have increased or remained stable for the period of 1999–2007. Some of the biggest increases were seen for musculoskeletal defects, cardiac and circulatory defects, genitourinary defects, eye and ear defects, and central nervous system defects.

Contaminants in Schools and Child Care Facilities

- The pesticides chlorpyrifos, *cis*-permethrin, and diazinon were detected in all dust samples collected at Ohio and North Carolina child care centers in 2000–2001. Chlorpyrifos and diazinon were also detected in all indoor air samples collected at these child care centers.

- Dibutyl phthalate was detected in all indoor air and dust samples collected at Ohio and North Carolina child care centers.

- Pyrethrin and pyrethroid insecticides accounted for the greatest volume of pesticide use in California schools overall from 2002 to 2007.

Environments and Contaminants

Introduction

Why is EPA tracking levels of contaminants and other aspects of children's environments in *America's Children and the Environment*?

Pollutants or contaminants that can affect the health of children can be found in air, water, food, and soil. This section describes contaminants in the air children breathe, the water they drink, and the food they eat. This section also addresses the conditions of children's environments by considering indoor environments, contaminated lands, and climate change. Trends over time can indicate the successes and shortcomings of efforts to reduce potential exposures and also identify opportunities for future action. Differences in the environmental conditions between geographic areas or demographic groups can inform more targeted actions.

What Environments and Contaminants topics are included in *America's Children and the Environment, Third Edition (ACE3)*?

Environments and Contaminants topics were selected for ACE3 based on: (1) research findings identifying environmental contaminants or characteristics that may have adverse effects on children's health; and (2) the availability of data suitable for constructing a national indicator. EPA obtained input from its Children's Health Protection Advisory Committee to assist in selecting topics from among the many contaminants and other aspects of the environment that may affect children's health. The ACE3 Environments and Contaminants indicators address the following topics:

- Criteria air pollutants
- Hazardous air pollutants
- Indoor environments
- Drinking water contaminants
- Chemicals in food
- Contaminated lands
- Climate change[i]

Data for all of the Environments and Contaminants indicators were obtained from surveys and databases maintained by U.S. government agencies. These include the Air Quality System (Environmental Protection Agency, EPA); National Air Toxics Assessment (EPA); National Health Interview Survey (National Center for Health Statistics); American Healthy Homes Survey (Housing and Urban Development, HUD); National Survey of Lead and Allergens in Housing

---

[i] Although climate change is addressed in this section, a climate change indicator is not currently presented. EPA is currently developing a new children's environmental health indicator for climate change. The new indicator will focus on the frequency of extreme heat events over time. EPA intends to complete development of this new indicator in 2014, and it will be made available at www.epa.gov/ace when completed.

(HUD); Safe Drinking Water Information System (EPA); Pesticide Data Program (U.S. Department of Agriculture); Comprehensive Environmental Response, Compensation, and Liability Information System (EPA); and the Resource Conservation and Recovery Act Information dataset (EPA). Although all of the data sources feature data collected across the United States, some are not designed to produce estimates describing the nation overall. These and other data limitations are described for each indicator presented. However, targeted samples can provide important insight into environmental conditions and lead to improved measurement over time.

Other environmental hazards that may potentially be of concern for children's health are not addressed in this section. Examples of these additional environmental hazards include contaminants in surface waters, ionizing radiation, and chemicals that may be present in parks and playgrounds.

What can we learn from the Environments and Contaminants indicators?

For some of the selected Environments and Contaminants topics, health-based standards have been established. By comparing data on contaminant levels against these standards, which often include a margin of safety, it is possible to determine the percentage of children living in areas where standards or targeted levels have been exceeded. For topics where health-based standards do not exist, indicator values may still summarize conditions over time or the conditions of different groups of children, such as by race/ethnicity or income level.

It is important to realize that children may be exposed to the same contaminant through a variety of sources and pathways. For example, children can be exposed to lead by ingesting dust, consuming drinking water, and breathing air that contains lead. Each Environments and Contaminants indicator shown here only informs our understanding of potential exposure from a single pathway. A separate Biomonitoring section of ACE3 presents indicators that report levels of selected chemicals measured in blood and urine samples. The biomonitoring approach provides an integrated measure of exposure from all possible sources and pathways. However, biomonitoring data are not available for all chemicals and contaminants represented in the Environments and Contaminants indicators. The Environments and Contaminants indicators and the Biomonitoring indicators are complementary in that they represent different types of information about children's potential environmental exposures.

What information is provided for each Environments and Contaminants topic?

An introduction section defines the topic and describes its relevance to children's health, including a discussion of potential health concerns associated with exposures to the contaminants or environmental conditions. The introduction is followed by a description of the indicators, including a summary of the data available and a brief description of how each indicator was calculated. One to three indicators, each a graphical presentation of the available data, are included for each topic. Most of the indicators present time series data. Where data over time are unavailable, the indicators present data for the most recent year available. Where possible, the indicators incorporate information on race/ethnicity and income level. Beneath each figure are explanatory bullet points highlighting key findings from the data presented in the figure, along with key data from any supplemental data tables. References are provided for each topic at the end of the report.

Data tables are provided in Appendix A. The tables include all indicator values depicted in the indicator figures, along with additional data of interest not shown in the figures. Metadata describing the data sources are provided in Appendix B. Documents providing details of how the indicators were calculated are available on the ACE website (www.epa.gov/ace).

Many of the topics presented in the Environments and Contaminants indicators are addressed in Healthy People 2020, which provides science-based, 10-year national objectives for improving the health of all Americans. Appendix C provides examples of the alignment of the Environments and Contaminants topics presented in ACE3 with objectives in Healthy People 2020.

How were the indicators calculated and presented?

Data files: The indicators were calculated using data files obtained from the government agency websites or from government agency staff.

Population age groups: Most of the indicators used data for children ages 17 years and younger. The indicators for indoor environments were restricted to younger ages because younger children have been specifically identified as more susceptible to the effects of tobacco smoke and lead exposure. The indicator for environmental tobacco smoke (E5) used data for children ages 0 to 6 years. The indicator for interior lead hazards (E6) used data for children ages 0 to 5 years.

Calculation of percentages: For most of the Environments and Contaminants topics, information on environmental contaminants/characteristics was used to identify counties where one or more environmental contaminants were above target levels established for the indicator. For example, the calculation of percentages in Indicator E1 involved identifying counties with at least one air pollutant measurement above the level of a National Ambient Air Quality Standard. The population of children in counties with an environmental contaminant above the target level was then calculated using census data, and divided by the total population of children to derive the indicator value as a percentage of all children in the United States.

For the indoor environments topics, survey data were obtained from representative samples of people (to estimate the percentage of children in homes with regular exposure to environmental tobacco smoke) and homes (to estimate the percentage of children in homes with lead hazards). Sample weights equal to the number of children in the U.S. population represented by each sampled child were applied to yield estimates representing the U.S. population of children. The indicator on chemicals in food reports the percentage of samples of selected foods with detectable levels of pesticides.

Statistical testing: Statistical analysis has been applied to the two indicators derived from probability-based sample data (the indoor environments indicators for environmental tobacco smoke and lead) to evaluate differences over time or between demographic groups. Statistical analysis has also been applied to the criteria pollutants data to evaluate trends over time. The remaining environment and contaminant indicators do not readily lend themselves to statistical analysis, due to the characteristics of the underlying databases.[ii]

[ii] Standard errors for the indoor environments indicator values, which are derived from survey data, are provided in a file available on the ACE website (www.epa.gov/ace). Standard errors could not be calculated for the remaining Environments and Contaminants indicators.

Criteria Air Pollutants

Air pollution contributes to a wide variety of adverse health effects. EPA has established national ambient air quality standards (NAAQS) for six of the most common air pollutants—carbon monoxide, lead, ground-level ozone, particulate matter, nitrogen dioxide, and sulfur dioxide—known as "criteria" air pollutants (or simply "criteria pollutants"). The presence of these pollutants in ambient air is generally due to numerous diverse and widespread sources of emissions. The primary NAAQS are set to protect public health. EPA also sets secondary NAAQS to protect public welfare from adverse effects of criteria pollutants, including protection against visibility impairment, or damage to animals, crops, vegetation, or buildings.

As required by the Clean Air Act,[1] EPA periodically conducts comprehensive reviews of the scientific literature on health and welfare effects associated with exposure to the criteria air pollutants.[2-7] The resulting assessments serve as the basis for making regulatory decisions about whether to retain or revise the NAAQS that specify the allowable concentrations of each of these pollutants in the ambient air.[8]

The primary standards are set at a level intended to protect public health, including the health of at-risk populations, with an adequate margin of safety. In selecting a margin of safety, EPA considers such factors as the strengths and limitations of the evidence and related uncertainties, the nature and severity of the health effects, the size of the at-risk populations, and whether discernible thresholds have been identified below which health effects do not occur. In general, for the criteria air pollutants, there is no evidence of discernible thresholds.[2-7]

The Clean Air Act does not require EPA to establish primary NAAQS at a zero-risk level, but rather at a level that reduces risk sufficiently so as to protect public health with an adequate margin of safety. In all NAAQS reviews, EPA gives particular attention to exposures and associated health risks for at-risk populations. Standards include consideration of providing protection for a representative sample of persons comprising at-risk populations rather than to the most susceptible single person in such groups. Even in areas that meet the current standards, individual members of at-risk populations may at times experience health effects related to air pollution.[9-13]

Childhood is often identified as a susceptible lifestage in the NAAQS reviews, because children's lungs and other organ systems are still developing, because they may have a preexisting disease (e.g., asthma), and because they may experience higher exposures due to their activities, including outdoor play.[14-17] Evaluating the effects of criteria air pollutants in children has been a central focus in several recent NAAQS reviews, including revisions of the lead,[18] ozone,[19] and particulate matter[20] standards to strengthen public health protection.

Some of the air quality standards are designed to protect the public from adverse health effects that can occur after being exposed for a short time, such as hours to days. Other standards are designed to protect people from adverse health effects that are associated with long-term exposures (months to years). For example, the standard for ozone is based on pollutant concentrations measured over a short-term period of eight hours. By contrast, the standard for lead considers average concentrations measured over a rolling three-month period. For fine particulate matter ($PM_{2.5}$), annual and 24-hour standards work together to provide protection against effects associated with long- and short-term exposures.

Health effects that have been associated with each of the criteria pollutants are summarized below. This information is drawn primarily from EPA's assessments of the scientific literature for the criteria pollutants.

Ozone

Ground-level ozone forms through the reaction of pollutants emitted by industrial facilities, electric utilities, and motor vehicles; chemicals that are precursors to ozone formation can also be emitted by natural sources, particularly trees and other plants.[2] Ground-level ozone can pose risks to human health, in contrast to the stratospheric ozone layer that protects the earth from harmful wavelengths of solar ultraviolet radiation. Short-term exposure to ground-level ozone can cause a variety of respiratory health effects, including inflammation of the lining of the lungs, reduced lung function, and respiratory symptoms such as cough, wheezing, chest pain, burning in the chest, and shortness of breath.[2,13,21] Ozone exposure can decrease the capacity to perform exercise.[2] Exposure to ozone can also increase susceptibility to respiratory infection. Exposure to ambient concentrations of ozone has been associated with the aggravation of respiratory illnesses such as asthma, emphysema, and bronchitis, leading to increased use of medication, absences from school, doctor and emergency department visits, and hospital admissions. Short-term exposure to ozone is associated with premature mortality.[2] Studies have also found that long-term ozone exposure may contribute to the development of asthma, especially among children with certain genetic susceptibilities and children who frequently exercise outdoors.[22-24] Long-term exposure to ozone can permanently damage lung tissue.

Particulate Matter

Particulate matter (PM) is a generic term for a broad class of chemically and physically diverse substances that exist as discrete particles (liquid droplets or solids) over a wide range of sizes. Particles originate from a variety of man-made stationary and mobile sources, as well as from natural sources such as forest fires. Particles may be emitted directly, or may be formed in the atmosphere by transformations of gaseous emissions such as oxides of sulfur (SO_x), oxides of nitrogen (NO_x), and volatile organic compounds (VOCs). The chemical and physical properties of PM vary greatly with time, region, meteorology, and the source of emissions. For regulatory purposes, EPA distinguishes between categories of particles based on size, and has established standards for fine and coarse particles. PM_{10}, in general terms, is an abbreviation for particles with an aerodynamic diameter less than or equal to 10 micrometers (μm), and represents inhalable particles small enough to penetrate deeply into the lungs (i.e., thoracic particles).[i] PM_{10} is composed of a coarse fraction referred to as $PM_{10-2.5}$ or as thoracic coarse particles

---

[i] For comparison, the diameter of PM_{10} particles is 1/7 the diameter of an average human hair or less.

(i.e., particles with an aerodynamic diameter less than or equal to 10 μm and greater than 2.5 μm) and a fine fraction referred to as $PM_{2.5}$ or fine particles (i.e., particles with an aerodynamic diameter less than or equal to 2.5 μm). Thoracic coarse particles are emitted largely as a result of mechanical processes and uncontrolled burning. Important sources include resuspended dust (e.g., resuspended by cars, wind, etc.), industrial processes, construction and demolition operations, residential burning, and wildfires. Fine particles are formed chiefly by combustion processes (e.g., from power plants, gas and diesel engines, wood combustion, and many industrial processes) and by atmospheric reactions of gaseous pollutants.

Although scientific evidence links harmful human health effects with exposures to both fine particles and thoracic coarse particles, the evidence is much stronger for fine particles than for thoracic coarse particles. Effects associated with exposures to both $PM_{2.5}$ and $PM_{10-2.5}$ include premature mortality, aggravation of respiratory and cardiovascular disease (as indicated by increased hospital and emergency department visits), and changes in sub-clinical indicators of respiratory and cardiac function. Such health effects have been associated with short- and/or long-term exposure to PM.[ii] Exposures to $PM_{2.5}$ are also associated with decreased lung function growth, exacerbation of allergic symptoms, and increased respiratory symptoms.[6] Children, older adults, individuals with preexisting heart and lung disease (including asthma), and persons with lower socioeconomic status are considered to be among the groups most at risk for effects associated with PM exposures.[6] Information is accumulating and currently provides suggestive evidence for associations between long-term $PM_{2.5}$ exposure and developmental effects such as low birth weight and infant mortality due to respiratory causes.[6]

Sulfur Dioxide

Fossil fuel combustion by electrical utilities and industry is the primary source of sulfur dioxide in the United States.[5] People with asthma are especially susceptible to the effects of sulfur dioxide.[5] Short-term exposures of asthmatic individuals to elevated levels of sulfur dioxide while exercising at a moderate level may result in breathing difficulties, accompanied by symptoms such as wheezing, chest tightness, or shortness of breath. Studies also provide consistent evidence of an association between short-term sulfur dioxide exposures and increased respiratory symptoms in children, especially those with asthma or chronic respiratory symptoms. Short-term exposures to sulfur dioxide have also been associated with respiratory-related emergency department visits and hospital admissions, particularly for children and older adults.[5]

Nitrogen Dioxide

Nitric oxide (NO) and nitrogen dioxide (NO_2) are emitted by cars, trucks, buses, power plants, and non-road engines and equipment. Emitted NO is rapidly oxidized into NO_2 in the atmosphere.[4] Exposure to nitrogen dioxide has been associated with a variety of health effects,

---

[ii] For $PM_{10-2.5}$, the evidence linking health effects to short-term (e.g., 24-hour) exposures is stronger than the evidence for effects of long-term exposures.

including respiratory symptoms, especially among asthmatic children, and respiratory-related emergency department visits and hospital admissions, particularly for children and older adults.[4]

Lead

Historically, the major source of lead emissions to the air was combustion of leaded gasoline in motor vehicles (such as cars and trucks). Following the elimination of leaded gasoline in the United States by the mid-1990s, the remaining sources of lead air emissions have been industrial sources, including lead smelting and battery recycling operations, and piston-engine small aircraft that use leaded aviation gasoline.[3] Lead accumulates in bones, blood, and soft tissues of the body. Exposure to lead can affect development of the central nervous system in young children, resulting in neurodevelopmental effects such as lowered IQ and behavioral problems.[3]

Carbon Monoxide

Gasoline-fueled vehicles and other on-road and non-road mobile sources are the primary sources of carbon monoxide (CO) in the United States.[7] Exposure to carbon monoxide reduces the capacity of the blood to carry oxygen, thereby decreasing the supply of oxygen to tissues and organs such as the heart. People with several types of heart disease already have a reduced capacity for pumping oxygenated blood to the heart, which can cause them to experience myocardial ischemia (reduced oxygen to the heart), often accompanied by chest pain (angina), when exercising or under increased stress. For these people, short-term CO exposure further affects their body's already compromised ability to respond to the increased oxygen demands of exercise or exertion. Thus people with angina or heart disease are identified as at greatest risk from ambient CO. Other potentially at-risk populations include those with chronic obstructive pulmonary disease, anemia, diabetes, and those in prenatal or elderly lifestages.[7]

The period of fetal development may be one of particular vulnerability for adverse health effects resulting from maternal exposure to some criteria air pollutants. This may occur if maternal exposure to air pollutants is transferred to the fetus during pregnancy; for example, lead and PM have both been shown to cross the placenta and accumulate in fetal tissue during gestation.[3,6] In addition to the findings noted above regarding associations of prenatal PM exposure and adverse birth outcomes (such as low birth weight), limited studies of prenatal exposure to criteria air pollutants have reported that exposure to PM and oxides of nitrogen and sulfur may increase the risk of developing asthma as well as worsen respiratory outcomes among those children that do develop asthma.[25-27] However, it is often difficult to distinguish the effects of prenatal and early childhood exposure because exposure to air pollutants is often very similar during both time periods.

Additional research indicates that exposure to pollution from traffic-related sources, a mix of criteria air pollutants and hazardous air pollutants, may pose particular threats to a child's respiratory system. Many studies have reported a correlation between proximity to traffic (or to traffic-related pollutants) and occurrence of new asthma cases or exacerbation of existing asthma and other respiratory symptoms, including reduced growth of lung function during childhood.[25,28-35] A report by the Health Effects Institute concluded that living close to busy roads appears to be an independent risk factor for the onset of childhood asthma.[36] The same report also concluded that the evidence was "sufficient" to infer a causal association between exposure to traffic-related pollution and exacerbations of asthma in children.[36] Some studies have suggested that traffic-related pollutants may contribute to the development of allergic disease, either by affecting the immune response directly or by increasing the concentration or biological activity of the allergens themselves.[37-39]

Many of the effects of criteria air pollutants on children can be reduced by limiting outdoor activities on high pollution days.[40] Such avoidance measures can have their own adverse impacts on children's health when they reduce opportunities for play and exercise.

The following three indicators provide different perspectives on children's exposures to criteria air pollutants. Indicator E1 summarizes the percentages of children over time living in counties where measured pollutant concentrations were above the levels of the short- and/or long-term standards for each of the criteria air pollutants.[iii] Indicator E2 provides additional detail on the frequency with which pollutant concentrations were above the levels of the ozone and 24-hour $PM_{2.5}$ standards in one year (2009). Indicator E3 focuses on the frequency with which children were exposed to good, moderate, or unhealthy daily air quality, based on EPA's Air Quality Index.

[iii] For standards with averaging times less than or equal to 24 hours, Indicator E1 includes counties where concentrations were above the level of the standards at least one day per year.

Indicator E1: Percentage of children ages 0 to 17 years living in counties with pollutant concentrations above the levels of the current air quality standards, 1999–2009

Indicator E2: Percentage of children ages 0 to 17 years living in counties with 8-hour ozone and 24-hour PM2.5 concentrations above the levels of air quality standards, by frequency of occurrence, 2009

About the Indicators: Indicators E1 and E2 present the percentage of children living in counties where measured ambient concentrations of criteria pollutants were greater than the levels of the Clean Air Act health-based standards at any time during a year. Indicator E1 presents results for each criteria pollutant for each year. Indicator E2 presents more detailed information on the frequency with which measured ambient ozone and fine particle ($PM_{2.5}$) concentrations were greater than the levels of the short-term standards for ozone and $PM_{2.5}$ in 2009. The air quality data used in these indicators are from an EPA database that compiles measurements of pollutants in ambient air from around the country each year.

Air Quality System

State and local environmental agencies that monitor air quality submit their data to EPA. EPA compiles the monitoring data in the national EPA Air Quality System (AQS) database.[iv] AQS contains some monitoring data from the late 1950s and early 1960s, but there is not an appreciable amount of data for lead until 1970, sulfur dioxide until 1971, nitrogen dioxide until 1974, carbon monoxide and ozone until 1975, and PM_{10} until 1987. AQS also contains monitoring data for $PM_{2.5}$ beginning in 1999; $PM_{2.5}$ was measured only infrequently prior to 1999. Indicators E1 and E2 are derived from analysis of air pollution data in AQS.

Air Quality Standards and Concentrations Above the Levels of the Standards

Under the Clean Air Act, EPA has established National Ambient Air Quality Standards (NAAQS) for carbon monoxide, lead, ground-level ozone, particulate matter, nitrogen dioxide, and sulfur dioxide. There are four basic elements of NAAQS that together serve to define each standard: the definition of the pollutant,[v] the averaging time (e.g., annual average or 24-hour average), the level, and the form of the standard (which defines the air quality statistic compared to the level of the standard in determining whether an area attains the standard—for example, the 24-hour $PM_{2.5}$ standard uses 98th percentile concentrations, averaged over three years). These elements must be considered collectively in evaluating the health and welfare protection afforded by the NAAQS.

[iv] Information on the AQS database is available at http://www.epa.gov/airdata/.

[v] In the development of NAAQS, the term "indicator" defines the chemical species or mixture that is to be measured in determining whether an area attains the standard. To avoid confusion with the way in which "indicator" is used throughout *America's Children and the Environment*, the term is not used in the following paragraphs, except to refer to the ACE criteria pollutant indicators E1, E2, and E3.

Indicators E1 and E2 consider the first three elements of a NAAQS: the definition of the pollutant, the averaging time, and the level of the standard. The indicators present percentages of children living in areas with pollutant concentrations above the level of the current standards, using the appropriate averaging time. The indicators do not consider the form of the standard, which often includes considerations for multiple years of air quality data (e.g., 3 years), adjustments for missing data and less-than daily monitoring, and consideration for the frequency and magnitude with which a standard level is exceeded. In considering the form of the NAAQS, these standards are defined to allow some days to be above the level of the standard while limiting the extent to which they are above the level of the standard. Furthermore, determinations of attainment with the NAAQS are generally based on air quality averaged over multiple years. Therefore, air quality in any one-year period, as presented in Indicators E1 and E2, cannot be used to characterize whether air quality does or does not meet the NAAQS. The analyses for Indicators E1 and E2 therefore differ from the analyses used by EPA for the designation of "nonattainment areas" (locations that have not attained the standard) for regulatory compliance purposes.[41] Nonetheless, looking at air quality within a given year, or across many individual years, provides important public health information.

For each of the years 1999–2009, Indicator E1 reflects comparisons of the monitoring data with the levels of the current NAAQS. The indicator for all years therefore incorporates the 2006 revision of the level of the 24-hour $PM_{2.5}$ standard[20] from 65 $\mu g/m^3$ to 35 $\mu g/m^3$; the 2008 revision of the level of the eight-hour ozone standard[19] from 0.08 ppm to 0.075 ppm;[vi] the 2008 revision of the level of the three-month standard[18] for lead from 1.5 $\mu g/m^3$ to 0.15 $\mu g/m^3$; the establishment of a new one-hour standard[42] for nitrogen dioxide with a level of 100 ppb, issued in 2010; and the establishment of a new one-hour standard[43] for sulfur dioxide with a level of 75 ppb, issued in 2010. Note that EPA promulgated a revised annual $PM_{2.5}$ standard in December 2012, which has not been incorporated into this analysis. Table 1 in the Methods documentation shows the criteria pollutant levels used for the purpose of this indicator to determine whether concentrations were above the standard level for each pollutant.[vii]

NAAQS are intended to provide public health protection, including providing protection for at-risk populations, with an adequate margin of safety.[viii] EPA's selection of the current standards

[vi] In January 2010, the EPA Administrator proposed to reconsider the ozone standard because she believed "that a standard set as high as 0.075 would not be considered requisite to protect public health with an adequate margin of safety, and that consideration of lower levels [was] warranted" (75 FR 2996, January 19, 2010). EPA is currently conducting the next statutorily mandated periodic review of the ozone standards, which the Agency plans to complete in 2014. See http://www.epa.gov/ttn/naaqs/standards/ozone/s_o3_index.html for more information on the current and previous ozone NAAQS reviews.

[vii] All criteria pollutants are included in Indicator E1, but for some pollutants with multiple primary standards (reflecting different averaging times), only a single standard is included. For CO only the 8-hour standard is included, because the 1-hour standard is rarely exceeded. For NO_2 only the 1-hour standard is included, because the annual standard is rarely exceeded.

[viii] The legislative history of section 109 of the Clean Air Act indicates that a primary standard is to be set at "the maximum permissible ambient air level... which will protect the health of an [sensitive] group of the population," and that for this purpose, "reference should be made to a representative sample of persons comprising the sensitive group rather than to a single person in such a group" S. Rep. No. 91-1196, 91st Cong., 2d Sess. 10 (1970).

for ozone, nitrogen dioxide, and sulfur dioxide were intended to protect against respiratory effects in at-risk populations, including children. EPA's selection of the current standards for particulate matter was based primarily on concerns for mortality and cardiovascular effects, as well as respiratory effects. EPA's selection of the current standard for lead was intended to reduce risks of neurodevelopmental effects in children. The standard for carbon monoxide is intended primarily to protect against potential effects in people with heart disease. The Clean Air Act does not require the EPA Administrator to establish a primary NAAQS at a zero-risk level or at background concentration levels, but rather at a level that reduces risk sufficiently so as to protect health with an adequate margin of safety. However, pollutant concentrations that are lower than the levels of the standards are not necessarily without risk for all individuals. No risk-free level of exposure has been determined for any of the criteria pollutants.

Data Presented in the Indicators

Indicator E1 presents the percentage of children living in counties with measured pollutant concentrations above the level of a NAAQS for any of the criteria pollutants, for each year from 1999–2009.[ix] The indicator begins with data for 1999 because, as noted above, this was the first year of widespread monitoring for $PM_{2.5}$. In addition to presenting data for each of the criteria pollutants separately, the indicator also presents the percentage of children living in counties with measured concentrations above the level of a NAAQS for any criteria air pollutant (i.e., exceedance of standard levels for one or more criteria air pollutants).

Indicator E1 does not differentiate between counties in which concentrations were above standard levels frequently or by a large margin, and areas in which concentrations were above standard levels only rarely or by a small margin. It also assumes that air pollutant concentrations are consistent throughout a county. Some pollutants, such as ozone and $PM_{2.5}$, tend to be well dispersed and generally have limited spatial variation within a county, whereas other pollutants such as lead might have higher concentrations within relatively smaller areas. The indicator is based on concentrations of individual pollutants compared with individual standard levels, and does not reflect any combined effect of exposure to multiple criteria pollutants.

All children living in all counties are considered in the indicator; however, many counties do not have air pollution monitors. Monitoring networks are typically designed to focus on areas that are expected to have higher concentrations or that have larger populations. If any of the unmonitored counties have concentrations above the levels of the NAAQS, Indicator E1 will understate the percentage of children living in counties with concentrations above standard levels. The indicator thus represents the percentage of all children who lived in counties with confirmed pollutant concentrations above the levels of the standards each year, where confirmation is provided by a valid monitor value in that year. The percentages of children in unmonitored counties in 2009 range from about 30% for ozone and $PM_{2.5}$ to about 50% for

---

[ix] For standards with averaging times less than or equal to 24 hours, Indicator E1 includes counties where concentrations were above the level of the standards at least one day per year.

PM_{10}, carbon monoxide, sulfur dioxide and nitrogen dioxide, and about 80% for lead.[x] These percentages have been fairly stable from 1999–2009, though there are some limited changes in monitoring from year to year. Those limited changes in monitoring mean that there are some small changes in data available for calculation of the indicator over time.

The supplemental data tables E1a and E1b show the percentage of children living in counties with concentrations above the levels of the air quality standards in 2009 by race/ethnicity (Table E1a) and family income (Table E1b).

Ambient concentrations were more frequently above the levels of the 8-hour ozone and the 24-hour $PM_{2.5}$ standards than the levels of the standards for other criteria pollutants. Indicator E2 provides information on the frequency with which concentrations were above the levels of these two standards in 2009. Counties were classified by the number of days during 2009 that measured pollutant concentrations were above the levels of the 8-hour ozone and 24-hour $PM_{2.5}$ standards. This indicator, therefore, shows the percentage of children living in counties in which concentrations were measured above the levels of these two short-term standards a few times, as well as the percentage in counties with more frequent measurements above the levels of the standards. The percentage of children in counties without monitors for these two pollutants in 2009 is also shown in Indicator E2. The data table for this indicator (Table E2) also provides the same information for each year 1999–2009, using the current level of the standards for each year's calculation.

Values in this indicator may be understated due to the fact that most monitors do not operate every day. Ozone monitors operate daily during the ozone season, which lasts from 6 to 7 months in most locations but can be between 5 and 12 months (based on ranges of dates when high temperatures associated with high ozone concentrations may occur). $PM_{2.5}$ monitors operate year round, but may collect measurements daily or every third or every sixth day. EPA requires areas that measure concentrations within 5% of the 24-hour $PM_{2.5}$ standard to monitor daily. Monitors for other criteria pollutants operate year round.

Statistical Testing

Statistical analysis has been applied to Indicator E1 to evaluate trends over time in the percentage of children living in counties with concentrations above the standard levels each year. These analyses use a 5% significance level, meaning that a conclusion of statistical significance is made only when there is no more than a 5% probability that the observed trend occurred by chance ($p \leq 0.05$). The statistical analysis of trends over time is dependent on how the annual values vary as well as on the number of annual values. For example, the statistical test is more likely to detect a trend when data have been obtained over a longer period. It should be noted that conducting statistical testing for multiple air quality standards increases the probability that some trends identified as statistically significant may actually have occurred by chance.

---

[x] EPA issued increased requirements for lead monitoring in December 2010.[44]

A finding of statistical significance is useful for determining that an observed trend was unlikely to have occurred by chance. However, a determination of statistical significance by itself does not convey information about the magnitude of the increase or decrease in indicator values. Furthermore, a lack of statistical significance means only that occurrence by chance cannot be ruled out. Thus, a conclusion about statistical significance is only part of the information that should be considered when determining the public health implications of trends.

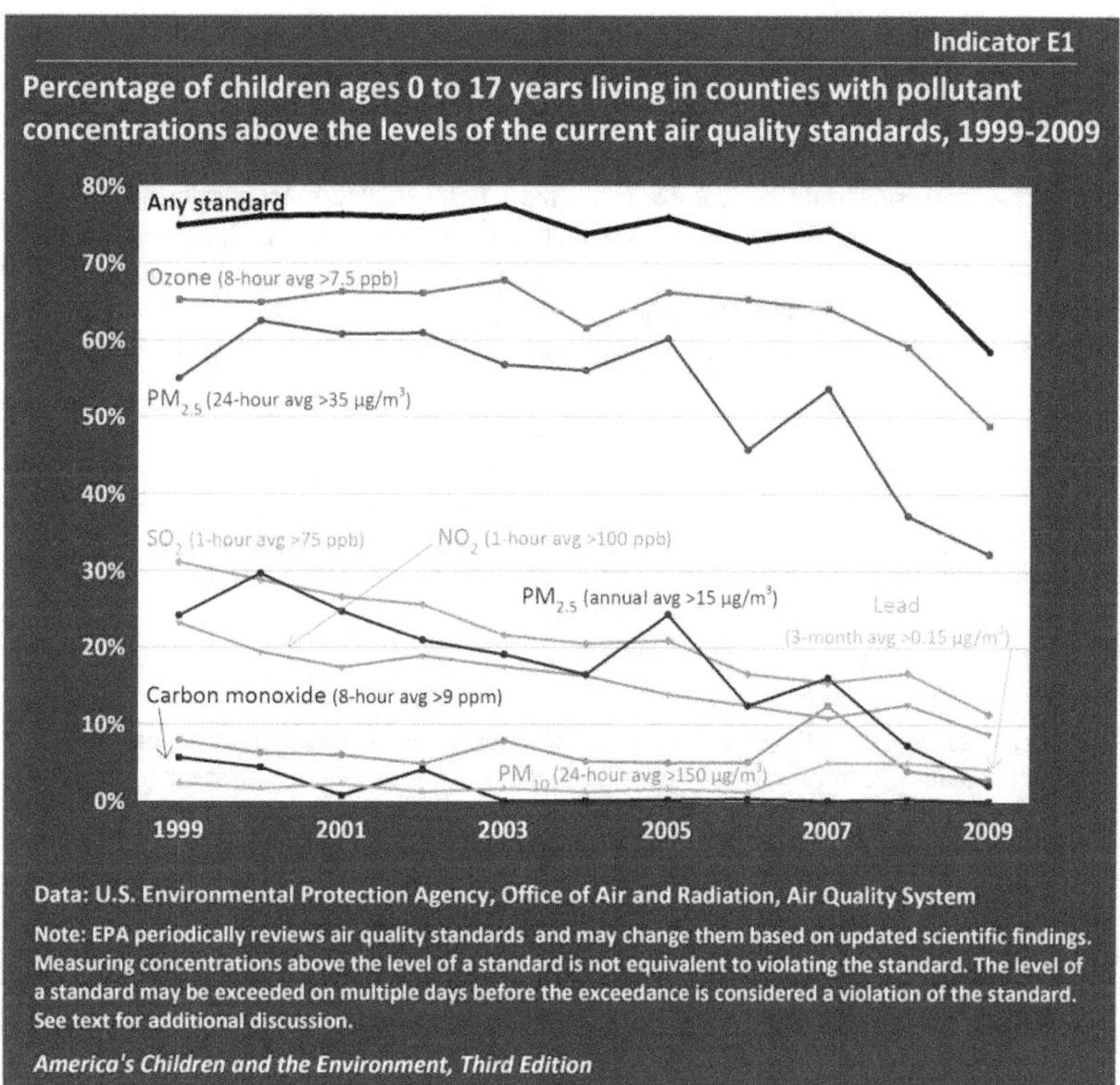

Indicator E1

Percentage of children ages 0 to 17 years living in counties with pollutant concentrations above the levels of the current air quality standards, 1999-2009

Data: U.S. Environmental Protection Agency, Office of Air and Radiation, Air Quality System

Note: EPA periodically reviews air quality standards and may change them based on updated scientific findings. Measuring concentrations above the level of a standard is not equivalent to violating the standard. The level of a standard may be exceeded on multiple days before the exceedance is considered a violation of the standard. See text for additional discussion.

America's Children and the Environment, Third Edition

Data characterization
- Data for this indicator are obtained from EPA's database of air quality monitoring measurements.
- Air pollution monitors are placed in locations throughout the country, with an emphasis on areas expected to have higher pollutant concentrations or that have larger populations. Not all counties in the United States have air pollution monitors, and the number of counties with monitors has changed over time.
- Monitors generally tend to stay in the same location over many years, but there may be some limited changes in the number or location of monitors providing data from year to year.

- From 1999 to 2009, the proportion of children living in counties with measured pollutant concentrations above the levels of one or more national ambient air quality standards decreased from 75% to 59%. This includes both concentrations above the level of any current short-term standard at least once during the year as well as average concentrations above the level of any current long-term standards.

- The decreasing trend over the years 1999–2009 was statistically significant.
- From 1999–2009, the percentage of children living in counties with measured ozone concentrations above the level of the current 8-hour ozone standard at least one day during the year decreased from 65% to 49%.
  - The decreasing trend for ozone over the years 1999–2009 was statistically significant.
- From 1999–2009, the percentage of children living in counties with measured $PM_{2.5}$ concentrations above the level of the current 24-hour $PM_{2.5}$ standard at least once per year decreased from 55% to 32%. Over the same years, the percentage of children living in counties with a measured annual average concentration above the level of the current annual $PM_{2.5}$ standard declined from 24% to 2%.
  - The decreasing trends for $PM_{2.5}$ were statistically significant.
- From 1999–2009, the percentage of children living in counties with measured sulfur dioxide concentrations above the level of the current one-hour standard for sulfur dioxide at least one day per year declined from 31% to 11%. Over the same years, the percentage of children living in counties with measured concentrations above the level of the current one-hour standard for nitrogen dioxide at least one day per year decreased from 23% to 9%.
  - The decreasing trends for both sulfur dioxide and nitrogen dioxide were statistically significant.
- In each year since 1999, between 1 and 5% of children lived in counties with measured ambient lead concentrations above the level of the current three-month standard for lead. In 2009, 8 counties with 4% of U.S. children reported concentrations above the level of the three-month standard for lead.
- In 2009, 3% of children lived in counties with measured PM_{10} concentrations above the level of the current 24-hour standard for PM_{10} at least one day per year, and no children lived in counties with measured concentrations above the level of the current standard for carbon monoxide.

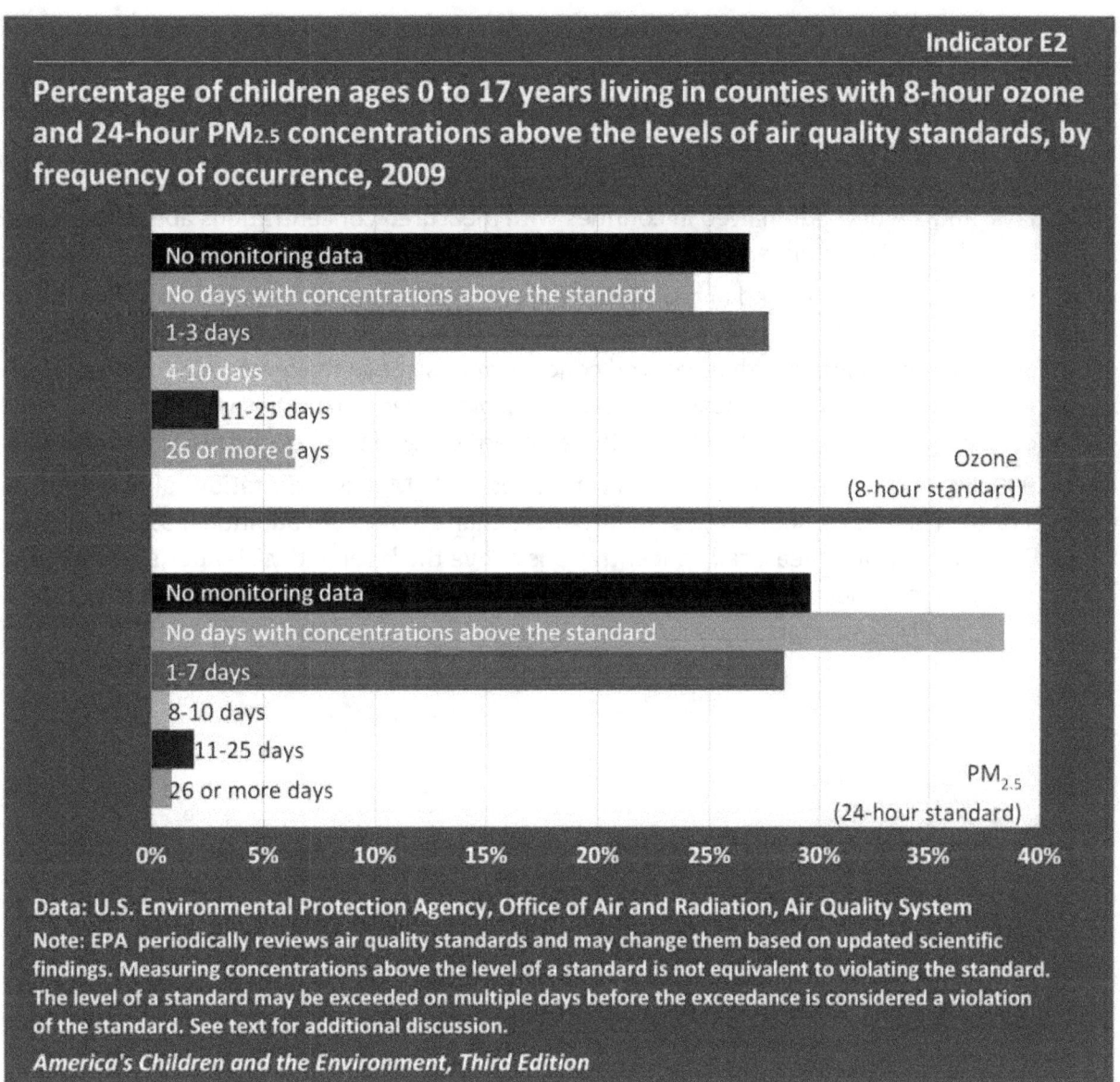

Indicator E2

Percentage of children ages 0 to 17 years living in counties with 8-hour ozone and 24-hour PM$_{2.5}$ concentrations above the levels of air quality standards, by frequency of occurrence, 2009

Ozone (8-hour standard):
- No monitoring data
- No days with concentrations above the standard
- 1-3 days
- 4-10 days
- 11-25 days
- 26 or more days

PM$_{2.5}$ (24-hour standard):
- No monitoring data
- No days with concentrations above the standard
- 1-7 days
- 8-10 days
- 11-25 days
- 26 or more days

Data: U.S. Environmental Protection Agency, Office of Air and Radiation, Air Quality System

Note: EPA periodically reviews air quality standards and may change them based on updated scientific findings. Measuring concentrations above the level of a standard is not equivalent to violating the standard. The level of a standard may be exceeded on multiple days before the exceedance is considered a violation of the standard. See text for additional discussion.

America's Children and the Environment, Third Edition

Data characterization
- Data for this indicator are obtained from EPA's database of air quality monitoring measurements.
- Air pollution monitors are placed in locations throughout the country, with an emphasis on areas expected to have higher pollutant concentrations or that have larger populations. Not all counties in the United States have air pollution monitors.
- Some air pollution monitors do not operate every day, so some days with pollutant concentrations above the levels of the air quality standards may not be identified.
- In 2009, 27% of children lived in counties with no monitoring data for ozone, and 30% lived in counties with no monitoring data for PM$_{2.5}$.

- In 2009, 6% of children lived in counties with measured ozone concentrations above the level of the 8-hour ozone standard on more than 25 days. An additional 3% of children lived in counties with measured concentrations above the level of the ozone standard between

11 and 25 days, and 12% of children lived in counties where concentrations were above the level of the standard between 4 and 10 days.

- In 2009, 1% of children lived in counties with measured $PM_{2.5}$ concentrations above the level of the 24-hour $PM_{2.5}$ standard on more than 25 days. An additional 2% of children lived in counties with measured concentrations above the level of this standard between 11 and 25 days, and 1% of children lived in counties with measured concentrations above the level of the 24-hour $PM_{2.5}$ standard between 8 and 10 days.

- In 1999, 23% of children lived in counties with measured ozone concentrations above the level of the current 8-hour ozone standard on more than 25 days. An additional 27% of children lived in counties with measured concentrations above the level of the ozone standard between 11 and 25 days, and 11% of children lived in counties where concentrations were above the level of the standard between 4 and 10 days. (See Table E2.)

- In 1999, 6% of children lived in counties with measured $PM_{2.5}$ concentrations above the level of the current 24-hour $PM_{2.5}$ standard more than 25 days. An additional 9% of children lived in counties with measured concentrations above the level of this standard 11 and 25 days, and 3% of children lived in counties with measured concentrations above the level of the 24-hour $PM_{2.5}$ standard between 8 and 10 days. (See Table E2.)

Indicator E3: Percentage of days with good, moderate, or unhealthy air quality for children ages 0 to 17 years, 1999–2009

About the Indicator: Indicator E3 presents data from EPA's Air Quality Index (AQI). The AQI produces a rating of the air quality for each county on each day, considering all monitoring results available on that day for carbon monoxide, ozone, nitrogen dioxide, particulate matter, and sulfur dioxide. Air quality in each county is considered to be "good," "moderate," or "unhealthy" based on comparison of the monitored pollutant concentrations to breakpoints defined by the AQI. The indicator is calculated by considering the number of children in counties with each rating for each day of the year, then summing the number of children for all days in the year.

Air Quality Index

EPA's Air Quality Index (AQI)[xi] represents air quality for each individual day and is widely reported in newspapers and other media outlets in metropolitan areas. The AQI is based on daily measurements of up to five of the six air quality criteria pollutants (carbon monoxide, ozone, nitrogen dioxide, particulate matter, and sulfur dioxide). The standard for lead is not included in the AQI because it requires averaging concentrations over a three-month period, and it can take several weeks to collect and analyze lead samples.

The specific pollutants considered in the AQI for each metropolitan area depend on the pollutants monitored in that area each day. Each pollutant concentration is given a value on a scale relative to the air quality standard for that pollutant. The daily AQI is based on the single pollutant with the highest index value that day. An AQI value of 100 corresponds to the level of the short-term (e.g., daily or hourly) NAAQS for each criteria pollutant. An AQI value of 50 is defined either as the level of the annual standard, if one has been established (e.g., $PM_{2.5}$, NO_2), or as a concentration equal to one-half the value of the short-term standard used to define an index value of 100 (e.g., CO).

EPA has divided the AQI scale into categories. Air quality is considered "good" (referred to as "code green") if the AQI is between 0 and 50, posing little or no risk. Air quality is considered "moderate" ("code yellow") if the AQI is between 51 and 100. Some pollutants at this level may present a moderate health concern for a small number of individuals. Air quality is considered "unhealthy for sensitive groups" if the AQI is between 101 and 150 (referred to as "code orange"). On code orange days, members of at-risk populations such as children may experience health effects, but the rest of the general population is unlikely to be affected. Air quality is considered "unhealthy" if the AQI is between 151 and 200 ("code red"). The general population may begin to experience health effects, and members of at-risk populations may experience more serious health effects. Values of 201 to 300 are designated as "very unhealthy" ("code purple"), while values of 301 to 500 are considered "hazardous" ("code maroon"). Decisions about the pollutant concentrations at which to set the various AQI

[xi] Available at http://www.airnow.gov/.

breakpoints that delineate the various AQI categories draw directly from the underlying health information that supports the reviews of the NAAQS.

For $PM_{2.5}$, the AQI values used in preparing Indicator E3 were calculated with a 24-hour concentration of 40 $\mu g/m^3$ used to define air quality as "unhealthy for sensitive groups" (i.e., an AQI value of 100), rather than the level of the current 24-hour $PM_{2.5}$ standard of 35 $\mu g/m^3$. As a consequence, Indicator E3 likely overstates the days with moderate air quality and understates the days with unhealthy air quality.[xii]

Data Presented in the Indicator

Indicator E3 is based on the reported AQI for counties in the United States. EPA calculates an AQI value each day in each county for five major air pollutants regulated by the Clean Air Act: ozone, particulate matter, carbon monoxide, sulfur dioxide, and nitrogen dioxide. The highest of these pollutant-specific AQI values is reported as the county's AQI value for that day.

Indicator E3 was developed by reviewing the AQI designation for each day for each county and weighting the daily designations by the number of children living in each county. The calculation, therefore, is a summation of the AQI values for all children in the United States, based on county of residence, for each day of the year. For example, the number of days of good air quality during the year is counted up for each child in the population based on the daily air quality in the county where they live. The overall indicator reports the percentage of children's days in each year considered to be of good (AQI 0–50; code green), moderate (AQI 51–100; code yellow), or unhealthy (AQI greater than 100; codes orange, red, purple, and maroon combined) air quality.[xiii] The percentage of children's days with no AQI value available (representing the absence of monitoring data) are also reported in Indicator E3.

Whereas Indicator E1 presents an annual analysis of counties in which concentrations were above the level of a standard for a pollutant, the AQI data used in Indicator E3 are based on the concentrations for all pollutants for which an AQI has been established in each county over the course of a year. The E3 method uses data on the air quality category for each day, rather than simply reporting whether a county ever exceeds the standard for each pollutant during the year. However, the AQI method has some limitations. The AQI is based on the single pollutant with the highest value for each day; it does not reflect any combined effect of multiple pollutants or the effects of pollutants that were not measured on a given day.

[xii] In December 2012, EPA promulgated a rule to change the AQI to use 35 $\mu g/m^3$ for defining the AQI value of 100 for $PM_{2.5}$. Prior to this rule, although the AQI had not formally been changed, an EPA guidance document[45] recommended use of 35 $\mu g/m^3$ for defining the AQI value of 100 for $PM_{2.5}$. States have generally been using 35 $\mu g/m^3$ in calculating and reporting their daily AQI values.

[xiii] As discussed above, an AQI value of 100 generally corresponds to the level of a short-term national ambient air quality standard. When AQI values are above 100, air quality is considered to be unhealthy—at first for certain sensitive groups of people (101 to 150), then for everyone as AQI values get higher.

Indicator E3 starts in 1999 because this was the first year of widespread monitoring for $PM_{2.5}$. The indicator uses a consistent set of pollutant concentrations to define good, moderate, or unhealthy air quality for all years shown, 1999–2009, but as noted above, the level of the current 24-hour standard for $PM_{2.5}$ has not been incorporated into calculation of the indicator.

Tables E3a and E3b show the percentage of children's days of exposure to good, moderate, or unhealthy air quality in 2009 by race/ethnicity (Table E3a) and family income (Table E3b). These calculations do not account for any possible variation in air quality within a county, and thus may not fully reflect the variability in air quality among children of different race/ethnicity and income.

Statistical Testing

Statistical analysis has been applied to Indicator E3 to evaluate trends over time in the percentage of children's days of with good, moderate, or unhealthy air quality. These analyses use a 5% significance level, meaning that a conclusion of statistical significance is made only when there is no more than a 5% probability that the observed trend occurred by chance ($p \leq 0.05$). The statistical analysis of trends over time is dependent on how the annual values vary as well as on the number of annual values. For example, the statistical test is more likely to detect a trend when data have been obtained over a longer period.

A finding of statistical significance is useful for determining that an observed trend was unlikely to have occurred by chance. However, a determination of statistical significance trend over time does not imply anything about the magnitude of the increase or decrease in indicator values. Furthermore, a lack of statistical significance means only that occurrence by chance cannot be ruled out. Thus, a conclusion about statistical significance is only part of the information that should be considered when determining the public health implications of trends.

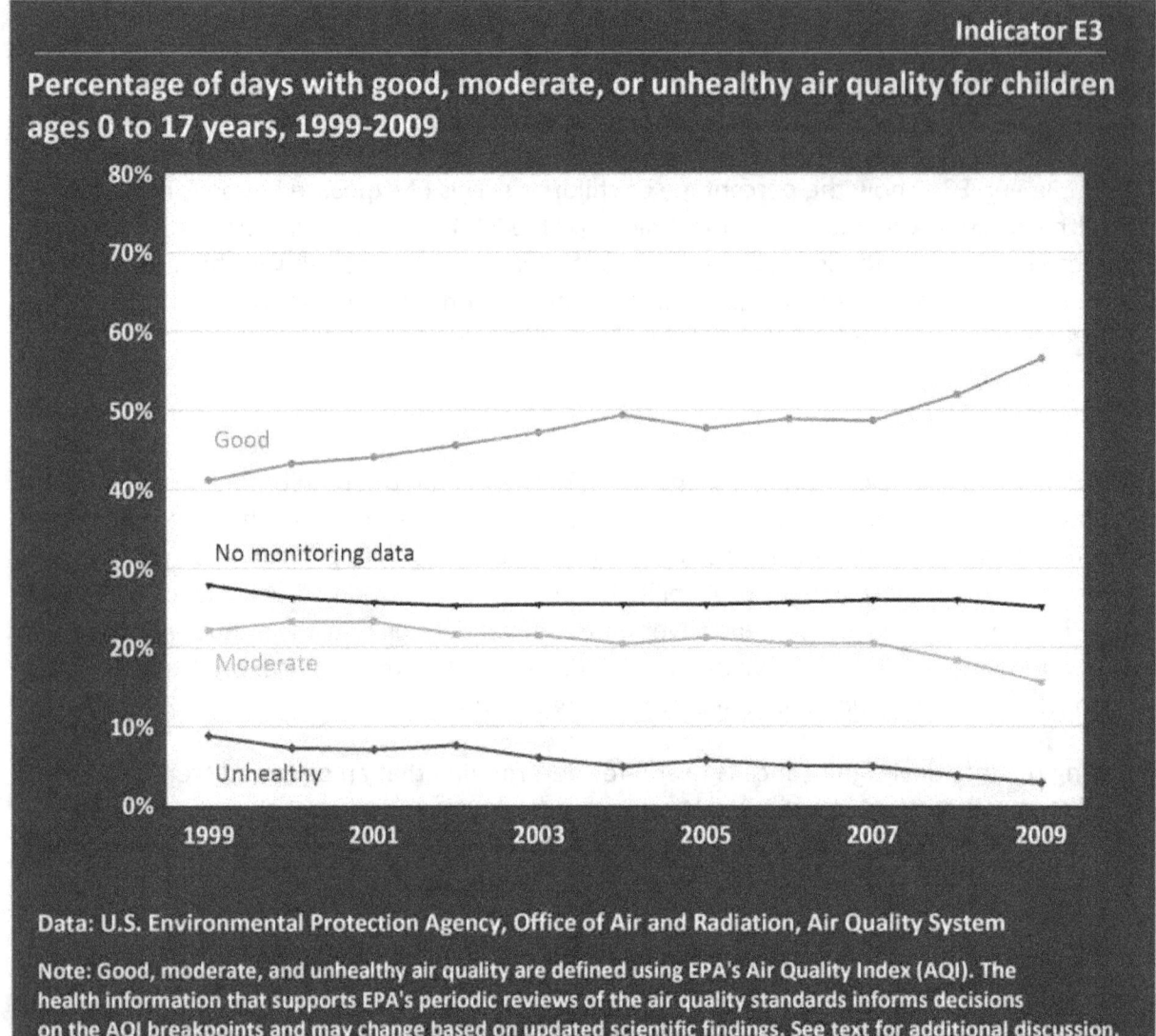

Percentage of days with good, moderate, or unhealthy air quality for children ages 0 to 17 years, 1999-2009

Data: U.S. Environmental Protection Agency, Office of Air and Radiation, Air Quality System

Note: Good, moderate, and unhealthy air quality are defined using EPA's Air Quality Index (AQI). The health information that supports EPA's periodic reviews of the air quality standards informs decisions on the AQI breakpoints and may change based on updated scientific findings. See text for additional discussion.

America's Children and the Environment, Third Edition

Data characterization
– Data for this indicator are obtained from EPA's database of daily Air Quality Index (AQI) values for each county in the United States.
– Air pollution monitors are placed in locations throughout the country, with an emphasis on areas expected to have higher pollutant concentrations or that have larger populations.
– AQI values are based on daily monitoring data for up to five criteria air pollutants. Some counties do not have monitors, and some monitors do not operate every day, so some days do not have AQI values.
– For this indicator, the available monitoring data are used to assign a value of "good," "moderate," "unhealthy," or "no monitoring data" for each day in each U.S. county.

- The percentage of children's days that were designated as having "unhealthy" air quality decreased from 9% in 1999 to 3% in 2009. The percentage of children's days with "good" air quality increased from 41% in 1999 to 57% in 2009. The percentage of children's days with "moderate" air quality was approximately constant at 21–23% from 1999 to 2007, and then decreased to 16% in 2009.

  - The 1999 to 2009 trends in "unhealthy," "good," and "moderate" air quality days were statistically significant.

Hazardous Air Pollutants

Hazardous air pollutants (HAPs) are air contaminants, frequently referred to as "air toxics," that are known or suspected to cause serious human health effects or adverse environmental effects.[1] The Clean Air Act identifies 187 substances as HAPs. Examples include benzene, trichloroethylene, mercury, chromium, and dioxin. The "criteria" air pollutants such as ozone and particulate matter are excluded from the HAPs list.[i]

HAPs are emitted into ambient air from a diverse range of facilities, businesses, and vehicles that are grouped into three general categories: major sources, area sources, and mobile sources. Major sources typically are large industrial facilities such as chemical manufacturing plants, refineries, and waste incinerators. These sources may release air toxics from equipment leaks, when materials are transferred from one location to another, or during discharge through emission stacks or vents. Area sources typically are smaller stationary facilities such as dry cleaners, auto body repair shops, and small manufacturing operations. Though emissions from individual area sources often are relatively small, collectively they can be of concern— particularly where large numbers of sources are located in heavily populated areas. Mobile sources include both on-road sources, such as cars, light trucks, large trucks, and buses, and non-road sources such as farm and construction equipment, lawn and garden equipment, marine engines, aircraft, and locomotives. Some HAPs are also emitted from natural sources such as volcanoes. Health effects associated with HAPs include cancer, asthma and other respiratory ailments, birth defects, reproductive effects, and neurodevelopmental effects.[2-9]

In some cases, health concerns are based on studies of workers exposed to high levels of particular HAPs on the job. For example, EPA has determined that HAPs such as benzene; 1,3-butadiene; chromium; nickel; and vinyl chloride are carcinogenic to humans, based on findings in occupational studies.[10-14] Similarly, toluene diisocyanate exposure has been associated with effects on the lung, and manganese exposure with neurological effects, in occupational studies.[15,16]

A limited number of HAPs have also been studied in human populations that have been exposed in their day-to-day lives. For examples, several studies have reported associations between formaldehyde exposure (usually indoors at home or at school) and childhood asthma.[3] In addition, a series of recent studies conducted in New York City reported that children of women who were exposed to increased levels of polycyclic aromatic hydrocarbons (PAHs, produced when gasoline and other materials are burned) during pregnancy are more likely to have experienced adverse effects on neurological development (such as reduced intelligence quotient (IQ) or behavioral problems[6,7]), as well as respiratory effects.[17-19]

---

[i] Lead is an exception: it is regulated as a criteria pollutant, and "lead compounds" are included on the list of HAPs. Note that criteria pollutants are discussed further in the *Criteria Air Pollutants* topic.

For the majority of HAPs, however, there are no human epidemiological studies, or very few, and concern for health effects is based on findings from animal studies. For example, many HAPs, such as PAHs,[20] acetaldehyde[21] and carbon tetrachloride[22] are considered likely to be carcinogenic to human based primarily on evidence from long-term laboratory animal studies.

Although many HAPs are of concern due to their potential to cause cancer, a substantial number of HAPs lack evidence of cancer—either because the relevant long-term studies have not been conducted, or because studies have been conducted and do not indicate carcinogenic potential. An example of a HAP that is not associated with cancer is acrolein; there are no appropriate human or animal studies with which to assess the carcinogenic potential of acrolein. However, acrolein has been identified as a HAP of particular concern for effects other than cancer.[23,24] Health concerns for acrolein include respiratory effects and irritation of the eyes, nose, and throat, based on animal studies and on short-term studies of small groups of humans intentionally exposed to high levels of acrolein.[25]

EPA relies on both monitoring and modeled data to characterize ambient air concentrations of HAPs, and to estimate potential human exposure and risk of adverse health effects associated with these toxics. EPA and state monitoring programs do not cover all the places where people live in the United States. For this reason, the following indicator relies on modeled data from the National Air Toxics Assessment.[26] The indicator presents the percentage of children living in census tracts with estimated HAP concentrations greater than benchmark comparison levels derived from health effects information.

In addition to their presence in ambient air, many HAPs also have indoor sources, and the indoor sources may frequently result in greater exposure than the presence of HAPs in ambient air. Sufficient data are not available to develop an indicator considering the combined exposure to HAPs from both indoor and outdoor sources; therefore the following indicator considers only levels of HAPs in ambient air.[ii]

[ii] Indoor sources of HAPs are further discussed in the *Indoor Environments* and *Contaminants in Schools and Child Care Facilities* topics, and in several of the biomonitoring topics.

Indicator E4: Percentage of children ages 0 to 17 years living in census tracts where estimated hazardous air pollutant concentrations were greater than health benchmarks in 2005

About the Indicator: Indicator E4 presents estimates of the percentage of children living in census tracts with ambient hazardous air pollutant (HAP) concentrations greater than benchmark values representing levels of concern for health effects. The HAP concentrations are computer model estimates for 2005, representing all identified sources of HAP emissions, including factories and motor vehicles. The health benchmarks are based on concerns for cancer and other adverse health effects that may be associated with HAP exposure.

National Air Toxics Assessment

EPA's National Air Toxics Assessment (NATA) provides estimated concentrations of 181 HAPs in ambient air for the year 2005. NATA is the most comprehensive resource on potential human exposure to and risk of adverse health effects from HAPs in the United States. Monitoring data are insufficient to characterize HAP concentrations across the country because of the limited number of monitors, and because concentrations of many HAPs may vary considerably within a metropolitan area or region.

Under NATA, EPA develops modeled estimates of ambient concentrations of HAPs using estimated emissions data from major, area, onroad mobile, and non-road mobile sources. These emissions data are collected and updated periodically, and are maintained in an emissions inventory. The original NATA was developed using emissions data for the year 1996. Since the initial release, EPA has developed additional estimates of ambient air concentrations of HAPs using updated emissions inventories for 1999, 2002, and 2005. NATA's computer modeling approach has the advantage of allowing estimation of HAP concentrations at locations throughout the United States, rather than in just those locations that have HAP monitors. However, compared with monitoring, the computer model requires estimating quantities of HAP emissions, estimating locations of HAP emissions sources, and modeling the dispersion of HAPs in the atmosphere after they have been emitted.

The most recent assessment developed estimated ambient concentrations of 179 air toxics for the year 2005. A computer model provided estimates for every census tract in the United States. The modeled estimates generally are consistent with the limited set of ambient air toxics monitoring data, although at many locations the model estimates for some HAPs are lower than measured concentrations.[27] The 2005 NATA estimates do not reflect any changes in emissions that may have occurred since 2005 due to new regulations, new technologies, changes in economic activity, or changes in the vehicle fleet and vehicle miles traveled.

Health Benchmarks for Hazardous Air Pollutants

Indicator E4 presents comparisons of modeled concentrations of HAPs in ambient air for 2005 with three health benchmark concentrations derived from scientific assessments conducted by EPA and other environmental agencies.[28] EPA uses the three benchmark risk levels to identify HAPs that are of priority concern.[29]

Two benchmarks reflect potential cancer risks, at levels of 1-in-100,000 risk and 1-in-10,000 risk. If a particular hazardous air pollutant is present in ambient air at a 1-in-100,000 benchmark concentration, for example, it is estimated that one additional case of cancer would occur in a population of 100,000 people exposed for a lifetime. The comparison to the cancer risk benchmark incorporates data for all HAPs considered carcinogenic to humans, likely carcinogenic to humans, or with suggestive evidence of carcinogenicity. The majority of HAPs included in the comparison to the cancer risk benchmarks are considered "carcinogenic to humans" or "likely carcinogenic to humans."[30]

The third benchmark concentration corresponds to the level at which exposure to the hazardous air pollutant is estimated to be of minimal risk for adverse non-cancer health effects; exposures above this benchmark may be associated with adverse health effects such as respiratory or neurological effects. Due to variation in human response to HAP exposure and uncertainty in the benchmark values, it is not necessarily the case that a person living in a location where this benchmark is exceeded will experience adverse effects. It is also possible that individuals may experience effects at levels below the benchmark level.

The health benchmarks are generally derived from laboratory animal studies, although for some HAPs they are derived from human epidemiological studies of workers exposed on the job. For some HAPs, even the animal studies are very limited and no benchmark has been derived. Health benchmarks were available to assess 87 HAPs as cancer-causing agents and 105 HAPs as agents that cause adverse health effects other than cancer. Some HAPs had benchmarks for both cancer and non-cancer health endpoints; a total of 141 air toxics were used in calculating the indicator.

Because they are typically based on studies of adults or mature laboratory animals, the three benchmarks generally reflect health risks to adults, rather than potential risks to children or risks in adulthood stemming from childhood exposure. Benchmarks for non-cancer effects incorporate assumptions that are based on adult respiratory physiology (i.e., breathing rates and lung structure); benchmarks for some HAPs would be lower if they were adjusted for children's respiratory physiology.[31]

Under a policy adopted in 2005, EPA adjusts risk estimates for certain carcinogens to account for increased risks associated with exposures during early life.[32] This adjustment has been applied to the cancer benchmarks for PAHs, acrylamide, benzidine, and ethyl carbamate. Benchmark values for other HAPs that are suspected carcinogens receive no adjustment for potential elevated risks from early-life exposures because they do not meet the criteria of the EPA policy or lack sufficient data to support application of the adjustment.

Further, the benchmarks reflect risks of continuous exposure over the course of a lifetime. Potential risks from higher concentrations experienced over a short amount of time (one day, one hour, or less) may in some cases trigger immediate responses, such as asthma attacks or effects on the central nervous system are not addressed by these benchmarks.

Finally, the benchmark values for HAP s are uncertain to varying degrees, due to data limitations and the lag in time between when new studies become available and the completion of updated assessments by EPA and other government agencies.

Data Presented in the Indicator

Indicator E4 presents the percentage of children living in census tracts where estimated 2005 HAP concentrations exceeded benchmark levels for cancer (at levels of 1-in-100,000 risk and 1-in-10,000 risk) and for other (non-cancer) adverse health effects. The indicator is calculated by comparing the estimated HAP concentrations for each U.S. census tract in 2005 to each of the benchmark concentrations. Census tracts are geographic areas within U.S. counties that vary in size and generally have 1,500 to 8,000 residents, with a typical population of 4,000 residents.

The comparison to the cancer risk benchmark sums up data for all carcinogenic HAPs. The comparison to the benchmark for other adverse health effects considers only individual HAPs; that is, a county is considered to exceed this benchmark if the modeled concentration for any single HAP exceeds the corresponding non-cancer benchmark for that HAP, but it does not consider adverse effects of combinations of HAPs.

Available information indicates that the NATA estimates of ambient HAP concentrations tend to be similar to or lower than actual HAP concentrations.[27] To the extent that underestimation occurs, the percentage of children living in census tracts exceeding the benchmark levels may be understated. In addition, the indicator does not differentiate between census tracts in which the benchmarks are exceeded by a large margin and those in which estimated HAP concentrations are just above the benchmark concentrations. The indicator presents results only for 2005, and does not compare results across assessment years, such as between 1999 and 2005, because each update of the assessment brings new improvements to methods. For example, improvements to the emissions estimation methodologies made in the 2005 assessment were not applied to the earlier versions, so the ambient concentration estimates are not entirely comparable between years.

Actual exposures may differ from ambient concentrations. Indoor concentrations of HAPs from outdoor sources may be slightly lower than ambient concentrations, although they can be significantly higher if any indoor sources are present.[33-36] Levels of some hazardous pollutants may be substantially higher inside cars and school buses,[37-39] and those higher levels would increase the risks.

In addition, this indicator only considers exposures to air toxics that occur by inhalation. For many air toxics, dietary exposures are also important. Air toxics that are persistent in the environment settle out of the atmosphere onto land and water, and then may accumulate in fish and other animals in the food web. For HAPs that are persistent in the environment and accumulate significantly in food, exposures through food consumption typically are greater than inhalation exposures. HAPs for which food chain exposures are important include mercury, dioxins, and PCBs.[40-42]

The comparison of ambient HAP concentrations In 2005 to the health benchmarks is not equivalent to an estimate of risk to the population from chronic HAP exposure. Actual risks to health depend on concentrations of HAPs in many environments over an extended period of time. Ambient concentrations will change over time as the mix of sources changes (e.g., due to businesses opening and closing), vehicle use changes (e.g., more cars and trucks traveling longer distances), and regulatory controls are applied. In addition, children spend most of their time indoors at home, at school, or at child care centers, and pollutant concentrations in indoor environments may be greater or lesser than the modeled ambient concentrations.

In addition to the indicator presented in the figure, which is based on where children live, the same statistics are calculated based on where children's schools are located (see data tables). Exposures at school are an important consideration, as children spend an average of 33 hours per week in school.[43] The data tables also provide indicator values by race/ethnicity and income, based on where children live.

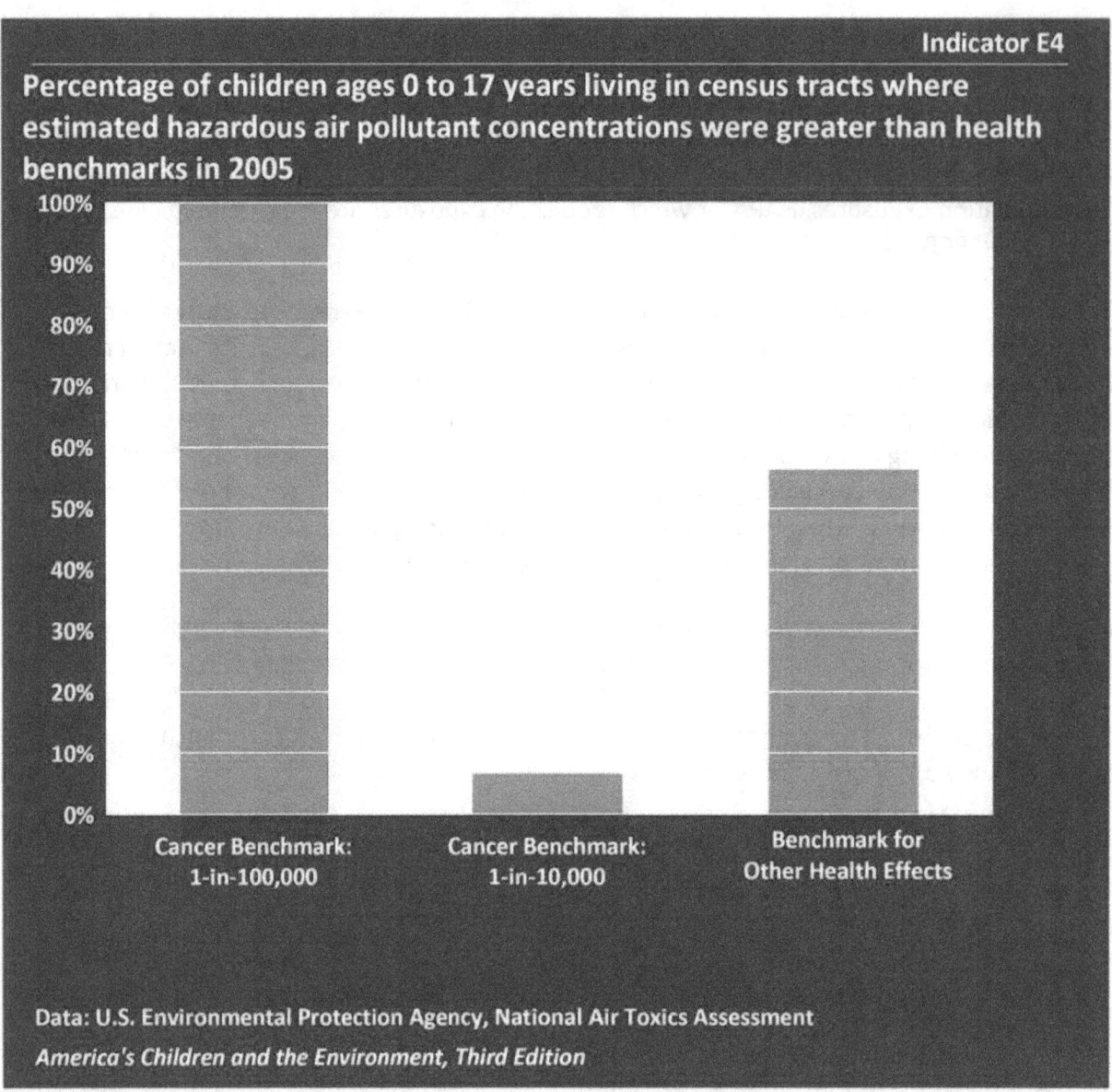

Indicator E4

Percentage of children ages 0 to 17 years living in census tracts where estimated hazardous air pollutant concentrations were greater than health benchmarks in 2005

Data: U.S. Environmental Protection Agency, National Air Toxics Assessment

America's Children and the Environment, Third Edition

Data characterization

–   Data for this indicator are obtained from EPA's National Air Toxics Assessment computer model predictions of hazardous air pollutant (HAP) concentrations in outdoor air.

–   The model produces estimates of HAP concentrations from emissions data for all census tracts in the United States (census tracts typically have about 4000 residents each).

■   In 2005, nearly all children lived in census tracts in which HAP concentrations combined to exceed the 1-in-100,000 cancer risk benchmark.

■   Seven percent of children lived in census tracts in which HAPs combined to exceed the 1-in-10,000 cancer risk benchmark. The pollutants that contributed most to this result were formaldehyde, benzene, acetaldehyde, carbon tetrachloride, and hexavalent chromium. Formaldehyde, benzene, and hexavalent chromium are considered to be carcinogenic to humans,[5,10,13] and acetaldehyde and carbon tetrachloride are considered likely to be carcinogenic to humans.[21,22]

- Fifty-six percent of children lived in census tracts in which at least one HAP exceeded the benchmark for health effects other than cancer. In almost all cases, this result was attributable to the pollutant acrolein, which is a respiratory irritant. More than 90% of acrolein emissions are from wood-burning fires and mobile sources such as cars, trucks, buses, planes, and construction equipment.

- Exposures to diesel particulate matter from diesel engine emissions are not included in this indicator due to uncertainty regarding the appropriate values to use as cancer benchmarks. Some studies have found that cancer risks from diesel particulate matter exceed those of the HAPs considered in this indicator.[44] Although EPA does not endorse any particular cancer benchmark value for diesel particulate matter, if the State of California's benchmark for diesel particulate matter were used in this analysis, 73% of children would live in census tracts where HAP estimates combined to exceed the 1-in-10,000 cancer risk benchmark.

- In 2005, all children's schools were located in census tracts where HAPs concentrations combined to exceed the 1-in-100,000 cancer risk benchmark. Six percent of children attended schools in census tracts where the HAPs concentrations exceeded the higher 1-in-10,000 cancer risk benchmark.

- Fifty-seven percent of children attended schools that were located in census tracts where at least one HAP exceeded the benchmark for health effects other than cancer.

Indoor Environments

Children spend most of their time in indoor environments, including homes, schools, child care facilities, and other buildings.[1] The chemicals found indoors or measured in indoor air are numerous and diverse. Hundreds of chemicals have been measured in indoor air, including multiple pesticides, fragrance-related compounds, polychlorinated biphenyls (PCBs), phthalates, combustion byproducts, carbon monoxide, benzene, formaldehyde, and other compounds.[2-4] Pollutants in indoor environments can come from many different sources, including combustion sources such as furnaces, gas stoves, fireplaces, and cigarettes; building materials and furnishings such as treated wood, paints, furniture, carpet, and fabrics; consumer goods such as electronics and toys; cleaning products, pesticides, and other products used for maintenance of the home or building; and products used for hobbies, science projects, arts and crafts projects, and other activities.

Children may also be routinely exposed to chemical contaminants that accumulate in dust, including lead, nicotine, pesticides, brominated flame retardants, phthalates, and perfluorinated chemicals.[3,5-9] Many pesticides and other chemicals that break down relatively quickly outdoors are much more persistent and long-lasting indoors, where they are less exposed to natural elements such as sunlight, moisture, and microorganisms that can accelerate the breakdown of chemicals.[10-12]

Infants and small children may have the highest exposure to house dust contaminants due to their frequent and extensive contact with floors, carpets, and other surfaces where dust gathers, as well as their frequent hand-to-mouth activity. However, children of all ages (as well as adults) are likely to be exposed to dust contaminants through hand-to-mouth activity[1,13] and other ingestion pathways, such as the settling of dust onto food and food preparation surfaces in the kitchen.

The indoor environments of personal cars and school buses are also important to children's exposure, as a child can spend up to an average of 84 minutes per day in a vehicle, depending on his or her age.[1] School bus cabins can have levels of fine particulate matter ($PM_{2.5}$) four times higher than levels in ambient air.[14] In addition, children riding school buses in urban areas are likely to be exposed to elevated levels of benzene, formaldehyde, and other pollutants in motor vehicle emissions. It is estimated that school buses commuting through congested urban areas may contribute up to 30% of a child's daily exposure to diesel engine-related pollutants.[15]

Adult smoking in personal cars can have a significant impact on children's environmental tobacco smoke exposures, as the air in smokers' cars tends to have significantly higher nicotine concentrations than that in non-smokers' cars.[16] Smoking in cars also leaves nicotine residues that may linger in dust and surfaces within smokers' cars, leading to continued exposure even after the practice of smoking within the car has ceased.[17]

Pollutants in indoor environments can also come from outside sources. For example, pollutants in outdoor air will penetrate to the indoor environment,[18,19] and contaminants from workplaces, streets, or lawns may be carried into the home on people's shoes or clothing.[20,21] Some contaminants in drinking water can enter indoor air through uses of hot water such as showering.[22,23] In areas where groundwater is contaminated, chemicals may enter indoor environments via vapor intrusion.[24,25] Radon, a gaseous radioactive element that causes lung cancer, is found in soils and can enter homes through cracks in the foundation and other entry points.[26]

Indoor air pollutants from biological sources such as mold; dust mites; pet dander (skin flakes); and droppings and body parts from cockroaches, rodents, and other pests or insects are commonly found in children's homes.[27-30] These contaminants are important because they can lead to allergic reactions, exacerbate existing asthma, and have been associated with the development of respiratory symptoms.[28-31]

Two indoor environmental contaminants for which there is extensive evidence of children's health effects are environmental tobacco smoke and lead. The following indicators present data on environmental tobacco smoke and lead dust hazards in children's homes, because they are well-established indoor hazards to children's health and because they have nationally representative data available for more than one point in time. Other indoor environmental hazards in children's homes generally lack nationally representative data necessary for development of indicators that can identify any changes over time. Unlike many outdoor pollutants, indoor pollutants are not regulated or systematically monitored in residential settings, and data collection for indoor pollutants is much more limited. Indicator E5 presents data on environmental tobacco smoke, based on national survey data of homes with young children where someone smokes regularly. Indicator E6 presents data on lead dust hazards in children's homes. Further information on these issues is provided in the following sections. In addition, indoor environments in children's schools and in child care facilities are discussed in the Supplementary Topics section of this report.

Environmental Tobacco Smoke

Environmental tobacco smoke (ETS), commonly referred to as secondhand smoke, is a complex mixture of gases and particles and includes smoke from burning cigarettes, cigars, and pipe tobacco (sidestream smoke), as well as exhaled mainstream smoke.[32] There are at least 250 chemicals in ETS that are known to be toxic or carcinogenic, including acrolein, ammonia, benzene, carbon monoxide, formaldehyde, hydrogen cyanide, nicotine, nitrogen oxides, and sulfur dioxide.[32,33] In 1992, EPA classified ETS as a known human carcinogen.[34] Children can be exposed to ETS in their homes or in places where people are allowed to smoke, such as some restaurants in some locations throughout the United States.

According to the U.S. Surgeon General, there is no safe level of exposure to ETS, and breathing even a small amount can be harmful to human health.[32] The Surgeon General has concluded that exposure to ETS causes sudden infant death syndrome (SIDS), acute lower respiratory infection, ear problems, and more severe asthma in children. Smoking by parents causes respiratory symptoms and slows lung growth in their children.[32] Young children appear to be more susceptible to the respiratory effects of ETS than are older children.[29,34,35]

The exposure of a pregnant woman to ETS can also be harmful to her developing fetus. The Surgeon General has determined that exposure of pregnant women to ETS causes a small reduction in mean birth weight, and that the evidence is suggestive (but not sufficient to infer causation) of a relationship between maternal exposure to environmental tobacco smoke during pregnancy and preterm delivery.[32] In addition, the Surgeon General concluded the evidence is suggestive but not sufficient to infer a causal relationship between prenatal and postnatal exposure to ETS and childhood cancer.[32]

Exposure to ETS in the home is influenced by adult behaviors, including the decisions to smoke at home and to allow visitors to smoke inside the home. Children living in homes with smoking bans have significantly lower levels of cotinine (a biological marker of exposure to ETS) in urine than children living in homes without smoking bans.[36] Household smoking bans can significantly decrease children's exposures to ETS, but do not completely eliminate them, especially in multi-unit housing where ETS from other apartments may infiltrate through seepage in walls or shared ventilation systems.[37-39] Furthermore, children may be exposed to toxic residues that remain from ETS in dust and on surfaces inside the home for weeks or months after smoke has cleared from the air.[6,40-43] These residues, referred to as "third-hand smoke," may be re-emitted into the gas phase or may react with other compounds to form secondary pollutants.[40,43] The risk of exposure to third-hand smoke may be particularly high for infants, due to their close proximity to contaminated objects such as blankets, carpets, and floor surfaces, and their frequent hand-to-mouth activity.[6]

Parental smoking status inside the home greatly affects children's exposures to ETS, but research suggests a difference in impact between maternal and paternal smoking. Maternal smoking is associated with higher cotinine levels in children, and maternal smoking appears to have a greater effect on lower respiratory illnesses than does paternal smoking.[32]

In recent years there has been a significant decline in children's exposures to ETS.[44] This reduction is in part attributable to a decline in the percentage of adults who smoke, and is likely related to increased restrictions on smoking at workplaces and other public places, as well as efforts to reduce the exposure of nonsmokers in homes.[44] In 2010, an estimated 19.3% of adults were current smokers, down from 24.7% in 1997.[45,46] In addition, the prevalence of smoke-free households increased from 43% of U.S. homes in 1992–1993 to 72% in 2003.[47] Children living in homes with smoking bans have significantly lower levels of cotinine than children living in homes without smoking bans.[36] Recent studies also suggest that smoking bans in workplaces and other public places can reduce the number of asthma-related emergency room visits and hospitalizations, including among children when legal bans lead to an increase in voluntary smoking bans in homes.[48,49] However, despite the increasing numbers of adults disallowing smoking in the home, approximately 34% of children live in a home with at least one smoker as of 2009.[50]

Lead in House Dust

The ingestion of lead-contaminated house dust, soil, and water is the primary pathway of current childhood exposure to lead.[51] Children have a greater risk of exposure to lead-contaminated dust than that of adults, due to their frequent and extensive contact with floors, carpets, and other surfaces where dust gathers, as well as their high rate of hand-to-mouth activity. Additionally, lead-contaminated dust particles are more readily absorbed into the body than soil or paint chips, and children's bodies absorb up to 10 times more ingested lead than adults do as a result of their less-developed gastrointestinal pathways.[52] Children living in homes with higher levels of lead-contaminated dust tend to have higher blood lead levels.[53-58]

Lead dust is composed of fine particles of soil, paint, and other settled industrial or automotive emissions from the outdoor and indoor air.[59] Residences with deteriorated lead-based paint tend to have higher levels of lead in house dust and the surrounding soil.[51,60] Deteriorated lead-based paint that is cracked, peeling, or chipped can be ingested directly by children or can mix with and contaminate house dust, which can also be ingested.[61] Normal wear as the result of cleaning activities or repeated surface friction can lead to further deterioration and the release of lead-based paint particles.[62] Any house built before 1978 may contain lead-based paint. As of the year 2000, approximately 38 million older housing units in the United States still contained lead-based paint.[51]

Home maintenance and renovation activities that disturb lead-based paint, such as sanding, scraping, cutting, and demolition, create hazardous lead dust and chips and have been associated with higher levels of lead dust and blood lead in children.[60,63] Beginning in April 2010, all contractors performing renovation, repair, and painting projects that disturb lead-based paint in pre-1978 homes and child-occupied facilities, such as child care facilities and preschools, must be certified and follow specific work practices to prevent lead contamination.[60] Lead-contaminated soil is another contributor to lead in house dust. Known sources of lead in soil include historical airborne emissions from leaded gasoline use, emissions from industrial sources such as smelters, and lead-based paint. Current sources of lead in ambient air in the United States include smelters, ore mining and processing, lead acid battery manufacturing, and coal combustion activities, such as electricity generation.[58] Lead-contaminated dust and soil from the outdoors can be transported into the home after becoming airborne via soil resuspension, or can be tracked into the home by occupants or family pets.[52]

The National Toxicology Program (NTP) has concluded that childhood lead exposure is associated with reduced cognitive function.[64] Children with higher blood lead levels generally have lower scores on IQ tests[55,65-70] and reduced academic achievement.[64] The NTP has also concluded that childhood lead exposure is associated with attention-related behavioral problems (including inattention, hyperactivity, and diagnosed attention-deficit/hyperactivity disorder) and increased incidence of problem behaviors (including delinquent, criminal, or antisocial behavior).[64]

Until recently, the Centers for Disease Control and Prevention (CDC) defined a blood lead level of 10 micrograms per deciliter (µg/dL) as "elevated." This definition was used to identify children for blood lead case management.[71,72] However, no level of lead exposure has been identified that is without risk of deleterious health effects.[58] CDC's Advisory Committee on Childhood Lead Poisoning Prevention (ACCLPP) recommended in January 2012 that the 97.5[th] percentile of children's blood lead distribution (currently 5 µg/dL) be defined as "elevated" for purposes of identifying children for follow up activities such as environmental investigations and ongoing monitoring.[73] CDC has adopted the ACCLPP recommendation.[74] CDC specifically notes that "no level of lead in a child's blood can be specified as safe,"[75] and the NTP has concluded that there is sufficient evidence for adverse health effects in children at blood lead levels less than 5 µg/dL.[64]

The current federal standards indicate that floor and window lead dust should not exceed 40 micrograms of lead per square foot ($\mu g/ft^2$) and 250 $\mu g/ft^2$, respectively, in order to protect children from developing "elevated" blood lead levels as formerly defined by the CDC. EPA is currently reviewing the lead dust standards to determine whether they should be lowered, based on indications from more recent epidemiological studies that the current standards may not be sufficiently protective of children.[76]

Childhood blood lead and house dust lead levels in the United States differ across groups in the population, such as those defined by socioeconomic status, race/ethnicity,[51,53,77] and geographic location. Children living in poverty and Black non-Hispanic children tend to have higher blood lead levels[53,78] and higher levels of lead-contaminated dust in the home than do White non-Hispanic children.[77] Blood lead levels tend to be higher for children living in older housing, because older housing units are more likely to contain lead-based paint.[77,79] Additionally, housing in the Northeast and Midwest has twice the prevalence of lead-based paint hazards compared with housing in the South and West,[59] because of the older housing stock in those areas.

> About the Indicator: Indicator E5 presents the percentage of children ages 0 to 6 years regularly exposed to environmental tobacco smoke (ETS) in the home. The data are from a national survey that collects health information from a representative sample of the population. The survey provides data on children exposed to ETS in the home on four or more days per week for the years 1994, 2005, and 2010. The focus is on children ages 6 years and under because these younger children have been specifically identified as more susceptible to the effects of tobacco smoke.

National Health Interview Survey

Comparable, nationally representative data on children living in homes where someone smokes regularly come from the National Health Interview Survey (NHIS) for 1994, 2005, and 2010. The NHIS is a large-scale household interview survey of a representative sample of the civilian noninstitutionalized U.S. population, conducted by the National Center for Health Statistics. In 1994, interviews were conducted with household adults representing about 5,450 children ages 0 to 6 years, and ETS exposure information was reported for about 5,390 of those children. In 2005, interviews were conducted with household adults representing about 10,100 children ages 0 to 6 years, and ETS exposure information was reported for about 7,800 of those children. In 2010, interviews were conducted with household adults representing about 9,350 children ages 0 to 6 years, and ETS exposure information was reported for about 6,900 of those children. Questions related to smoking in the home are included in the NHIS only in selected years. In 1994, the NHIS asked, "Does anyone who lives here smoke cigarettes, cigars, or pipes anywhere inside this home?" Similarly, in 2005 and 2010, the NHIS asked, "In a usual week, does ANYONE who lives here, including yourself, smoke cigarettes, cigars, or pipes anywhere inside this home?" If the answer was positive, participants were asked how many days per week smoking usually occurred anywhere inside the home. The NHIS also included questions about smoking in the home in the 1998 survey, but the questions used in 1998 provide data that are not directly comparable to the 1994, 2005, and 2010 data.

Data Presented in the Indicator

Indicator E5 presents data from NHIS for the percentage of children ages 0 to 6 years living in homes where someone smokes on a regular basis (defined as four days or more per week). Studies have found that questionnaire data on smoking in the home are relatively accurate in predicting serum levels of cotinine (a metabolite of nicotine used as a marker of ETS exposure) in children,[80,81] and researchers have used these data to associate ETS exposure with adverse effects on childhood lung function and other health outcomes.[32] However, comparisons of questionnaire data with measures of serum cotinine in children suggest that questionnaires may underestimate actual exposure to ETS, particularly in multi-unit housing or in cases where visitors and other non-family members may smoke in the home.[32,39,82-84]

While the indicator provides information on the presence and number of days per week of smoking in the home, it does not indicate the intensity of smoking (e.g., the number of cigarettes smoked in the home per day). Furthermore, children exposed to ETS at home fewer than four days per week are not included in this indicator, but may also experience adverse health effects since no level of exposure to ETS is without a risk to health.

We focus on children ages 0 to 6 years because these younger children have been specifically identified as more susceptible to the effects of tobacco smoke and are targeted by the indicator used in the federal government's Healthy People 2010 initiative.[85] Children ages 6 years and under also have less control over their environment and are likely to spend more time in close proximity to adult caregivers.[32] Children of all ages, however, may be affected by exposure to ETS.

The indicator presents data on children's exposures to ETS in the home for 1994, 2005, and 2010, based on family income level. Additional information regarding ETS exposures for different race/ethnicity groups is presented in Table E5a.

Statistical Testing

Statistical analysis has been applied to the 2010 data to evaluate differences in indicator values between demographic groups. These analyses use a 5% significance level, meaning that a conclusion of statistical significance is made only when there is no more than a 5% probability that the observed difference occurred by chance ($p \leq 0.05$). A finding of statistical significance depends on the numerical difference in the indicator value between two groups, the number of observations in each group, and various aspects of the survey design. For example, the statistical test is more likely to detect a difference between two groups when data have been obtained from a larger number of people in those groups. It should be noted that when statistical testing is conducted for differences among multiple demographic groups (for example, considering both race/ethnicity and income level), the large number of comparisons involved increases the probability that some differences identified as statistically significant may actually have occurred by chance.

A finding of statistical significance is useful for determining that an observed difference was unlikely to have occurred by chance. However, a determination of statistical significance by itself does not convey information about the magnitude of the difference in indicator values or the potential difference in risk of associated health outcomes. Furthermore, a lack of statistical significance means only that occurrence by chance cannot be ruled out. Thus a conclusion about statistical significance is only part of the information that should be considered when determining the public health implications of differences in indicator values.

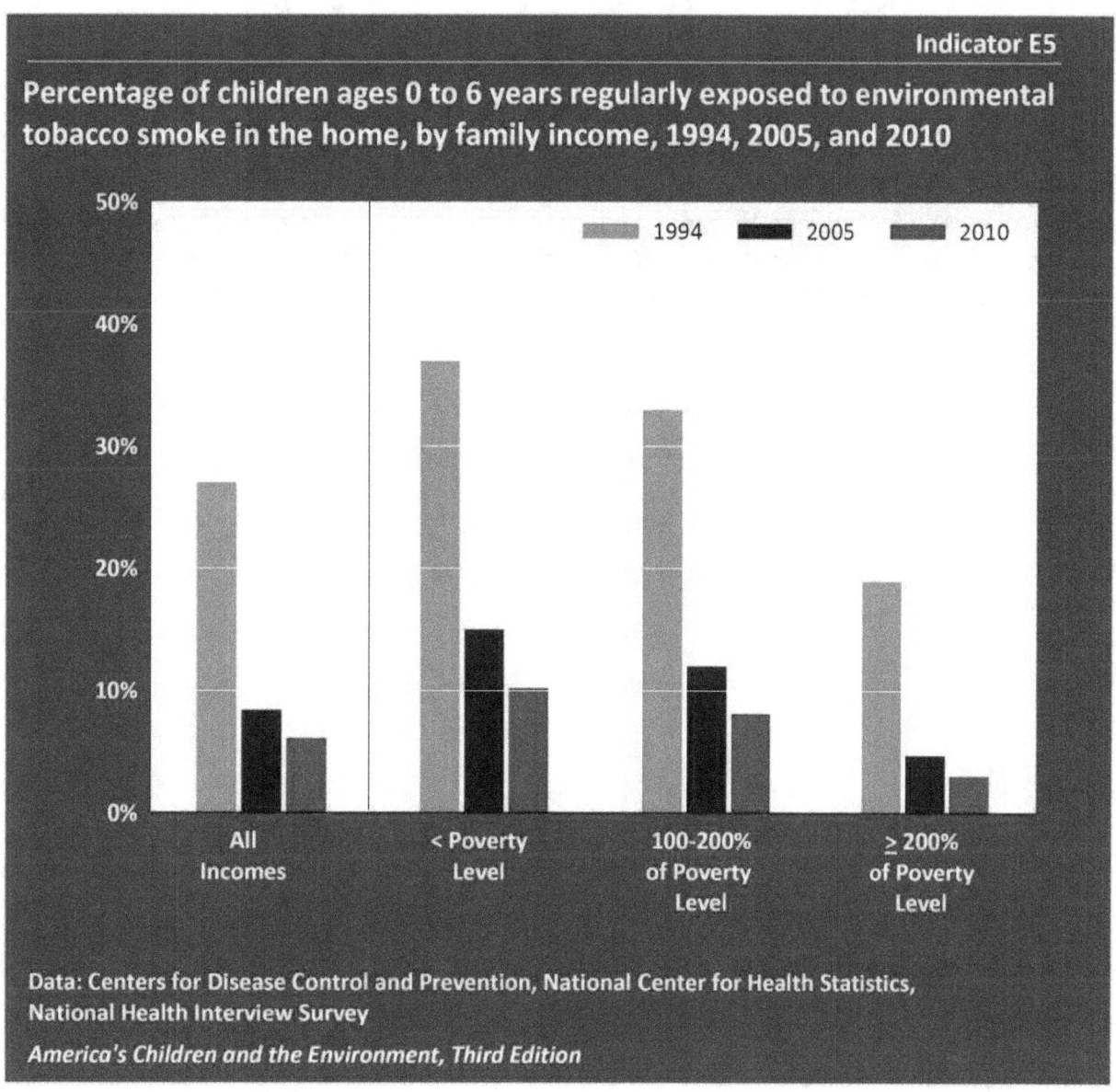

Percentage of children ages 0 to 6 years regularly exposed to environmental tobacco smoke in the home, by family income, 1994, 2005, and 2010

Legend: 1994, 2005, 2010

X-axis categories: All Incomes | < Poverty Level | 100-200% of Poverty Level | ≥ 200% of Poverty Level

Data: Centers for Disease Control and Prevention, National Center for Health Statistics, National Health Interview Survey

America's Children and the Environment, Third Edition

Data characterization

- Data for this indicator are obtained from an ongoing annual survey conducted by the National Center for Health Statistics.
- Survey data are representative of the U.S. civilian noninstitutionalized population.
- In 1994, 2005, and 2010, an adult survey participant in each sampled household was asked whether any resident smokes inside the home and the number of days per week that smoking occurred.

- In 2010, 6% of children ages 0 to 6 years lived in homes where someone smoked regularly, compared with 27% in 1994.

- Children living in homes with family incomes below the poverty level were more likely than their peers at higher income levels to be living in homes where someone smoked regularly. In 2010, 10% of children below the poverty level lived in homes where someone smoked regularly, compared with 8% of children in homes with incomes between 100–200% of poverty level, and 3% of children in homes with incomes at least twice the poverty level.

- The differences between children in homes with family incomes below the poverty level and children in homes with family incomes at or above the poverty level were statistically significant.

- In 2010, 20% of White non-Hispanic children below poverty lived in homes where someone smoked regularly, compared with 10% of Black non-Hispanic children and 2% of Hispanic children living below poverty. (See Table E5a.) These differences were statistically significant.

About the Indicator: Indicator E6 shows the percentage of children ages 0 to 5 years who lived in homes with interior lead-based paint hazards. The data are from two nationally representative surveys of homes conducted in 1998–1999 and 2005–2006. The surveys involved collection of dust, soil, and paint samples from homes and measurement of the lead levels in these samples. The focus of the indicator is on children ages 0 to 5 years, due to the elevated exposures that occur during early childhood and the sensitivity of the developing brain to the effects of lead.

NSLAH/AHHS

The United States Department of Housing and Urban Development (HUD) has conducted two nationally representative surveys of housing in the United States to assess children's potential household exposure to lead and other contaminants. The American Healthy Homes Survey (AHHS) was conducted from 2005–2006 to update the National Survey of Lead and Allergens in Housing (NSLAH), which was conducted from 1998–1999. AHHS also included measurements of arsenic, pesticides, and mold; however, these substances were not measured in the earlier NSLAH.

Samples of paint, dust, and soil were taken from 831 total housing units (184 units with children ages 0 to 5 years) in NSLAH, and 1,131 total housing units (206 units with children ages 0 to 5 years) in AHHS. The lead sampling components of AHHS were designed to be very similar to NSLAH so that results of the two studies could be compared.

Lead-Based Paint Hazards

Samples collected from the housing units surveyed in NSLAH and AHHS were analyzed to determine their lead content. HUD then compared these measured lead levels to federal guidelines to identify homes with lead-contaminated dust, deteriorated lead-based paint, and lead-contaminated soil hazards.

EPA has established Residential Lead Hazard Standards under Title X of the Toxic Substances Control Act (TSCA), section 403, for identifying lead-based paint hazards in all housing built before 1978. These standards were adopted by HUD under the Lead Safe Housing Act, which applies to all federally owned or assisted housing in the United States. According to these regulations, a lead-based paint hazard is the presence of deteriorating lead-based paint, lead-contaminated dust, or lead-contaminated soil above federal standards.

For lead-contaminated dust, there are separate standards for dust on the floor and dust on windowsills. Floor dust samples should not have more than 40 micrograms of lead per square foot ($\mu g/ft^2$) and window dust samples should not have more than 250 $\mu g/ft^2$.[61,86]

Additionally, current federal standards qualify a significantly deteriorated lead-based paint hazard as the deterioration of an area of lead-based paint greater than 20 square feet (exterior) and 2 square feet (interior) for large-surface items, such as walls and doors; or damage to more than 10% of the total surface area of small-surface components—such as windowsills, baseboards, and trim—with lead-based paint.

The level of deterioration is an important variable in determining exposure. The presence of lead-based paint alone is not necessarily indicative of a significant hazard; except during renovations, maintenance, and similar disturbances, intact lead-based paint is believed to pose very little risk to occupants.[87] However, deteriorated lead-based paint that is cracked, peeling, or chipped can be ingested directly by children or can contaminate house dust that can be inhaled or ingested by children.[61]

Data Presented in the Indicator

Indicator E6 presents the percentage of children ages 0 to 5 years who lived in homes with interior lead-based paint hazards, using data from NSLAH and AHHS and three hazard definitions.

The first hazard definition, "interior lead dust," presents the percentage of children ages 0 to 5 years living in homes with a lead dust hazard, based on the number of homes with dust containing levels of lead that exceeded the levels defined by EPA's Residential Lead Hazard Standards (established under Title X of TSCA, section 403). The second hazard definition, "interior deteriorated lead-based paint," displays the percentage of children ages 0 to 5 years who lived in homes with significantly deteriorated lead-based paint indoors as defined by EPA's Residential Lead Hazard Standards. The last definition, "either interior lead dust or interior deteriorated lead-based paint," represents the percentage of children living in homes with an interior dust hazard, a deteriorated lead-based paint hazard, or both.

This indicator represents the potential for children's indoor exposure to lead based solely on the percentage of children ages 0 to 5 years living in homes with levels of lead-based paint and dust above federal standards. The indicator does not represent differences in paint lead levels, paint deterioration levels, or the amount of lead in the dust above the standards. It also does not account for the possibility that children living in homes with levels of lead-based paint and dust below federal standards may still have some exposure to lead. Furthermore, while this indicator focuses on children ages 0 to 5 years, older children may also experience health effects from exposure to lead.

Survey records identify the race/ethnicity and income level of survey respondents; however, estimates of lead hazards in the home for children ages 0 to 5 years broken out by race/ethnicity and income are not statistically reliable, due to the relatively small number of homes in each group. Therefore, the indicator provides data only for all children ages 0 to 5 years combined.

Statistical Testing

Statistical analysis has been applied to Indicator E6 to evaluate differences over time in the indicator values (for example, percentage of children living in homes with lead-contaminated dust). These analyses use a 5% significance level, meaning that a conclusion of statistical significance is made only when there is no more than a 5% probability that the observed difference occurred by chance ($p \leq 0.05$). The statistical analysis depends on the numerical difference in the indicator value over time, the number of observations in each time period, and various aspects of the survey design. For example, the statistical test is more likely to detect a change over time when data have been obtained from a larger number of people in each time period.

A finding of statistical significance is useful for determining that an observed difference was unlikely to have occurred by chance. However, a determination of statistical significance by itself does not convey information about the magnitude of the difference in indicator values or the potential difference in risk of associated health outcomes. Furthermore, a lack of statistical significance means only that occurrence by chance cannot be ruled out. Thus a conclusion about statistical significance is only part of the information that should be considered when determining the public health implications of changes over time.

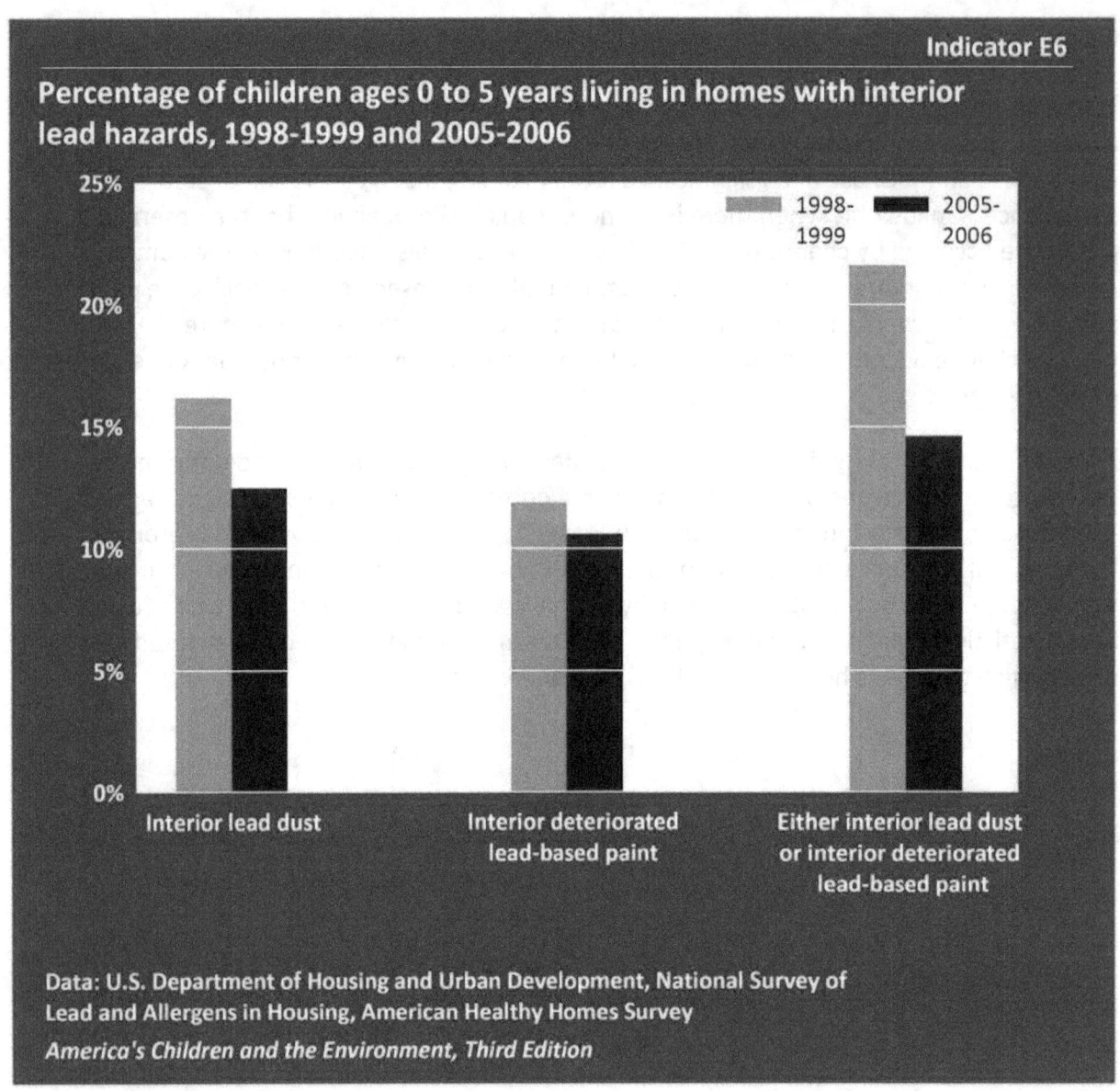

Indicator E6

Percentage of children ages 0 to 5 years living in homes with interior lead hazards, 1998-1999 and 2005-2006

Data: U.S. Department of Housing and Urban Development, National Survey of Lead and Allergens in Housing, American Healthy Homes Survey

America's Children and the Environment, Third Edition

Data characterization

- Data for this indicator are obtained from two surveys of U.S. homes conducted by the Department of Housing and Urban Development.
- Surveyed homes were representative of permanently occupied, non-institutional housing units in the United States in which children may live. Only surveyed homes with children ages 0 to 5 years were included in calculation of this indicator.
- Lead was measured in samples of paint and dust collected from the surveyed homes.

- In 2005–2006, 13% of children ages 0 to 5 years lived in homes with an interior lead dust hazard, compared with 16% in 1998–1999.
- In 2005–2006, 11% of children ages 0 to 5 years lived in homes with an interior deteriorated lead-based paint hazard, compared with 12% in 1998–1999.

- In 2005–2006, 15% of children ages 0 to 5 years lived in homes with either an interior lead dust hazard or an interior deteriorated lead-based paint hazard, compared with 22% in 1998–1999.
- Changes in percentages between the two surveys were not statistically significant.

Drinking Water Contaminants

Drinking water sources may contain a variety of contaminants that, at elevated levels, have been associated with increased risk of a range of diseases in children, including acute diseases such as gastrointestinal illness, developmental effects such as learning disorders, endocrine disruption, and cancer.[1-3] Because children tend to take in more water relative to their body weight than adults do, children are likely to have higher exposure to drinking water contaminants.

Drinking water sources include surface water, such as rivers, lakes, and reservoirs;[4] and groundwater aquifers, which are subsurface layers of porous soil and rock that contain large collections of water.[5] Groundwater and surface water are not isolated systems and are continually recharged by each other as well as by rain and other natural precipitation.[6]

Several types of drinking water contaminants may be of concern for children's health. Examples include microorganisms, (e.g., *E. coli*, *Giardia*, and noroviruses), inorganic chemicals (e.g., lead, arsenic, nitrates, and nitrites), organic chemicals (e.g., atrazine, glyphosate, trichloroethylene, and tetrachloroethylene), and disinfection byproducts (e.g., chloroform). EPA and the Food and Drug Administration (FDA) are both responsible for the safety of drinking water. FDA regulates bottled drinking water, while EPA regulates drinking water provided by public water systems. EPA sets enforceable drinking water standards for public water systems, and unless otherwise specified, the term "drinking water" in this text refers to water provided by these systems. The drinking water standards include maximum contaminant levels and treatment technique requirements for more than 90 chemical, radiological, and microbial contaminants, designed to protect people, including sensitive populations such as children, against adverse health effects.[2,7] Microbial contaminants, lead, nitrates and nitrites, arsenic, disinfection byproducts, pesticides, and solvents are among the contaminants for which EPA has set health-based standards.

Microbial contaminants include bacteria, viruses, and protozoa that may cause severe gastrointestinal illness.[2] Children are particularly sensitive to microbial contaminants, such as *Giardia*, *Cryptosporidium*, *E. coli*, and noroviruses, because their immune systems are less developed than those of most adults.[8-14]

Drinking water is a known source of lead exposure among children in the United States, particularly from corrosion of pipes and other elements of the drinking water distribution systems.[15-17] Exposure to lead via drinking water may be particularly high among very young children who consume baby formula prepared with drinking water that is contaminated by leaching lead pipes.[15] The National Toxicology Program has concluded that childhood lead exposure is associated with reduced cognitive function, reduced academic achievement, and increased attention-related behavioral problems.[18]

Fertilizer, livestock manure, and human sewage can be significant contributors of nitrates and nitrites in groundwater sources of drinking water.[19,20] High levels of nitrates and nitrites can cause the blood disorder methemoglobinemia (blue baby syndrome)[21-23] and have been associated with thyroid dysfunction in children[24,25] and pregnant women.[24,26,27] Moderate deficits in maternal thyroid hormone levels during early pregnancy have been linked to reduced childhood IQ scores and other neurodevelopmental effects, as well as unsuccessful or complicated pregnancies.[28]

Arsenic enters drinking water sources from natural deposits in the earth, which vary widely from one region to another, or from agricultural and industrial sources where it is used as a wood preservative and a component of fertilizers, animal feed, and a variety of industrial products.[29] Population studies of health effects associated with arsenic exposure have been conducted primarily in countries such as Bangladesh, Taiwan, and Chile, where arsenic levels in drinking water are generally much higher than in the United States due to high levels of naturally occurring arsenic in groundwater.[30] Long-term consumption of arsenic-contaminated water has been associated with the development of skin conditions and circulatory system problems, as well as increased risk of cancer of the bladder, lungs, skin, kidney, nasal passages, liver, and prostate.[29,31] In many cases, long-term exposure to arsenic begins during prenatal development or childhood, which increases the risk of mortality and morbidity among young adults exposed to arsenic long-term.[32] A review of the literature concluded that epidemiological studies of associations between exposure to arsenic and some adverse health outcomes pertinent to children's health have mixed findings. These include studies of associations between high levels of exposure to arsenic and abnormal pregnancy outcomes, such as spontaneous abortion, still-births, reduced birth weight, and infant mortality, as well as associations between early-life exposure to arsenic and increased incidence of childhood cancer and reduced cognitive function.[33]

Water can contain microorganisms such as parasites, viruses, and bacteria; the disinfection of drinking water to reduce water-borne infectious disease is one of the major public health advances of the 20th century.[34] The method by which infectious agents are removed or chemically inactivated depends on the type and quality of the drinking water source and the volume of water to be treated. Surface water systems are more exposed than groundwater systems to weather and runoff; therefore, they may be more susceptible to contamination.[4,35] Surface and groundwater systems use filtration and other treatment methods to physically remove particles. Disinfectants, such as chlorine and chloramine, ultraviolet radiation, and ozone are added to drinking water provided by public water systems to kill or neutralize microbial contaminants.[36] However, this process can produce disinfection byproducts, which form when chemical disinfectants react with naturally occurring organic matter in water.[37] The most common of these disinfection byproducts are chloroform and other trihalomethanes. Consumption of drinking water from systems in the United States and other industrialized countries with relatively high levels of disinfection byproducts has been associated with bladder cancer and developmental effects in some studies.[38-41] Some individual epidemiological studies have reported associations between the presence of disinfection byproducts in drinking water and increased risk of birth defects, especially neural tube defects and oral clefts; however, recent articles reviewing the body of literature determined that the evidence is too limited to make conclusions about a possible association between exposure to disinfection byproducts and birth defects.[38,42-45]

Some of the most widely used agricultural pesticides in the United States, such as atrazine and glyphosate, are also drinking water contaminants.[46,47] Pesticides can enter drinking water sources as runoff from crop production in agricultural areas and enter groundwater through abandoned wells on farms.[48] Some epidemiological studies have reported associations between prenatal exposure to atrazine and reduced fetal growth.[49-52]

The use of glyphosate, an herbicide used to kill weeds, has increased dramatically in recent years because of the growing popularity of crops genetically modified to survive glyphosate treatment.[53] Previous safety assessments have concluded that glyphosate does not affect fertility or reproduction in laboratory animal studies.[54,55] However, more recent studies in laboratory animals have found that male rats exposed to high levels of glyphosate, either during prenatal or pubertal development, may suffer from reproductive problems, such as delayed puberty, decreased sperm production, and decreased testosterone production.[56,57] Very few epidemiological human studies have investigated effects of glyphosate exposure on reproductive endpoints. In contrast to the results of animal studies, one such epidemiological study of women living in regions with different levels of exposure to glyphosate found no associations between glyphosate exposure and delayed time to pregnancy.[58]

A variety of other chemical contaminants can enter the water supply after use in industry.[47] Examples include trichloroethylene and tetrachloroethylene (also known as perchloroethylene), which are solvents widely used in industry as degreasers, dry cleaning agents, paint removers, chemical extractors, and components of adhesives and lubricants.[59-61] Potential health concerns from exposure to trichloroethylene, based on limited epidemiological data and evidence from animal studies, include decreased fetal growth and birth defects, particularly cardiac birth defects.[61] A study conducted in Massachusetts reported associations between birth defects and maternal exposure to drinking water contaminated with high levels of tetrachloroethylene around the time of conception.[62] An additional study reported that older mothers or mothers who had previously miscarried, and who were exposed to high levels of tetrachloroethylene in contaminated drinking water, had a higher risk of delivering a baby with reduced birth weight.[63] However, other studies did not find associations between maternal exposure to tetrachloroethylene and pregnancy loss, gestational age, or birth weight.[64,65] Studies in laboratory animals indicate that mothers exposed to high levels of tetrachloroethylene can have spontaneous abortion, and their fetuses can suffer from altered growth and birth defects.[60]

EPA has not determined whether standards are necessary for some drinking water contaminants, such as personal care products. Personal care products, such as cosmetics, sunscreens, and fragrances; and pharmaceuticals, including prescription, over-the-the counter, and veterinary medications, can enter water systems after use by humans or domestic animals[66] and have been measured at very low levels in drinking water sources.[67] Many concentrated animal feeding operations treat livestock with hormones and antibiotics, and can be one significant source of pharmaceuticals in water.[35] Other major sources of pharmaceuticals in water are human waste, manufacturing plants and hospitals, and other human activities such as showering and swimming.[66] Any potential health implications of long-term exposure to levels of pharmaceuticals and personal care products found in drinking water are unclear.

Manganese is a naturally occurring mineral that can enter drinking water sources from rocks and soil or from human activities.[68] While manganese is an essential nutrient at low doses, chronic exposure to high doses may be harmful, particularly to the nervous system. Many of the reports on adverse effects from manganese exposure are based on inhalation exposures in occupational settings. Fewer studies have examined health effects associated with oral exposure to manganese.[68] However, some recent epidemiological studies have reported associations between long-term exposure to high levels of manganese in drinking water during prenatal development or childhood and intellectual impairment; decreased non-verbal memory, attention, and motor skills; hyperactivity; and other behavioral effects.[69-73] Most studies on the health effects of manganese have been conducted in countries where manganese exposure is generally higher than in the United States. However, two individual studies conducted in specific areas of relatively high manganese contamination in the United States reported associations between prenatal or childhood manganese exposure and problems with general intelligence, memory, and behavior.[74,75] Although there is no health-based regulatory standard for manganese in drinking water, EPA has set a voluntary standard for manganese as a guideline to assist public water systems in managing their drinking water for aesthetic considerations, such as taste, color and odor.[7]

Perchlorate is a naturally occurring and man-made chemical that has been found in surface and groundwater in the United States.[76-78] Perchlorate is used in the manufacture of fireworks, explosives, flares, and rocket fuel.[78] Perchlorate was detected in just over 4% of public water systems in a nationally representative monitoring study conducted from 2001–2005.[78] Some infant formulas have been found to contain perchlorate, and the perchlorate content of the formula is increased if it is prepared with perchlorate-contaminated water.[79-82] Exposure to elevated levels of perchlorate can inhibit iodide uptake into the thyroid gland, possibly disrupting the function of the thyroid and potentially leading to a reduction in the production of thyroid hormone.[83,84] As noted above, thyroid hormones are particularly important for growth and development of the central nervous system in fetuses and infants.

In January 2009, EPA issued an interim health advisory level to help state and local officials manage local perchlorate contamination issues in a health-protective manner, in advance of a final EPA regulatory determination.[78,85] In February 2011, EPA decided to develop a federal drinking water standard for perchlorate, based on the concern for effects on thyroid hormones and the development and growth of fetuses, infants, and children.[78] The process for developing the standard will include receiving input from key stakeholders as well as submitting any formal rule to a public comment process.

The two indicators that follow use the best nationally representative data currently available to characterize the performance of water systems in meeting EPA's health-based drinking water standards and in reporting monitoring results over time. Indicator E7 estimates the percentage of children served by community water systems that did not meet all applicable health-based drinking water standards. Indicator E8 estimates the percentage of children served by systems with violations of drinking water monitoring and reporting requirements. Monitoring and reporting violations occur when a water system does not monitor, does not report monitoring results, or was late in reporting results.[86] Such violations in monitoring and reporting may mean that some health-based violations were not reported; this could cause the percentages shown in Indicator E7 to be underestimated.

Indicator E7: Estimated percentage of children ages 0 to 17 years served by community water systems that did not meet all applicable health-based drinking water standards, 1993–2009

Indicator E8: Estimated percentage of children ages 0 to 17 years served by community water systems with violations of drinking water monitoring and reporting requirements, 1993–2009

About the Indicators: Indicators E7 and E8 estimate the percentage of children served by community water systems that did not meet all health-based drinking water standards or failed to adhere to monitoring and reporting requirements. The data are from an EPA database that compiles drinking water violations reported by public water systems. Indicator E7 shows the estimated percentage of children served by community water systems that did not meet health-based drinking water standards in each year from 1993 to 2009. Indicator E8 shows the estimated percentage of children served by community water systems that did not adhere to monitoring and reporting requirements in each year.

SDWIS/FED

EPA's Safe Drinking Water Information System, Federal Version (SDWIS/FED) provides information on violations of drinking water standards. Public drinking water systems in the United States are required to monitor the presence of certain individual contaminants at specific time intervals and locations to assess whether they are complying with drinking water standards. These standards include Maximum Contaminant Levels (MCLs), which are numerical limits on how much of a contaminant may be present in drinking water; as well as mandatory treatment techniques and processes, such as those intended to prevent microbial contamination of drinking water. When a violation of a drinking water standard is detected, the public water system is required to report the violation to the state, which in turn reports to the federal government. All health-based violations are compiled in SDWIS/FED. SDWIS/FED was created in 1995 and includes data from various precursor database systems that have violation and inventory data going back to 1976. SDWIS/FED also reports the number of people served by each water system.

Health-Based Drinking Water Standard Violations

Indicator E7 presents statistics on violations of drinking water standards grouped into several categories:

- The "Surface water treatment" category includes violations of requirements in the Surface Water Treatment Rule and Interim Enhanced Surface Water Treatment Rule that specify the type of treatment and maintenance activities that systems must use to prevent microbial contamination of drinking water.

- The "Chemical and radionuclide" category includes violations of the MCLs for organic and inorganic chemicals, such as atrazine, glyphosate, trichloroethylene, tetrachloroethylene, arsenic, cadmium, and mercury, in addition to radionuclide contaminants, such as radium and uranium.
- The "Lead and copper" category includes violations of treatment technique requirements for systems to control the corrosiveness of their water.[2]
- The "Total coliforms" category covers all violations of the MCL for total coliform bacteria, which is an indicator of the presence of various fecal pathogens, including *E.Coli*.[87,88]
- The "Nitrate/nitrite" category takes account of all violations of the MCLs for nitrates and nitrites.

The "Disinfectants and disinfection byproducts" category covers violations of standards for several disinfectants—chlorine, chloramine, and chlorine dioxide—and disinfectant byproducts—total trihalomethanes, haloacetic acids, chlorite, and bromate.[89]

Monitoring and Reporting Violations

Indicator E8 presents statistics on violations of monitoring and reporting requirements. Monitoring and reporting violations occur when a water system does not monitor, does not report monitoring results, or was late in reporting results.[86] All monitoring and reporting violations are compiled from SDWIS/FED.

Data Presented in the Indicators

Indicator E7 estimates the percentage of children ages 0 to 17 years served by community water systems that did not meet all applicable health-based drinking water standards between 1993 and 2009. The indicator is calculated by identifying all community water systems with violations in SDWIS/FED each year by state, then summing the number of people served by those systems with violations. Census data for the number of children in each state are then used to adjust these estimates of the total population served to estimate the percentage of children served by systems with violations in relation to all children served by community water systems.

Indicator E8 estimates the percentage of children ages 0 to 17 years served by community water systems with violations of drinking water monitoring and reporting requirements. This indicator is based on data reported to SDWIS/FED for violations between 1993 and 2009. Violations of monitoring and reporting requirements for Indicator E8 were grouped into the same categories as in Indicator E7, except for the Nitrate/nitrite category.

For the most part, the indicator represents comparisons with a consistent set of standards over the years 1993–2009, with some exceptions. Revisions to the surface water treatment standard were finalized in 2002.[89] A revised standard for radionuclides went into effect in 2003, and for arsenic (included in the chemical and radionuclide category) in 2006.[90] A new standard for disinfection byproducts was implemented in 2002 for larger drinking water systems, and in

2004 for smaller systems.[91] The revisions to the surface water treatment standard were significant enough to warrant a break in the trend lines for this category in Indicators E7 and E8 between 2001 and 2002. The break in the "any violation" trend line between 2001 and 2002 is due to both the revision of the surface water standard and the implementation of the new disinfection byproducts standard for large systems beginning in 2002. Revisions to other standards had only minimal impacts on the indicator values. As new and revised drinking water standards take effect, water system compliance with all applicable health-based standards signifies higher levels of public health protection over time.

Violations of health-based standards (as represented in Indicator E7) may be under-reported as a result of monitoring and reporting violations. An EPA audit of drinking water data from 2002–2004 found that only 62% of health-based standards violations were reported to SDWIS.[86] Therefore, the data on systems reporting no violations of health based standards include a number of systems that have not gathered or reported all of the required data needed to make this determination.

Indicators E7 and E8 provide information about the extent to which contaminants in community water systems reach levels that may be of concern for children. However, the indicators do not provide a direct measure of children's exposure to drinking water contaminants and do not give an indication about how drinking water violations are related to health risks. A violation of a health-based standard represents a potential concern for children's health, but the importance of any violation depends on the particular contaminant, the magnitude and duration of the violation, and the extent of the violation within a system. Indicator E7 does not reflect the extent to which a standard has been exceeded or the extent to which a water system's distribution system may have been affected by a violation. The indicator does not take into account the duration of a violation within any calendar year. However, a violation that continues over an extended period of time is included in the indicator for each calendar year in which it occurs. A large water system with a single violation of short duration may significantly affect the indicator value for a single year.

The ability to examine children's potential exposure to contaminated drinking water is limited by the type of information collected and stored in the SDWIS/FED database. States are not required to report the actual contaminant levels measured to SDWIS/FED; instead, they report when standards are not met. As a result, SDWIS/FED data cannot be used to analyze national or local trends in contaminant concentrations, or to provide comparisons to the current health-based standards across all years shown.[i] EPA is working with states to develop a new drinking water data system that will compile and make available actual measurements of contaminant levels.

---

[i] EPA requires community water systems to provide annual drinking water quality reports to their customers. These reports summarize the contaminants measured in each system's drinking water over the course of a year, providing much more detail than the information reported to SDWIS. The drinking water quality reports for many systems can be found at: http://water.epa.gov/lawsregs/rulesregs/sdwa/ccr/index.cfm.

Indicators E7 and E8 are based on drinking water provided to residences served by community water systems. Community water systems are public water systems that serve water to the same residential population year-round.[92] The indicators do not account for all sources of children's drinking water. Some drinking water comes from other types of public water systems, including those that may not serve residences, or may not operate year-round (e.g., schools, factories, office buildings, and hospitals that have their own water systems; gas stations and campgrounds); and bottled water.[ii] [93-95]

In addition, many homes are not served by community water systems and instead obtain their drinking water from individual residential wells.[93,96] EPA does not have the authority under the Safe Drinking Water Act to regulate wells that serve fewer than 25 persons or 15 service connections. Thus, the SDWIS/FED database does not contain data on non-public water systems, such as privately owned household wells, that are not required to monitor or report the quality of drinking water to EPA.[94,97] In 2000, approximately 15% of the total U.S. population was served by non-public water systems[97] and more than 90,000 new domestic wells are installed every year.[98] Separate data collection activities have found that the contaminants in untreated groundwater are generally at lower levels than the MCL; however, more than 20% of wells sampled by the U.S. Geological Survey between 1991 and 2004 contained at least one contaminant at a level of potential health concern.[99] Approximately 4% of the 2,167 sampled wells exceeded the nitrate MCL, and 7% exceeded the arsenic MCL.[99] Nitrate concentrations above the MCL were more frequently detected in agricultural regions than any other land-use setting.[99] Groundwater-sourced wells in rural and agricultural regions may be at an increased risk for nitrate and nitrite contamination due to local fertilizer use and animal waste runoff.[100]

[ii] Bottled water is regulated by the Food and Drug Administration.

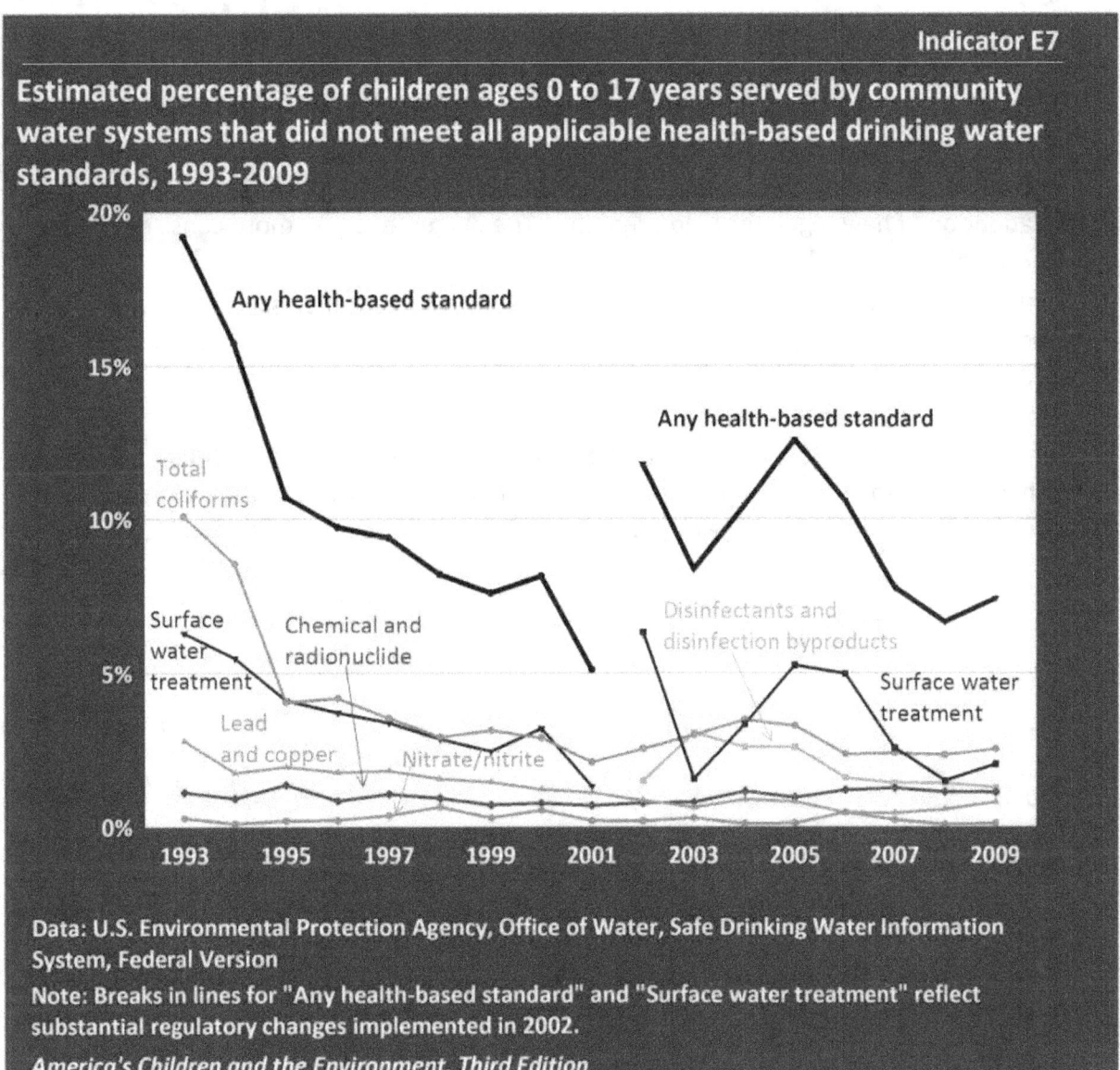

Indicator E7

Estimated percentage of children ages 0 to 17 years served by community water systems that did not meet all applicable health-based drinking water standards, 1993-2009

Data: U.S. Environmental Protection Agency, Office of Water, Safe Drinking Water Information System, Federal Version

Note: Breaks in lines for "Any health-based standard" and "Surface water treatment" reflect substantial regulatory changes implemented in 2002.

America's Children and the Environment, Third Edition

Data characterization
- Data for this indicator are obtained from EPA's database to which states are required to report public water system violations of national drinking water standards.
- All violations of health-based standards are supposed to be reported to the database; however, it is known that not all violations are reported and the magnitude of underreporting is not known.
- Some drinking water standards have been changed over time to increase the level of public health protection; therefore, as noted on the figure, some types of violations in more recent years are not strictly comparable to violations in earlier years.
- Non-public drinking water systems, such as private wells, are not represented in the database. In 2000, about 15% of the U.S. population was served by non-public water systems.

- The estimated percentage of children served by community drinking water systems that did not meet all applicable health-based standards declined from 19% in 1993 to about 5% in 2001. Since 2002, this percentage has fluctuated between 7% and 13%, and was 7% in 2009.

- The estimated percentage of children served by community drinking water systems that did not meet surface water treatment standards varied substantially from 2002–2007, following the adoption of new regulatory requirements. The percentage was more consistent from 2007–2009, and was 2% in 2009.

- Total coliforms indicate the potential presence of harmful bacteria associated with infectious illnesses. The estimated percentage of children served by community drinking water systems that did not meet the health-based standard for total coliforms was about 10% in 1993 and about 3% in 2009.

- A new standard for disinfection byproducts was adopted in 2001. The estimated percentage of children served by community water systems that had violations of the disinfection byproducts standard has declined steadily from 3% in 2003 to about 1% in 2009.

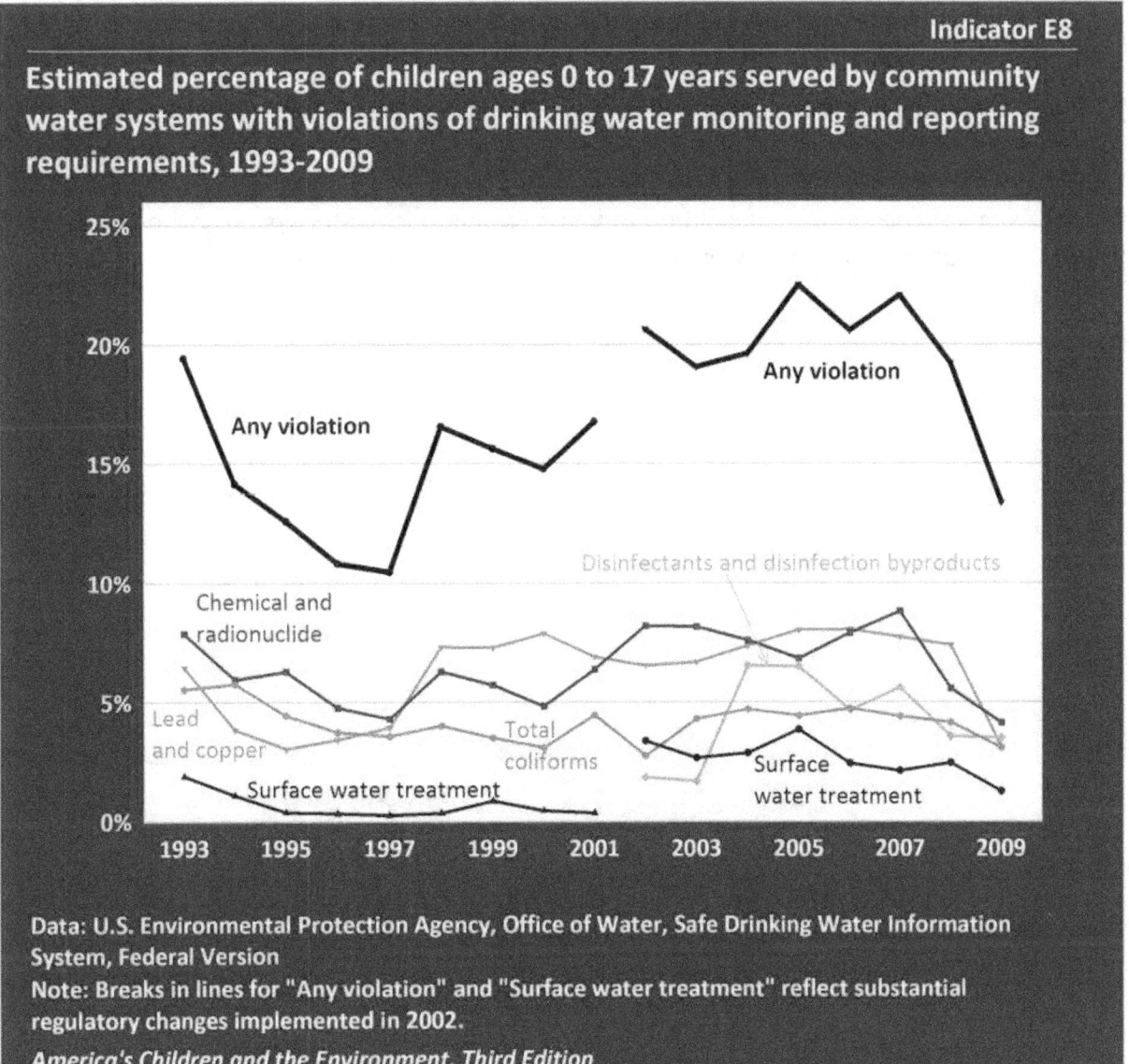

Indicator E8

Estimated percentage of children ages 0 to 17 years served by community water systems with violations of drinking water monitoring and reporting requirements, 1993-2009

Data: U.S. Environmental Protection Agency, Office of Water, Safe Drinking Water Information System, Federal Version

Note: Breaks in lines for "Any violation" and "Surface water treatment" reflect substantial regulatory changes implemented in 2002.

America's Children and the Environment, Third Edition

Data characterization
- Data for this indicator are obtained from EPA's database to which states are required to report public water system violations of national drinking water standards.
- Not all violations of monitoring and reporting requirements are reported to the database, and the magnitude of underreporting is not known.
- Some drinking water standards have been changed over time to increase the level of public health protection; therefore, as noted on the figure, some types of violations in more recent years are not strictly comparable to violations in earlier years.
- Non-public drinking water systems, such as private wells, are not represented in the database. In 2000, about 15% of the U.S. population was served by non-public water systems.

- Between 1993 and 2009, the estimated percentage of children served by community water systems that had at least one monitoring and reporting violation fluctuated between about 11% and 23%, and was 13% in 2009.

- In 1993, approximately 6% of children served by community water systems lived in an area with significant monitoring and reporting violations for lead and copper. This figure dropped to about 3% in 2009.

- The estimated percentage of children served by community water systems with a chemical and radionuclide monitoring violation has varied between 4 and 9%, and was 4% in 2009.

Chemicals in Food

Children's diets are an important pathway for exposure to some environmental chemicals and other contaminants. Children may be at a greater risk for exposures to contaminants because they consume more food relative to their body weight than do adults. Additionally, children's dietary patterns are often less varied than those of adults, suggesting that there are greater opportunities for continuous exposure to a foodborne contaminant than in adults.[1]

Food contamination can come from multiple sources, including antibiotics and hormones in meat and dairy products, as well as microbial contamination that can lead to illness. An estimated 48 million Americans suffer from foodborne illnesses each year,[2] and children under age five have the highest incidence of most of these infections.[3] Microbial contamination of food is monitored and regulated by a number of federal agencies, including the Department of Agriculture and the Food and Drug Administration.[i] In addition, a wide variety of chemicals from man-made sources may be found in or on foods, typically at low levels. Chemicals in foods may come from application of pesticides to crops, from transport of industrial chemicals in the environment, or from chemicals used in food packaging products. A number of persistent environmental contaminants tend to accumulate in all types of animals, and are frequently found in meat, poultry, fish, and dairy products. Other chemicals, such as perchlorate and a variety of pesticides, are often found in fruits, vegetables, and other agricultural commodities. Some chemicals in food, such as mercury and perchlorate, have naturally occurring as well as man-made sources. The health risks from chemicals in food are dependent on both the actual level of a chemical in the food as well as the amount of the food consumed by individuals.

Following this text, an indicator is presented for organophosphate pesticides in selected foods. Many chemicals of concern in food lack sufficient data to generate reliable, nationally representative indicators, particularly for children. Selected chemicals of concern for children's health that are frequently found in foods are summarized below. Further details can be found in the Biomonitoring section of this report for several of these chemicals, including methylmercury, polychlorinated biphenyls (PCBs), polybrominated diphenyl ethers (PBDEs), phthalates, perfluorochemicals (PFCs), and perchlorate.

Methylmercury

Mercury is a naturally occurring element that is released to the environment from a variety of sources, including the combustion of coal, the use of mercury in industrial processes, and from breakage of products such as mercury thermometers and fluorescent lighting, as well as from natural sources such as volcanoes. Mercury may enter water bodies through direct release or through emissions to the atmosphere that are subsequently deposited to surface waters.

---

[i] More information on microbial contaminants in food is available at http://www.fda.gov/Food/ResourcesForYou/Consumers/ucm103263.htm, http://fsrio.nal.usda.gov/pathogen-detection-and-monitoring, and http://www.fsis.usda.gov/fact_sheets/Foodborne_Illness_&_Disease_Fact_Sheets/index.asp.

Bacteria in water bodies convert the deposited mercury into methylmercury.[4] Methylmercury can be absorbed by small aquatic organisms that then are consumed by predators, including fish.[5] As each organism builds up methylmercury in its own tissues, and as smaller fish are eaten by larger fish, concentrations of methylmercury can accumulate, particularly in large fish with longer lifespans[6-8] such as sharks and swordfish.[9]

EPA has determined that methylmercury is known to have neurotoxic and developmental effects in humans.[10] This conclusion is based on severe adverse effects observed in exposed populations in two high-dose mercury poisoning events in Japan and Iraq. Some other studies of populations with prenatal exposure to methylmercury through regular consumption of fish have reported more subtle adverse effects on childhood neurological development.[11-15] Although ingestion of methylmercury in fish may be harmful, other compounds naturally present in many fish (such as high quality protein and other essential nutrients) are extremely beneficial.

In particular, fish are an excellent source of omega-3 fatty acids, which are nutrients that contribute to the healthy development of infants and children.[16] Pregnant women are advised to seek dietary sources of these fatty acids, including many species of fish. However, the levels of both methylmercury and omega-3 fatty acids can vary considerably by fish species. Thus, the type of fish, as well as portion sizes and frequency of consumption are all important considerations for health benefits of fish and the extent of methylmercury exposure.[16] For this reason, EPA and the U.S. Food and Drug Administration (FDA) provide advisory information on fish consumption to females who are pregnant, breastfeeding, or of childbearing age, and to young children. The advisory encourages consumption of up to 12 ounces per week of a variety of fish and shellfish that are lower in mercury, or, in the absence of a local advisory, consumption of up to 6 ounces per week of fish caught from local waters and no other fish that week. EPA and FDA also recommend that these categories of women and young children avoid consuming shark, swordfish, tile fish, or king mackerel, because these species may contain high levels of methylmercury.[17] Fish that are high in omega-3 fatty acids and low in mercury are expected to offer the greatest health benefits.[9,16,18] EPA and FDA are currently working to update the fish consumption advisory to incorporate the most current science regarding the health benefits of fish consumption and the risks from methylmercury in fish. In 2011, the Departments of Agriculture and Health and Human Services jointly released the 2010 Dietary Guidelines for Americans, which recommended that pregnant or breastfeeding women should consume 8–12 ounces of seafood per week, but avoid consumption of the same high-mercury-containing fish identified in EPA and FDA's advisory.[19] More information regarding current fish advisories, and links to lists of fish and shellfish typically containing lower levels of mercury, can be found at http://water.epa.gov/scitech/swguidance/fishshellfish/fishadvisories/general.cfm#tabs-4. Tribal and state-specific fish advisories can be found at http://fishadvisoryonline.epa.gov/General.aspx.

Polychlorinated biphenyls

Polychlorinated biphenyls (PCBs) are a group of persistent chemicals used in electrical transformers and capacitors for insulating purposes, in gas pipeline systems as a lubricant, and in caulks and other building materials. The manufacture, sale, and use of PCBs were generally banned by law in 1979, although EPA regulations have authorized their continued use in certain existing electrical equipment. Due to their persistent nature, large reservoirs of previously released PCBs remain in the environment. PCBs accumulate in fat tissue, so they are commonly found in foods derived from animals. Consumption of fish is a common source of PCB exposure, but other foods with lower PCB levels that are consumed more frequently, including meat, dairy, and poultry products, also contribute to PCB exposure.[20,21] A study by the U.S. Department of Agriculture found that levels of certain PCBs in beef and chicken declined between 2002 and 2008, while levels in turkey and pork remained relatively constant during the same years.[22] Exposure to PCBs remains widespread;[23,24] however, declining environmental levels of PCBs suggest that children today are exposed to lower levels of PCBs compared with children in previous generations.[20,25-28]

Prenatal exposure to PCBs has been associated with adverse effects on children's neurological development and impaired immune response, primarily through studies of populations that consume fish regularly.[29-31] Although there is some inconsistency in the epidemiological literature, several reviews of the literature have found that the overall evidence supports a concern for effects of PCBs on children's neurological development.[29,30,32-34] The Agency for Toxic Substances and Disease Registry has determined that "Substantial data suggest that PCBs play a role in neurobehavioral alterations observed in newborns and young children of women with PCB burdens near background levels."[20] Some studies have also detected associations between childhood exposure and adverse health effects.[30,35-37] In addition to PCBs, many other organochlorine chemicals, including dioxins, dibenzofurans, and organochlorine pesticides, are persistent and bioaccumulative and are frequently found in foods derived from animals.[38]

Polybrominated diphenyl ethers

Polybrominated diphenyl ethers (PBDEs) are a class of flame retardants used in many applications, including furniture foam, small appliances, and electronic products. PBDEs are intended to slow the ignition and rate of fire growth. Of three forms of PBDEs once used in the United States (pentaBDE, octaBDE, and decaBDE), only the decaBDE form, used primarily in televisions, personal computers, and other electrical appliances, is still in production. Manufacturers of decaBDE have agreed to phase out all uses of the chemical by the end of 2013.[39] However, products manufactured prior to the elimination of the pentaBDE and octaBDE forms in 2004, and products manufactured prior to the phaseout of decaBDE in 2013, can remain in use and contribute to the presence of PBDEs in the environment.

Like PCBs, PBDEs are persistent in the environment, accumulate in fat tissue, and have been found in a variety of foods, including fish, meat, poultry, and dairy products as well as breast milk.[40-48] Exposure studies have concluded that the presence of PBDEs in house dust and in foods are both important contributors to PBDE exposures for people of all ages, and that exposures from house dust are generally greater than those from food.[46,47,49-54] PBDE toxicity to the developing nervous system as well as endocrine disruption have been identified as areas of potential concern.[40,55-59]

Bisphenol A

Bisphenol A (BPA) is an industrial chemical used in the production of epoxy resins used as inner liners of metallic food and drink containers to prevent corrosion. BPA is also used in the production of polycarbonate plastics that may be used in food and drink containers. The primary route of human exposure to BPA is through diet, when BPA migrates from food and drink containers, particularly when a container is heated.[60-62]

Much of the scientific interest in BPA is related to published research suggesting that BPA may be an endocrine disrupting chemical.[63,64] Endocrine disruptors act by interfering with the biosynthesis, secretion, action, or metabolism of naturally occurring hormones.[63-65] BPA has demonstrated developmental effects in laboratory animals at high doses, though the effects of lower doses similar to typical human exposure levels are the subject of scientific debate.[61,66-70] Based on a critical review of the existing scientific literature, in 2008 the National Toxicology Program (NTP) determined that there was "some concern" (the midpoint on a five-level scale ranging from "negligible" to "serious")[ii] for effects of BPA on the brain, behavior, and prostate gland in fetuses, infants, and children.[61] Although there is uncertainty regarding the effects in humans of BPA at typical exposure levels, several retailers and manufacturers have begun phasing out baby products such as bottles and sippy cups that contain BPA. Several states have also introduced legislation to ban or limit BPA in food containers and consumer products. Additional studies by both government and non-government entities are being conducted to provide additional information and address uncertainties about the safety of BPA.

Phthalates

Phthalates are a class of chemicals commonly used to increase the flexibility of plastics in a wide array of consumer products, and have been used in food packaging.[71-74] Some phthalates have been found at higher levels in fatty foods such as dairy products, fish, seafood, and oils, which are most likely to absorb phthalates.[74] Phthalates in a mother's body can enter her breast milk. Ingestion of that breast milk and infant formula containing phthalates may also contribute to infant phthalate exposure.[75] Certain phthalates are suspected endocrine disruptors, and have shown a number of reproductive and developmental effects in laboratory animal studies[76-85] as well as some reported associations in human epidemiological studies.[86-89]

Perfluorochemicals

Perfluorochemicals (PFCs) are a group of chemicals used in a variety of consumer products, including food packaging, and in the production on nonstick coatings on cookware.[90,91] Long-chain PFCs, including perfluorooctane sulfonic acid (PFOS) and perfluorooctanoic acid (PFOA),

---

[ii] More information on NTP concern levels is available at http://www.niehs.nih.gov/news/media/questions/sya-bpa.cfm.

have already been or will be phased out by the chemical industry by 2015, although the persistence of these chemicals means that they will remain in the environment for several years despite reductions in emissions. While the routes of human exposure to PFCs are not fully understood, two recent studies have identified food consumption as the primary exposure pathway.[92,93] PFC-treated food-contact packaging, such as microwave popcorn bags, may be a source of PFC exposure.[94,95] Heating these materials may cause PFCs to migrate into food, or into the air where they may be inhaled.[iii] Meats may also be contaminated with PFCs due to exposure of source animals to air, water, and feed contaminated with PFCs.[95-97] PFCs have also been detected in some plant-based foods.[93] Studies in laboratory animals have demonstrated reproductive and developmental toxicity of PFCs.[98,99] Some human health studies have reported associations between prenatal exposure to PFCs and a number of adverse birth outcomes,[100-103] while other studies have not.[104,105]

Perchlorate

Perchlorate is a naturally occurring and man-made chemical that has been detected in surface water and groundwater in the United States.[106-109] Perchlorate is used in the manufacture of fireworks, explosives, flares, and rocket propellant.[107,109] Perchlorate has been detected in human breast milk, dairy products, as well as in leafy vegetables and other produce.[108,110-115] Infant formulas have been found to contain perchlorate, and the perchlorate content of the formula is increased if it is prepared with perchlorate-contaminated water.[116-118]

Exposure to high doses of perchlorate has been shown to inhibit iodide uptake into the thyroid gland, thus possibly disrupting the function of the thyroid and potentially leading to a reduction in the production of thyroid hormone.[107,119,120] Thyroid hormones are particularly important for growth and development of the central nervous system in fetuses and infants.[121] Due to the sensitivities of the developing fetus, perchlorate exposures among pregnant women, especially those with preexisting thyroid disorders or iodide deficiency, carry the potential for risk of adverse health effects.

Organophosphate Pesticides

Agricultural crops are frequently treated with pesticides to control insects and other pests that may affect crop growth. Some of the most prevalent classes of pesticides used in growing food crops are the carbamates, pyrethroids, and the organophosphates. After crops are harvested, they may retain residues of these pesticides. Apples, corn, oranges, rice, and wheat are among the agricultural commodities consumed in large amounts by children.

[iii] The U.S. Food and Drug Administration recently worked with several manufacturers to remove grease-proofing agents containing C8 perfluorinated compounds from the marketplace. These manufacturers volunteered to stop distributing products containing these compounds in interstate commerce for food-contact purposes as of October 1, 2011. For more information, see http://www.fda.gov/Food/FoodIngredientsPackaging/FoodContactSubstancesFCS/ucm308462.htm.

Organophosphates are one class of pesticides that are of concern for children's health. Examples of organophosphate pesticides include chlorpyrifos, azinphos methyl, methyl parathion, and phosmet. These pesticides are frequently applied to many of the foods important in children's diets, and certain organophosphate pesticide residues can be detected in small quantities on these foods. Organophosphates can interfere with the proper function of the nervous system when exposure is sufficiently high.[1,122] Childhood is a period of increased vulnerability, because many children may have low capacity to detoxify organophosphate pesticides through age 7 years.[123] Recent studies have reported an association between prenatal organophosphate exposure and childhood attention deficit/hyperactivity disorder (ADHD) in U.S. communities with relatively high exposures to organophosphate pesticides,[124] as well as with exposures found within the general US population.[125] Other recent studies have reported associations between prenatal organophosphate pesticide exposures and a variety of neurodevelopmental deficits in childhood, including reduced IQ, perceptual reasoning, and memory.[126-128] Since 1999, EPA has imposed restrictions on the use of the organophosphate pesticides azinphos methyl, chlorpyrifos, and methyl parathion on certain food crops and around the home, due largely to concerns about potential exposures of children.[129-131]

The 1996 Food Quality Protection Act required EPA to identify and assess the extent of dietary pesticide exposure in the United States, and to determine whether there was a "reasonable certainty of no harm" to vulnerable populations including infants and children.[132] The U.S. Department of Agriculture's Pesticide Data Program (PDP) provides data annually on pesticide residues in food, with a specific focus on foods often consumed by children.[133] Other researchers have supplemented the PDP with their own analyses. A recent study measured pesticide residues in 24-hour duplicate food samples of fruits, vegetables, and juices served to children, and found that 14% of the samples contained at least one organophosphate pesticide.[134] Additional pesticide residue data are available from FDA's pesticide residue monitoring program.[135] A number of pesticide residues, along with a variety of other chemicals in food, are also measured in FDA's Total Diet Study.[136] When pesticide residue data are combined with dietary consumption surveys, it can be possible to estimate pesticide exposure from dietary intake.

Indicator E9 presents the percentage of samples of two fruits and two vegetables analyzed by the USDA PDP that have detectable residues of organophosphate pesticides. This indicator allows for a general comparison of the frequency of organophosphate detection over time for four foods typically consumed by children, although data are not available on each fruit or vegetable for every year.

About the Indicator: Indicator E9 presents the percentage of sampled apples, carrots, grapes, and tomatoes that were found to contain detectable residues of organophosphate pesticides from 1998–2009. These foods were selected because they are frequent components of children's diets, and because data for these foods were available for multiple years. The data are from an analysis of pesticide residues in foods conducted annually by the U.S. Department of Agriculture.

Pesticide Data Program

The U.S. Department of Agriculture (USDA) collects data on pesticide residues in food annually. USDA's Pesticide Data Program (PDP), initiated in 1991, focuses on measuring pesticide residues in foods that are important parts of children's diets, including apples, apple juice, bananas, carrots, grapes, green beans, orange juice, peaches, pears, potatoes, and tomatoes.

Samples are collected from food distribution centers in 10 states across the country.[137] The PDP has a statistical design in which food samples are randomly selected from the national food distribution system and reflect what is typically available to the consumer, including both domestic and imported foods.[137] Different foods are sampled each year. In its history, the PDP has tested for more than 440 different pesticides.[133] In 2009, the PDP analyzed fruit and vegetables for 309 pesticides and related chemicals. Prior to analysis, the PDP processes samples by following the preparations an average individual would use before consuming an item. This includes washing fruits and vegetables, as well as removing inedible portions of a food item. For example, tomatoes and grapes are washed with the stems and other materials removed, while apples are washed and the stems and cores are removed.

Data Presented in the Indicator

Indicator E9 displays the percentage of apple, grape, carrot, and tomato samples with detectable organophosphate pesticide residues reported by the PDP from 1998–2009. These four foods were selected as those that were sampled by the PDP in at least five years from 1998–2009 and are among the 20 most-consumed foods identified in an analysis by EPA.[138] Other foods not shown here may have either greater or lesser frequencies of organophosphate pesticide residue detection than the four foods presented in this indicator.

The 43 organophosphates that were sampled in every one of the years 1998–2009 are included in calculation of the indicator; 53 other organophosphates that were added to or dropped from the program in these years are excluded so that the chart represents a consistent set of pesticides for all years shown. Some aspects of trends in organophosphate residues could be missed by the indicator if any organophosphates other than the 43 considered in the indicator had substantial changes in use on the four selected foods during the years 1998–2009. For example, a decrease in the percentage of detections of organophosphate residues may reflect an actual decrease in the use of organophosphate pesticides, but can definitively represent only a decrease in the residues of the 43 OPs included in the indicator; it does not account for potential substitution with other organophosphates or other types of pesticides.

The indicator also defines "detectable" based on the ability to measure residues in the PDP in 1998, so that introduction of more sensitive measurement techniques over time does not affect the indicator and allows for direct comparison of data collected in previous years with those collected today. This means that some produce samples analyzed in recent years with improved detection technology would, for purposes of indicator calculation, be considered to have non-detectable organophosphate residues based on comparison with the older, higher limit of detection.[iv]

The fruits and vegetables shown in this indicator were each sampled in five to seven years between 1998 and 2009. Gaps in the percentage of residue detections from year to year thus represent missing information, rather than an absence of organophosphate residues.

This indicator is a surrogate for children's exposure to pesticides in foods: If the frequency of detectable levels of pesticides in foods decreases, it is likely that exposures will decrease. However, the indicator does not account for many additional factors that affect the risk to children. For example, some organophosphates pose greater risks to children than others do, and residues on some foods may pose greater risks than residues on other foods due to differences in amounts consumed. The indicator also does not distinguish between residue levels that are barely detectable and those that are much higher, which would pose a greater concern for children's health. Finally, exposures to organophosphate pesticides may also occur by pathways other than the diet, such as ingestion of pesticides present in house dust and drinking water.

[iv] An alternate analysis of the data that considered all detectable residues, without holding the limit of detection constant at 1998 levels, resulted in percentages of food samples with detectable organophosphate pesticide residues very similar to those shown in the indicator.

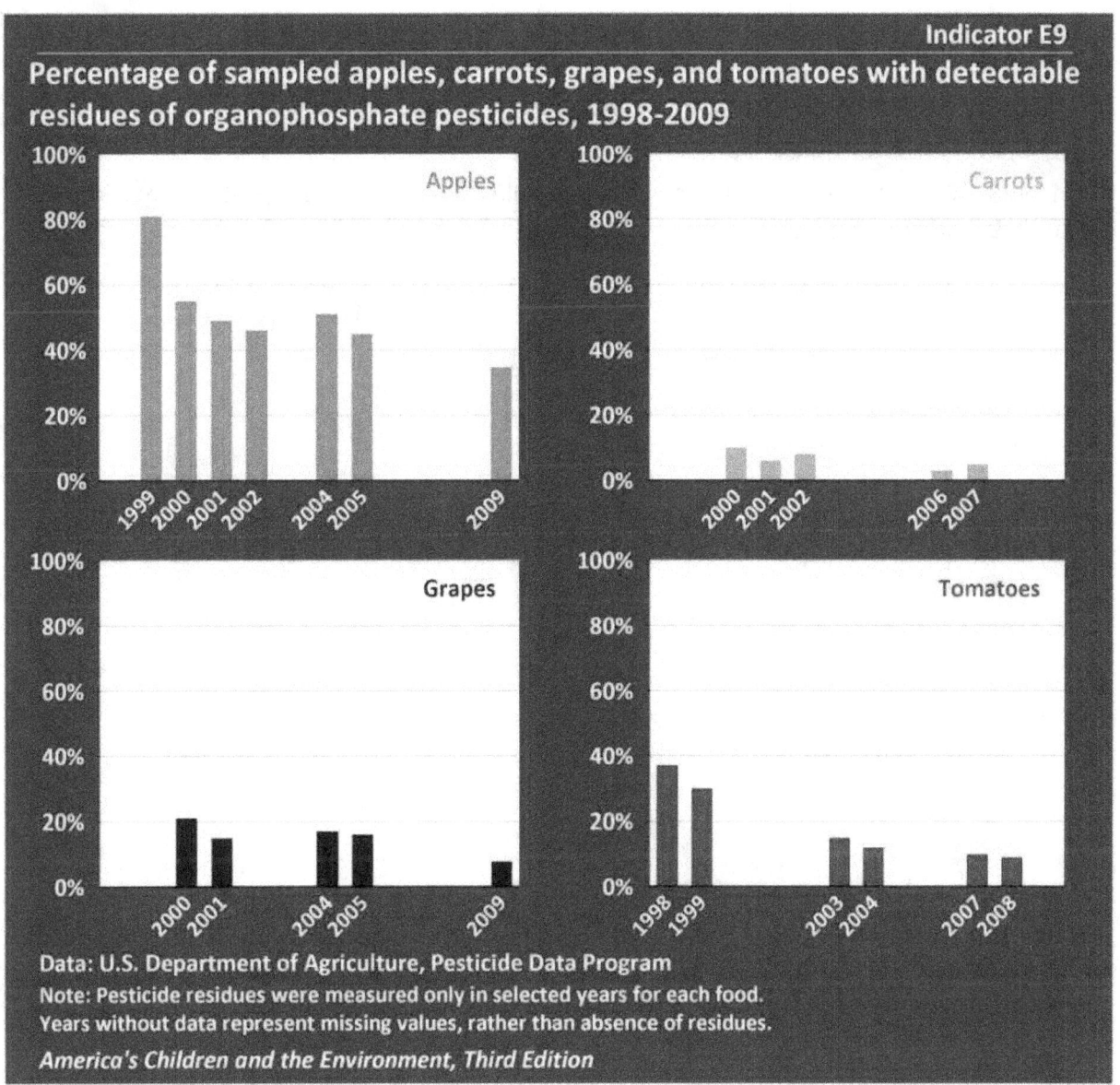

Indicator E9

Percentage of sampled apples, carrots, grapes, and tomatoes with detectable residues of organophosphate pesticides, 1998-2009

Data: U.S. Department of Agriculture, Pesticide Data Program
Note: Pesticide residues were measured only in selected years for each food.
Years without data represent missing values, rather than absence of residues.
America's Children and the Environment, Third Edition

Data characterization
- Data for this indicator are obtained from a U.S. Department of Agriculture program that measures pesticide residues in food samples collected from 10 states.
- Food samples are randomly selected from the national food distribution system and reflect what is typically available to the consumer.
- The types of foods sampled change over time; so, for example, data for pesticide residues on apples are not available every year.
- The indicator is calculated using the measurement sensitivity as of 1998 for each year shown; more sensitive measurement techniques have been incorporated over time.

- In 1999, 81% of sampled apples had detectable organophosphate pesticide residues. In 2009, 35% had detectable residues.

- In 2000, 10% of sampled carrots had detectable organophosphate pesticide residues. In 2007, 5% had detectable residues.

- In 2000, 21% of sampled grapes had detectable organophosphate pesticide residues. In 2009, 8% had detectable residues.

- In 1998, 37% of sampled tomatoes had detectable organophosphate pesticide residues. In 2008, 9% had detectable residues.

Contaminated Lands

Accidents, spills, leaks, and improper disposal and handling of hazardous materials and wastes have resulted in tens of thousands of contaminated sites across the United States. The nature of the contaminants and the hazards they present vary greatly from site to site. These contaminants include industrial solvents, petroleum products, metals, residuals from manufacturing processes, pesticides, and radiological materials, as well as certain naturally occurring substances such as asbestos. Contaminated lands can threaten human health and the environment, in addition to hampering economic growth and the vitality of local communities.

The presence of contaminated soils in a particular location may or may not have health consequences. Soils, unlike air and water, are not intentionally inhaled, absorbed, or ingested. Contaminants diffuse more slowly through soil than through air or water, so contaminants are rarely distributed uniformly across a contaminated site. Soils are a concern if children are playing, attending school, or residing on or near to contaminated land. People and pets may track contaminated soils and dusts into homes where infants and toddlers are playing. Some contaminants may harm or penetrate the skin, and by touching or playing in soil children may come into direct contact with them. Children may ingest soils through hand-to-mouth play or by eating without first washing their hands after having touched contaminated soil. Soil dust may be carried on the wind and inhaled into the lungs, where it can be very damaging. The optimal approach to minimizing risks to children from contaminated soils is to prevent these exposures.

In addition, contaminated land may contribute to pollution of ground water, surface water, ambient air, and foods, creating additional potential human exposure routes. For example, consumption of fish caught at or near a contaminated site may increase risk of exposure to contaminants from the site. The same is true of drinking water from contaminated ground- or surface water sources. When drinking water sources are affected at EPA-tracked contaminated sites, an alternate water supply may need to be provided, in some cases permanently.

Cleanup of contaminated lands may be conducted by EPA, other federal agencies, states, tribes, municipalities, or the party responsible for the contamination. As of September 2011, EPA's programs for assessing and cleaning up contaminated lands track roughly 22 million acres of land across the United States, or nearly 1% of the entire U.S. land mass.[1] EPA and its partners conduct work on contaminated lands through federally mandated programs such as the Superfund and Corrective Action programs. The Superfund program, implemented under the Comprehensive Environmental Response, Compensation, and Liability Act (CERCLA), aims to clean up some of the most hazardous and highly polluted inactive commercial, industrial, and residential properties in the country. The Corrective Action program, implemented under the Resource Conservation and Recovery Act (RCRA), aims to control and clean up releases at operating hazardous waste treatment, storage, and disposal facilities. EPA is also responsible for other programs that focus on management of contaminated lands, including Brownfields, underground storage tanks, and RCRA waste management and minimization programs.

EPA prioritizes sites for cleanup using information from initial investigations regarding possible threats to human health or the environment. EPA's primary concern is to protect people from the most contaminated lands and to clean up these sites to a standard that is protective, and that state, local, or tribal governments and communities deem appropriate based on the future uses of the individual site. EPA and partner agencies work to contain possible routes for exposure as soon as possible.[2,3]

When a potential pathway for exposure is identified, a process is normally initiated for the pathway to be minimized or eliminated. For Superfund sites and for hazardous waste facilities requiring Corrective Action, EPA or authorized state regulators assess contaminated media, exposure pathways, risks from complete pathways, and the significance of any risks. If no significant human health risks are identified, a determination is made that the site has all human health protective measures in place. If significant human health risks are or may be present, regulators choose site-specific controls (e.g., fencing, caps, containment walls) and cleanup activities (e.g., excavation, groundwater treatment) necessary to reduce the risks.

If additional contamination or previously unrecognized pathways of exposure are identified, a site that is designated as having all human health protective measures in place may lose that designation until pathways of exposure are controlled.

When a site is designated as having all human health protective measures in place, known pathways of exposure have been controlled, although additional cleanup work may remain. These sites pose a reduced risk to children compared with most sites that have not yet been designated as having all human health protective measures in place. However, there can be a number of reasons why a site has not yet achieved that designation. For example, some sites have not yet been adequately assessed, and it is thus unknown whether these sites pose significant risk to human health.

This approach to managing potential exposures is based on identified presence of contaminants and potential exposure pathways because there is often an absence of information identifying actual children's exposures; however, there are notable exceptions where EPA and other federal and state agencies have addressed documented exposures.[4-10]

Children who have been exposed to contaminants do not all experience the same health outcomes. The magnitude and duration of an exposure, the pathway of exposure (ingestion, inhalation, dermal), the stage of development at which a child is exposed, and differences in genetic susceptibility all influence the variation in outcome from exposure. Even after exposure characteristics and genetic factors have been taken into consideration, variation remains in risks experienced by different individuals and different communities as a consequence of exposures to contaminants. This variation may in part be explained through socio–cultural and socioeconomic factors that have been associated with physical and psychological health, including family income, unemployment, nutrition, education, housing and infrastructure, race, gender, class, access to health services, social cohesion, participation in local decision-making, exercise, and health-related behaviors (e.g., smoking, drug abuse).[11-22]

Of the many sociological determinants of health, the relationships between race/ethnicity and health status and between lower levels of income and less optimal health are among the most documented.[23-26] Because these factors are related to many of the other sociological determinants, they are frequently used as proxies for a larger set of factors. For these reasons, the following indicators of children living in proximity to contaminated lands focus on differences by race/ethnicity and family income level.

Indicator E10: Percentage of children ages 0-17 years living within one mile of Superfund and Corrective Action sites that may not have all human health protective measures in place, 2009

Indicator E11: Distribution by race/ethnicity and family income of children living near selected contaminated lands in 2009, compared with the distribution by race/ethnicity and income of children in the general U.S. population

About the Indicators: Indicators E10 and E11 present information about children living within one mile of Superfund sites or RCRA Corrective Action sites that may not have had all human health protective measures in place as of October 1, 2009. Site boundaries were estimated and a computer mapping tool was used to identify all land areas within one mile of each of these sites. Data from the 2000 U.S. Census were then used to estimate the population of children living within these areas. Indicator E10 provides information about U.S. children living within one mile of these selected sites, including the percentage of children in proximity by race, ethnicity, and family income. Indicator E11 compares the race/ethnicity profile of children living within one mile of these selected sites with the profile for all children living in the United States.

Corrective Action and Superfund Sites

EPA's Office of Solid Waste and Emergency Response manages the RCRA Corrective Action Program and the Superfund Program, and maintains inventories of sites in each program. The Comprehensive Environmental Response, Compensation, and Liability Information System (CERCLIS) database provides information on Superfund sites, and the Resource Conservation and Recovery Act Information (RCRAInfo) database provides information on RCRA Corrective Action sites. As of October 1, 2009 there were 1,653 Corrective Action and Superfund sites, totaling more than 10 million acres, that may not have had all human health protective measures in place.[3] Of the 3,746 Corrective Action sites at that time, 1,297 fell into this category. Of the 1,727 Superfund sites (which includes both sites that are on the National Priorities List and sites that are not on the NPL but for which the Superfund program has some responsibilities), a total of 356 fell into this category. The location and extent of each site are characterized by the latitude and longitude of a single point within that site, and the area (total acres) of the site, obtained from the official documentation for each site.[i] A map displaying the distribution of these sites across the country and their prevalence in urbanized areas is available in the Methods document for this topic (available at www.epa.gov/ace).

Some of the largest sites that EPA oversees are federal facilities. Among the sites that may not have had all human health protective measures in place in 2009, 47 Corrective Action sites and 62 Superfund sites are federal facilities.

---

[i] Actual boundaries of the sites are available in digital form for only a few sites.

Estimating Site Areas and Children's Proximity

For purposes of indicator calculation, the actual land area within each site was approximated using the latitude/longitude and acreage information. A circle whose area equaled the site's acreage was drawn around each site's latitude/longitude identification point. It is important to note that these areas are not the actual site boundaries, and are not expected to reflect the actual area of contamination. Contamination will likely be determined by factors such as the release of waste, the contours of the land, and groundwater flow. Sites also have hotspots (areas with high levels of contamination) and areas that have been remediated or were never contaminated. The site boundaries are therefore likely to overestimate the area of a site that is contaminated. Nonetheless, approximating the area of a site with a circle is a reasonable assumption that provides the best available information for this analysis.

To identify land areas in proximity to the selected contaminated lands, a one-mile buffer was drawn around the circle representing each site. Data on total child population, and population by race and ethnicity, were collected from the 2000 Census for children living in Census blocks whose center point was within the one-mile buffer boundary. Information on family income levels (percentage above and below poverty level, by race and ethnicity) was extrapolated for these blocks from Census block group data. Data from the 2000 census were used in order to obtain necessary population race/ethnicity and income statistics at the local level; this information is not available in the 2009 census estimates.[ii]

Data Presented in the Indicators

Each indicator presents a characterization of the population of children living within one mile of Superfund or RCRA Corrective Action sites that may not have had all human health protective measures in place as of October 1, 2009. Indicator E10 shows the percentage of children living within one mile of a site, by race/ethnicity and family income. Indicator E11 shows the proportion of children of each race and ethnicity among those living in proximity to the selected sites, compared with the race/ethnicity proportions among all children in the United States. This comparison is also made for children living in homes with incomes below poverty level. Tables of values for these indicators at the state level are available in the Appendix to this document.

Data for seven race/ethnicity groups are presented in the indicators: White, Black, Asian, American Indian or Alaska Native (AIAN), Native Hawaiian or Other Pacific Islander (NHOPI), All Other Races, and Hispanic. The "All Other Races" category includes all other races not specified, together with those individuals who report more than one race. Children of Hispanic ethnicity may be of any race. Data presented by race do not include any designation of ethnicity; for example, the indicator value labeled "Black" includes both Hispanic and non-Hispanic Black children, and children who are Black and Hispanic are included in the indicator values for both

---

[ii] A greater percentage of children were living in poverty in 2009 than in 2000; therefore, these calculations will understate the proportion of children below poverty living in proximity to the selected contaminated lands in 2009.

"Black" and "Hispanic" children. Three family income categories are presented in the indicators: all incomes, below the poverty level, and greater than or equal to the poverty level.

Designation of sites that may not have all human health protective measures in place were made for the first time in 2009; trend data are not reported because these designations were not analyzed for purposes of this report in earlier years.[iii]

For purposes of these indicators, proximity to a site is used as a surrogate for potential exposure to contaminants found at these sites. The indicators do not imply any specific relationship between childhood illness and a child's proximity to a Superfund or Corrective Action site. Information on amounts of environmental contamination, which would be a source of exposure to children, is generally available for these sites, but information on the extent to which children are actually exposed is not generally available. Because of the ways in which children can be exposed to land contaminants and the potential for certain contaminants to move into groundwater or to vaporize through soil, the proximity to contaminated sites may increase the potential for exposure and the possible health consequences, but proximity to a site does not mean that there will always be exposure. Nor does proximity to a site represent risks of adverse health effects. The risk of exposure posed to children varies significantly across all the different types of contaminated sites and the different activities of children on or near the sites. Many sites do not pose risks outside of property boundaries.

These indicators present a high-end approximation of children at risk from the Corrective Action and Superfund sites that may not have all human health protective measures in place, but do not include children near the much larger universe of Brownfield sites, leaking underground storage tanks, and sites addressed solely by state, tribal, and local authorities or private companies. While the indicators include those RCRA Corrective Action sites assumed to have the most potential for contamination, these sites represent only a subset of waste treatment, storage, or disposal facilities currently regulated by EPA. The indicators also do not capture the proportion of children living near contaminated sites that are yet to be identified. Access to uncontrolled contamination remains the greatest risk of potential exposure, and risks are most likely to have been greatest prior to intervention by EPA and partner agencies. The ultimate cleanup of these sites best assures reduced health risks for children by eliminating the possibility of exposure and promotes the health of their communities since cleanup opens the way for sustainable redevelopment and revitalization opportunities.

[iii] These data cannot be compared to Indicator E9 from previous editions of *America's Children and the Environment*. Previous versions considered only Superfund sites; represented each site as a single point, rather than an area; and did not consider the status of human health protective measures put in place at the sites.

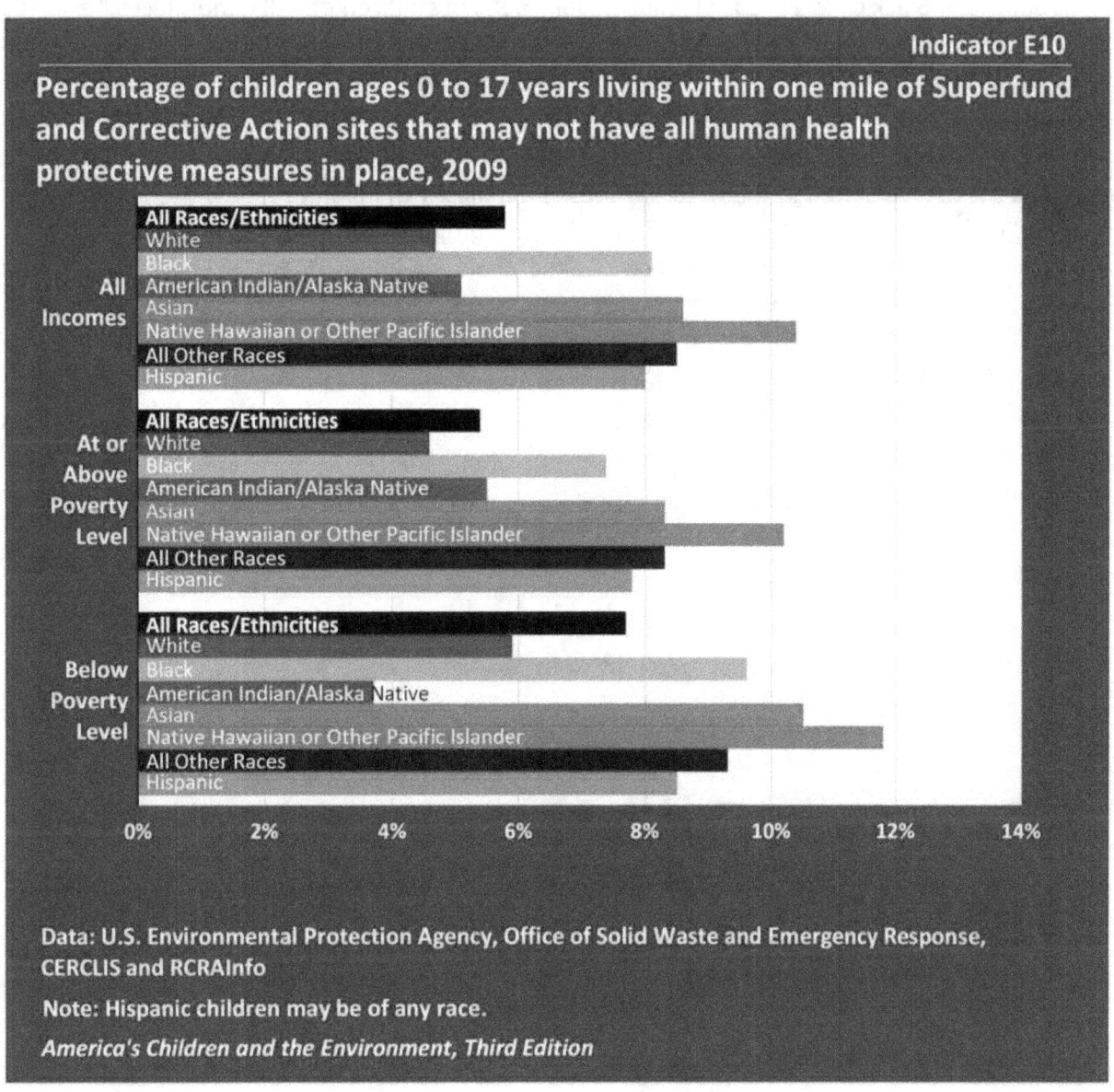

Indicator E10

Percentage of children ages 0 to 17 years living within one mile of Superfund and Corrective Action sites that may not have all human health protective measures in place, 2009

Data: U.S. Environmental Protection Agency, Office of Solid Waste and Emergency Response, CERCLIS and RCRAInfo

Note: Hispanic children may be of any race.

America's Children and the Environment, Third Edition

Data characterization
- Data on Superfund and RCRA Corrective Action sites are reported by EPA regional offices and states, and compiled in EPA's databases of information on contaminated sites.
- Information for each site includes the site name, state in which the site is located, latitude, longitude, estimated acreage, and site status.
- Areas of known or suspected contamination may be less than the total acreage at each site.

- Approximately 6% of all children in the United States lived within one mile of a Corrective Action or Superfund site that may not have had all human health protective measures in place as of 2009.

- About 8% of Black children, 9% of Asian children, 9% of children of "All Other Races," and 10% of Native Hawaiian and Other Pacific Islander (NHOPI) children lived in proximity to the designated sites. About 8% of Hispanic children, who may be of any race, lived in proximity to the sites. In contrast, about 5% of White children and 5% of American Indian/Alaska Native children lived in proximity to the designated sites.

- About 8% of all children in the United States in families with incomes below the poverty level lived within one mile of the designated sites, compared with about 5% of children above the poverty level. The proportion of children below the poverty level in proximity to the designated sites was generally greater than the proportion for those above poverty level for each race and ethnicity; the only exception to this pattern was for American Indian and Alaskan Native (AIAN) children.

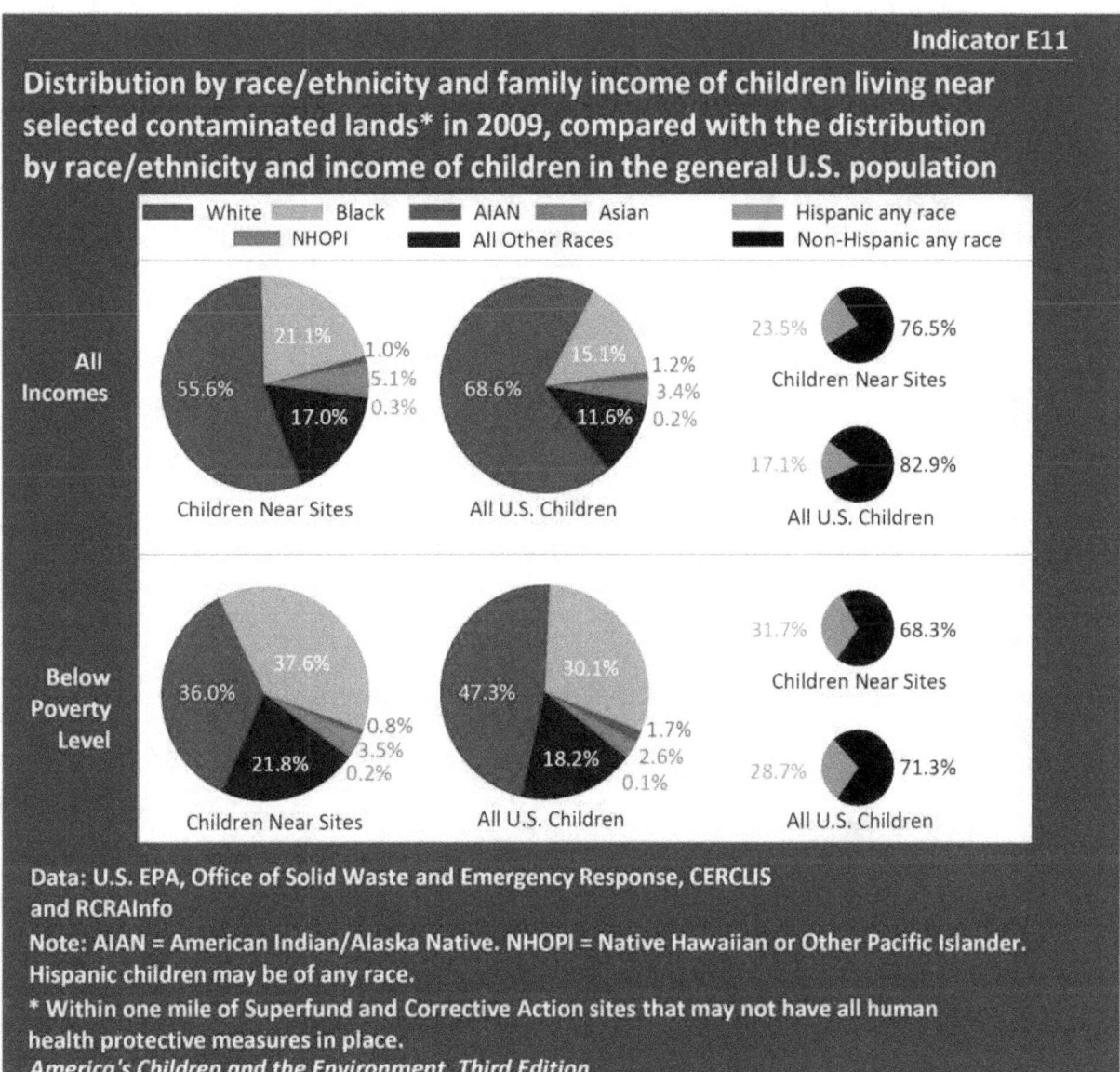

Indicator E11

Distribution by race/ethnicity and family income of children living near selected contaminated lands* in 2009, compared with the distribution by race/ethnicity and income of children in the general U.S. population

Legend: White, Black, AIAN, Asian, Hispanic any race, NHOPI, All Other Races, Non-Hispanic any race

All Incomes — Children Near Sites: 55.6%, 21.1%, 1.0%, 5.1%, 0.3%, 17.0%
All Incomes — All U.S. Children: 68.6%, 15.1%, 1.2%, 3.4%, 0.2%, 11.6%
All Incomes — Children Near Sites: 23.5%, 76.5%
All Incomes — All U.S. Children: 17.1%, 82.9%

Below Poverty Level — Children Near Sites: 36.0%, 37.6%, 0.8%, 3.5%, 0.2%, 21.8%
Below Poverty Level — All U.S. Children: 47.3%, 30.1%, 1.7%, 2.6%, 0.1%, 18.2%
Below Poverty Level — Children Near Sites: 31.7%, 68.3%
Below Poverty Level — All U.S. Children: 28.7%, 71.3%

Data: U.S. EPA, Office of Solid Waste and Emergency Response, CERCLIS and RCRAInfo
Note: AIAN = American Indian/Alaska Native. NHOPI = Native Hawaiian or Other Pacific Islander. Hispanic children may be of any race.
* Within one mile of Superfund and Corrective Action sites that may not have all human health protective measures in place.
America's Children and the Environment, Third Edition

Data characterization
- Data on Superfund and RCRA Corrective Action sites are reported by EPA regional offices and states, and compiled in EPA's databases of information on contaminated sites.
- Information for each site includes the site name, state in which the site is located, latitude, longitude, estimated acreage, and site status.
- Areas of known or suspected contamination may be less than the total acreage at each site.

- Approximately 21% of all children living within one mile of a Corrective Action or Superfund site that may not have had all human health protective measures in place were Black, while 15% of children in the United States as a whole are Black. Black children account for about 30% of all U.S. children in homes below poverty level; among children below poverty level living within one mile of a designated site, about 38% were Black.

- The percentages of Asian children, Hispanic children, and children of "All Other Races" among children living close to the designated sites were also greater than the percentages of these children in the entire U.S. population, considering all incomes and considering only those in homes with incomes below poverty level.

Climate Change

Climate change refers to any significant change in climate variables including temperature, precipitation, or wind that lasts for decades or longer. It may include changes in variability of average weather conditions or extreme weather conditions. Both human activities and natural factors contribute to climate change. Human activities, such as burning fossil fuels; cutting down forests; and developing land for farms, cities, and roads, release heat-trapping greenhouse gases into the atmosphere. Natural causes, such as changes in the Earth's orbit, the sun's intensity, the circulation of the ocean and the atmosphere, and volcanic activity, contribute to climate change in a variety of ways.[1]

Climate change may increase children's exposure to extreme temperatures, polluted air and water, extreme weather events, wildfires, infectious disease, allergens, pesticides, and other chemicals. These exposures may affect children's health in a number of direct and indirect ways. It is important to note that climate change will likely result in a mix of both positive and negative health impacts. For example, warmer summers may increase the number of heat-related injuries and deaths, while warmer winters may result in fewer cases of cold-related injuries and deaths.[2] The effects of climate change will also vary from one location to another and will likely change over time as climate change continues.[2,3] Furthermore, the human health risks from climate change may be affected strongly by changes in health care advances and accessibility, public health infrastructure, and technology.[2,4-6]

Direct effects of extreme temperatures are one area of concern, as climate change is expected to increase the number and intensity of hot days, hot nights, and heat waves in the United States.[5,7,8] Heat exposure can result in heat rashes, heat stroke, heat exhaustion, and even death; children may be especially at risk because they often spend more time outside than adults do.[2,9] Children's bodies are less effective at adapting to heat compared with those of adults.[10] Also, children may not feel the need to drink as urgently, which can lead to severe dehydration and electrolyte imbalance.[10,11] Humidity can further exacerbate heat stress in children.[10,11] Infants may be especially vulnerable to heat events in part because they depend on adults for care and are unable to communicate thirst and discomfort.[6,12,13] Caregivers can help protect children from heat-related health effects.[14]

Many factors can modify the impact of heat exposure, including geographic location, income level, and the built environment.[15] Studies have shown that the temperature at which mortality and morbidity (e.g., respiratory hospital admissions) can occur from heat exposure varies based on location.[16-18] Extreme heat exposure may have a greater impact on populations living in regions that experience high temperatures less frequently, such as the Northwest and Midwest United States. In warmer climates such as those in the South and Southwest United States, the population may be acclimated to heat and area infrastructure is better designed to accommodate high temperatures.[13,19] A higher income allows families to adapt more easily to meet the challenges of climate change compared with lower-income families, because they can afford the use of air conditioners and other cooling methods to create a more ideal and comfortable environment.[3]

The urban built environment can both exacerbate and alleviate the effects of heat. For example, high concentrations of buildings in urban areas cause what is known as the urban heat island effect: generating as well as absorbing and releasing heat, resulting in urban centers that are several degrees warmer than surrounding areas. Expanding the area of parks and green spaces and increasing the density of trees in and around cities can help to reduce this effect.[6]

Warmer winters may have the effect of decreasing the number of cold-related deaths and injuries.[2,15] It is difficult to estimate the net changes in mortality due to climate change; however, a recent assessment by the United States Global Change Research Program concluded that increases in heat-related mortality due to climate change are unlikely to be compensated by decreases in cold-related mortality.[8]

High temperatures, heat waves, and associated stagnant air masses can increase levels of air pollution, specifically ground level ozone, fine particulate matter ($PM_{2.5}$), nitrogen oxides, and sulfur oxides.[2,6,8,9] These air pollutants can be harmful for children: they may contribute to the development of new cases of asthma, aggravate preexisting cases of asthma, cause decrements to lung function, increase respiratory symptoms such as coughing and wheezing, and increase hospital admissions and emergency room visits for respiratory diseases.[20-35] Because children may spend a lot of time outdoors, often while exerting themselves for sports or play, they can be especially vulnerable to the impacts of poor air quality.[8]

Climate change is likely to change the timing, frequency, and intensity of extreme weather events, including heat waves, hurricanes, heavy rainfall, droughts, high coastal waters, and storm surges.[5,36] These events can cause traumatic injury and death, as well as emotional trauma. Extreme weather events are also associated with increased risk of food- and water-borne illnesses as sanitation, hygiene, and safe food and water supplies are often compromised after these types of events.[2] One study found that periods of heavy rainfall were associated with increased emergency room visits for gastrointestinal illness among children.[37] Heavy rainfall may result in flooding, which can lead to contamination of water with dangerous chemicals, heavy metals, or other hazardous substances from storage containers or from preexisting chemical contamination already in the environment.[2,36] Elevated temperatures and low precipitation are also projected to increase the size and severity of wildfires. This can lead to increased eye and respiratory illnesses and injuries, which include burns and smoke inhalation.[2] Extreme weather events can be especially dangerous for children because they are dependent on adults for care and protection.[7]

A number of infectious diseases may be affected by climate change. The combined effects of increased temperature and precipitation are projected to cause increases in some water-, food-, and vector-borne illnesses. In general, increased temperature results in higher replication, transmission, persistence, habitat range, and survival of bacterial pathogens (the effect on viral pathogens is less clear), and produces a greater number of water- and food-borne parasitic infections.[5,6,8] Climate change is also expected to expand or shift the habitat and range of disease-carrying organisms, such as mosquitoes, ticks, and rodents.[5] Changes in the geographic distribution of disease-carrying organisms may alter the spread of vector-borne diseases such as Lyme disease, West Nile virus and Dengue fever.[5] Children may be at greater risk for these types of infectious diseases as they spend more time outdoors compared with adults, where they might contact disease-carrying organisms, and they have less-developed immune systems.[14]

Climate change, including changes in carbon dioxide (CO_2) concentrations and temperature, may affect the growth and distribution of allergen-producing vegetation such as weeds, grasses, and trees. Climate change has already caused an earlier onset of the U.S. spring pollen season and a lengthened ragweed season.[15,38] The aeroallergens (e.g., pollen) themselves might be changed in terms of production, distribution, dispersion, and allergic potency.[2,6,15] Exposure to weed and grass pollen has been associated with exacerbation of children's asthma, emergency room visits, and hospitalizations.[39-41]

Through various indirect pathways, climate change may lead to increasing levels and/or frequencies of childhood exposure to harmful contaminants.[6,14] Changes in temperature, rainfall, and crop practices related to climate change are likely to affect exposure to pathogens, pesticides, and other chemicals in a number of ways. Broader geographic distribution of pests and increased growth of invasive weeds will likely lead to greater use of pesticides.[6,8] Increased precipitation and increased variability in precipitation are likely to increase pathogen and contaminant levels in lakes and other surface waters.[2,42] The distribution of chemicals in the environment is likely to change: for example, an increase in ice melts caused by a warming climate may release some past emissions of globally transported chemicals, such as polychlorinated biphenyls (PCBs) and mercury, that have been trapped in polar ice.[43,44] Increasing concentrations of these chemicals in the atmosphere, and subsequent deposition to land and water, have the potential to increase concentrations of these chemicals in fish and other foods derived from animals. Warmer water temperatures may also increase the release of chemical contaminants from sediments, increasing their uptake in fish.[2] Climate change may result in children spending more time indoors. Buildings that are tightly sealed in response to adverse weather conditions may result in increased exposure to contaminants from poor ventilation and higher concentrations of indoor pollutants such as radon, environmental tobacco smoke, and formaldehyde.[45]

Children are expected to be especially sensitive to the effects of climate change for a number of reasons. Young children and infants are particularly vulnerable to heat-related illness and death.[6] Compared with adults, children have higher breathing rates, spend more time outside, and have less developed respiratory tracts—all making children more sensitive to air pollutants. Additionally, children have immature immune systems, meaning that they can experience more serious impacts from infectious diseases.[8] The greatest impacts are likely to fall on children in poor families, who lack the resources, such as adequate shelter and access to air conditioning, to cope with climate change.[8]

EPA is currently developing a new children's environmental health indicator for climate change. The new indicator will focus on the frequency of extreme heat events over time. EPA intends to complete development of this new indicator in 2014, and it will be made available at www.epa.gov/ace when completed.

Biomonitoring

Introduction

What is biomonitoring?

In the field of human exposure assessment, biomonitoring refers to the measurement of chemicals in human body fluids and tissues, such as blood, urine, breast milk, saliva, and hair. Measurements of the levels of pollutants in children's bodies provide direct information about their exposures to environmental contaminants. Measurements in women who may become pregnant, currently are pregnant, or currently are breastfeeding provide information about exposures that may affect conception, the fetus, or the developing child.

Biomonitoring measurements provide an estimate of the amount of a chemical absorbed into the body from all pathways of exposure (for example, ingestion of drinking water, inhalation of air), and thus give a cumulative estimate of the chemical burden that a person carries in their body, sometimes referred to as a body burden. Biomonitoring can characterize differences in exposure among groups within a population, and can characterize changes in population exposure over time. Biomonitoring is an increasingly important element of epidemiological research when evaluating whether chemical exposures are associated with adverse health effects in humans.

What environmental chemicals are included in the Biomonitoring indicators for *America's Children and the Environment, Third Edition (ACE3)*?

Biomonitoring topics were selected for ACE3 based on: (1) research that indicates an association between exposure and children's health or suggests a potential association between exposure and children's health; (2) significant public interest; and (3) the nature of the biomonitoring data available (for example, range of ages for which data are available and frequency of detection). EPA obtained input from its Children's Health Protection Advisory Committee to assist in selecting topics from among the many chemicals with biomonitoring data available. The ACE3 Biomonitoring indicators address the following topics:

- Lead
- Mercury
- Cotinine (a marker for environmental tobacco smoke exposure)
- Perfluorochemicals (PFCs)
- Polychlorinated biphenyls (PCBs)
- Polybrominated diphenyl ethers (PBDEs)
- Phthalates
- Bisphenol A (BPA)
- Perchlorate

For many of the chemicals addressed in this section, scientific findings have reported associations between children's health and the mother's exposure during pregnancy. For this reason, indicators for several of these topics present data for women of child-bearing age—defined here as ages 16 to 49 years.

What data sources were used to develop the Biomonitoring indicators?

Biomonitoring data are generated by collecting samples of blood, urine or other biological specimens from a group of individuals, then measuring the concentrations of selected chemicals in those specimens. There are many scientific research efforts that collect biomonitoring data in the United States, but only the National Health and Nutrition Examination Survey (NHANES) conducted by the National Center for Health Statistics (NCHS) measures chemicals in the blood and urine[1] of a nationally representative sample of the U.S. population. NHANES was therefore identified as the most suitable data source for all Biomonitoring indicators presented in ACE3. Summary statistics for more than 200 chemicals measured in NHANES are reported in the *Fourth National Report on Human Exposure to Environmental Chemicals*,[1] and data files containing individual measurements for each chemical are available from the NHANES website.[2]

Because NHANES is an ongoing, continuous survey that provides data over a number of years using a consistent sample design and consistent methods of measurement, the biomonitoring levels can be compared over time and across demographic groups. However, because of the highly clustered sample design of the survey, multiple NHANES cycles should be combined to yield sample sizes necessary for certain types of statistical analysis. NHANES is not designed to provide detailed estimates for populations that are highly exposed to particular environmental chemicals. In addition, military personnel and people who reside in institutions are excluded from NHANES.

For most of the environmental chemicals currently measured in NHANES, data are available starting in 1999 or more recently; measurement of lead and cotinine began earlier. Availability of NHANES biomonitoring data for children varies by chemical and type of sample, and, in general, biomonitoring data for young children are quite limited. For environmental chemicals measured in urine, NHANES collects data from survey participants ages 6 years and older. For most environmental chemicals measured in blood, NHANES collects data from survey participants ages 12 years and older. Exceptions apply to three chemicals presented in this section: measurements of lead and mercury in blood are conducted for all survey participants ages 1 year and older, and measurements of cotinine in blood are conducted for all participants ages 3 years and older. NHANES does not measure chemicals in breast milk, an important route of exposure for infants, or in target organs where chemicals affect the body.

What can we learn from biomonitoring indicators?

Biomonitoring indicators in ACE3 provide summaries of biomonitoring measurements in blood or urine specimens obtained from a nationally representative target population—either children within a specified age range, or women of child-bearing age. For chemicals that are persistent in the human body, biomonitoring measurements may be reflective of exposures that have occurred over several months or years. For chemicals that are cleared from the body more rapidly, a biomonitoring measurement may typically reflect exposures that have occurred within the previous few days.

The Biomonitoring indicators prepared for ACE3 focus primarily on presenting biomonitoring data collected over multiple years to evaluate whether there are any changes over time. The biomonitoring indicator values are also compared across various race/ethnicity, income, or age groups.

When health benchmarks are available, biomonitoring data may provide insights about the percentage of a population at risk for adverse health effects; however, in most cases information on health risks associated with levels of chemicals in blood or urine typical for the general population is limited. For some chemicals, such as lead and cotinine, there is an extensive body of literature demonstrating that adverse effects can occur in children with levels of exposure commonly experienced in the general population. However, biomonitoring by itself does not reveal whether any adverse effects have occurred in an individual or in the population.

Biomonitoring indicators present data for one chemical at a time, but biomonitoring studies have found that individuals have multiple chemicals in their bodies.[3-5] While the evidence is still developing for the links between exposures to environmental chemicals and disease, a wide variety of chemicals may act together to produce common adverse outcomes.[6] Thus, even small biological alterations caused by exposure to a single chemical in isolation may have important effects when combined with exposure to other chemicals. The ACE3 Biomonitoring indicators do not reflect this context of simultaneous or sequential exposure to multiple chemicals.

An important limitation of biomonitoring is that, by itself, it provides few clues as to the source(s) of exposure. Data on environmental sources of the chemical are necessary to separate contributions from air, water, food, and/or contaminated soil or dust.

What information is provided for each Biomonitoring topic?

For each topic, an introduction section explains the potential relevance of the chemical to children's health, including a discussion of typical exposure pathways and scientific findings concerning possible adverse health effects.

The introduction section is followed by a description of the indicators, including a summary of the data available from NHANES for the specific chemical or chemical group and information on how each indicator was calculated. One or two indicators, each presented as a graphical representation of the available data, are included for each topic. Where data are available for a sufficient number of years (at least three NHANES two-year cycles), the indicator presents a time series. When time series data are not available, the indicator shows a comparison of the most current biomonitoring data by race/ethnicity and income level.

All indicator figures present median (50th percentile) values; some time series figures also provide 95th percentile values. The median is the value in the middle of the chemical's distribution: half of the measured population has levels of the chemical in their urine or blood that are greater than the median, and half has levels below the median. The 95th percentile is a value representing the upper range of levels: 5% of the specified group has levels of the chemical in their urine or blood that are greater than the 95th percentile. This value therefore can be thought of as representing a high level relative to the rest of the population, but not a maximum level.[i]

Beneath each figure is a description of the data source and explanatory bullet points highlighting key findings from the data presented in the figure, along with key data from any supplemental data tables. References are provided for each topic at the end of the report.

Data tables are provided in Appendix A. The tables include all indicator values depicted in the indicator figures, along with additional data of interest not shown in the figures. Metadata describing the data sources are provided in Appendix B. Documents providing details of how the indicators were calculated are available on the ACE website (www.epa.gov/ace).

Many of the topics presented in the biomonitoring indicators are addressed in Healthy People 2020, which provides science-based, 10-year national objectives for improving the health of all Americans. Appendix C provides examples of the alignment of the biomonitoring topics presented in ACE3 with objectives in Healthy People 2020.

What race/ethnicity groups are used in reporting indicator values?

For each topic in the Biomonitoring section, indicator values are provided for defined race/ethnicity groups—either in the indicator figures or in the data tables—for the following races/ethnicities:

- White non-Hispanic
- Black non-Hispanic
- Mexican-American
- All Other Races/Ethnicities

Values are provided for "Mexican-American" ethnicity rather than "Hispanic" ethnicity because in all years up to 2006, NHANES was designed to provide statistically reliable estimates for Mexican-Americans rather than all Hispanics.[ii]

---

[i] Frequently, a small portion of the population may appear to have much higher levels of an environmental chemical compared with everyone else. In these cases, percentiles in the lower portion of the distribution (below the median) are generally less variable than those well above the median. In NHANES, estimates above the 95th percentile are generally very uncertain due to the sample size, the survey design, and (for many measurements) the substantially skewed distributions.

[ii] NHANES now oversamples Hispanics instead of Mexican-Americans, beginning with NHANES 2007–2008.

The "All Other Races/Ethnicities" category includes all other races and ethnicities not specified, together with those individuals who report more than one race. The limits of the sample design and sample size often preclude statistically reliable estimates for smaller race/ethnicity groups.[iii]

What income groups are used in reporting indicator values?

The ACE3 Biomonitoring indicators present values for income groups defined on the basis of the federal poverty level. Poverty level is defined by the federal government, and is based on income thresholds that vary by year, family size and composition. In 2010, for example, the poverty threshold was $22,113 for a household with two adults and two related children.[7] The Biomonitoring indicators (in figures and/or data tables) provide data separately for individuals in families with incomes below poverty level, and those in families with incomes at or above poverty level.

How were the indicators calculated and presented?

Data files: All indicators were calculated from publicly available data files obtained from the NHANES website. Files include values for the biomonitoring measurement, and information on the sampled individual's age, sex, race/ethnicity, and income level (that is, the family income divided by the poverty level). Each individual observation also has a sample weight that is used in calculating population statistics; the weight equals the number of people in the U.S. population represented by the particular observation.

Population age groups: Indicators of biomonitoring data in children used all data available for children ages 17 years and younger, except for lead where the indicator focuses on children ages 5 years and younger. Indicators of biomonitoring data in women of child-bearing age used all available data for women ages 16 to 49 years. As noted above, indicators for women of child-bearing age are included in ACE3 when there are concerns for children's health associated with the mother's exposure during pregnancy. Adjustments were applied in calculating the population distribution of women ages 16 to 49 years to incorporate birth rates specific to age and race/ethnicity.[8] These adjustments give greater weight to women of ages more likely to give birth, and reduce the contribution to the calculated indicator values of women of ages less likely to give birth (e.g., those ages 40 to 49 years). Without the birth rate adjustment, the indicator values would be calculated as if all women ages 16 to 49 years are equally likely to give birth.[iv]

Please see http://www.cdc.gov/nchs/nhanes/nhanes2007-2008/sampling_0708.htm/ .

[iii] Separate estimates for Asians may be feasible for some biomonitoring measures in the future, as NHANES started oversampling Asians in the 2011-2012 cycle.

[iv] The adjustment involves calculating age- and race/ethnicity-specific birth rates. Birth rates (i.e., average number of births per woman annually) are derived for each single year of age (age 16, age 17, etc.) separately for each race/ethnicity group (White non-Hispanic, Black non-Hispanic, Mexican-American, and "All Other Races/Ethnicities"). The standard NHANES sample weight for each observation is then multiplied by the calculated

Calculation of 50th and 95th percentiles over specified time periods: For all ACE3 Biomonitoring indicators, the 50th and 95th percentile values were selected as the indicator statistics to represent the central tendency and upper end of the exposure distribution. Where data are available for at least three 2-year NHANES survey periods, the indicator presentation focuses on how the measured values have changed over time. If data are available for only one or two NHANES survey periods, the indicator presentation focuses on demographic comparisons.

The 50th and 95th percentiles were also calculated for different population groups (defined by race/ethnicity or income) for all chemicals considered in the indicators. A single two-year NHANES cycle frequently will not include enough sampled individuals to provide statistically reliable estimates for all population groups of interest. Four-year data sets were used to ensure that there were a sufficient number of observations for each population group, using the two most current two-year NHANES cycles reported for each chemical. All calculations incorporated the NHANES sample weights.

Statistical considerations in presenting and characterizing the indicators: Statistical analysis has been applied to the ACE3 Biomonitoring indicators to evaluate trends over time in indicator values (for example, median concentration of lead in blood), or differences in indicator values between demographic groups.[v] These analyses use a 5% significance level, meaning that a conclusion of statistical significance is made only when there is no more than a 5% probability that the observed trend or difference occurred by chance ($p \leq 0.05$).

The statistical analysis of trends over time for an ACE3 Biomonitoring indicator is dependent on how the indicator values vary over time, the number of NHANES survey cycles with data included in the analysis, the number and variability of measurements in each survey cycle, and various aspects of the survey design. The evaluation of trends over time incorporates data from each survey cycle within the time period reported (for example, 2001–2002, 2003–2004, 2005–2006, 2007–2008, and 2009–2010). A finding of statistical significance for differences in indicator values between demographic groups depends on the magnitude of the difference, the number and variability of measurements in each group, and various aspects of the survey design. For example, if two groups from the U.S. population have different median levels of a chemical in blood or urine, the statistical test is more likely to detect a difference when samples have been obtained from a larger number of people in those groups. Similarly, if there is low

birth rate corresponding to the age and race/ethnicity of the sampled woman. This produces a birth rate-adjusted weight that is applied in the same manner as standard NHANES sample weights. There may be multiple ways to implement an adjustment to the data that accounts for birth rates by age. The National Center for Health Statistics has not fully evaluated the method used in ACE, or any other method intended to accomplish the same purpose, and has not used any such method in its publications. NCHS and EPA are working together to further evaluate the birth rate adjustment method used in ACE and alternative methods.

[v] The approach used in ACE3 focuses on identifying statistical trends and differences in the 50th and 95th percentiles of the NHANES biomonitoring data. Other approaches to analyzing trends in the NHANES biomonitoring data may focus on different summary statistics, such as the geometric mean or the percentage of the population exceeding some designated level. Assessment of trends in other summary statistics (such as the geometric mean) will not necessarily lead to the same conclusions as assessments of trends in the 50th and 95th percentiles.

variability in measured levels of the chemical within each group, then a difference between groups is more likely to be detected. It should be noted that when statistical testing is conducted for differences among multiple demographic groups (for example, considering both race/ethnicity and income level), or for multiple chemicals, the large number of comparisons involved increases the probability that some differences identified as statistically significant may actually have occurred by chance.

A finding of statistical significance is useful for determining that an observed trend or difference was unlikely to have occurred by chance. However, a determination of statistical significance by itself does not convey information about the magnitude of the difference in chemical concentrations or the potential difference in the risk of associated health outcomes. Furthermore, a lack of statistical significance means only that occurrence by chance cannot be ruled out. Thus, a conclusion about statistical significance is only part of the information that should be considered when determining the public health implications of trends or differences in indicator values.

In some cases, calculated indicator values have substantial uncertainty. Uncertainty in these estimates is assessed by looking at the relative standard error (RSE), a measure of how large the variability of the estimate is in relation to the estimate (RSE = standard error divided by the estimate).[vi] The estimate should be interpreted with caution if the RSE is at least 30% but is less than 40%; a notation is provided for such estimates in the indicator figures and tables. If the RSE is greater than 40%, the estimate is considered to have very large uncertainty and is not reported.[vii]

[vi] Standard errors for all Biomonitoring indicator values are provided in a file available on the ACE website (www.epa.gov/ace).

[vii] The RSE itself may also be uncertain for some estimates, particularly for values based on small samples, such as values stratified by race/ethnicity or income and chemicals measured in a subsample of NHANES participants (rather than all NHANES participants). Degrees of freedom is a statistical measure that provides an indication of this uncertainty. Estimates with between 7 and 11 degrees of freedom have a notation stating that they should be interpreted with caution. Estimates with fewer than 7 degrees of freedom were considered unreliable and are not reported.

Lead

Lead is a naturally occurring metal used in the production of fuels, paints, ceramic products, batteries, solder, and a variety of consumer products. The use of leaded gasoline and lead-based paint was eliminated or restricted in the United States beginning in the 1970s, resulting in substantial reductions in exposure to lead. However, children continue to be exposed to lead due to the widespread distribution of lead in the environment. For example, children are exposed to lead through the presence of lead-based paint in many older homes, the presence of lead in drinking water distribution systems, and current use of lead in the manufacture of some products.

In the United States, the major current source of early childhood lead exposure is lead-contaminated house dust.[1,2] Exposure to lead in house dust tends to be highest for young children, due to their frequent and extensive contact with floors, carpets, window areas, and other surfaces where dust gathers, as well as their frequent hand-to-mouth activity. A major contributor to lead in house dust is deteriorated or disrupted lead-based paint.[3-5] Housing units constructed before 1950 are most likely to contain lead-based paint, but any housing unit constructed before 1978 may also contain lead-based paint.[6] As of 2000, approximately 15.5 million housing units in the United States had one or more lead dust hazards on either floors or windowsills.[7] New lead dust hazards occur when lead in house paint is released during home renovation and remodeling activities.[8,9]

Two other contributors to lead in house dust are lead-contaminated soil and airborne lead.[10-13] Known sources of lead in soil include historical airborne emissions of leaded gasoline, emissions from industrial sources such as smelters, and lead-based paint.[14,15] Current sources of lead in ambient air in the United States include smelters, ore mining and processing, lead acid battery manufacturing, and coal combustion activities such as electricity generation.[15]

Lead-contaminated house dust is not the only source of childhood lead exposure. Direct contact with lead-contaminated soil,[13] ingestion of lead-based paint chips,[16] and inhalation of lead in ambient air also contribute to childhood lead exposure. Drinking water is an additional known source of lead exposure among children in the United States, particularly from corrosion of pipes and other elements of the drinking water distribution systems.[5,17,18] Exposure to lead via drinking water may be particularly high among very young children who consume baby formula prepared with drinking water that is contaminated by leaching lead pipes.[17] Although childhood exposure to lead in the United States typically occurs through contact with contaminated environmental media; children may also be exposed through lead-contaminated toys;[5,19] jewelry;[20] tobacco smoke;[21] imported candies, spices, and condiments;[5,22] and imported folk remedies.[23,24]

Compared with adults, children's bodies typically absorb a much greater fraction of a given amount of ingested lead. Once absorbed, most of the lead is stored in bones, where it can stay many years, while other lead goes into the blood and can be eliminated more quickly. Elimination of lead from the body usually occurs through urine or feces.[25]

Childhood blood lead levels in the United States differ across groups in the population, such as those defined by socioeconomic status and race/ethnicity.[26] Children living in poverty and Black non-Hispanic children tend to have higher blood lead levels[27] and higher levels of lead-contaminated dust in the home[6] than do other children. Blood lead levels tend to be higher for children living in older housing, most likely because older housing units are more likely to contain lead-based paint.[6,28] Blood lead levels may vary by nutritional status: conditions such as iron deficiency have been associated with higher blood lead levels in children.[15] In addition, some children who have immigrated to the United States may have been exposed to lead in their previous countries of residence. Foreign birth place and recent foreign residence have both been positively associated with the risk of elevated blood lead levels among immigrant children in the United States.[27,29]

Childhood blood lead levels in the United States have declined substantially since the 1970s. The decline in blood lead levels is due largely to the phasing out of lead in gasoline between 1973 and 1995,[30] and to the reduction in the number of homes with lead-based paint hazards.[7] Some decline was also a result of regulations reducing lead levels in drinking water, as well as legislation limiting the amount of lead in paint and restricting the content of lead in solder, faucets, pipes, and plumbing, and the elimination of lead-soldered cans for food use.[5] In the United States, lead content is banned or limited in many products, including food and beverage containers, ceramic ware, toys, Christmas trees, polyvinyl chloride pipes, vinyl mini-blinds, and playground equipment.[5] However, because trace levels of lead may be present in these products, normal use may still result in lead exposure.[5]

The National Toxicology Program (NTP) has concluded that childhood lead exposure is associated with reduced cognitive function.[31] Children with higher blood lead levels generally have lower scores on IQ tests[32-38] and reduced academic achievement.[31] In addition to the effects on IQ and school performance, research on the effects of lead has increasingly been addressing the effects of lead on behavior. The NTP has concluded that childhood lead exposure is associated with attention-related behavioral problems (including inattention, hyperactivity, and diagnosed attention-deficit/hyperactivity disorder) and increased incidence of problem behaviors (including delinquent, criminal, or antisocial behavior).[31] Studies have reported that lead exposure in children may contribute to decreased attention,[38-43] hyperactivity-impulsivity,[44] and increased likelihood of attention-deficit/hyperactivity disorder.[44-52] Other adverse behavioral outcomes that have been associated with childhood lead exposure in some studies include conduct disorders,[53,54] increased risks of juvenile delinquency and antisocial behaviors,[55-57] higher total arrest rates, and arrest rates for violent crimes in early adulthood.[58,59] Socioeconomic status may also modify the effect of lead on these cognitive and behavioral changes, resulting in stronger effects in children with lower socioeconomic status.[60,61]

Mothers who are exposed to lead can transfer lead to the fetus during pregnancy and to the child while breast feeding.[62,63] The NTP has concluded that there is "limited evidence" that prenatal lead exposure is associated with cognitive and behavioral effects in children.[31] The Centers for Disease Control and Prevention (CDC) has recently published guidelines for screening pregnant and lactating mothers for possible lead exposure to better protect the fetus.[64]

Many studies of the effects of lead focus on outcomes in children ages 5 years and younger. This focus reflects scientific thinking that early childhood is when children tend to experience peak exposures to lead, and also when they are most biologically susceptible to the effects of lead. Increased susceptibility to the neurodevelopmental effects of lead in the first three years of life is expected because this period is characterized by major growth and developmental events in the nervous system.[15] However, lead is toxic to individuals of all ages, and children older than 5 years may also be susceptible to the neurodevelopmental effects of lead. Blood lead measurements at various ages in early childhood have been found to be strongly correlated with cognitive deficits,[36] and some analyses have found that effects are more strongly associated with blood lead levels at school age (i.e., 5- to 6-year-old children) compared with levels measured earlier in life.[65,66]

Childhood lead exposures may also have lifelong effects. For instance, high childhood blood lead concentrations are associated with significant region-specific brain volume loss in adults, with greater effects seen in males.[67,68] Childhood blood lead concentrations are also inversely associated with intellectual functioning in young adulthood.[69] In addition, lead stored in bones has the potential to be released into the bloodstream later in life. Such is the case with pregnant women, breastfeeding women, and elderly persons, as blood lead levels are comparatively elevated in these populations.[25,70,71] Finally, childhood exposures to lead may contribute to a variety of neurological disorders and neurobehavioral effects in later life. [25,71-73]

Until recently, CDC defined a blood lead level of 10 micrograms per deciliter (µg/dL) as "elevated"; this definition was used to identify children for blood lead case management.[72,74] However, no level of lead exposure has been identified that is without risk of deleterious health effects.[15] CDC's Advisory Committee on Childhood Lead Poisoning Prevention (ACCLPP) recommended in January 2012 that the 97.5[th] percentile of children's blood lead distribution (currently 5 µg/dL) be defined as "elevated" for purposes of identifying children for follow-up activities such as environmental investigations and ongoing monitoring.[75] CDC has adopted the ACCLPP recommendation.[76] CDC specifically notes that "no level of lead in a child's blood can be specified as safe,"[1] and the NTP has concluded that there is sufficient evidence for adverse health effects in children at blood lead levels less than 5 µg/dL.[31]

The following two indicators use the best nationally representative data available on blood lead levels over time in children. Indicators B1 and B2 present blood lead concentrations for children ages 1 to 5 years.

Indicator B1: Lead in children ages 1 to 5 years: Median and 95th percentile concentrations in blood, 1976–2010

Indicator B2: Lead in children ages 1 to 5 years: Median concentrations in blood, by race/ethnicity and family income, 2007–2010

About the Indicators: Indicators B1 and B2 present concentrations of lead in blood of U.S. children ages 1 to 5 years. The data are from a national survey that collects blood specimens from a representative sample of the population every two years, and then measures the concentration of lead in the blood. Indicator B1 presents concentrations of lead in blood over time. Indicator B2 shows how blood lead levels differ by race/ethnicity and family income.

NHANES

The National Health and Nutrition Examination Survey (NHANES) provides nationally representative biomonitoring data for lead. NHANES is designed to assess the health and nutritional status of the civilian noninstitutionalized U.S. population and is conducted by the National Center for Health Statistics, part of the Centers for Disease Control and Prevention (CDC). NHANES conducts interviews and physical examinations with approximately 10,000 people in each two-year year survey cycle. CDC's National Center for Environmental Health measures concentrations of environmental chemicals in blood and urine samples collected from NHANES participants. Summaries of the measured values for more than 200 chemicals are provided in the *Fourth National Report on Human Exposure to Environmental Chemicals*.[77]

Lead

Indicators B1 and B2 present levels of lead in children's blood. Blood lead levels are reflective of relatively recent exposure and, to a varying extent across individuals, may also incorporate contributions of long-term lead exposures.[15] All values are reported as micrograms of lead per deciliter of blood (µg/dL).

Concentrations of lead in the blood of children have been measured in NHANES beginning with the 1976–1980 survey cycle (referred to as NHANES II). For 2009–2010, NHANES collected lead biomonitoring data for 8,793 individuals ages 1 year and older, including 836 children ages 1 to 5. Lead was detected in 100% of all individuals sampled. The median blood lead level among all NHANES participants in 2009–2010 was 1.1 µg/dL and the 95th percentile was 3.3 µg/dL.

Data Presented in the Indicators

Indicator B1 presents median and 95th percentile concentrations of lead in blood over time for children ages 1 to 5 years, using NHANES data from 1976–2010.

Indicator B2 presents current median concentrations of lead in blood for children ages 1 to 5 years of different races/ethnicities and levels of family income, using NHANES data from 2007–2008 and 2009–2010.

The data from two NHANES cycles are combined to increase the statistical reliability of the estimates for each race/ethnicity and income group, and to reduce any possible influence of geographic variability that may occur in two-year NHANES data. The current 95th percentiles of blood lead by race/ethnicity and income are presented in the data tables.

Four race/ethnicity groups are presented in Indicator B2: White non-Hispanic, Black non-Hispanic, Mexican-American, and "All Other Races/Ethnicities." The "All Other Races/Ethnicities" category includes all other races and ethnicities not specified, together with those individuals who report more than one race. The limits of the sample design and sample size often prevent statistically reliable estimates for smaller race/ethnicity groups. The data are also tabulated across three income categories: all incomes, below the poverty level, and greater than or equal to the poverty level.

The sensitivity of measurement techniques has improved over the years spanned by Indicator B1, allowing increased detection of lower blood lead levels. These improvements do not affect the comparability of the median or 95th percentiles over time, since between 92 and 100% of children have had detectable levels of lead in each NHANES cycle.

Additional information on how median and 95th percentile blood lead levels vary among different age groups for children ages 1 to 17 years is presented in a supplementary data table. Another data table provides median blood lead levels for the same race/ethnicity and income groups in 1991–1994, for comparison with the more current data presented in Indicator B2.

The indicators focus on ages 1 to 5 years because this age range has been the focus for research, data collection, and intervention due to the elevated exposures that occur during early childhood and the sensitivity of the developing brain to the effects of lead. Blood lead data for school-age children, whose neurological development is also affected by lead exposure, are included in the data tables for this indicator.

Please see the Introduction to the Biomonitoring section for an explanation of the terms "median" and "95th percentile," and information on the statistical significance testing applied to these indicators.

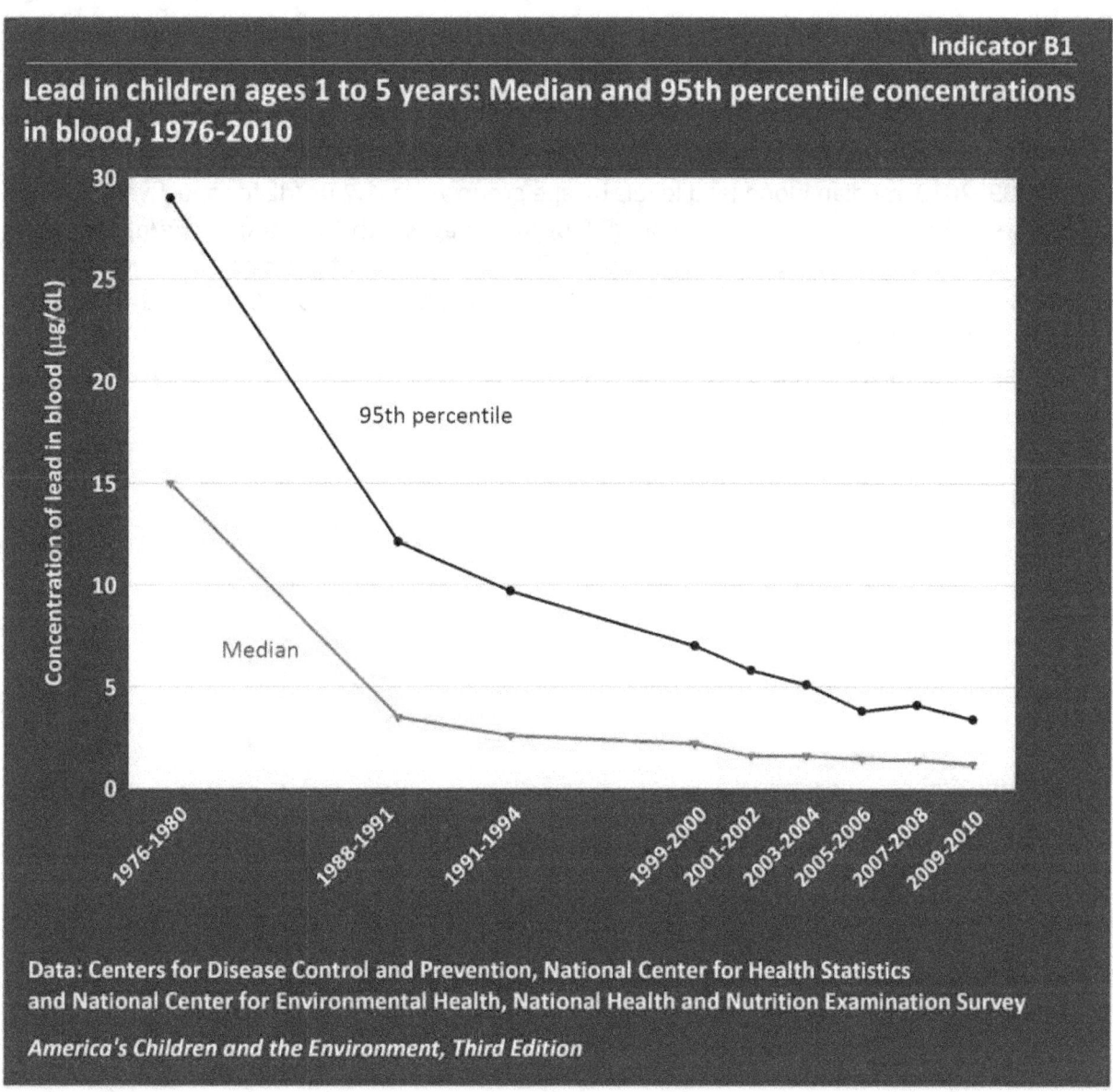

Indicator B1

Lead in children ages 1 to 5 years: Median and 95th percentile concentrations in blood, 1976-2010

Data: Centers for Disease Control and Prevention, National Center for Health Statistics and National Center for Environmental Health, National Health and Nutrition Examination Survey

America's Children and the Environment, Third Edition

Data characterization

- Data for this indicator are obtained from an ongoing continuous survey conducted by the National Center for Health Statistics.
- Survey data are representative of the U.S. civilian noninstitutionalized population.
- Lead is measured in blood samples obtained from individual survey participants.

- The median concentration of lead in the blood of children between the ages of 1 and 5 years dropped from 15 µg/dL in 1976–1980 to 1.2 µg/dL in 2009–2010, a decrease of 92%.

- The concentration of lead in blood at the 95th percentile in children ages 1 to 5 years dropped from 29 µg/dL in 1976–1980 to 3.4 µg/dL in 2009–2010, a decrease of 88%.

- The largest declines in blood lead levels occurred from the 1970s to the 1990s, following the elimination of lead in gasoline. The data show continuing declines in blood lead levels from

1999–2000 through 2009–2010, when the primary focus of lead reduction efforts has been on lead-based paint in homes.

- These decreasing trends were all statistically significant, including the trend in both the median and 95th percentile over the most recent 12 years (from 1999–2000 to 2009–2010).

- In 2009–2010, median blood lead levels by age group were: 1.2 µg/dL for age 1 year and age 2 years; 1.1 µg/dL for ages 3 to 5 years; 0.8 µg/dL for ages 6 to 10 years; 0.7 µg/dL for ages 11 to 15 years; and 0.7 µg/dL for ages 16 to 17. The 95th percentile blood lead levels were 4.2, 3.5, 2.8, 2.1, 1.7, and 1.4 µg/dL, respectively, for ages 1, 2, 3 to 5, 6 to 10, 11 to 15, and 16 to 17 years. (See Table B1a.)

  - The differences among age groups in median and 95th percentile blood lead levels were statistically significant.

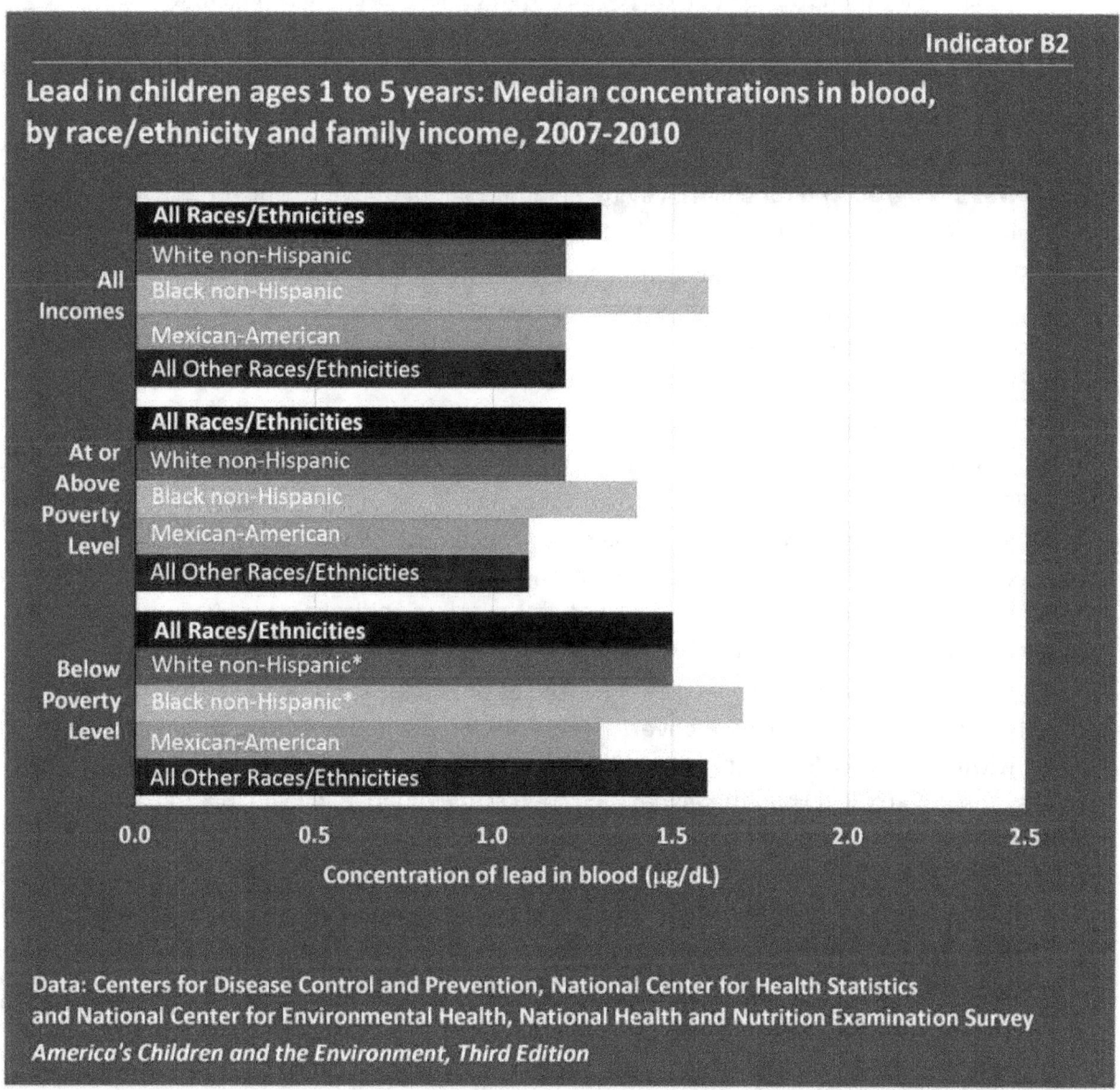

Indicator B2

Lead in children ages 1 to 5 years: Median concentrations in blood, by race/ethnicity and family income, 2007-2010

Data: Centers for Disease Control and Prevention, National Center for Health Statistics and National Center for Environmental Health, National Health and Nutrition Examination Survey
America's Children and the Environment, Third Edition

*The estimate should be interpreted with caution because the standard error of the estimate is relatively large: the relative standard error, RSE, Is at least 30% but is less than 40% (RSE = standard error divided by the estimate), or the RSE may be underestimated.

Data characterization
- Data for this indicator are obtained from an ongoing continuous survey conducted by the National Center for Health Statistics.
- Survey data are representative of the U.S. civilian noninstitutionalized population.
- Lead is measured in blood samples obtained from individual survey participants.

- The median blood lead level in children ages 1 to 5 years in 2007–2010 was 1.3 µg/dL. The median blood lead level in Black non-Hispanic children ages 1 to 5 years in 2007–2010 was

1.6 µg/dL, higher than the level of 1.2 µg/dL in White non-Hispanic children, Mexican-American children, and children of "All Other Races/Ethnicities."

- The median blood lead level in Black non-Hispanic children was statistically significantly higher than the median level for each of the remaining race/ethnicity groups.

- The median blood lead level for children living in families with incomes below the poverty level was 1.5 µg/dL, and for children living in families at or above the poverty level it was 1.2 µg/dL, a difference that was statistically significant.

- The 95th percentile blood lead level among all children ages 1 to 5 years was 3.9 µg/dL. The 95th percentile blood lead level in Black non-Hispanic children ages 1 to 5 years in 2007–2010 was 5.8 µg/dL, compared with 3.5 µg/dL for White non-Hispanic children and children of "All Other Races/Ethnicities," and 3.3 µg/dL for Mexican-American children. (See Table B2a.)

  - The 95th percentile blood lead level in Black non-Hispanic children was statistically significantly higher than the 95th percentile for each of the remaining race/ethnicity groups.

- Among children ages 1 to 5 years in families with incomes below poverty level, the 95th percentile blood lead was 4.7 µg/dL, and among those in families at or above the poverty level, it was 3.3 µg/dL, a difference that was statistically significant after accounting for differences by age, sex, and race/ethnicity. (See Table B2a.)

- The 95th percentile blood lead levels in children ages 1 to 5 years were higher for those in families with incomes below the poverty level compared with those at or above the poverty level within each race/ethnicity group. Black non-Hispanic children in families with incomes below the poverty level had the highest 95th percentile blood lead level, 6.8 µg/dL, which was 60% higher than for Black non-Hispanic children with families at or above the poverty level. (See Table B2a.)

  - The differences in 95th percentile blood lead levels between income groups were statistically significant for Black non-Hispanic children and children of "All Other Races/Ethnicities." The difference was also statistically significant for Mexican-American children after accounting for differences by age and sex.

- Between 1991–1994 and 2007–2010, median blood lead levels among Black non-Hispanic children ages 1 to 5 years declined 63%: from 4.3 µg/dL to 1.6 µg/dL. Over the same time period, median blood lead levels among Mexican-American children ages 1 to 5 years declined 61%: from 3.1 µg/dL to 1.2 µg/dL, and median blood lead levels among White non-Hispanic children ages 1 to 5 years declined 48%: from 2.3 µg/dL to 1.2 µg/dL. The differences over time were statistically significant for each race/ethnicity. (See Table B2b.)

Mercury

Mercury is a metal that is liquid at room temperature. There are three major forms of mercury: 1) organic mercury; 2) non-elemental forms of inorganic mercury; and 3) elemental mercury. Organic mercury, predominantly in the form of methylmercury, is found primarily in fish. Non-elemental forms of inorganic mercury are found primarily in batteries, some disinfectants, and some health products and creams. Lastly, elemental mercury is found in thermometers, fluorescent bulbs, dental amalgam fillings, switches in certain automobiles (used for convenience lighting in hoods and trunks, mostly in vehicles manufactured prior to 2003), and other sources.[1,2]

Mercury is released from its natural form in the earth's crust as a result of both human activities and natural processes. Coal-burning power plants are the largest source of mercury emissions in the United States.[3] Other sources of mercury emissions include the combustion of waste and industrial processes that use mercury.[3,4] When released into the atmosphere, either from human activities or from non-human sources, such as volcanoes, mercury can travel long distances on global air currents and can be deposited on land and water far from its original source.[4,5] In addition to these mercury emissions, there is concern that an increase in ice melts caused by a warming climate may release some past mercury emissions that have been trapped in polar ice.[6] Moreover, mercury deposited on the surface in the Arctic vaporizes each spring when the sunlight returns, causing increased concentrations in the atmosphere.[7,8]

Human exposure to elemental and inorganic mercury can occur at work, from accidental mercury releases, through the use of products containing mercury, through ritual and folk medicine uses of mercury, as well as dental restorations with mercury-silver amalgams.[4,9,10] Sources of childhood exposure to elemental and inorganic mercury in the home include the tracking of mercury into the home from the workplace by parents, mercury-containing devices in the home, and very rarely from intentionally heating mercury in the home for the purpose of extracting gold.[11] In schools, the most common sources of exposure are elemental and inorganic mercury stored in science laboratories, and mercury from broken instruments such as thermometers; less common sources are certain mercury-containing gymnasium floors manufactured between 1960 and 1980 found in some schools.[11,12] The adverse health effects of elemental and inorganic mercury exposure in childhood have not been extensively studied. However, inhaling high concentrations of elemental mercury vapor can lead to lung problems, neurobehavioral effects, mood changes, and tremors.[9] Although elemental mercury vapor emissions from dental amalgams are a major source of mercury exposure in the U.S. general population, two prospective clinical trials in children have found no evidence of adverse effects on IQ, memory, attention, or other neurological functions.[13-15]

Thimerosal is an organic mercury-containing preservative that is used in some vaccines to prevent contamination and growth of harmful bacteria in vaccine vials. The presence of thimerosal in many vaccines administered to infants led to concerns about possible effects on children's neurological development, including a hypothesis that mercury in vaccines could be a contributing factor to the incidence of autism. The Institute of Medicine has rejected the hypothesis of a causal relationship between thimerosal-containing vaccines and autism.[16] In addition, two recent studies have concluded that prenatal and infant exposure to thimerosal-containing vaccines is not related to increased risk of autism.[17,18] Since 2001, thimerosal has not been used in routinely administered childhood vaccines, with the exception of some influenza vaccines.[19]

Methylmercury is another form of organic mercury, which may form when mercury is deposited into water systems such as oceans, rivers, lakes, and wetlands; the mercury is converted by bacteria and other microorganisms into methylmercury. Methylmercury then bioaccumulates up the aquatic food web; fish that live long and feed on other fish (i.e., predatory fish) can accumulate high levels of methylmercury. The concentration of methylmercury in the larger fish at the top of the food chain can reach levels a million times higher than in the water.[20] Consuming fish is the main way that people are exposed to methylmercury. This includes fish commercially distributed in stores and restaurants as well as those that people catch for consumption by their families and communities. Each person's exposure depends on the amount of methylmercury in the fish that they eat and how often they eat fish. These exposure levels are of particular importance for women of child-bearing age because of the potential for prenatal exposure: methylmercury easily crosses the placenta and blood-brain barrier.[15] As such, the prenatal period is considered the most sensitive period of exposure.[15]

EPA has determined that methylmercury is known to have neurotoxic and developmental effects in humans.[4] This determination was based on effects in people prenatally exposed to extremely high levels of methylmercury during accidental mercury poisoning events in Japan and Iraq. Severe adverse health effects observed in the prenatally exposed population included cerebral palsy, intellectual disability (mental retardation), deafness, and blindness.[15,21,22]

Prospective cohort studies have been conducted in island populations where frequent fish consumption leads to methylmercury exposure in pregnant women at levels much lower than in the poisoning incidents but much greater than those typically observed in the United States. These studies are designed to investigate possible associations of prenatal methylmercury exposure with more subtle adverse neurodevelopmental effects than those observed in the poisoning incidents. However, the expected beneficial impacts of prenatal fish consumption on neurodevelopment can make it more difficult to detect such outcomes. Prenatal exposure to mercury in these studies is represented by measurement of total mercury in blood or hair samples obtained from a woman during pregnancy or at delivery. Results from such studies in New Zealand and the Faroe Islands[15,23-28] suggested that increased prenatal mercury exposure due to maternal fish consumption was associated with decrements in attention, language, memory, motor speed, and visual-spatial function (like drawing) during childhood. These associations were not seen in initial results reported from a study in the Seychelles Islands.[29] Further analyses of the Seychelles study population did find associations between prenatal mercury exposure and some neurodevelopmental deficits, after researchers had accounted for the developmental benefits of fish consumption.[30-32]

More recent studies have been conducted in Massachusetts and New York City, with maternal blood mercury levels within the range of typical levels in the U.S. general population.[33-35] In Massachusetts, total mercury in blood samples collected during the second trimester of pregnancy was associated with reduced cognitive development in testing conducted at age 3 years, after adjusting for the positive effects of fish/seafood consumption during pregnancy.[34] In the New York study, total cord blood mercury was associated with decreased IQ scores in testing conducted at age 4 years, after adjusting for the positive effects of fish/seafood consumption during pregnancy.[33]

Findings of neurodevelopmental effects from early childhood methylmercury exposure are more limited than for prenatal exposure, with several studies reporting mixed findings.[25,36-39] Animal and epidemiological studies suggest that early life exposure to methylmercury (including prenatal exposures) may also affect cardiovascular,[40,41] immune,[15,42,43] and reproductive health.[15]

Although ingestion of methylmercury in fish may be harmful, other compounds naturally present in many fish (such as high quality protein and other essential nutrients) are beneficial. In particular, fish are an excellent source of omega-3 fatty acids, which are nutrients that contribute to the healthy development of infants and children.[44] Pregnant women are advised to seek dietary sources of these fatty acids, including many species of fish. However, the levels of both methylmercury and omega-3 fatty acids can vary considerably by fish species. Thus, the type of fish, as well as portion sizes and frequency of consumption, are all important considerations for health benefits of fish and the extent of methylmercury exposure.

For these reasons, EPA and the U.S. Food and Drug Administration (FDA) issued a fish consumption advisory in 2004 that advises young children and pregnant females to consume up to 12 ounces a week of lower-mercury fish and shellfish, such as shrimp, canned light tuna, salmon, pollock, and catfish, but to avoid any consumption of high-mercury-containing fish, such as shark, swordfish, tile fish, or king mackerel.[45] EPA and FDA are currently working to update the fish consumption advisory to incorporate the most current science regarding the health benefits of fish consumption and the risks from methylmercury in fish. In 2011, the Departments of Agriculture and Health and Human Services jointly released the *2010 Dietary Guidelines for Americans*, which recommended that pregnant or breastfeeding women should consume 8–12 ounces of seafood per week, but avoid consumption of the same high-mercury-containing fish identified in the EPA-FDA advisory.[46] In addition, many state health departments provide advice regarding healthy sources of fish that are lower in mercury. Web links to state advice regarding fish consumption can be found at http://www.epa.gov/waterscience/fish/states.htm (for an example, see Washington state's "Eat Fish, Choose Wisely" available at http://www.doh.wa.gov/ehp/oehas/fish/fishchart.htm). State advisories may address both store-bought fish and fish caught by individuals in local lakes, rivers, and coastal waters.

Because methylmercury exposure in pregnant women is a concern for children health, studies have measured the level of mercury in women's bodies. Mercury can be measured in blood and is often called "blood mercury." In most cases, total blood mercury is reported, and the measurements do not distinguish methylmercury in blood from the other forms of mercury. In the United States, and in populations where most mercury exposure comes from fish consumption, the majority of total blood mercury is from methylmercury. Among women 16 to 49 years of age in the United States, levels of mercury in blood tend to be highest for Native American, Pacific Islander, Asian American, and multi-racial women.[47-49] A survey of adults in New York City found that blood mercury levels were three times higher than the national levels. Asian Americans in this study had higher blood mercury levels than other race/ethnicity groups.[50] Among women ages 16 to 49 years in the United States, blood mercury levels are higher for those who eat fish more often or in higher quantities.[51,47] Asian American populations have been identified as high consumers of seafood compared with White non-Hispanics or Black non-Hispanics.[50]

For women of all races, blood mercury levels tend to be higher in those women with higher family incomes.[48,50,52] Fish consumption rates are highest among women with relatively high family incomes, and this higher rate of fish consumption leads to increased blood mercury levels.[48,52] Concentrations of total mercury in blood among women also seem to vary with geographic region, and potentially by coastal region. Based on data from 1999–2004, blood mercury levels for women ages 16 to 49 years were higher in the Northeastern region of the United States compared with other regions.[48] Estimated mercury intake from fish consumption also follows this observed pattern. Women living in coastal regions had blood mercury levels higher than those living in noncoastal regions, and among coastal populations, the highest blood mercury levels were reported for the Atlantic and Pacific coastal regions, followed by the Gulf Coast and Great Lakes regions, respectively. Furthermore, subsistence populations (individuals who sustain a portion of their diets by catching and eating fish from local waters), or those who consume fish as a large portion of their diet because of taste preference or in the pursuit of health benefits, may have elevated blood mercury levels, depending on the source and species of fish.[4]

The indicator that follows uses the best nationally representative data currently available on blood mercury levels over time for women of child-bearing age. Indicator B3 presents median and 95th percentile blood mercury levels for women ages 16 to 49 years.

Indicator B3: Mercury in women ages 16 to 49 years: Median and 95th percentile concentrations in blood, 1999–2010

About the Indicator: Indicator B3 presents concentrations of mercury in blood of U.S. women ages 16 to 49 years. The data are from a national survey that collects blood specimens from a representative sample of the population every two years, and then measures the concentration of mercury in the blood. The indicator presents concentrations of mercury in blood over time. The focus on women of child-bearing age is based on concern for potential adverse effects in children born to women who have been exposed to mercury.

NHANES

The National Health and Nutrition Examination Survey (NHANES) provides nationally representative biomonitoring data for mercury. NHANES is designed to assess the health and nutritional status of the civilian noninstitutionalized U.S. population and is conducted by the National Center for Health Statistics, part of the Centers for Disease Control and Prevention (CDC). Interviews and physical examinations are conducted with approximately 10,000 people in each two-year survey cycle. CDC's National Center for Environmental Health measures concentrations of environmental chemicals in blood and urine samples collected from NHANES participants. Summaries of the measured values for more than 200 chemicals are provided in the *Fourth National Report on Human Exposure to Environmental Chemicals*.[53]

Mercury

Indicator B3 presents levels of mercury in blood of women of child-bearing age. Organic, inorganic, and total mercury can be measured in blood.[i] The concentration of total mercury in blood is a marker of exposure to methylmercury in populations where fish consumption is the predominant source of mercury exposure. Previous analysis shows that, in general, methylmercury accounts for a large percentage of total mercury in blood among women of child-bearing age in the United States.[47] Total blood mercury is generally representative of methylmercury exposures in the past few months.[54,55] All values are reported as micrograms of mercury per liter of blood (µg/L).

Concentrations of total blood mercury have been measured in all NHANES participants ages 1 to 5 years and all female participants ages 16 to 49 years beginning with the 1999–2000 survey cycle. Starting with the 2003–2004 survey cycle, NHANES measured blood mercury in all participants ages 1 year and older.[56] Separate measurements of inorganic blood mercury have been reported starting with the 2003–2004 NHANES survey cycle.

---

[i] NHANES also measures mercury levels in participant's urine samples, which is considered a more robust determinant of body burden of mercury from long-term exposure, particularly for inorganic mercury.

For 2009–2010, NHANES collected mercury biomonitoring data for 8,793 individuals ages 1 year and older, including 1,871 women ages 16 to 49 years. Mercury was detected in 81% of all individuals sampled. The frequency of mercury detection was 83% in women ages 16 to 49 years.[ii] The median blood mercury level among all NHANES participants in 2009–2010 was 0.8 µg/L and the 95th percentile was 5.1 µg/L.

Birth Rate Adjustment

Indicator B3 uses measurements of mercury in blood of women ages 16 to 49 years to represent the distribution of mercury exposures to women who are pregnant or may become pregnant. However, blood mercury levels increase with age,[56] and women of different ages have a different likelihood of giving birth. For example, in 2003–2004, women aged 27 years had a 12% annual probability of giving birth, and women aged 37 years had a 4% annual probability of giving birth.[57] A birth rate-adjusted distribution of women's mercury levels is used in calculating this indicator,[iii] meaning that the data are weighted using the age-specific probability of a woman giving birth.[58]

Data Presented in the Indicators

Indicator B3 presents median and 95th percentile concentrations of mercury in blood over time for women ages 16 to 49 years, using NHANES data from 1999–2010.

Additional information showing how median and 95th percentile blood mercury levels vary by race/ethnicity and family income for women ages 16 to 49 years is presented in supplemental data tables for these indicators. Data tables also display the median and 95th percentile blood mercury levels for children ages 1 to 5 years over time and the median and 95th percentile blood mercury levels for children ages 1 to 17 years for 2007–2010.

Please see the Introduction to the Biomonitoring section for an explanation of the terms "median" and "95th percentile," a description of the race/ethnicity and income groups used in the ACE3 biomonitoring indicators, and information on the statistical significance testing applied to these indicators.

[ii] The percentage for women ages 16 to 49 years is calculated with the birth rate adjustment described below.
[iii] There may be multiple ways to implement an adjustment to the data that accounts for birth rates by age. The National Center for Health Statistics has not fully evaluated the method used in ACE, or any other method intended to accomplish the same purpose, and has not used any such method in its publications. NCHS and EPA are working together to further evaluate the birth rate adjustment method used in ACE and alternative methods.

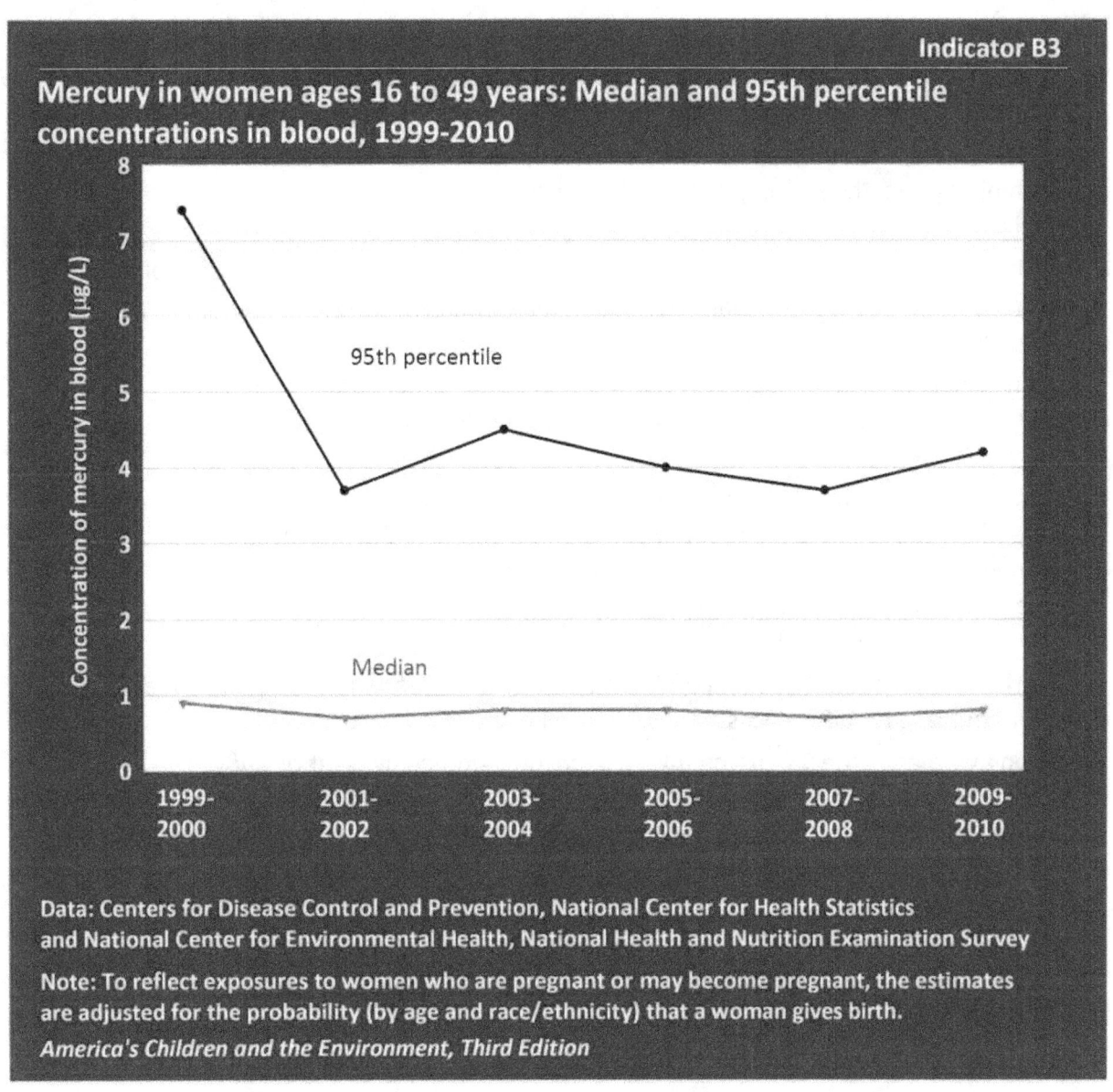

Indicator B3

Mercury in women ages 16 to 49 years: Median and 95th percentile concentrations in blood, 1999-2010

Data: Centers for Disease Control and Prevention, National Center for Health Statistics and National Center for Environmental Health, National Health and Nutrition Examination Survey

Note: To reflect exposures to women who are pregnant or may become pregnant, the estimates are adjusted for the probability (by age and race/ethnicity) that a woman gives birth.

America's Children and the Environment, Third Edition

Data characterization
- Data for this indicator are obtained from an ongoing continuous survey conducted by the National Center for Health Statistics.
- Survey data are representative of the U.S. civilian noninstitutionalized population.
- Mercury is measured in blood samples obtained from individual survey participants.

- The median concentration of total mercury in the blood of women ages 16 to 49 years has shown little change between 1999–2000 and 2009–2010, and was 0.8 µg/L in 2009–2010.
- Among women in the 95th percentile of exposure, the concentration of total mercury in blood decreased from 7.4 µg/L in 1999–2000 to 3.7 µg/L in 2001-2002. From 2001–2002 to 2009–2010, the 95th percentile of total blood mercury remained between 3.7 and 4.5 µg/L.

- The decrease in the 95th percentile levels of blood mercury between 1999–2000 and 2001–2002 was statistically significant. From 2001-2002 to 2009-2010 there was no statistically significant change.

- In 1999–2000, the 95th percentile total mercury level was 8 times the median level. For the remaining years, the 95th percentile total mercury levels were about 5 times the median levels.

- For the years 2007–2010, women of "All Other Races/Ethnicities" had median blood mercury levels of 1.3 µg/L, compared with median mercury levels for the remaining race/ethnicity groups of 0.6–0.8 µg/L. (See Table B3a.)

  - The median blood mercury level in women of "All Other Races/Ethnicities" was statistically significantly higher than the median level for each of the remaining race/ethnicity groups.

- Among women in the 95th percentile of exposure, differences in total mercury in blood were observed across race/ethnicity groups. For the years 2007–2010, White non-Hispanic women had a blood mercury level of 3.7 µg/L, Black non-Hispanics had 2.9 µg/L, Mexican-American women had 2.3 µg/L, and women in the "All Other Races/Ethnicities" group had 6.7 µg/L. (See Table B3b.)

  - The differences between race/ethnicity groups were statistically significant after accounting for differences by income level and age.

- Among women in the 95th percentile of exposure, women living at or above the poverty level had higher blood levels of total mercury (4.0 µg/L) compared with women living below poverty level (2.9 µg/L), a difference that was statistically significant. (See Table B3b.)

- The median and 95th percentile values for women of child-bearing age were about 2 to 4 times those of children ages 1 to 5 years. (See Table B3 and Table B3c.)

- Among children ages 1 to 5 years in the 95th percentile of exposure, the concentration of total mercury in blood showed a decreasing trend from 2.3 µg/L in 1999–2000 to 1.3 µg/L in 2009–2010. The median blood mercury level for children ages 1 to 5 years stayed relatively constant for the same time period. (See Table B3c.)

  - The decreasing trend in 95th percentile blood mercury levels in children was statistically significant. There was no statistically significant change in median blood mercury levels in children.

- Among children ages 1 to 17 years, median and 95th percentile blood mercury levels generally increased with age in 2007–2010, with higher blood mercury levels among children ages 6 years and older. Children ages 16 to 17 years had a median level of mercury in blood of 0.5 µg/L and a 95th percentile of 2.8 µg/L. (See Table B3d.)

  - The differences by age group were statistically significant at both the median and the 95th percentile.

Cotinine

Environmental tobacco smoke (ETS), commonly referred to as secondhand smoke, is a complex mixture of gases and particles and includes smoke from burning cigarettes, cigars, and pipe tobacco (sidestream smoke), as well as exhaled mainstream smoke.[1] There are at least 250 chemicals in ETS that are known to be toxic or carcinogenic, including acrolein, ammonia, benzene, carbon monoxide, formaldehyde, hydrogen cyanide, nicotine, nitrogen oxides, and sulfur dioxide.[1,2] In 1992, EPA classified ETS as a known human carcinogen.[3] Children can be exposed to ETS in their homes or in places where people are allowed to smoke, such as some restaurants in some locations throughout the United States.

According to the U.S. Surgeon General, there is no safe level of exposure to ETS, and breathing even a small amount can be harmful to human health.[1] The Surgeon General has concluded that exposure to ETS causes sudden infant death syndrome (SIDS), acute lower respiratory infection, ear problems, and more severe asthma in children. Smoking by parents causes respiratory symptoms and slows lung growth in their children.[1] Young children appear to be more susceptible to the respiratory effects of ETS than are older children.[3-5] It is also possible that early-life exposures to ETS may lead to adverse health effects in adulthood. Exposure to ETS in childhood has been reported to be associated with early emphysema in adulthood among nonsmokers.[6]

The exposure of a pregnant woman to ETS can also be harmful to her developing fetus. The Surgeon General has determined that exposure of pregnant women to ETS causes a small reduction in mean birth weight and the evidence is suggestive (but not sufficient to infer causation) of a relationship between maternal exposure to environmental tobacco smoke during pregnancy and preterm delivery.[1] In addition, the Surgeon General concluded the evidence is suggestive but not sufficient to infer a causal relationship between prenatal and postnatal exposure to ETS and childhood cancer.[1]

Exposure to ETS in the home is influenced by adult behaviors, including the decisions to smoke at home and to allow visitors to smoke inside the home. Children living in homes with smoking bans have significantly lower levels of cotinine (a biological marker of exposure to ETS) in urine than children living in homes without smoking bans.[7] Household smoking bans can significantly decrease children's exposures to ETS, but do not completely eliminate them.[8]

In recent years there has been a significant decline in children's exposures to ETS.[9] This reduction is in part attributable to a decline in the percentage of adults who smoke. In 2010, an estimated 19.3% of adults were current smokers, down from 24.7% in 1997.[10,11] In addition, the prevalence of smoke-free households increased from 43% of U.S. homes in 1992–1993 to 72% in 2003.[12] However, despite the increasing numbers of adults disallowing smoking in the home, approximately 34% of children live in a home with at least one smoker as of 2009.[13] The enactment of smoking bans in restaurants, bars, and other public places has led to a decrease in ETS exposure for both children and adults.[14] Recent studies suggest that smoking bans can reduce the number of asthma-related emergency room visits and hospitalizations and reduce asthmatic symptoms, including persistent wheeze, wheeze-medication use, and chronic night cough in children.[15-18]

Cotinine is considered the best biomarker of exposure to tobacco smoke for both active smokers and those exposed to ETS.[19] The two indicators that follow use the best nationally representative data currently available on blood cotinine levels over time for women of child-bearing age and children. Indicator B4 presents median and 95th percentile blood serum levels of cotinine for children ages 3 to 17 years. Indicator B5 presents median and 95th percentile blood serum levels of cotinine for women ages 16 to 49 years.

Indicator B4: Cotinine in nonsmoking children ages 3 to 17 years: Median and 95th percentile concentrations in blood serum, 1988–2010

Indicator B5: Cotinine in nonsmoking women ages 16 to 49 years: Median and 95th percentile concentrations in blood serum, 1988–2010

About the Indicators: Indicators B4 and B5 present concentrations of cotinine in blood serum of U.S. children ages 3 to 17 years and women ages 16 to 49 years. Cotinine is a marker of exposure to environmental tobacco smoke (ETS). The data are from a national survey that collects blood specimens from a representative sample of the population every two years, and then measures the concentration of cotinine in the blood serum. Indicator B4 presents concentrations of cotinine in children's blood serum over time and Indicator B5 presents concentrations of cotinine in women's blood serum over time. The focus on both children and women of child-bearing age is based on concern for potential adverse effects in children exposed to ETS and in children born to women who have been exposed to ETS.

NHANES

The National Health and Nutrition Examination Survey (NHANES) provides nationally representative biomonitoring data for cotinine. NHANES is designed to assess the health and nutritional status of the civilian noninstitutionalized U.S. population and is conducted by the National Center for Health Statistics, part of the Centers for Disease Control and Prevention (CDC). Interviews and physical examinations are conducted with approximately 10,000 people in each two-year survey cycle. CDC's National Center for Environmental Health measures concentrations of environmental chemicals in blood and urine samples collected from NHANES participants. Summaries of the measured values for more than 200 chemicals are provided in the *Fourth National Report on Human Exposure to Environmental Chemicals*.[19]

Environmental Tobacco Smoke (ETS) and Cotinine

Indicators B4 and B5 present blood serum levels of cotinine as a marker of exposure to ETS. Nicotine is a distinctive component of tobacco that is found in large amounts in tobacco smoke, including ETS. Once nicotine enters the body, it is rapidly broken down in a matter of a few hours into other chemicals. Cotinine is a primary breakdown product of nicotine, and has a longer half-life. This characteristic makes cotinine a better indicator than nicotine of an individual's exposure to ETS.[20-22]

Measurement of cotinine in blood serum is a marker for exposure to ETS in the previous few days.[23] Some studies have shown that, given the same exposure to tobacco smoke, cotinine levels may differ by race/ethnicity and sex, and there may be genetic differences in the rate at which cotinine is removed from the body.[1,24-28]

These indicators present cotinine levels for non-tobacco-users only. Children and women who were active smokers, as indicated by a relatively high serum cotinine level, were excluded from these statistics. For these analyses, individuals with a serum cotinine level greater than 10 nanograms of cotinine per milliliter of serum (ng/mL) are considered active smokers, and all individuals with cotinine levels below 10 ng/mL are considered nonsmokers.[19] Active smokers will almost always have serum cotinine levels above 10 ng/mL, and sometimes those levels will be higher than 500 ng/mL.[19,29] Nonsmokers who are exposed to typical levels of ETS have serum cotinine levels of less than 1 ng/mL, whereas those nonsmokers with heavy exposure to ETS will have serum cotinine levels between 1 and 10 ng/mL.[19]

Concentrations of cotinine in blood serum have been measured in NHANES participants ages 4 years and older for the 1988–1991 and 1991–1994 survey cycles, and then for ages 3 years and older beginning with the 1999–2000 survey cycle.

For 2009–2010, NHANES collected cotinine biomonitoring data for 6,678 nonsmoking individuals ages 3 years and older, including 2,191 children ages 3 to 17 years and 1,395 women ages 16 to 49 years. Cotinine was detected in about 67% of all nonsmoking individuals sampled. The frequency of cotinine detection was 71% in children ages 3 to 17 years and 66% in women ages 16 to 49 years.[i] The median blood serum cotinine level for all nonsmoking NHANES participants in 2009–2010 was 0.03 ng/mL and the 95th percentile was 1.3 ng/mL.

Birth Rate Adjustment

Indicator B5 uses measurements of cotinine in blood serum of women ages 16 to 49 years to represent the distribution of ETS exposures to women who are pregnant or may become pregnant. For example, in 2003–2004, women aged 27 years had a 12% annual probability of giving birth, and women aged 37 years had a 4% annual probability of giving birth.[30] A birth rate-adjusted distribution of women's cotinine levels is used in calculating this indicator,[ii] meaning that the data are weighted using the age-specific probability of a woman giving birth.[31]

Data Presented in the Indicators

Indicator B4 presents median and 95th percentile concentrations of cotinine in blood serum over time as a marker of exposure to ETS among non-smoking children ages 3 to 17 years, using NHANES data from 1988–2010.

Indicator B5 presents median and 95th percentile concentrations of cotinine in blood serum over time as a marker of exposure to ETS among non-smoking women ages 16 to 49 years, using NHANES data from 1988–2010.

---

[i] The percentage for women ages 16 to 49 years is calculated with the birth rate adjustment described below.
[ii] There may be multiple ways to implement an adjustment to the data that accounts for birth rates by age. The National Center for Health Statistics has not fully evaluated the method used in ACE, or any other method intended to accomplish the same purpose, and has not used any such method in its publications. NCHS and EPA are working together to further evaluate the birth rate adjustment method used in ACE and alternative methods.

Although the sensitivity of measurement techniques has improved over the years spanned by Indicators B4 and B5, allowing increased detection of lower serum cotinine levels over time, these improvements do not affect the comparability of the median or 95th percentiles over time since the majority of children and women have had detectable levels of cotinine in each NHANES cycle.

Additional information showing how median and 95th percentile blood serum levels of cotinine vary by race/ethnicity, family income, and age for children ages 3 to 17 years is presented in the supplemental data tables for these indicators. Data tables also show how median and 95th percentile blood serum levels of cotinine vary by race/ethnicity and family income for women ages 16 to 49 years.

NHANES does not provide cotinine measurements for children under the age of 3 years (or under age 4 years prior to 1999), who may be especially sensitive to the effects of ETS exposure.

Please see the Introduction to the Biomonitoring section for an explanation of the terms "median" and "95th percentile," a description of the race/ethnicity and income groups used in the ACE3 biomonitoring indicators, and information on the statistical significance testing applied to these indicators.

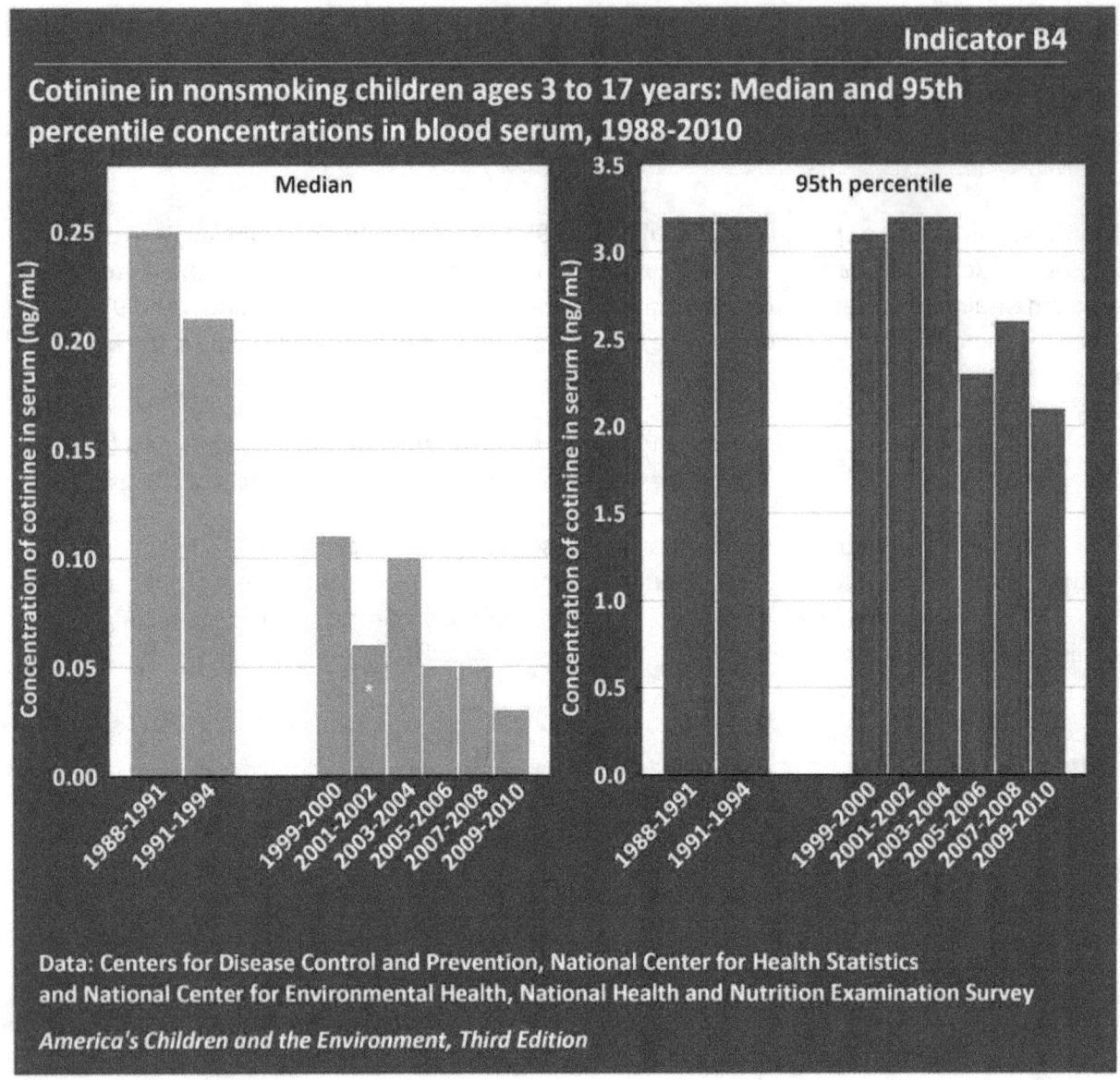

*The estimate should be interpreted with caution because the standard error of the estimate is relatively large: the relative standard error, RSE, is at least 30% but is less than 40% (RSE = standard error divided by the estimate), or the RSE may be underestimated.

Data characterization
- Data for this indicator are obtained from an ongoing continuous survey conducted by the National Center for Health Statistics.
- Survey data are representative of the U.S. civilian noninstitutionalized population.
- Cotinine is measured in blood samples obtained from individual survey participants.

- The median level of cotinine measured in blood serum of nonsmoking children ages 3 to 17 years dropped from 0.25 ng/mL in 1988–1991 to 0.03 ng/mL in 2009–2010, a decrease of 88%. This decreasing trend was statistically significant.

- Cotinine values at the 95th percentile decreased by 34% from 1988–1991 to 2009–2010. This trend was also statistically significant.

- Children at the 95th percentile of cotinine levels had much higher levels than those at the median. In 1988–1991, the 95th percentile cotinine level (3.2 ng/mL) was 13 times the median level (0.25 ng/mL); in 2009–2010, the 95th percentile cotinine level (2.1 ng/mL) was 70 times the median level (0.03 ng/mL).

- In every time period measured, children at the 95th percentile had higher levels of cotinine in their blood than women at corresponding levels. (Compare with Indicator B5.)

- Eighty-seven percent of nonsmoking children ages 4 to 17 years had detectable levels (at or above 0.05 ng/mL) of cotinine in 1988–1991. Forty percent of nonsmoking children ages 3 to 17 years had levels at or above 0.05 ng/mL of cotinine in 2009–2010, although improvements in laboratory methods made it possible to detect cotinine at lower concentrations starting with the 2001–2002 survey cycle. (Data not shown.)

- In 2007–2010, median concentrations of cotinine in blood for nonsmokers were approximately 0.11 ng/mL for Black non-Hispanic children, 0.04 ng/mL for White non-Hispanic children, and 0.02 ng/mL for Mexican-American children. The differences between these race/ethnicity groups were statistically significant. (See Table B4a.)

- In 2007–2010, the median concentration of cotinine in blood serum for nonsmoking children living below the poverty level (0.14 ng/mL) was about 5 times the median for nonsmoking children living at or above the poverty level (0.03 ng/mL). The differences between income groups were statistically significant. (See Table B4a.)

- In 2007–2010, 95th percentile concentrations of cotinine in blood for nonsmokers were 2.9 ng/mL for White non-Hispanic children and 2.6 ng/mL for Black non-Hispanic children, while Mexican-American children had levels that were more than 3 times lower (0.8 ng/mL). (See Table B4b.)

  - The differences between levels for Mexican-American children and both White non-Hispanic and Black non-Hispanic children were statistically significant.

- For the years 2007–2010, median levels of cotinine in younger nonsmoking children ages 3 to 5 years and 6 to 10 years were 0.06 and 0.05 ng/mL, respectively, compared with 0.03 and 0.04 ng/mL in older nonsmoking children ages 11 to 15 years and 16 to 17 years, respectively. (See Table B4c.)

  - The differences between the levels for the four age groups were not statistically significant.

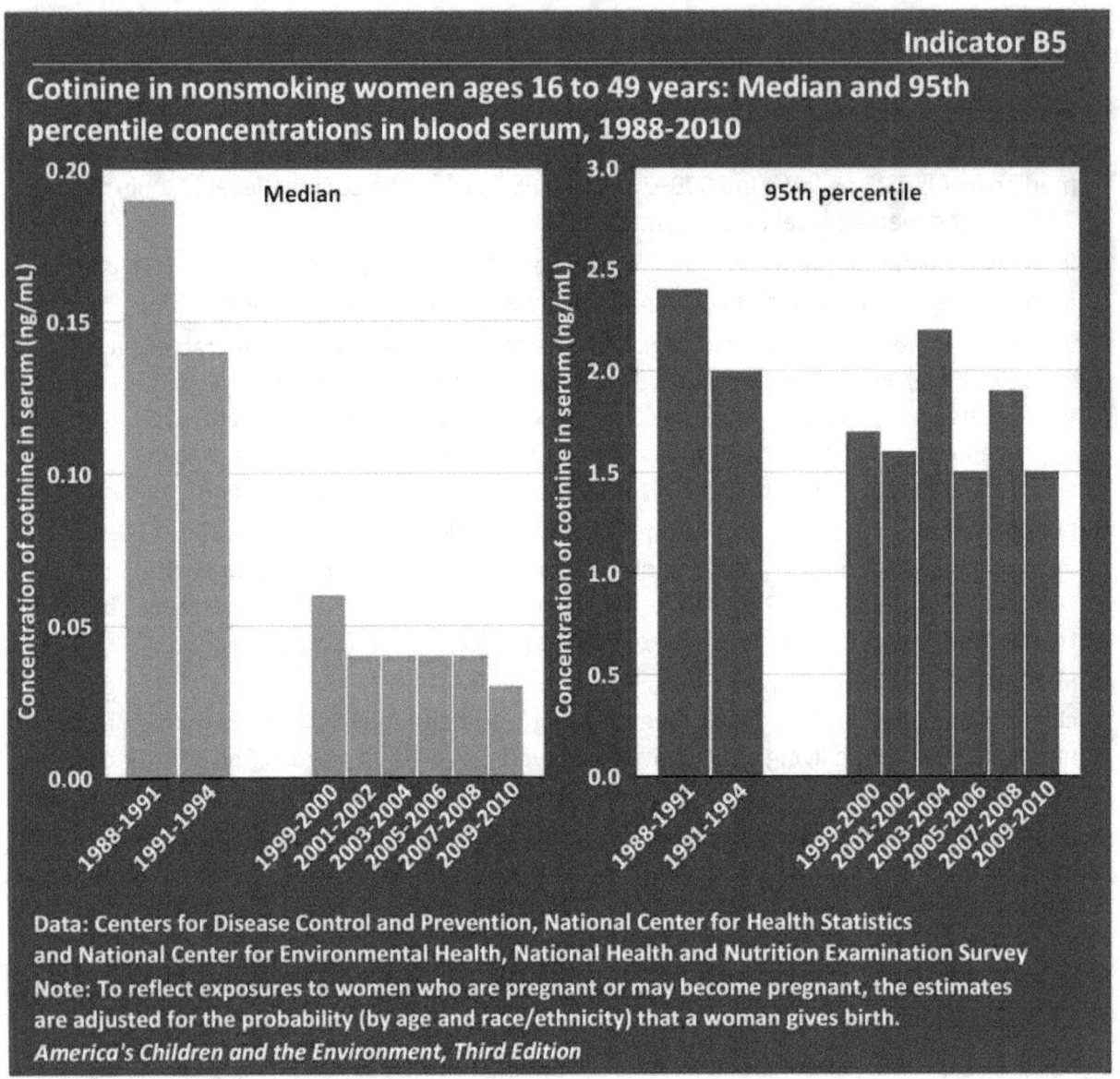

Indicator B5

Cotinine in nonsmoking women ages 16 to 49 years: Median and 95th percentile concentrations in blood serum, 1988-2010

Data: Centers for Disease Control and Prevention, National Center for Health Statistics and National Center for Environmental Health, National Health and Nutrition Examination Survey
Note: To reflect exposures to women who are pregnant or may become pregnant, the estimates are adjusted for the probability (by age and race/ethnicity) that a woman gives birth.
America's Children and the Environment, Third Edition

Data characterization
- Data for this indicator are obtained from an ongoing continuous survey conducted by the National Center for Health Statistics.
- Survey data are representative of the U.S. civilian noninstitutionalized population.
- Cotinine is measured in blood samples obtained from individual survey participants.

- The median level of cotinine measured in blood serum of nonsmoking women of child-bearing age dropped from 3.2 ng/mL in 1988–1991 to 2.1 ng/mL in 2009–2010, a decrease of 86%. This decreasing trend was statistically significant.

- Cotinine values at the 95th percentile decreased by 35% from 1988–1991 to 2009–2010. This trend was also statistically significant.

- Women at the 95th percentile cotinine levels had much higher levels than those at the median. In 1988–1991, the 95th percentile cotinine level (2.3 ng/mL) was 11 times the median level (0.21 ng/mL); in 2009–2010, the 95th percentile cotinine level (1.5 ng/mL) was 50 times the median level (0.03 ng/mL).

- In 2007–2010, median concentrations of cotinine in blood for nonsmoking women were approximately 0.1 ng/mL for Black non-Hispanic women and 0.03 ng/mL for White non-Hispanic women and Mexican-American women. (See Table B5a.)

  - The difference between Black non-Hispanic women and Mexican-American women was statistically significant. The difference between Black non-Hispanic and White non-Hispanic women was not statistically significant after adjusting for age and income differences.

- Cotinine values at the 95th percentile were more than twice as high for nonsmoking women living below the poverty level (3.5 ng/mL) as for nonsmoking women living at or above the poverty level (1.4 ng/mL) in 2007–2010. The differences between income groups were statistically significant. (See Table B5b.)

Perfluorochemicals (PFCs)

Perfluorochemicals (PFCs) are a group of synthetic chemicals that have been used in many consumer products.[1] The structure of these chemicals makes them very stable, hydrophobic (water-repelling), and oleophobic (oil-repelling). These unique properties have led to extensive use of PFCs in surface coating and protectant formulations for paper and cardboard packaging products; carpets; leather products; and textiles that repel water, grease, and soil. PFCs have also been used in fire-fighting foams and in the production of nonstick coatings on cookware and some waterproof clothes.[1] Due in part to their chemical properties, some PFCs can remain in the environment and bioconcentrate in animals.[2-8] Data from human studies suggest that some PFCs can take years to be cleared from the body.[9-13]

The PFCs with the highest production volumes in the United States have been perfluorooctane sulfonic acid (PFOS) and perfluorooctanoic acid (PFOA).[1] PFOS and PFOA are also two of the most frequently detected PFCs in humans.[14] Other PFCs include perfluorohexane sulfonic acid (PFHxS), which is a member of the same chemical category as PFOS; and perfluorononanoic acid (PFNA), which is a member of the same chemical category as PFOA.[15] Chemicals within a given PFC chemical category share similar chemical structures and uses. Although some studies have addressed PFHxS and PFNA specifically, the majority of scientific research has focused on PFOS and PFOA.[15]

In 2000, one of the principal perfluorochemical manufacturers, 3M, began phasing out the production of PFOA, PFOS, and PFOS-related compounds. The 3M phaseout of PFOS and PFHxS was completed in 2002, and its phaseout of PFOA was completed in 2008.[16] In 2006, to address PFOA production by other manufacturers, EPA launched the 2010/15 PFOA Stewardship Program, with eight companies voluntarily agreeing to reduce emissions and product content of PFOA, PFNA, and related chemicals by 95% no later than 2010. The industry participants also committed to work toward eliminating emissions and product content of these chemicals by 2015.[17] However, the fact that some of these chemicals may be persistent in the environment and have a long half-life in humans means that they may continue to persist in the environment and in people for many years, despite reductions in emissions.[2-13] EPA is currently evaluating the potential need for regulation of PFCs using the authorities of the Toxic Substances Control Act.[15]

The major sources of human exposure to PFCs are poorly understood, but may include food, water, indoor and outdoor air, breast milk, and dust.[4] Two recent studies pointed to food consumption as the primary pathway of exposure to PFOS and PFOA for Americans and Europeans.[18,19] PFC-treated food-contact packaging, such as microwave popcorn bags,[i] has been a

---

[i] The U.S. Food and Drug Administration recently worked with several manufacturers to remove grease-proofing agents containing C8 perfluorinated compounds from the marketplace. These manufacturers volunteered to stop distributing products containing these compounds in interstate commerce for food-contact purposes as of October 1, 2011. For more information, see
http://www.fda.gov/Food/FoodIngredientsPackaging/FoodContactSubstancesFCS/ucm308462.htm.

source of PFC exposure.[20] Meat and dairy products may also be contaminated with PFCs due to exposure of source animals to air, water, and feed contaminated with PFCs,[21-23] although a recent study reported that PFCs were undetected in nearly all milk samples tested in the United States.[24] In some areas, such as those near industrial facilities that either make or use PFCs, these contaminants have been found at high levels in drinking water, groundwater, and/or surface water.[25-31] PFCs have been detected in human breast milk.[32-36] PFCs have been measured in house dust as well, with some perfluorochemicals, such as PFOS, PFOA, PFHxS, perfluorobutane sulfonic acid (PFBuS), and perfluorohexanoic acid (PFHxA), found to be present in the majority of dust samples examined.[37-40] Infants and small children may be more highly exposed to the PFCs present in house dust than adults are, due to their frequent and extensive contact with floors, carpets, and other surfaces where dust gathers, as well as their frequent hand-to-mouth activity.[18,19,41,42] Children could have increased exposure to PFCs in carpet and carpet protectants, due to the amount of time they spend lying, crawling, and playing on carpet.[15,41] Limited available data on levels of PFCs in children's blood suggest that the blood serum levels of most PFCs are higher in children ages 3 to 11 years compared with other age groups.[43,44]

Some PFCs have been widely detected in pregnant women and in umbilical cord blood, suggesting that the developing fetus can be exposed to PFCs while in the womb. However, findings between studies vary. For example, PFOS and PFOA were detected in 99–100% of blood samples collected from both pregnant and non-pregnant women in 2003–2004.[45] Additionally, PFOS and PFOA were detected in 99% and 100% of umbilical cord blood samples, respectively, collected from newborns in Baltimore.[46] In another study conducted in Japan, the level of PFOS circulating in a pregnant woman's blood was highly correlated with the level in cord blood. However, PFOA was detected in maternal samples but was not detected in umbilical cord samples in the Japanese study.[47] Even though studies suggest that the correlation between maternal and fetal exposure may vary, the ubiquitous presence of PFOS, PFOA, and other PFCs in blood of women of child-bearing age and in umbilical cord blood may indicate that fetal exposure to these chemicals is widespread.[45,46,48]

Some human health studies have found associations between prenatal exposure to PFOS or PFOA and a range of adverse birth outcomes, such as low birth weight, decreased head circumference, reduced birth length, and smaller abdominal circumference.[49-52] However, there are inconsistencies in the results of these studies, and two other studies did not find an association between prenatal PFC exposure and birth weight.[53,54] The participants in all of these studies had PFC blood serum levels comparable to levels in the general population. Animal studies echo these findings, though typically at levels much higher than what humans are normally exposed to. Developmental and reproductive effects, including reduced birth weight, decreased gestational length, structural defects, delays in postnatal growth and development, increased neonatal mortality, and pregnancy loss have all been associated with prenatal rodent exposure to PFOS and PFOA.[55-65]

Findings from a limited number of studies suggest that exposure to PFOS or PFOA may have negative impacts on human thyroid function. However, there are inconsistencies in the findings between these studies. Some studies have found that PFC exposures are associated with alterations in thyroid hormone levels, as well as an increased risk of thyroid disease in the general public and in workers with occupational exposures.[66-68] However, a recent study of pregnant women with exposures comparable to those in the general population found that increasing levels of PFOS, PFOA, and PFHxS were not associated with differences in thyroid hormone levels.[69] The results from animal studies have been more consistent. Multiple animal studies have found that thyroid hormone levels are altered in animals exposed to PFOS.[57,62,63,65,70-74] One of these studies also found that PFOA-treated rats have altered thyroid hormone levels.[71] The health risks associated with maternal thyroid hormone disruption during pregnancy may make this a cause for concern. Moderate deficits in maternal thyroid hormone levels during early pregnancy have been linked to reduced childhood IQ scores and other neurodevelopmental effects, as well as unsuccessful or complicated pregnancies.[75]

Both animal and some human studies have found an association between PFCs exposure and cholesterol and/or triglyceride levels, although physiological differences between humans and experimental animals may cause lipid levels to vary in opposite directions.[76] Structurally, PFCs resemble fatty acids and can bind to receptors that play key roles in lipid metabolism and fat production.[77] In animal studies involving various species, PFCs are associated with decreased serum levels of these lipids;[64,65,73] in contrast some human studies show an increase in blood lipid levels with increased presence of PFCs, including PFOS, PFOA, PFHxS, and PFNA, while other human studies show no change in lipid levels with PFC exposure.[77-84] This could be a potential concern for children, because the mother's body provides a source of cholesterol and triglycerides to the developing fetus. Cholesterol and fatty acids support cellular growth, differentiation, and adipose accumulation during fetal development.[49,85] Finally, although human studies have not looked at the associations between PFC exposure and the immune system, animal studies have found an association between PFOS and PFNA exposure (in utero and in adulthood) and immune suppression, including alterations in function and production of immune cells and decreased lymphoid organ weights.[86-88]

The indicator that follows uses the best nationally representative data currently available on blood serum levels of perfluorochemicals over time for women of child-bearing age. Indicator B6 presents median blood serum levels of PFOS, PFOA, PFHxS, and PFNA for women ages 16 to 49 years.

About the Indicator: Indicator B6 presents concentrations of perfluorochemicals (PFCs) in blood serum of U.S. women ages 16 to 49 years. The data are from a national survey that collects blood specimens from a representative sample of the population every two years, and then measures the concentration of PFCs in the blood serum. The indicator presents concentrations of PFCs in blood serum over time. The focus on women of child-bearing age is based on concern for potential adverse effects in children born to women who have been exposed to PFCs.

NHANES

The National Health and Nutrition Examination Survey (NHANES) provides nationally representative biomonitoring data for PFCs. NHANES is designed to assess the health and nutritional status of the civilian noninstitutionalized U.S. population and is conducted by the National Center for Health Statistics, part of the Centers for Disease Control and Prevention (CDC). Interviews and physical examinations are conducted with approximately 10,000 people in each two-year survey cycle. CDC's National Center for Environmental Health measures concentrations of environmental chemicals in blood and urine samples collected from NHANES participants. Summaries of the measured values for more than 200 chemicals are provided in the *Fourth National Report on Human Exposure to Environmental Chemicals*.[2]

Perfluorinated Compounds

Indicator B6 presents blood serum levels of four important PFCs: perfluorooctane sulfonic acid (PFOS), perfluorooctanoic acid (PFOA), perfluorohexane sulfonic acid (PFHxS), and perfluorononanoic acid (PFNA). These four PFCs were chosen because they are commonly detected in humans, and the bulk of PFCs health effects research in both humans and laboratory animals has focused on these contaminants—especially PFOS and PFOA.

PFCs bind to proteins in the serum of blood. Because PFCs remain in the human body for years, blood serum levels of PFCs are reflective of long-term exposures to these contaminants. Serum accounts for about half the weight of whole blood, so the blood serum concentration of PFCs is about twice the concentration of PFCs in whole blood.[89] The blood serum PFC levels for this indicator are given in nanograms of PFC per milliliter of blood serum (ng/mL).[ii]

Concentrations of 12 different PFCs, including PFOS, PFOA, PFHxS, and PFNA, have been measured in blood serum from a representative subset of NHANES participants ages 12 years

---

[ii] Most persistent organic pollutants (POPs) are lipophilic, meaning that they accumulate in fatty tissues; however, this is not the case for PFCs, which are both hydrophobic (water-repelling), and oleophobic (oil-repelling). They instead bind to proteins in the serum of blood. While blood levels of lipophilic POPs are commonly lipid-adjusted, the PFC measurements in blood are not.

and older beginning with the 1999–2000 survey cycle, although PFCs were not measured in the 2001–2002 cycle.

In 2007–2008, NHANES collected PFCs biomonitoring data for 2,100 individuals ages 12 years and older, including 495 women ages 16 to 49 years. The four selected PFCs were detected in 99% to 100% of the individuals sampled in NHANES 2007–2008. The median and 95th percentile of blood serum PFC levels for all NHANES participants in 2007–2008 were 14 ng/mL and 41 ng/mL, respectively, for PFOS; 4 ng/mL and 10 ng/mL, respectively, for PFOA; 2 ng/mL and 10 ng/mL, respectively, for PFHxS; 2 ng/mL and 4 ng/mL, respectively, for PFNA.

Birth Rate Adjustment

Indicator B6 uses measurements of PFCs in blood serum of women ages 16 to 49 years to represent the distribution of PFC exposures to women who are pregnant or may become pregnant. However, women of different ages have a different likelihood of giving birth. For example, in 2003–2004, women aged 27 years had a 12% annual probability of giving birth, and women aged 37 years had a 4% annual probability of giving birth.[90] A birth rate-adjusted distribution of women's PFC levels is used in calculating this indicator,[iii] meaning that the data are weighted using the age-specific probability of a woman giving birth.[91]

Data Presented in the Indicator

Indicator B6 presents median concentrations of PFOS, PFOA, PFHxS, and PFNA in blood serum over time for women ages 16 to 49 years, using NHANES data from 1999–2008 (excluding the years 2001–2002).

Additional information on the 95th percentile blood serum levels of PFOS, PFOA, PFHxS, and PFNA for women ages 16 to 49 years is presented in the supplemental data tables for this indicator, along with information showing how median and 95th percentile blood serum levels of PFCs in women of child-bearing age vary by race/ethnicity and family income.

Please see the Introduction to the Biomonitoring section for an explanation of the terms "median" and "95th percentile," a description of the race/ethnicity and income groups used in the ACE3 biomonitoring indicators, and information on the statistical significance testing applied to these indicators.

[iii] There may be multiple ways to implement an adjustment to the data that accounts for birth rates by age. The National Center for Health Statistics has not fully evaluated the method used in ACE, or any other method intended to accomplish the same purpose, and has not used any such method in its publications. NCHS and EPA are working together to further evaluate the birth rate adjustment method used in ACE and alternative methods.

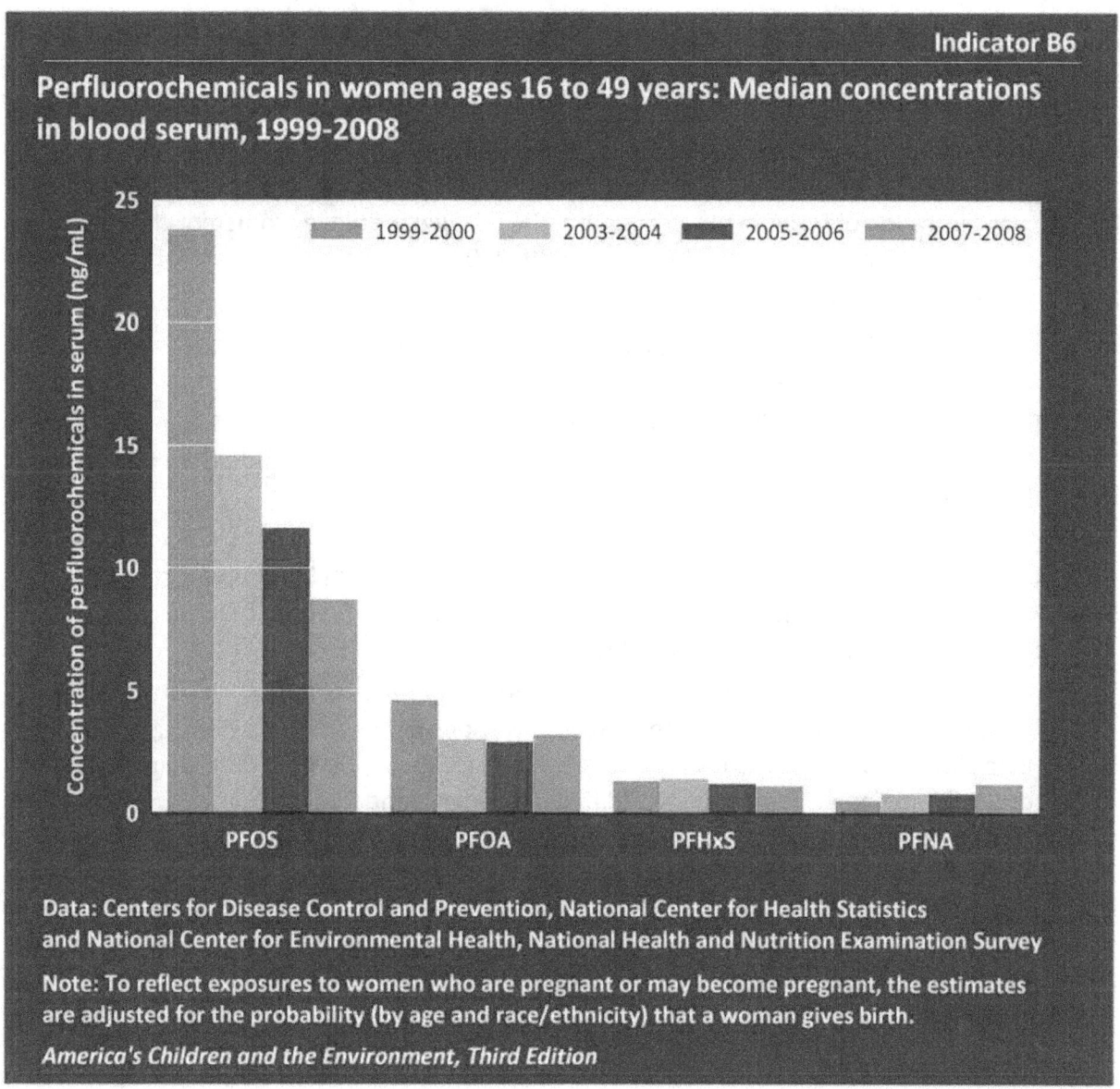

Indicator B6

Perfluorochemicals in women ages 16 to 49 years: Median concentrations in blood serum, 1999-2008

Data: Centers for Disease Control and Prevention, National Center for Health Statistics and National Center for Environmental Health, National Health and Nutrition Examination Survey

Note: To reflect exposures to women who are pregnant or may become pregnant, the estimates are adjusted for the probability (by age and race/ethnicity) that a woman gives birth.

America's Children and the Environment, Third Edition

Data characterization
- Data for this indicator are obtained from an ongoing continuous survey conducted by the National Center for Health Statistics.
- Survey data are representative of the U.S. civilian noninstitutionalized population.
- PFCs are measured in blood samples obtained from individual survey participants.

- Between 1999–2000 and 2007–2008, median blood serum levels of PFOS in women of child-bearing age declined from 24 ng/mL in 1999–2000 to 9 ng/mL in 2007–2008. Median blood serum levels of PFOA in women of child-bearing age declined from 5 ng/mL in 1999–2000 to 3 ng/mL in 2007–2008. These decreasing trends were statistically significant.

- The median blood serum levels of PFHxS and PFNA were lower than those of PFOS and PFOA in women of child-bearing age. Median levels of PFHxS have remained relatively constant over time. Between 1999–2000 and 2007–2008, median blood serum levels of PFNA showed an increasing trend, from 0.5 ng/mL in 1999–2000 to 1.2 ng/mL in 2007–2008.

- The increasing trend in median PFNA levels was statistically significant.

- The concentration of PFOS in blood serum at the 95th percentile in women of child-bearing age showed a decreasing trend from 50 ng/mL in 1999–2000 to 23 ng/mL in 2007–2008. The concentration of PFOA in blood serum at the 95th percentile in women of child-bearing age remained relatively constant between 1999–2000 and 2007–2008. (See Table B6a.)

  - The decreasing trend in 95th percentile PFOS levels was statistically significant.

- For the years 2005–2008, women of child-bearing age living at or above poverty level had higher median and 95th percentile concentrations of PFOS and PFOA in their blood serum compared with women living below poverty level. (See Tables B6b and B6c.)

  - The differences between income groups were statistically significant after adjustment for differences in race/ethnicity and age.

- For the years 2005–2008, median concentrations of PFOA were higher in White non-Hispanic women of child-bearing age (3.5 ng/mL) compared with Black non-Hispanic women (2.7 ng/mL), Mexican-American women (2.3 ng/mL), and women of "All Other Races/Ethnicities" (2.4 ng/mL). (See Table B6b.)

  - These differences in median PFOA concentrations by race/ethnicity were statistically significant. The difference in median concentrations between White non-Hispanic and Black non-Hispanic women was no longer statistically significant after accounting for other demographic characteristics (differences in age and income).

- In 2005–2008, median and 95th percentile concentrations of PFOS were lower in Mexican-American women of child-bearing age at 7.4 ng/mL and 17.3 ng/mL, respectively, compared with White non-Hispanic women at 11.4 ng/mL and 28.4 ng/mL, respectively, Black non-Hispanic women at 11.2 ng/mL and 25.7 ng/mL, respectively, and women of "All Other Races/Ethnicities" at 8.3 ng/mL and 24.9 ng/mL, respectively. (See Tables B6b and B6c.)

  - These differences were statistically significant both with and without adjustment for other demographic characteristics, with the following exceptions: the difference between Mexican-American women and women of "All Other Races/Ethnicities" was statistically significant at the median only after accounting for differences by age and income; and was not statistically significant at the 95th percentile.

Polychlorinated Biphenyls (PCBs)

Polychlorinated biphenyls (PCBs) are a family of industrial chemicals that were produced in the United States from 1929 to 1979 and used primarily as insulating fluids in capacitors, transformers, and other electrical equipment.[1] PCBs were also used as plasticizers in many paints, plastics, and rubber products, and had numerous applications in industry and building construction.[1]

Each PCB has a common structure of a biphenyl molecule with 1 to 10 chlorine atoms attached; each possible variant is called a congener. In theory, there could be as many as 209 PCB congeners; however, a smaller number of congeners were found in manufactured PCB mixtures, and measurements of PCBs in the environment and in human blood samples typically target a subset of the congeners.[2,3]

The PCB congeners are sometimes separated into two categories, "dioxin-like" or "non-dioxin-like," that are defined by structural differences and that act by different toxicological mechanisms.[4,5] The dioxin-like PCBs are structurally and toxicologically related to the chemical 2,3,7,8- tetrachlorodibenzo-p-dioxin, which has been studied very extensively in toxicological and epidemiological research. However, both categories have been associated with adverse health outcomes and it is unknown which congeners are the most potent, particularly for outcomes most relevant to children's health.[2,4]

Manufacture, sale, and use of PCBs was generally banned in the United States in 1979,[6] but EPA regulations have authorized their continued use in certain equipment manufactured prior to the ban. Due to their persistent nature, PCBs remain widely distributed in the environment, and they are also present at many Superfund sites.[2] The persistent nature of PCBs and their distribution through the food chain has resulted in continuing human exposure. However, dietary intake of PCBs and levels measured in blood serum have declined since the ban.[2,7,8] Measured levels of PCBs in human blood decreased by an estimated 87% from 1973–2003,[7] and levels of PCB-153, one of the major PCB congeners, also showed significant decline from the late 1980s to 2002.[8] Although levels of PCBs in environmental samples have declined from their peak, the rate of decline has slowed in recent years.[9,10]

A large body of health effects research comes from children born to mothers who were exposed to high concentrations of a mixture of PCBs and polychlorinated dibenzofurans (a class of dioxin-like chemicals) in accidental poisoning incidents in Taiwan and Japan. These prenatally exposed children exhibited a number of adverse health effects, including neurodevelopmental effects such as cognitive deficits, developmental delays, effects on motor skills, behavioral effects, immunological effects, and skin alterations ranging from irritation to chloracne,[2] a potentially serious inflammatory condition.[11-16]

Following the poisoning incidents, several studies have been conducted to examine the effects of PCBs at more typical exposure levels. Many of these studies have linked early-life exposure to PCBs with neurodevelopmental effects, such as lowered intelligence, and behavioral deficits, including inattention and impulsive behavior.[17-23] The observed effects have been most frequently associated with exposure in the womb resulting from the mother having eaten food contaminated with PCBs,[24-29] but some studies have detected relationships between adverse effects and PCB exposure during infancy and childhood.[22,29-31] Although there is some inconsistency in the epidemiological literature, several reviews of the literature have concluded that the overall evidence supports a concern for effects of PCBs on children's neurological development.[16,30,32-34] The Agency for Toxic Substances and Disease Registry has determined that "Substantial data suggest that PCBs play a role in neurobehavioral alterations observed in newborns and young children of women with PCB burdens near background levels."[2] Research on dioxin-like chemicals in general also supports a concern for neurodevelopmental effects from the dioxin-like PCBs.[35] Similar outcomes have been observed in experimental animal studies, including behavioral changes and learning deficits in rats and monkeys exposed to PCBs in their diets.[2,36]

Prenatal PCB exposures have also been associated with immunological effects, such as increased infections, in multiple epidemiological studies,[37-43] with supporting evidence from the literature on effects of dioxin-like chemicals.[35] Possible other effects of exposure to PCBs—with limited or inconclusive evidence—include preterm birth and low birthweight,[16] as well as effects on the timing of puberty in both boys and girls.[44] PCBs are also considered "reasonably anticipated to be human carcinogens," based on experimental animal studies.[45]

Biomonitoring data in U.S. children under 12 years of age is limited. One study of 6- to 9-year-old girls from 2005–2007 in California and Ohio showed a median level of PCB-153, the congener with the highest concentration, of 7.4 nanograms per gram of lipid (ng/g lipid). The same congener was measured in a nationally representative sample of the U.S. population ages 12 years and older in 2003–2004. The median level of PCB-153 in males and females ages 12 to 19 years was 5.4 ng/g lipid and for adults ages 20 years and over the median level was 24.2 ng/g lipid.[46,47]

Due to the continued presence of PCBs in fish, especially salmon, meat, poultry, dairy products, and breast milk,[48] dietary intake is an important pathway of exposure for PCBs.[2] In infants, dietary intake is important since PCBs accumulate in the mother's body over many years and are stored in the fat in breast milk,[35] and breast-feeding infants are exposed to the PCBs in the milk.[49] PCBs can also cross the placental barrier to transfer prenatally from mother to fetus, and PCBs have been measured in cord blood.[2,50]

Recent findings also suggest that the presence of PCBs in indoor dust and indoor air may constitute an important exposure pathway for some portion of the population.[51-54] The importance of PCBs in indoor environments may be greater for toddlers than for adults and children of other ages, because toddlers tend to have more contact with house dust.[51] A study of homes with unusually high indoor air concentrations of PCBs found that a PCB-containing wood flooring finish applied in the 1950s and 1960s can be a major contributor to current elevations of PCBs in blood for people living in those homes.[53] PCBs have been found in caulk in some schools and other buildings constructed or renovated before the late 1970s, which may contribute significantly to indoor air and dust levels of PCBs in those buildings.[55,56] Many schools have lighting systems containing PCBs that were produced before PCBs were banned. While well-contained lighting systems pose little risk, the PCB-containing ballasts are only expected to last 10–15 years. Existing ballasts from before the ban are past their life expectancy and are at a greater risk for leaks and fires, resulting in a greater risk of PCB exposure.[57] Finally, the inadvertent presence of PCBs has been found in pigments that are currently manufactured for use in paints, inks, textiles, paper, cosmetics, leather, and other materials.[58,59]

Blood levels of PCBs generally increase with age, because these chemicals are persistent.[60,61] However, the decline in levels of PCBs in the environment and in foods over the past three decades, suggests that young people today are exposed to lower levels of PCBs through the diet than were previous generations.[2,7,9,10]

Although environmental levels of PCBs have been declining, there are concerns that some past PCB emissions trapped in polar ice may be released to the environment in coming years with increasing ice melts.[62,63] Furthermore, environments where heavy PCB contamination previously occurred continue to be remediated, which may dislodge or expose additional PCBs. The Hudson River, contaminated with 1.3 million pounds of PCBs between 1947 and 1977, is undergoing remediation to remove PCB-contaminated sediments. After Phase 1 of the remediation in 2009 there was a short-term increase in the PCB levels in fish samples, but more recent samples from 2010 did not have increased PCB concentrations.[64,65]

The following indicator presents the best nationally representative data on PCB levels in women of child-bearing age. Indicator B7 presents median blood serum levels of PCBs for women ages 16 to 49 years. Although data are available only for two two-year survey periods at this time, the data provide a baseline that will be updated with PCB measurements over time from subsequent survey cycles. No indicator is presented for PCBs in children due to the limited availability of data.

Indicator B7: PCBs in women ages 16 to 49 years: Median concentrations in blood serum, by race/ethnicity and family income, 2001–2004

About the Indicator: Indicator B7 presents concentrations of PCBs in blood serum of U.S. women ages 16 to 49 years. The data are from a national survey that collects blood specimens from a representative sample of the population every two years, and then measures the concentration of PCBs in the blood serum. The indicator presents comparisons of PCBs in blood serum for women of different race/ethnicities, and for women of different income levels. The focus on women of child-bearing age is based on concern for potential adverse effects in children born to women who have been exposed to PCBs.

NHANES

The National Health and Nutrition Examination Survey (NHANES) provides nationally representative biomonitoring data for PCBs. NHANES is designed to assess the health and nutritional status of the civilian noninstitutionalized U.S. population and is conducted by the National Center for Health Statistics, part of the Centers for Disease Control and Prevention (CDC). Interviews and physical examinations are conducted with approximately 10,000 people in each two-year survey cycle. CDC's National Center for Environmental Health measures concentrations of environmental chemicals in blood and urine samples collected from NHANES participants. Summaries of the measured values for more than 200 chemicals are provided in the *Fourth National Report on Human Exposure to Environmental Chemicals*.[46]

PCB Congeners

Indicator B7 presents blood serum levels of PCBs in women of child-bearing age. There are 209 possible PCBs, referred to as "congeners," which are defined by the number of chlorine atoms (from 1 to 10) and their position in the chemical structure. Most of these congeners were not present in the manufactured PCB mixtures and have not been measured in environmental or human samples.

PCB concentrations are measured in blood serum. PCBs are lipophilic, meaning that they tend to accumulate in fat. Serum PCB concentrations are measured and expressed on a lipid-adjusted basis, as these values better represent the amount of PCBs stored in the body compared with unadjusted values.[46] The indicator uses lipid-adjusted concentrations, meaning that the concentration of PCBs in serum is divided by the concentration of lipid in serum. The resulting units are nanograms of PCB per gram of lipid (ng/g lipid) in serum.[i]

Concentrations of PCBs in blood serum have been measured in a representative subset of NHANES participants ages 12 years and older beginning with the 1999–2000 survey cycle.

---

[i] Serum levels of PCBs can also be reported without lipid adjustment. Both the lipid-adjusted values and the unadjusted "whole weight" values are reported in CDC's *Fourth National Report on Human Exposure to Environmental Chemicals*.

NHANES sampled for 34 PCB congeners in 2001–2002, and added 4 congeners in 2003–2004 for a total of 38 congeners. Indicator B7 uses NHANES data on four specific congeners: PCBs 118, 138, 153, and 180. These four congeners are generally found at higher levels in the environment—and in human blood samples—than other PCB congeners. This combination of congeners has been frequently used to represent PCB exposure in the epidemiological studies described above that identified children's health concerns for PCBs. PCBs 118, 138, 153, and 180 were detected in the majority of samples for women ages 16 to 49 years in 2001–2002, and in virtually all samples for this population group in 2003–2004.

Indicator B7 was calculated by summing together the measured values of the 4 selected congeners for each woman 16 to 49 years sampled in NHANES; this approach is commonly used in studies assessing levels of PCBs in human blood samples and environmental samples.[2,30] If the congener was not detected in a sample, a default value below the detection limit was assigned for purposes of calculating the summed total.[ii] This assumption has a small impact on the indicator values, because all four congeners were detected in most samples in the combined four-year (2001–2004) data set.

In 2001–2004, a sum of measured PCBs 118, 138, 153, and 180 is available from NHANES for 4,205 individuals ages 12 years and older, including 1164 women ages 16 to 49 years. The four selected PCBs were detected in 81% of the individuals sampled in NHANES 2001-2004,[iii] and in 71% of women ages 16 to 49 years.[iv] The median sum of the four PCB congeners in blood serum among all NHANES participants in 2001-2004 was 71 ng/g lipid and the 95th percentile sum was 316 ng/g lipid.

Birth Rate Adjustment

Indicator B7 uses measurements of PCBs in the blood of women ages 16 to 49 years to represent the distribution of PCB exposures to women who are pregnant or may become pregnant. However, women of different ages have a different likelihood of giving birth. For example, in 2003–2004, women aged 27 years had a 12% annual probability of giving birth, and women aged 37 years had a 4% annual probability of giving birth.[66] A birth rate-adjusted distribution of women's PCB levels is used in calculating this indicator,[v] meaning that the data are weighted using the age-specific probability of a woman giving birth.[67]

[ii] The default value used for non-detect samples is equal to the limit of detection divided by the square root of 2.
[iii] In 2003–2004, PCBs were detected in 100% of the individuals sampled. The detection frequency was lower in 2001–2002 due to use of less sensitive measurement techniques.
[iv] The percentage for women ages 16 to 49 years is calculated with the birth rate adjustment described below.
[v] There may be multiple ways to implement an adjustment to the data that accounts for birth rates by age. The National Center for Health Statistics has not fully evaluated the method used in ACE, or any other method intended to accomplish the same purpose, and has not used any such method in its publications. NCHS and EPA are working together to further evaluate the birth rate adjustment method used in ACE and alternative methods.

Data Presented in the Indicator

Indicator B7 presents median concentrations of PCBs in blood serum, computed as the sum of PCBs 118, 138, 153, and 180, for women ages 16 to 49 years of different races/ethnicities and levels of family income, using NHANES data from 2001–2002 and 2003–2004.

Data from 1999–2000 are not included in the indicator because less sensitive measurement techniques were used in those years, and PCB levels could not be determined in a large proportion of the blood samples. Improvements in measurement sensitivity were achieved in 2001–2002, with further improvements in 2003–2004 resulting in the detection of PCBs in a majority of samples.[61] The data from the 2001–2002 and 2003–2004 NHANES cycles are combined to increase the statistical reliability of the estimates for each race/ethnicity and income group, and to reduce any possible influence of geographic variability that may occur in two-year NHANES data. No time series is shown because data from only two NHANES cycles are too limited to depict possible changes over time.

Four race/ethnicity groups are presented in Indicator B7: White non-Hispanic, Black non-Hispanic, Mexican-American, and "All Other Races/Ethnicities." The "All Other Races/Ethnicities" category includes all other races and ethnicities not specified, together with those individuals who report more than one race. The limits of the sample design and sample size often prevent statistically reliable estimates for smaller race/ethnicity groups. The data are also tabulated across three income categories: all incomes, below the poverty level, and greater than or equal to the poverty level.

Additional information on how 95th percentile blood serum levels of PCBs vary by race/ethnicity and family income for women ages 16 to 49 years is presented in the supplemental data tables for this indicator

Please see the Introduction to the Biomonitoring section for an explanation of the terms "median" and "95th percentile," along with information on the statistical significance testing applied to these indicators.

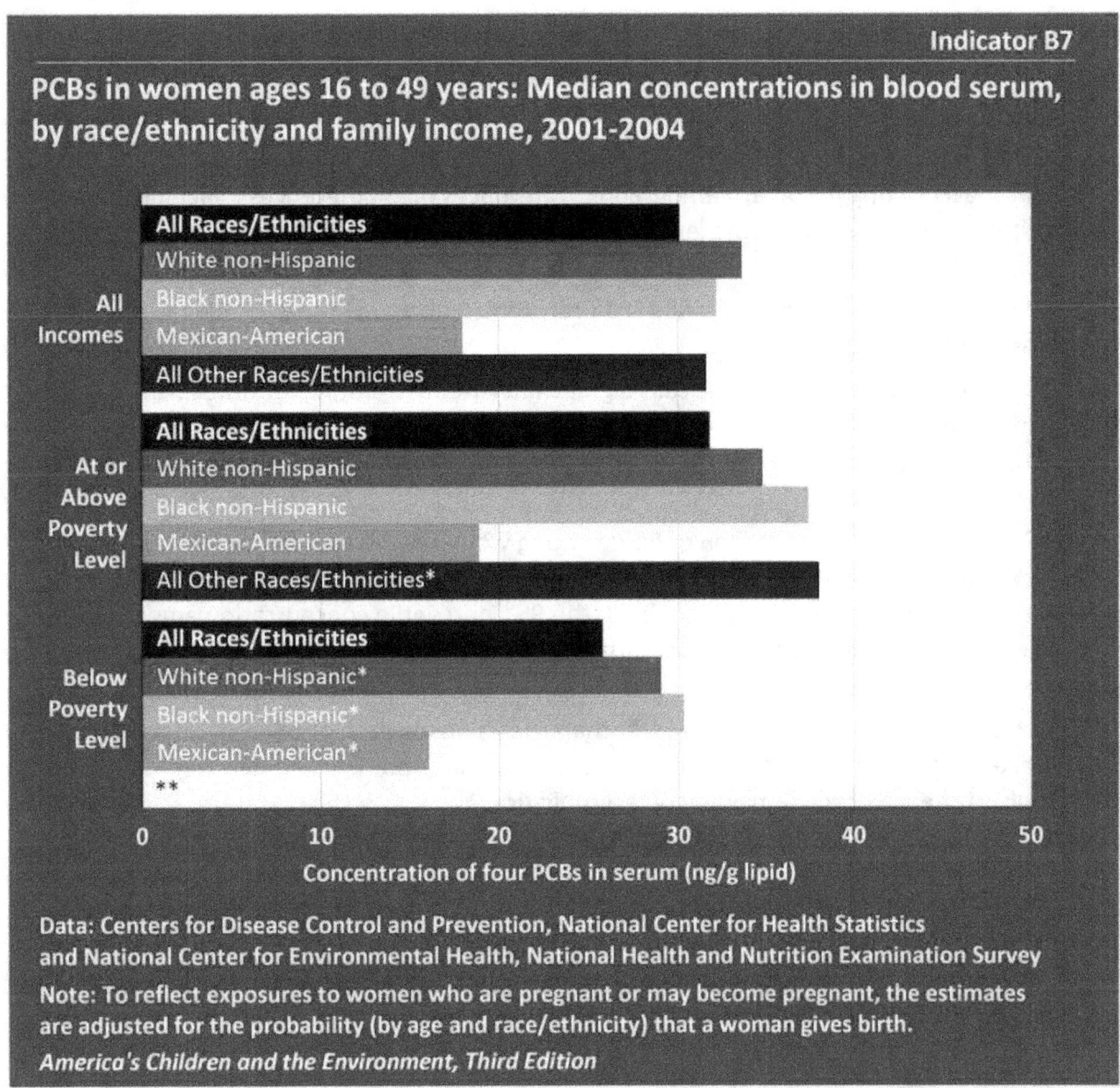

Indicator B7

PCBs in women ages 16 to 49 years: Median concentrations in blood serum, by race/ethnicity and family income, 2001-2004

Data: Centers for Disease Control and Prevention, National Center for Health Statistics and National Center for Environmental Health, National Health and Nutrition Examination Survey

Note: To reflect exposures to women who are pregnant or may become pregnant, the estimates are adjusted for the probability (by age and race/ethnicity) that a woman gives birth.

America's Children and the Environment, Third Edition

*The estimate should be interpreted with caution because the standard error of the estimate is relatively large: the relative standard error, RSE, is at least 30% but is less than 40% (RSE = standard error divided by the estimate), or the RSE may be underestimated.

** The estimate is not reported because it has large uncertainty: the relative standard error, RSE, is 40% or greater (RSE = standard error divided by the estimate), or the RSE cannot be reliably estimated.

Data characterization

- Data for this indicator are obtained from an ongoing continuous survey conducted by the National Center for Health Statistics.
- Survey data are representative of the U.S. civilian noninstitutionalized population.
- PCBs are measured in blood samples obtained from individual survey participants.

- In 2001–2004, the median level of PCBs in blood serum among women ages 16 to 49 years (the sum of PCBs 118, 138, 153 and 180) was 30 ng/g lipid.
- Median PCB levels were higher for women with higher incomes than for women with lower incomes, consistently for all race/ethnicity groups.
  - After accounting for other demographic differences (i.e., differences in age profile), the differences between income levels for each race/ethnicity group were not statistically significant except for the differences for White non-Hispanic women.
- Median PCB levels were lower among Mexican-American women than among women of any other race/ethnicity group.
  - These differences were statistically significant. After accounting for other demographic differences (i.e., differences in income or age profile), the differences remained statistically significant except for that between Mexican-American women and women of "All Other Races/Ethnicities."
- The 95th percentile concentration of PCBs among women ages 16 to 49 years was 106 ng/g lipid. Among women of "All Other Races/Ethnicities," the 95th percentile PCB concentration was substantially higher, at 245 ng/g lipid; the 95th percentile concentration among Mexican-American women was substantially lower at 49 ng/g lipid. (See Table B7a.)
  - These differences were statistically significant: the 95th percentile for women of "All Other Races/Ethnicities" was greater than the value for each of the remaining race/ethnicities; and the 95th percentile for Mexican-American women was less than the value for each of the remaining race/ethnicities.

Polybrominated Diphenyl Ethers (PBDEs)

Polybrominated diphenyl ethers (PBDEs) are a group of brominated flame retardant chemicals that have been incorporated into a variety of manufactured products, including foam cushioning used in furniture and plastics used in televisions and computers. Flame retardants are intended to slow the rate of ignition and fire growth, allowing more time for people to escape from a fire or extinguish it.

All PBDEs have a common structure of a diphenyl ether molecule, which may have from 1–10 bromine atoms attached; each particular PBDE variant is referred to as a congener. In theory, there could be as many as 209 PBDE congeners, but a much smaller number of congeners are commonly found in the commercial PBDE mixtures and in measurements of PBDEs in humans and the environment.

Three commercial PBDE mixtures have been used in manufactured products since the 1960s and 1970s, when these chemicals came into use,[1] with each mixture made up of congeners with varying degrees of bromination. The commercial pentabromodiphenyl ether (pentaBDE) and octabromodiphenyl ether (octaBDE) mixtures have not been manufactured or imported in the United States since 2004. The pentaBDE mixture, made up primarily of four- and five-bromine congeners, was used almost entirely in flexible polyurethane foam in furniture, mattresses, carpet padding, and automobile seats; and the octaBDE mixture, made up primarily of seven- and eight-bromine congeners, was used in acrylonitrile-butadiene-styrene (ABS) plastic for certain electric and electronic devices.

A third product, the commercial decabromodiphenyl ether (decaBDE) mixture, is still manufactured and used in the United States. The decaBDE mixture, made up almost entirely of the 10-bromine congener, has been used primarily in high-impact polystyrene (HIPS) plastic that was frequently used to make the back part of television sets, and in other electronic devices. However, there are indications that the use of decaBDE in electronic devices has declined in recent years, particularly since restrictions on the use of decaBDE in electronics were implemented in Europe beginning in 2008. DecaBDE is also used as a flame retardant on certain types of textiles; in electrical products, including uses in vehicles and airplanes; and in certain building materials. The major U.S. importers and manufacturers of decaBDE have announced that this mixture will be phased out by the end of 2013.[2] As use of PBDEs is reduced, they are being replaced by other flame-retardant chemicals or by materials that are inherently resistant to fire. EPA has conducted an assessment of alternatives to commercial pentaBDE,[3] and is conducting a similar assessment of alternatives to commercial decaBDE.[4]

PBDEs can be released into the environment at various points in their lifecycle, from their production and application to consumer products to their release from discarded products in landfills. Since PBDEs are not chemically bound to the products in which they are used, they can easily migrate into the surrounding air, dust, soil, and water. Although production and use of the commercial PBDE mixtures has been phased out (pentaBDE and octaBDE) or will soon be phased out (decaBDE), it is likely that PBDE congeners will continue to be present in the environment for many years. This is because products previously manufactured with PBDEs (e.g., sofas) will stay in use for many years. PBDEs will continue to be released from these products while they are in use, and these releases may continue when the products are disposed of or recycled. PBDEs are persistent in the environment, so even if there were no further releases they would continue to be detected for many years.

Exposure studies, focusing on selected PBDE congeners that were most predominant in the commercial mixtures or that are frequently measured in environmental samples, have concluded that the presence of PBDEs in house dust and in foods are both important contributors to PBDE exposures for people of all ages, and that exposures from house dust are generally greater than those from food.[1,5-11]

Studies conducted in multiple locations have consistently found PBDEs in U.S. house dust at levels greater than those found in other countries.[12-14] This is likely due to greater use of PBDE-containing products in homes in the United States than in other countries. Within the United States, the highest levels of three frequently measured PBDE congeners in dust have been found in California. The three congeners were all components of commercial pentaBDE mixtures, and the authors of these studies observed that the elevated levels may be due to California requirements for flame resistance in residential furniture that are not applicable in other states.[13,15] A study conducted in adults found a stronger association between direct contact with PBDE-containing materials and PBDE blood levels than between PBDE-contaminated house dust and PBDE blood levels.[16]

A second pathway of exposure to PBDEs is through diet. PBDEs are generally persistent chemicals that accumulate in fat tissue, so they are commonly found in foods derived from animals.[17-19] Information about how PBDEs enter the food web is limited, but release from manufacture of the PBDEs or of PBDE-containing products; release of PBDEs from products while they are in use; and release from products when disposed of or recycled are all likely contributors to PBDEs in the environment. PBDEs have been measured in a variety of supermarket foods, with the highest levels found in fish and other foods of animal origin.[18,20] A California study found associations between pork and poultry consumption and the levels of PBDEs measured in blood of children ages 2 to 5 years.[13]

Levels of PBDEs measured in blood are substantially greater in North America than in Europe and Asia, a difference that appears to be due to the higher levels of PBDEs in house dust in North America.[7,12,21,22] Studies comparing archived and current samples of blood and pooled serum from various locations in the United States have shown marked increases in PBDE levels since the late 1970s.[23,24]

Early-life exposures to PBDEs may be elevated in a number of ways. A child's exposure to PBDEs begins in the prenatal period, as PBDEs have been measured in cord blood, fetal blood, and placental tissue[25-27] and continues in early infancy due to the presence of PBDEs in breast milk.[11,22,28-31] Levels of PBDEs in breast milk are higher in North America than elsewhere,[12] and estimated intakes of PBDEs are substantially greater for a breastfeeding infant than exposures that occur during other life stages.[6,22]

Exposures are also elevated for young children up to age 7 years. While few studies have measured concentrations of PBDEs in young children, one large study conducted in Australia found that levels of PBDEs in blood are greatest for children ages 2 to 5 years, compared with older children and adults.[32] A study of 20 young children (ages 1.5 to 4 years) in various locations throughout the United States found that their PBDE blood levels were consistently higher than those of their mothers.[33] A study of California children ages 2 to 5 years found PBDE blood levels generally greater than those measured in California residents ages 12 to 60 years.[13,15] A study of 7-year-old Mexican-American children living in California reported PBDE blood levels that were three times the levels found in their mothers during pregnancy.[34]

The elevated exposures observed for young children are likely due to increased exposure to house dust, based on several studies that have estimated exposures based on measured levels of PBDEs in house dust, air, and food.[6,7,9,35] Infants and small children may have the highest exposure to PBDEs in house dust due to their frequent and extensive contact with floors, carpets, and other surfaces where dust gathers, as well as their frequent hand-to-mouth activity.[36] However, children of all ages (as well as adults) are likely to be exposed to dust contaminants through hand-to-mouth activity and other ingestion pathways, such as the settling of dust onto food and food preparation surfaces in the kitchen, as well as inhalation and absorption of PBDEs through the skin.[1,9,22]

Concerns about the health effects of PBDEs are based largely on laboratory animal studies, along with findings of the limited number of human epidemiological studies that have been conducted to date. A primary concern from the animal studies is for effects on the developing brain and nervous system, including effects on learning, memory, and behavior.[37-39] A study of children in New York City found significant associations between children's prenatal exposure to PBDEs and performance on IQ tests at ages up through 6 years.[40] A second epidemiological study conducted in the Netherlands found that prenatal exposure to PBDEs was associated with reduced scores on some tests of neurological development and improved scores on other tests at ages 5 to 6 years.[41]

PBDEs are suspected endocrine disruptors.[37] Endocrine disruptors act by interfering with the biosynthesis, secretion, action, or metabolism of naturally occurring hormones.[42,43] Given the importance of hormones in human physiology, there is concern in the scientific community over the potential for endocrine disruptors to adversely affect children's health, particularly in reproduction, early and adolescent development, and behavior.

Animal and human studies indicate that PBDEs may alter circulating levels of thyroid hormones.[37,44-47] Moderate deficits in maternal thyroid hormone levels during early pregnancy have been linked to reduced childhood IQ scores and other neurodevelopmental effects, as well as unsuccessful or complicated pregnancies.[48] Animal studies have found that PBDE exposure at critical stages of fetal development reduced levels of male hormones or caused other changes relevant to male reproductive development.[37,47,49-51] An epidemiological study of boys born in Denmark and Finland found that increased levels of PBDEs in breast milk were associated with an increased risk of cryptorchidism (undescended testes),[52] an effect that may be related to hormone disruption during critical stages of development.[53,54] Also, a study of Mexican immigrant women in California found effects on fertility (increased time to pregnancy) with increasing PBDE levels; this finding may be related to hormonal activity of PBDEs.[55]

The following indicator presents the best nationally representative data on PBDE levels in women of child-bearing age. Indicator B8 presents median blood serum levels of PBDEs for women ages 16 to 49 years. Although data are available only for a single two-year period at this time, the data provide a baseline that will be updated with PBDE measurements over time from subsequent survey cycles.

About the Indicator: Indicator B8 presents concentrations of PBDEs in blood serum of U.S. women ages 16 to 49 years. The data are from a national survey that collects blood specimens from a representative sample of the population every two years, and then measures the concentration of PBDEs in the blood serum. The indicator presents comparisons of PBDEs in blood serum for women of different race/ethnicities, and for women of different income levels. The focus on women of child-bearing age is based on concern for potential effects in children born to women who have been exposed to PBDEs.

NHANES

The National Health and Nutrition Examination Survey (NHANES) provides nationally representative biomonitoring data for PBDEs. NHANES is designed to assess the health and nutritional status of the civilian noninstitutionalized U.S. population and is conducted by the National Center for Health Statistics, part of the Centers for Disease Control and Prevention (CDC). Interviews and physical examinations are conducted with approximately 10,000 people in each two-year survey cycle. CDC's National Center for Environmental Health measures concentrations of environmental chemicals in blood and urine samples collected from NHANES participants. Summaries of the measured values for more than 200 chemicals are provided in the *Fourth National Report on Human Exposure to Environmental Chemicals*.[56]

PBDE Congeners

Indicator B8 presents blood serum levels of PBDEs in women of child-bearing age. There are 209 possible PBDEs, referred to as "congeners," which are defined by the number of bromine atoms (from 1 to 10) and their position in the chemical structure. Each congener is assigned a specific brominated diphenyl ether (BDE) number, such as BDE-47 (a tetrabromodiphenyl ether congener – four bromine atoms). Most of these congeners have not been detected in the manufactured PBDE mixtures and have not been measured in environmental or human samples.

PBDE concentrations are measured in blood serum. PBDEs are lipophilic, meaning that they tend to accumulate in fat. Serum PBDEs concentrations are measured and expressed on a lipid-adjusted basis, as these values better represent the amount of PBDEs stored in the body compared with unadjusted values.[56] The indicator uses lipid-adjusted concentrations, meaning that the concentration of PBDEs in serum is divided by the concentration of lipid in serum. The resulting units are nanograms of PBDE per gram of lipid (ng/g lipid) in serum.[i]

---

[i] Serum levels of PBDEs can also be reported without lipid adjustment. Both the lipid-adjusted values and the unadjusted "whole weight" values are reported in CDC's *Fourth National Report on Human Exposure to Environmental Chemicals*.

Concentrations of PBDEs in blood serum have been measured in a representative subset of NHANES participants ages 12 years and older in the 2003–2004 survey cycle. NHANES sampled for 10 PBDE congeners in 2003–2004, including those most frequently measured in environmental and human samples. These include BDEs 17, 28, 47, 66, 85, 99, 100, 153, 154, and 183. Most of these 10 congeners were components of the pentaBDE mixture that was used in polyurethane foam for furniture, mattresses, and automotive seating. Some of the congeners measured in NHANES were components of the octaBDE mixture, used in plastics for some household electric devices. The primary congener comprising the decaBDE formulation, BDE-209, was not measured in NHANES in 2003–2004.

Indicator B8 was calculated by summing together the measured values of the 10 congeners for each woman 16 to 49 years sampled in NHANES; this approach is commonly used in studies assessing levels of PBDEs in human blood samples and environmental samples.[1] Data are insufficient at this time to assess and quantify differences in toxicity of the measured PBDE congeners, or to inform approaches other than an unweighted summation of the 10 congeners.

If a congener was not detected in a particular blood sample, a default value below the detection limit was assigned for purposes of calculating the summed total for the sampled individual.[ii] This assumption has a small impact on the reported blood levels of PBDEs, because almost all women sampled had values well above the detection limit for at least some congeners. BDEs 47, 100 and 153 were each detected in more than 90% of women ages 16 to 49 years.

In 2003–2004, a sum of the 10 measured PBDEs is available from NHANES for 2,040 individuals ages 12 years and older, including 540 women ages 16 to 49 years. One or more PBDE congeners were detected in 99% of the individuals sampled in NHANES 2003–2004, and in 99% of women ages 16 to 49 years.[iii] The median sum of the ten PBDE congeners among NHANES participants in 2003–2004 was 38 ng/g lipid and the 95th percentile sum was 307 ng/g lipid.

Birth Rate Adjustment

Indicator B8 uses measurements of PBDEs in blood of women ages 16 to 49 years to represent the distribution of PBDE exposures to women who are pregnant or may become pregnant. However, women of different ages have a different likelihood of giving birth. For example, in 2003–2004, women aged 27 years had a 12% annual probability of giving birth, and women aged 37 years had a 4% annual probability of giving birth.[57] A birth rate-adjusted distribution of women's PBDE levels is used in calculating this indicator,[iv] meaning that the data are weighted using the age-specific probability of a woman giving birth.[58]

[ii] The default value used for non-detect samples is equal to the limit of detection divided by the square root of 2.

[iii] The percentage for women ages 16 to 49 years is calculated with the birth rate adjustment described below.

[iv] There may be multiple ways to implement an adjustment to the data that accounts for birth rates by age. The National Center for Health Statistics has not fully evaluated the method used in ACE, or any other method intended to accomplish the same purpose, and has not used any such method in its publications. NCHS and EPA are working together to further evaluate the birth rate adjustment method used in ACE and alternative methods.

Data Presented in the Indicator

Indicator B8 presents median concentrations of PBDEs in blood serum for women ages 16 to 49 years of different races/ethnicities and levels of family income, using NHANES data from 2003–2004.[v]

Three race/ethnicity groups are presented in Indicator B8: White non-Hispanic, Black non-Hispanic, and Mexican-American. The data are also tabulated across three income categories: all incomes, below the poverty level, and greater than or equal to the poverty level.

Additional information on how median blood serum levels of PBDEs vary by race/ethnicity and family income for children ages 12 to 17 years is presented in a supplemental data table for this indicator.

Please see the Introduction to the Biomonitoring section for an explanation of the term "median" and information on the statistical significance testing applied to this indicator.

---

[v] Unlike other biomonitoring indicators in this report, 95th percentile PBDE levels are not provided in a supplementary table. This is because most 95th percentile PBDE values do not meet ACE statistical reliability criteria. There is more uncertainty in 95th percentile estimates for PBDEs than for other chemicals because data are only available for two years (2003–2004) at this time. Similarly, separate values are not provided considering both race/ethnicity and income simultaneously, nor are values provided for the "All Other Races/Ethnicities" category, because (with data from only one NHANES cycle available at this time) such estimates lack statistical reliability.

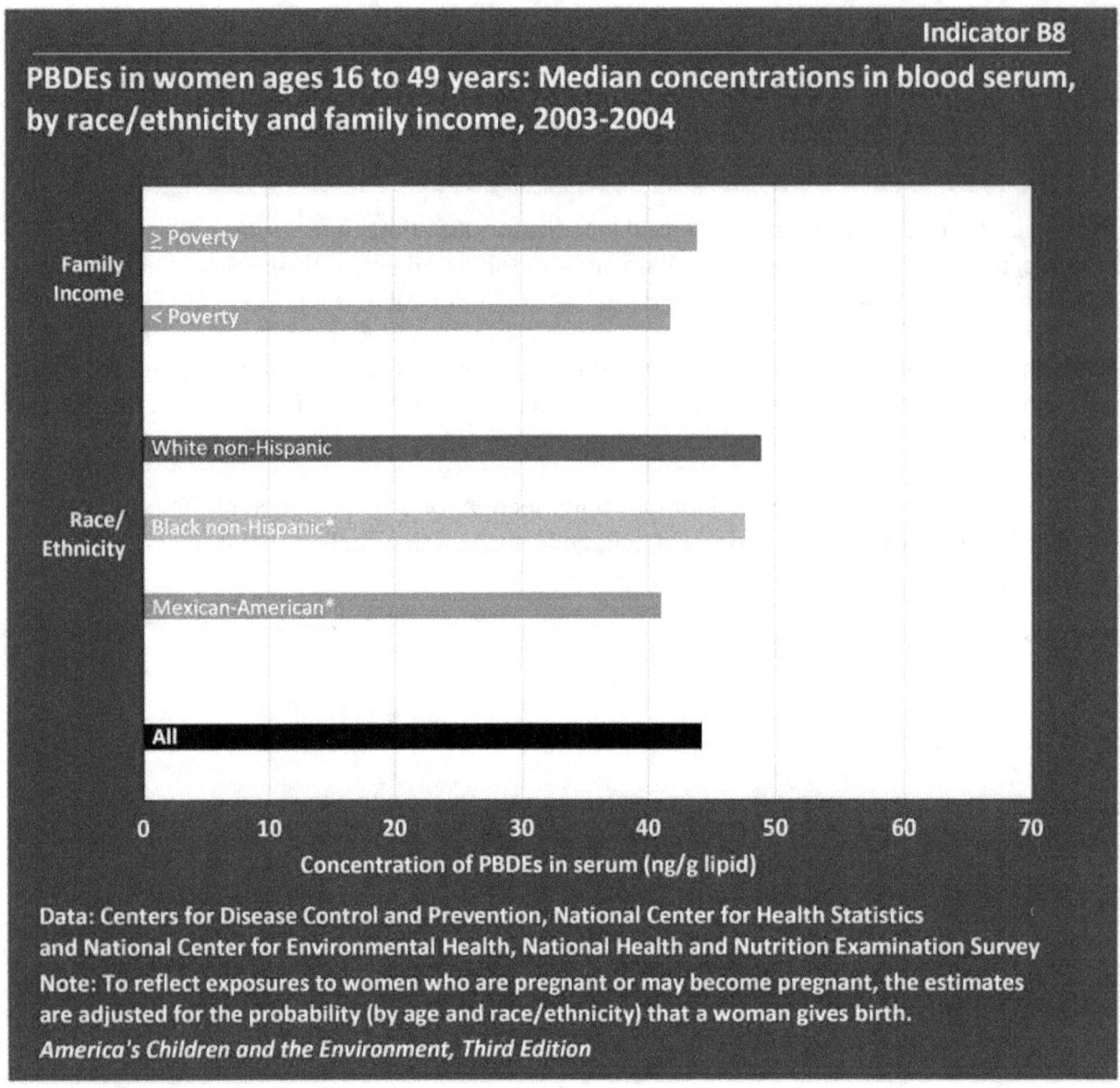

Indicator B8

PBDEs in women ages 16 to 49 years: Median concentrations in blood serum, by race/ethnicity and family income, 2003-2004

Data: Centers for Disease Control and Prevention, National Center for Health Statistics and National Center for Environmental Health, National Health and Nutrition Examination Survey

Note: To reflect exposures to women who are pregnant or may become pregnant, the estimates are adjusted for the probability (by age and race/ethnicity) that a woman gives birth.

America's Children and the Environment, Third Edition

*The estimate should be interpreted with caution because the standard error of the estimate is relatively large: the relative standard error, RSE, is at least 30% but is less than 40% (RSE = standard error divided by the estimate), or the RSE may be underestimated.

Data characterization
- Data for this indicator are obtained from an ongoing continuous survey conducted by the National Center for Health Statistics.
- Survey data are representative of the U.S. civilian noninstitutionalized population.
- PBDEs are measured in blood samples obtained from individual survey participants.

- The median concentration of PBDEs in blood serum of women ages 16 to 49 years was 44 ng/g lipid in 2003–2004.

- White non-Hispanic women and Black non-Hispanic women had the highest median PBDE levels at 49 and 48 ng/g lipid, respectively.
  - The differences by race/ethnicity were generally not statistically significant without accounting for differences by age and income across race/ethnicity groups. After accounting for differences by age and income across race/ethnicity groups, PBDE levels in White non-Hispanic women were statistically significantly greater than levels in Black non-Hispanic women. Also after adjustment, PBDE levels in Black non-Hispanic women were statistically significantly greater than levels in Mexican-American women.
- Among women of child-bearing age, there was little difference in median PBDE concentrations in blood serum between income groups.
- The median concentration of PBDEs in children ages 12 to 17 years overall was 53 ng/g lipid. The median concentration of PBDEs for children with family incomes below the poverty level was 63 ng/g lipid, and 50 ng/g lipid for children at or above poverty level. (See Table B8a.)
  - The difference in median PBDE concentration between the income groups was not statistically significant.
- Among children ages 12 to 17, White non-Hispanic children and Black non-Hispanic children had the lowest median PBDE levels at 48 and 50 ng/g lipid. Mexican-American children had median PBDE levels of 63 ng/g lipid. (See Table B8a.)
  - These differences by race/ethnicity were not statistically significant.

Phthalates

Phthalates are a class of manufactured chemicals commonly used to increase the flexibility of plastics in a wide array of consumer products. More than 470 million pounds of phthalates are produced or imported in the United States each year.[1]

By far the most common use of phthalates is in the production of polyvinyl chloride (PVC) products.[2] PVC is the second most commonly used plastic in the world, and is present in pipes and tubing, construction materials, packaging, electrical wiring, and thousands of consumer goods.[3,4] Phthalates are or have been used in wall coverings, tablecloths, floor tiles, furniture upholstery, carpet backings, shower curtains, garden hoses, rainwear, pesticides, some toys, shoes, automobile upholstery, food packaging, medical tubing, and blood storage bags.[5-8] Phthalates are not strongly bound in these products and can therefore leach out.[4-10] Some phthalates are also present in cosmetics, nail polish, hair products, skin care products, and some medications.[4,6,7,11,12]

The Consumer Product Safety Improvement Act of 2008 (CPSIA) banned the use of three phthalates in toys and child care articles at concentrations greater than 0.1 percent: di-2-ethylhexyl phthalate (DEHP), dibutyl phthalate (DBP), and butyl benzyl phthalate (BBzP). CPSIA also restricts the use of di-isononyl phthalate (DINP), di-isodecyl phthalate (DIDP), and di- n-octyl phthalate (DnOP) in toys that can be mouthed and child care articles. The Consumer Product Safety Commission has also appointed a Chronic Hazard Advisory Panel to examine the cumulative health risks of phthalates and phthalate substitutes, and to recommend whether to continue the ban of DINP, DIDP, and DnOP and whether any other phthalates or phthalate substitutes should be banned.[1] As use of phthalates is reduced, they are being replaced by other chemicals, such as di-isononylcyclohexane-1,2-dicarboxylate (DINCH) and di(2-ethylhexyl) terephthalate (DEHT), that also increase the flexibility of plastics.[13-15] EPA is planning to conduct an assessment of alternatives to several phthalates.[1]

For most phthalates, the major route of exposure is food ingestion.[4,16-19] However, personal care product use and inhalation are major routes of exposure for certain phthalates.[4-8,11,20] Some phthalates have been found at higher levels in fatty foods such as dairy products, fish, seafood, and oils.[8] Phthalates in a mother's body can enter her breast milk. Ingestion of that breast milk and infant formula containing phthalates may also contribute to infant phthalate exposure.[21] The phthalates that may be present in dust can be ingested by infants and children through hand-to-mouth activities.[10,22] Finally, infants and small children can be exposed to phthalates by sucking on toys and objects made with phthalate-containing plastics.[10]

Other minor routes of phthalate exposure include inhalation, drinking contaminated water, and absorption through the skin.[16,17] Phthalates can be released in small amounts to the air people breathe inside homes or schools from the many consumer products that contain them.[7,20] People living near phthalate-producing factories or hazardous waste sites may be exposed to phthalates released into the air or ground water where they live.[5,7,8] Individuals may be exposed to phthalates during the use of many personal care products containing phthalates, such as hair products, cosmetics, and lotions.[11,23,24] Phthalates in these products may be absorbed through contact with the skin or may be inhaled if some of the product is present in the air.[5] In addition, certain medical devices, such as intravenous tubing or flexible bags containing blood, medications, or nutritional products, contain phthalates. These can be a source of phthalate exposure to children and women of child-bearing age when the tubing or bags are used to administer medications, nutritional products, or blood to the individual. This can be a very significant route of exposure, especially for premature infants in intensive care units.[25-27]

Phthalate exposures, assessed from urinary concentrations of phthalate metabolites (i.e., breakdown products), appear to be higher for children compared with adolescents and adults. Studies of phthalate metabolites in children's urine are limited, but the few that have been published have found children's urinary phthalate metabolite levels to be higher than levels in adults and to decrease with age (i.e., younger children had more phthalate metabolites in their urine than older children did).[28-30] The exception is monoethyl phthalate (MEP), a metabolite of diethyl phthalate, which has been found to be present in higher levels in adult urine compared with children's urine.[28] Levels of MEP are most likely associated with use of consumer products that contain diethyl phthalate, such as detergents, soaps, cosmetics, shampoos, and perfumes.[5,28]

Some phthalates are suspected endocrine disruptors.[31-35] Endocrine disruptors act by interfering with the biosynthesis, secretion, action, or metabolism of naturally occurring hormones.[32,36] Given the importance of hormones in human physiology, there is concern in the scientific community over the potential for endocrine disruptors to adversely affect children's health, particularly in reproduction, development, and behavior. Male laboratory animals exposed to high doses of some phthalates have been known to display elements of "phthalate syndrome," which includes infertility, decreased sperm count, cryptorchidism (undescended testes), hypospadias (malformation of the penis in which the urethra does not open at the tip of the organ), and other reproductive tract malformations.[4] A number of animal studies have reported associations between exposure to certain phthalates and changes in male hormone production, altered sexual differentiation, and changes to reproductive organs, including hypospadias.[37-45] These findings in animal studies, although typically occurring at exposure levels much higher than what the general population may be exposed to, suggest a potential concern for health effects in children as well. The National Research Council has concluded that prenatal exposure to certain phthalates produces reproductive tract abnormalities in male rats, and also concluded that the same effects could plausibly occur in humans.[4]

There are only a limited number of human studies looking at the relationship between phthalate exposure and hormonal and reproductive health changes. In one study, prenatal exposure to some phthalates at typical U.S. population levels was associated with changes in physical measures of the distance between the anus and the genitals (anogenital distance) in male infants.[46,47] A shorter anogenital distance has been associated with decreased fertility in animal experiments[48,49] and a recent human study reported that a shorter anogenital distance in men was associated with decreased semen quality and low sperm count.[50] Another study reported an association between increased concentrations of phthalate metabolites in breast milk and altered reproductive hormone levels in newborn boys. The same study did not find an association between breast milk phthalate metabolite concentrations and cryptorchidism.[51]

Exposure to some phthalates has been associated with neurodevelopmental problems in children in some studies. Two studies of a group of New York City children ages 4 to 9 years reported associations between prenatal exposure to certain phthalates and behavioral deficits, including effects on attention, conduct, and social behaviors.[52,53] Studies conducted in South Korea of children ages 8 to 11 years reported that children with higher levels of certain phthalate metabolites in their urine were more inattentive and hyperactive, displayed more symptoms of attention-deficit/hyperactivity disorder (ADHD), and had lower IQ compared with those who had lower levels.[54,55] The exposure levels in these studies are comparable to typical exposures in the U.S. population.

A handful of studies have reported associations between prenatal exposure to some phthalates and preterm birth, shorter gestational length, and low birth weight;[56-59] however, one study reported phthalate exposure to be associated with longer gestational length and increased risk of delivery by Cesarean section.[60]

Finally, some researchers have hypothesized that phthalate exposure in homes may contribute to asthma and allergies in children. Two research groups have conducted studies, primarily in Europe, and reported associations between surrogates for potential phthalate exposure in the home and risk of asthma and allergies in children.[61] Examples of the exposure indicators and outcomes considered in these studies include an association between some phthalates in surface dust and increased risk of runny nose, eczema, and asthma,[62] and increased risk of bronchial obstruction associated with the presence of PVC in the home.[63]

The two indicators that follow use the best nationally representative data currently available on urinary phthalate metabolite levels over time for women of child-bearing age and children. The indicators focus on three important phthalates: di-2-ethylhexyl phthalate (DEHP), dibutyl phthalate (DBP), and butyl benzyl phthalate (BBzP). These three phthalates were chosen because their metabolites are commonly detected in humans and their potential connection to adverse children's health outcomes is supported by the scientific literature summarized in the following paragraphs.

DEHP is currently the only phthalate plasticizer used in PVC medical devices such as blood bags and plastic tubing. DEHP is also currently used in flooring, wallpaper, and raincoats and has been used in toys, auto upholstery, and food packaging.[64] DBP is used primarily in latex adhesives, cellulose plastics, dyes, and cosmetics and other personal care products.[65] The largest use of BBzP is in the production of PVC flooring materials, but it is also used in the manufacture of automotive materials, artificial leather, and food conveyor belts.[66,67]

In 2006, the National Toxicology Program (NTP) concluded that there is "concern" for effects on reproductive tract development in male infants less than one year old exposed to DEHP. In addition, the NTP also concluded that there is "some concern" (the midpoint on a five-level scale ranging from "negligible" to "serious" concern) [i] for effects on reproductive tract development in male children older than one year old exposed to DEHP, and also that there is "some concern" for effects of prenatal DEHP exposure on reproductive tract development in males. Concern was greater for males exposed to high levels of DEHP in the womb or early in life. These conclusions were based primarily on findings from animal studies, as human data are limited and were determined to be insufficient for evaluating the reproductive effects of DEHP.[64] Some studies have also reported associations of DEHP exposure with increased risk of asthma and bronchial obstruction, increased risk of ADHD symptoms, and altered pregnancy durations.[55,56,58,60,62,63] Human health studies have reported associations between exposures to DBP and altered reproductive hormone levels in newborn boys, and shifts in thyroid hormone levels in pregnant women.[51,68] Signs of feminization in young boys (as measured by reduced anogenital distance), altered hormone levels in newborn boys, and increased risk of rhinitis and eczema are health effects that have been associated with BBzP exposure in some studies.[46,47,51,62] The exposure levels in these studies are comparable to typical exposures in the U.S. population. It is important to note that while the following indicators present data on individual phthalate metabolites, evidence suggests that exposures to multiple phthalates may contribute to common adverse outcomes. The National Research Council has concluded that multiple phthalates may act cumulatively to adversely impact male reproductive development.[4]

Indicator B9 presents median concentrations of metabolites of DEHP, DBP, and BBzP in urine for women ages 16 to 49 years. Indicator B10 presents median metabolite levels of the same phthalates (DEHP, DBP, and BBzP) in urine for children ages 6 to 17 years.

---

[i] More information on NTP concern levels is available at http://www.niehs.nih.gov/news/media/questions/sya-bpa.cfm.

Indicator B9: Phthalate metabolites in women ages 16 to 49 years: Median concentrations in urine, 1999–2008

Indicator B10: Phthalate metabolites in children ages 6 to 17 years: Median concentrations in urine, 1999–2008

About the Indicators: Indicators B9 and B10 present concentrations of phthalate metabolites in urine of U.S. women ages 16 to 49 years and children ages 6 to 17 years. The data are from a national survey that collects urine specimens from a representative sample of the population every two years, and then measures the concentration of phthalate metabolites in the urine. Indicator B9 presents concentrations of phthalate metabolites in women's urine over time and Indicator B10 presents concentrations of phthalate metabolites in children's urine over time. The focus on both women of child-bearing age and children is based on concern for potential adverse effects in children born to women who have been exposed to phthalates and in children exposed to phthalates.

NHANES

The National Health and Nutrition Examination Survey (NHANES) provides nationally representative biomonitoring data for several phthalates. NHANES is designed to assess the health and nutritional status of the civilian noninstitutionalized U.S. population and is conducted by the National Center for Health Statistics, part of the Centers for Disease Control and Prevention (CDC). Interviews and physical examinations are conducted with approximately 10,000 people in each two-year cycle. CDC's National Center for Environmental Health measures concentrations of environmental chemicals in blood and urine samples collected from NHANES participants. Summaries of the measured values for more than 200 chemicals are provided in the *Fourth National Report on Human Exposure to Environmental Chemicals*.[69]

Phthalate Metabolites

Indicators B9 and B10 present urinary metabolite levels of three important phthalates: di-2-ethylhexyl phthalate (DEHP), dibutyl phthalate (di-n-butyl phthalate and di-isobutyl phthalate) (DBP), and butyl benzyl phthalate (BBzP).

In NHANES and many research studies, biomonitoring of phthalates is conducted by measuring phthalate metabolites in urine rather than the phthalates themselves. This is because phthalates may be present in the sampling and laboratory equipment used to study human exposure levels, and contamination of samples could occur. Also phthalate metabolism is so rapid that the parent phthalate may not appear in urine.[5-8,16,70,71] Furthermore, the phthalate metabolites, and not the parent phthalates, are generally considered to be the biologically active molecules.[5-8,16,72] Unlike other contaminants that have a tendency to accumulate in the human body, phthalates are metabolized and excreted quickly, with elimination half-lives on the order of hours.[5-8,71] Therefore, phthalate metabolites measured in humans are indicative of recent exposures. All values are reported as micrograms of phthalate metabolites per liter of urine (µg/L).

Concentrations of phthalate metabolites, including those for DEHP, DBP, and BBzP, have been measured in urine from a representative subset of NHANES participants ages 6 and older beginning with the 1999–2000 survey cycle. For DEHP, three metabolites are included: mono-2-ethylhexyl phthalate (MEHP), mono-(2-ethyl-5-oxohexyl) phthalate (MEOHP), and mono-(2-ethyl-5-hydroxyhexyl) phthalate (MEHHP).[ii] The urinary levels of MEHP, MEOHP, and MEHHP are summed together, as is common in phthalates research, to provide a more complete picture of an individual's total DEHP exposure than is given by any individual metabolite.[57,73-75] The primary urinary metabolites of DBP are mono-n-butyl phthalate (MnBP) and mono-isobutyl phthalate (MiBP). The urinary levels of MnBP and MiBP were measured together for the NHANES 1999–2000 survey cycle, but for the following years were measured separately. Indicators B9 and B10 present the combined urinary levels of MnBP and MiBP for each survey cycle. The primary urinary metabolite of BBzP is mono-benzyl phthalate (MBzP).

Calculation of the DEHP metabolite and DBP metabolite indicator values involves summing together separate measured values (3 metabolites of DEHP, and 2 metabolites of DBP in the survey cycles following 1999–2000). If a metabolite included in the sum was not detected in a sample, a default value below the detection limit was assigned for purposes of calculating the summed total.[iii]

In 2007–2008, NHANES collected phthalates biomonitoring data for 2,604 individuals ages 6 years and older, including 571 women ages 16 to 49 years and 690 children ages 6 to 17 years. DEHP metabolites were detected in about 67% of all individuals sampled. The frequency of DEHP metabolites detection was 75% in women ages 16 to 49 years[iv], and 69% in children ages 6 to 17 years. DBP metabolites and BBzP metabolite were detected in 98% of all individuals sampled. The frequency of DBP metabolites detection was 98% in women ages 16 to 49 years, and 99% in children ages 6 to 17 years. The frequency of BBzP metabolite detection was 97% in women ages 16 to 49 years, and 99% in children ages 6 to 17 years. The median and 95th percentile of phthalate levels in urine for all NHANES participants in 2007–2008 were 35 µg/L and 406 µg/L, respectively, for DEHP; 29 µg/L and 147 µg/L, respectively, for DBP; and 12 µg/L and 82 µg/L, respectively, for BBzP. The widespread detection of phthalate metabolites, combined with the fact that phthalates have short half-lives, indicates that phthalate exposure is widespread and relatively continuous.

[ii] A fourth DEHP metabolite, mono(2-ethyl-5-carboxypentyl) phthalate (MECPP), is now measured in NHANES but was not measured prior to 2003–2004. At least one other DEHP metabolite has been measured in laboratory studies but is not measured in NHANES.

[iii] The default value used for non-detect samples is equal to the limit of detection divided by the square root of 2.

[iv] The percentage for women ages 16 to 49 years is calculated with the birth rate adjustment described below.

Individual Variability in Urinary Measurements

NHANES data for phthalates are based on measurements made using a single urine sample for each person surveyed. Due to normal changes in an individual's urinary output throughout the day, this variability in urinary volume, among other factors related to the measurement of chemicals that do not accumulate in the body,[76] may mask differences between individuals in levels of phthalates. Since phthalates do not appear to accumulate in bodily tissues, the distribution of NHANES urinary phthalate levels may overestimate high-end exposures (e.g., at the 95th percentile) as a result of collecting one-time urine samples.[71,77,78] Many studies account for differences in hydration levels by reporting the chemical concentration per gram of creatinine. Creatinine is a byproduct of muscle metabolism that is excreted in urine at a relatively constant rate, independent of the volume of urine, and can in some circumstances partially account for the measurement variability due to changes in urinary output.[79] However, urinary creatinine concentrations differ significantly among different demographic groups, and are strongly associated with an individual's muscle mass, age, sex, diet, health status (specifically renal function), body mass index, and pregnancy status.[80,81] Thus, these indicators present the unadjusted phthalate concentrations so that any observed differences in concentrations between demographic groups are not due to differences in creatinine excretion rates. These unadjusted urinary levels from a single sample may either over- or underestimate urinary levels for a sampled individual. However, for a representative group, it can be expected that a median value based on single samples taken throughout the day will provide a good approximation of the median for that group. Furthermore, due to the large number of subjects surveyed, we expect that differences in the concentrations of phthalates that might be attributed to the volume of the urine sample would average out within and across the various comparison groups.

Birth Rate Adjustment

Indicator B9 uses measurements of phthalate metabolites in urine of women ages 16 to 49 years to represent the distribution of phthalate exposures to women who are pregnant or may become pregnant. However, women of different ages have a different likelihood of giving birth. For example, in 2003–2004, women aged 27 years had a 12% annual probability of giving birth, and women aged 37 years had a 4% annual probability of giving birth.[82] A birth rate-adjusted distribution of women's phthalate metabolite levels is used in calculating this indicator,[v] meaning that the data are weighted using the age-specific probability of a woman giving birth.[83]

---

[v] There may be multiple ways to implement an adjustment to the data that accounts for birth rates by age. The National Center for Health Statistics has not fully evaluated the method used in ACE, or any other method intended to accomplish the same purpose, and has not used any such method in its publications. NCHS and EPA are working together to further evaluate the birth rate adjustment method used in ACE and alternative methods.

Data Presented in the Indicators

Indicator B9 presents median concentrations of DEHP, DBP, and BBzP metabolites in urine over time for women ages 16 to 49 years, using NHANES data from 1999–2008.

Indicator B10 presents median concentrations of DEHP, DBP, and BBzP metabolites in urine over time for children ages 6 to 17 years, using NHANES data from 1999–2008.

Additional information on the 95th percentile levels of urinary phthalates and how median levels of phthalate metabolites vary by race/ethnicity and family income for women ages 16 to 49 years is presented in the supplemental data tables for this indicator. Data tables also display the 95th percentile levels of phthalate metabolites and how median levels of phthalate metabolites vary by race/ethnicity, family income and age for children ages 6 to 17 years.

NHANES only provides phthalate metabolite data for children ages 6 years and older, which means that the indicator is not able to capture the exposure of premature infants, some of whom may have high levels of phthalate exposure due to the use of medical equipment containing phthalates; or young children, whose play and mouthing behaviors may increase their exposure to phthalates in toys and house dust.

Please see the Introduction to the Biomonitoring section for an explanation of the terms "median" and "95th percentile," a description of the race/ethnicity and income groups used in the ACE3 biomonitoring indicators, and information on the statistical significance testing applied to these indicators.

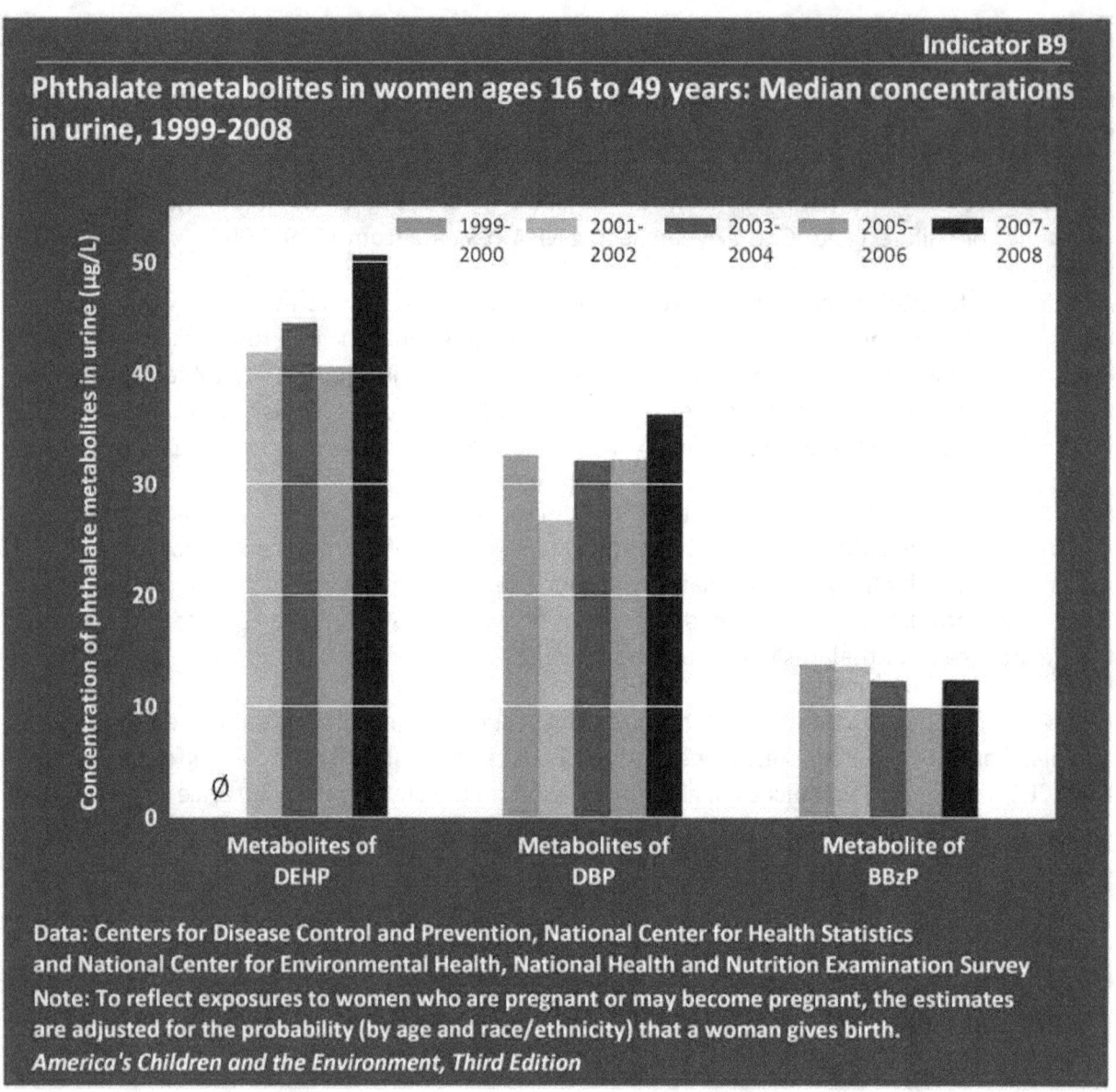

Indicator B9

Phthalate metabolites in women ages 16 to 49 years: Median concentrations in urine, 1999-2008

Data: Centers for Disease Control and Prevention, National Center for Health Statistics and National Center for Environmental Health, National Health and Nutrition Examination Survey

Note: To reflect exposures to women who are pregnant or may become pregnant, the estimates are adjusted for the probability (by age and race/ethnicity) that a woman gives birth.

America's Children and the Environment, Third Edition

Ø The estimate is not reported because the metabolites MEOHP and MEHHP were not measured in 1999–2000.

Data characterization

–   Data for this indicator are obtained from an ongoing continuous survey conducted by the National Center for Health Statistics.
–   Survey data are representative of the U.S. civilian noninstitutionalized population.
–   Phthalate metabolites are measured in urine samples obtained from individual survey participants.

■   From 2001–2002 to 2007–2008, the median level of DEHP metabolites in urine of women ages 16 to 49 years varied between 41 µg/L and 51 µg/L, and was 51 µg/L in 2007–2008. There was no statistically significant trend in median DEHP metabolites over 2001–2002 to 2007–2008.

- From 1999–2000 to 2007–2008, the median level of DBP metabolites in urine of women ages 16 to 49 years varied between 27 µg/L and 36 µg/L, and was 36 µg/L in 2007–2008.
- From 1999–2000 to 2007–2008, the median level of BBzP metabolites in urine of women ages 16 to 49 years varied between 10 µg/L and 14 µg/L, and was 12 µg/L in 2007–2008.
- From 2001–2002 to 2007–2008, the concentration of DEHP metabolites in the 95th percentile varied between 462 µg/L and 578 µg/L, and was 567 µg/L in 2007–2008. There was an increasing trend in the 95th percentile concentration of DBP metabolites in women of child-bearing age, from 128 µg/L in 2001–2002 to 160 µg/L in 2007–2008. From 1999–2000 and 2007–2008, the concentration of BBzP metabolite varied between 68 µg/L and 100 µg/L, and was 70 µg/L in 2007–2008. (See Table B9a.)
    - The increasing trend for DBP metabolites at the 95th percentile from 2001–2002 to 2007–2008 was statistically significant after accounting for differences by age, race/ethnicity, and income.
- The concentrations of DEHP metabolites in the 95th percentile ranged from 10 to 14 times higher than the median levels presented in this graph. The concentrations of DBP metabolites and BBzP metabolite in urine at the 95th percentile ranged from 4 to 7 times higher than the median levels presented in this graph. (See Table B9 and B9a.)
- For the years 2005–2008, Black non-Hispanic women of child-bearing age had higher median concentrations of all the phthalate metabolites shown here compared with White non-Hispanic women, Mexican-American women, and women of "All Other Races/Ethnicities," although these differences were frequently not statistically significant. (See Table B9b.)
- Median levels of urinary phthalate metabolites varied by family income. Women living below the poverty level had higher concentrations of phthalate metabolites in their urine compared with women living at or above the poverty level. (See Table B9b.)
    - The difference between income groups was statistically significant for the DBP metabolites after accounting for differences by race/ethnicity or age profile above and below poverty. The difference between income groups for the BBzP metabolite was statistically significant before accounting for race/ethnicity and age. The difference between income groups was not statistically significant for the DEHP metabolites.

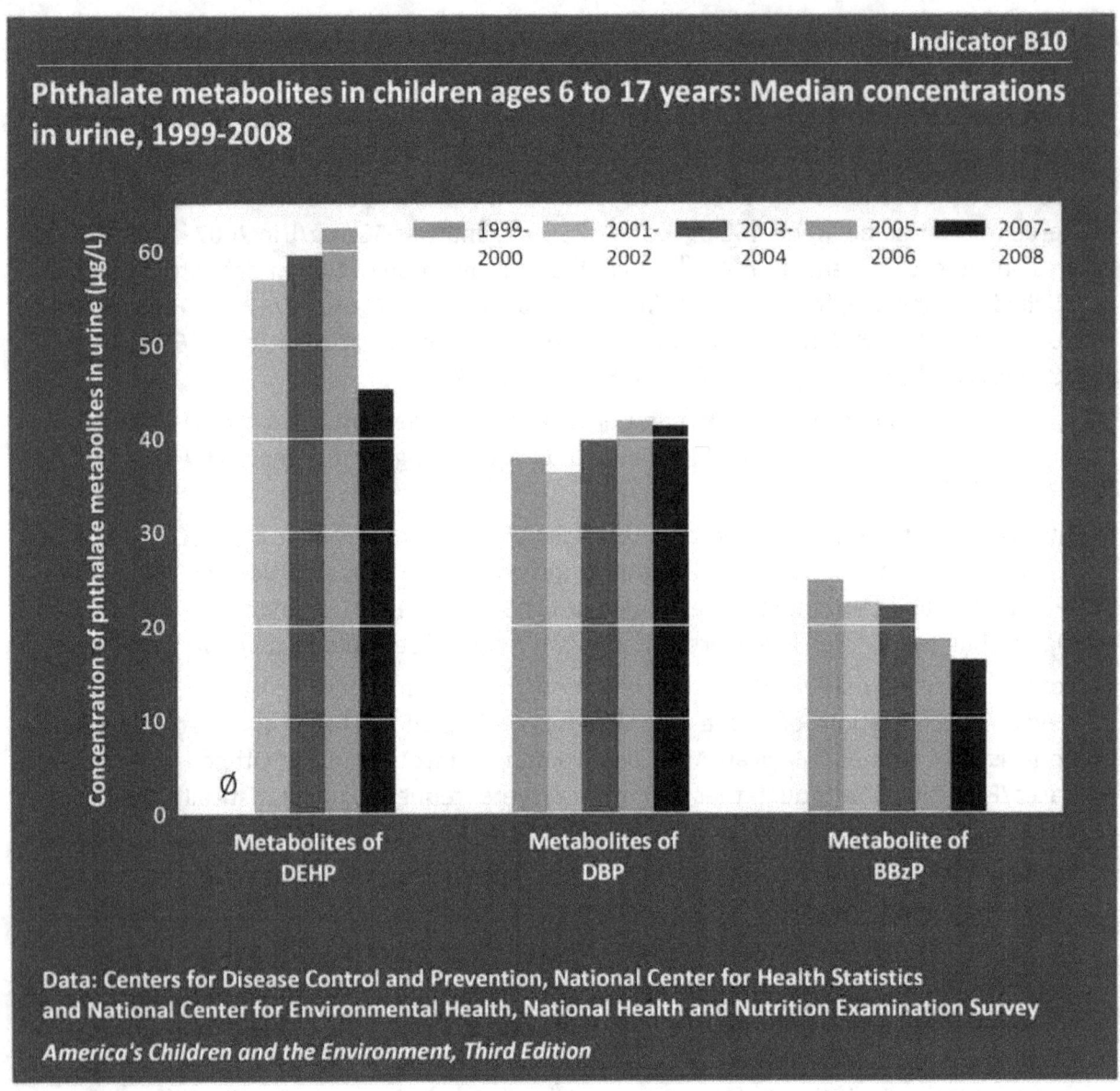

Indicator B10

Phthalate metabolites in children ages 6 to 17 years: Median concentrations in urine, 1999-2008

Legend: 1999-2000 | 2001-2002 | 2003-2004 | 2005-2006 | 2007-2008

Data: Centers for Disease Control and Prevention, National Center for Health Statistics and National Center for Environmental Health, National Health and Nutrition Examination Survey

America's Children and the Environment, Third Edition

Ø The estimate is not reported because the metabolites MEOHP and MEHHP were not measured in 1999–2000.

Data characterization
- Data for this indicator are obtained from an ongoing continuous survey conducted by the National Center for Health Statistics.
- Survey data are representative of the U.S. civilian noninstitutionalized population.
- Phthalate metabolites are measured in urine samples obtained from individual survey participants.

- From 2001–2002 to 2007–2008, the median level of DEHP metabolites in urine of children ages 6 to 17 years varied between 45 µg/L and 62 µg/L, and was 45 µg/L in 2007–2008.

- From 1999–2000 to 2007–2008, the median level of DBP metabolites in urine of children ages 6 to 17 years varied between 36 µg/L and 42 µg/L, and was 41 µg/L in 2007–2008.

- The median level of BBzP metabolite in urine of children ages 6 to 17 years decreased from 25 μg/L in 1999–2000 to 16 μg/L in 2007–2008. This decreasing trend was statistically significant.

- At the 95th percentile, there was an increasing trend in the concentration of DEHP metabolites in children, from 387 μg/L in 1999-2000 to 564 μg/L in 2007–2008. From 1999–2000 to 2007–2008, the concentration of DBP metabolites varied between 166 μg/L and 191 μg/L, and was 191 μg/L in 2007–2008. The concentration of BBzP metabolites varied between 104 μg/L and 151 μg/L, and was 107 μg/L in 2007–2008. (See Table B10a.)

  - The increasing trend for DEHP metabolites from 1999–2000 to 2007–2008 was statistically significant.

- Among children ages 6 to 17 years, the concentration of DEHP metabolites in urine at the 95th percentile ranged from 7 to 12 times higher than the median levels presented in this graph. The concentrations of metabolites of DBP and BBzP in the 95th percentile ranged from 4 to 7 times higher than the median levels. (See Table B10 and B10a.)

- Children living below the poverty level had higher median concentrations of DBP metabolites detected in their urine compared with children living at or above the poverty level. Median concentrations of DEHP metabolites and BBzP metabolite were similar among children living below the poverty level and children living at or above the poverty level. (See Table B10b.)

  - The difference between income groups for DBP metabolites was statistically significant.

- For the years 2005–2008, Mexican-American children had lower median concentrations of all the phthalate metabolites shown here compared with White non-Hispanic children and Black non-Hispanic children. (See Table B10b.)

  - Testing for differences by race/ethnicity found that Mexican-American children had statistically significantly lower median concentrations of phthalate metabolites as follows: for DEHP and BBzP, lower than both White non-Hispanic and Black non-Hispanic children; for DBP, lower than Black non-Hispanic children, and lower than White non-Hispanic children after accounting for differences in age, sex, and income.

- Children ages 6 to 10 years had higher median levels of phthalate metabolites in their urine compared to adolescents ages 16-17 years. These differences were relatively small for DEHP metabolites and DBP metabolites but greater for BBzP metabolite. (See Table B10c.)

  - The age group differences for BBzP were statistically significant.

Bisphenol A (BPA)

Bisphenol A (BPA) is a high-volume industrial chemical used in the production of epoxy resins and polycarbonate plastics. Polycarbonate plastics may be encountered in many products, notably food and drink containers, while epoxy resins are frequently used as inner liners of metallic food and drink containers to prevent corrosion. The use of BPA in food contact materials is regulated by the U.S. Food and Drug Administration.[1,2] BPA also serves as a coating on some types of thermal paper that are often used as receipts from cash registers, automatic teller machines, and other similar devices. It is used in the polyvinyl chloride (PVC) industries as well as in metal foundries where it is used to make casts and moldings. The primary route of human exposure to BPA is believed to be through diet, when BPA migrates from food and drink containers.[3-5] Migration is more likely to occur when the container is heated or washed.[5,6] Other possible sources of BPA exposure include air, dust, water, and dental sealants.[3,5]

Biomonitoring studies demonstrate that BPA exposure is prevalent in the United States, with detectable levels of BPA present in 93% of tested urine samples.[7] Because BPA is metabolized quickly in the body,[8] the high frequency of detection indicates that exposures are occurring regularly within the U.S. population. Exposures to BPA of infants and children up to age 6 years are estimated to be greater than BPA exposures in older children and adults.[3]

Much of the scientific interest in BPA is related to published research suggesting that BPA may be an endocrine disrupting chemical.[9,10] Endocrine disruptors act by interfering with the biosynthesis, secretion, action, or metabolism of naturally occurring hormones.[9,11] Given the importance of hormones in human physiology, there is concern in the scientific community over the potential for endocrine disruptors to adversely affect children's health, particularly in reproduction, early and adolescent development, and behavior.[9] BPA is described as a "weakly estrogenic" chemical, because its affinity for binding to estrogen receptors is approximately 10,000-fold weaker than natural estrogen.[12]

Recent attention to the developmental effects of BPA is based on several laboratory studies and a better understanding of the mechanisms by which BPA exerts an estrogenic effect.[10,13-15] In animal studies, exposure to high levels of BPA during pregnancy or lactation resulted in reduced birth weight, slowed growth, reduced survival, and delayed time to the onset of puberty in offspring.[16-19] Animal studies have also found that low-dose BPA exposure was associated with insulin resistance.[20,21] In addition, one study found that low-dose BPA exposure in pregnant animals was associated with symptoms similar to gestational diabetes, suggesting that BPA exposures may have adverse effects in pregnant women.[22] Other studies have found relationships between prenatal or early-life BPA exposure and neurological effects as well as the development of breast and prostate cancer in adult animals.[23-25] The effects of low-dose exposure to BPA in lab animals are debated within the scientific community, with some researchers finding no developmental effects, while others have identified behavioral and neural effects, abnormal urinary tract development, development of lesions in the prostate gland, and early onset of puberty in females.[3,23,26-36] Differences in reported results on the timing of puberty between low and high dose studies may be a result of dose differences, study design, or species of animal.[3] Based on a critical review of the existing scientific literature, in 2008 the National Toxicology Program (NTP) determined that there was "some concern" (the midpoint on a five-level scale ranging from "negligible" to "serious")[i] for effects of BPA on the brain, behavior, and prostate gland in fetuses, infants, and children; "minimal concern" for effects on the mammary gland and onset of puberty in females; and "negligible concern" for fetal or neonatal mortality, birth defects, or reduced birth weight and growth.[3]

Epidemiological data on the effects of BPA in human populations are limited. Studies of the U.S. general population have reported that adults with higher recent BPA exposure (as represented by urinary BPA concentrations) are more likely to have coronary heart disease, diabetes, immune dysfunction, and liver enzyme abnormalities.[37-39] Some of these associations are postulated to be due to non-estrogenic effects of BPA,[38] although there is limited understanding of the mechanisms by which BPA exposure may lead to an adverse health effect. Studies of workers in China reported an association between exposure to high levels of BPA and an increased risk of self-reported sexual dysfunction,[40,41] and that BPA exposure to pregnant workers was associated with decreased offspring birthweight.[42] A study of children in Ohio reported an association between prenatal BPA exposure, at levels typical for the general population, and aggression and hyperactivity in 2-year-old children.[43] Similar associations between behavioral effects and BPA exposure have been seen in animal studies.[3,44,45] However, another study of prenatal BPA exposure conducted in New York City found no association between prenatal BPA exposure and social behavior deficits in children at ages 7 to 9 years.[46] In 2009, the National Institutes of Health announced that it would spend $30 million over two years to better understand the link between low-dose BPA exposure and human health effects.

Studies have shown that detectable levels of BPA are present in human urine samples from all age groups including infants, toddlers, children and adults.[3,47-52] BPA has been identified in the blood of pregnant women,[53] and also can cross the placenta, potentially exposing the fetus.[54] Previous studies have identified higher levels of BPA in the urine of children ages 6 to 11 years compared with adults,[47,49,50] and found that consumption of soda and school lunches was also associated with higher urinary BPA concentrations.[50] Infants and young children also have a higher estimated daily intake of BPA compared with adults.[3,48] Although less information is available on BPA levels in infants than in older children, one study found that premature infants in intensive care units had greater urinary BPA concentrations than those observed in other infants or even older children, though the route of exposure for the premature infants is unclear.[55] Some laboratory animal studies have found that younger animals are less effective at metabolizing BPA than older animals are; while it has been proposed that such findings may apply to human infants and developing fetuses, this hypothesis is debated in the scientific literature.[3,36,52,56-61] One important part of ongoing research is to better understand how BPA is absorbed, distributed, metabolized, and excreted by the body, and how those processes change with age and with route of exposure.[56-58,60-62] Interpretation of these data will allow us to

---

[i] More information on NTP concern levels is available at http://www.niehs.nih.gov/news/media/questions/sya-bpa.cfm.

understand how environmental exposure equates to the internal dose routinely measured in biomonitoring studies.

The two indicators that follow use the best nationally representative data currently available on urinary BPA levels over time for women of child-bearing age and children. Indicator B11 presents median and 95th percentile concentrations of BPA in urine for women ages 16 to 49 years. Indicator B12 presents median and 95th percentile concentrations of BPA in urine for children ages 6 to 17 years.

Indicator B11: Bisphenol A in women ages 16 to 49 years: Median and 95th percentile concentrations in urine, 2003–2010

Indicator B12: Bisphenol A in children ages 6 to 17 years: Median and 95th percentile concentrations in urine, 2003–2010

About the Indicators: Indicators B11 and B12 present concentrations of bisphenol A (BPA) in urine of U.S. women ages 16 to 49 years and children ages 6 to 17 years. The data are from a national survey that collects urine specimens from a representative sample of the population every two years, and then measures the concentration of total BPA in the urine. Indicator B11 presents concentrations of BPA in women's urine over time and Indicator B12 presents concentrations of BPA in children's urine over time. The focus on both women of child-bearing age and children is based on concern for potential adverse effects in children born to women who have been exposed to BPA and in children exposed to BPA.

NHANES

The National Health and Nutrition Examination Survey (NHANES) provides nationally representative biomonitoring data for BPA. NHANES is designed to assess the health and nutritional status of the civilian noninstitutionalized U.S. population and is conducted by the National Center for Health Statistics, part of the Centers for Disease Control and Prevention (CDC). Interviews and physical examinations are conducted with approximately 10,000 people in each two-year survey cycle. CDC's National Center for Environmental Health measures concentrations of environmental chemicals in blood and urine samples collected from NHANES participants. Summaries of the measured values for more than 200 chemicals are provided in the *Fourth National Report on Human Exposure to Environmental Chemicals*.[63]

Bisphenol A and its Metabolites

Indicators B11 and B12 present urinary levels of BPA in women of child-bearing age and children. The reported measurements of BPA in urine represent "total BPA," which includes both free BPA and non-estrogenic metabolites of BPA (only free BPA is considered active based on measures of estrogenicity). Measured levels in the U.S. population may be composed predominantly of these metabolites,[64] but total BPA levels reflect previous exposure to the biologically active form of BPA and there is debate in the scientific community over the potential for conversion of non-estrogenic metabolites back to free BPA in various tissues.[65] Recent work has also highlighted the potential for conversion of non-estrogenic metabolites of BPA to the active form when crossing the placenta, increasing the relevance of total BPA measurements to children's health.[54,66] All values are reported as micrograms of BPA per liter of urine (µg/L).

Concentrations of BPA in urine have been measured in a representative subset of NHANES participants ages 6 years and older beginning with the 2003–2004 survey cycle. In 2009–2010,

NHANES collected BPA biomonitoring data for 2,749 individuals ages 6 years and older, including 608 women ages 16 to 49 years and 727 children ages 6 to 17 years. BPA was detected in about 90% of all individuals sampled. The frequency of BPA detection was 92% in women ages 16 to 49 years,[ii] and 92% in children ages 6 to 17 years. The median and 95th percentile BPA levels in urine for all NHANES participants in 2009–2010 were 2 μg/L and 10 μg/L, respectively. The widespread detection of BPA, combined with the fact that BPA has a short half-life, indicates that BPA exposure is widespread and relatively continuous.

Individual Variability in Urinary Measurements

NHANES data for BPA are based on measurements made using a single urine sample for each person surveyed. Due to normal changes in an individual's urinary output throughout the day, this variability in urinary volume, among other factors related to the measurement of chemicals that do not accumulate in the body,[67] may mask differences between individuals in levels of BPA. Since BPA does not appear to accumulate in bodily tissues, the distribution of NHANES urinary BPA levels may overestimate high-end exposures (e.g., at the 95th percentile) as a result of collecting one-time urine samples.[8,68,69] Many studies account for differences in hydration levels by reporting the chemical concentration per gram of creatinine. Creatinine is a byproduct of muscle metabolism that is excreted in urine at a relatively constant rate, independent of the volume of urine, and can in some circumstances partially account for the measurement variability due to changes in urinary output.[70] However, urinary creatinine concentrations differ significantly among different demographic groups, and are strongly associated with an individual's muscle mass, age, sex, diet, health status (specifically renal function), body mass index, and pregnancy status.[71,72] Thus, these indicators present the unadjusted BPA concentrations so that any observed differences in concentrations between demographic groups are not due to differences in creatinine excretion rates. These unadjusted urinary levels from a single sample may either over- or underestimate urinary levels for a sampled individual. However, for a representative group, it can be expected that a median value based on single samples taken throughout the day will provide a good approximation of the median for that group. Furthermore, due to the large number of subjects surveyed, we expect that differences in the concentrations of BPA that might be attributed to the volume of the urine sample would average out within and across the various comparison groups.

Birth Rate Adjustment

Indicator B11 uses measurements of BPA in urine of women ages 16 to 49 years to represent the distribution of BPA exposures to women who are pregnant or may become pregnant. However, women of different ages have a different likelihood of giving birth. For example, in 2003–2004, women aged 27 had a 12% probability of giving birth, and women aged 37 had a 4% probability of giving birth.[73] A birth rate-adjusted distribution of women's BPA levels is used

---

[ii] The percentage for women ages 16 to 49 years is calculated with the birth rate adjustment described below.

in calculating this indicator,[iii] meaning that the data are weighted using the age-specific probability of a woman giving birth.[74]

Data Presented in the Indicators

Indicators B11 presents median and 95th percentile concentrations of BPA in urine over time for women ages 16 to 49 years, using NHANES data from 2003–2010.

Indicator B12 presents median and 95th percentile concentrations of BPA in urine over time for children ages 6 to 17 years, using NHANES data from 2003–2010.

Additional information showing how the median and 95th percentile levels of BPA in urine vary by race/ethnicity and family income for women ages 16 to 49 years is presented in supplemental data tables for these indicators. Data tables also display information showing how the median and 95th percentile levels of BPA in urine vary by race/ethnicity, family income, and age for children ages 6 to 17 years.

Please see the Introduction to the Biomonitoring section for an explanation of the terms "median" and "95th percentile," a description of the race/ethnicity and income groups used in the ACE3 biomonitoring indicators, and information on the statistical significance testing applied to these indicators.

[iii] There may be multiple ways to implement an adjustment to the data that accounts for birth rates by age. The National Center for Health Statistics has not fully evaluated the method used in ACE, or any other method intended to accomplish the same purpose, and has not used any such method in its publications. NCHS and EPA are working together to further evaluate the birth rate adjustment method used in ACE and alternative methods.

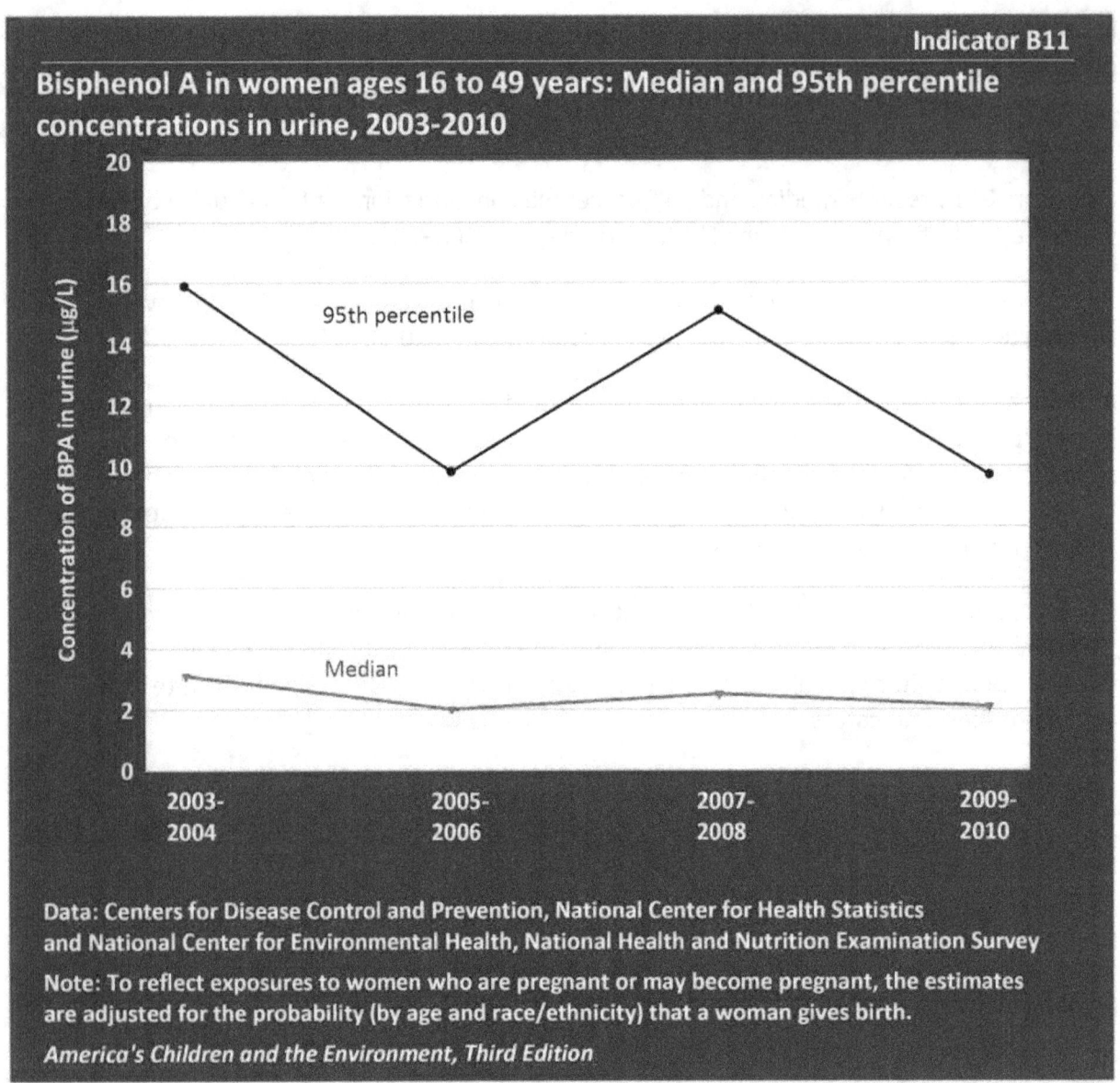

Indicator B11

Bisphenol A in women ages 16 to 49 years: Median and 95th percentile concentrations in urine, 2003-2010

Data: Centers for Disease Control and Prevention, National Center for Health Statistics and National Center for Environmental Health, National Health and Nutrition Examination Survey

Note: To reflect exposures to women who are pregnant or may become pregnant, the estimates are adjusted for the probability (by age and race/ethnicity) that a woman gives birth.

America's Children and the Environment, Third Edition

Data characterization

–   Data for this indicator are obtained from an ongoing continuous survey conducted by the National Center for Health Statistics.
–   Survey data are representative of the U.S. civilian noninstitutionalized population.
–   BPA is measured in urine samples obtained from individual survey participants.

■   From 2003–2004 to 2009–2010, the median concentration of BPA in urine among women ages 16 to 49 years varied between 2 µg/L and 3 µg/L. There was no statistically significant trend in median BPA levels over the years shown.

■   From 2003–2004 to 2009–2010, the concentrations of BPA in urine at the 95th percentile varied between 10 µg/L and 16 µg/L, and was 10 µg/L in 2009–2010. There was no statistically significant trend in 95th percentile concentrations of BPA over the years shown.

- Between 2003–2004 and 2009–2010, the concentrations of BPA in the 95th percentile ranged from 5 to 6 times the median levels for women ages 16 to 49 years.

- In 2007–2010, the median concentration of BPA in urine of Black non-Hispanic women was about 4 µg/L, which was higher than the median concentrations in White non-Hispanic women, Mexican-American women, and women of "All Other Races/Ethnicities." The differences between Black non-Hispanic women and women in other race/ethnicity groups were statistically significant. (See Table B11a.)

- Women living below the poverty level had higher median concentrations of BPA in urine than those living at or above poverty level, a difference that was statistically significant. (See Table B11a.)

- Among White non-Hispanic women and women of "All Other Races/Ethnicities," those with family incomes below poverty level had higher median concentrations of BPA in urine than those at or above poverty level. The differences between the income groups were statistically significant. (See Table B11a.)

- Higher concentrations of BPA were observed in the urine of women below the poverty level at the 95th percentile (15 µg/L) compared with women at or above the poverty level (11 µg/L). This difference was statistically significant after adjustment for differences in age and race/ethnicity. (See Table B11b.)

- White non-Hispanic women at the 95th percentile (10 µg/L) had lower concentrations of BPA in urine than Black non-Hispanic women (15 µg/L) and Mexican-American women (15 µg/L). These differences were statistically significant after adjustment for differences in age and income. (See Table B11b.)

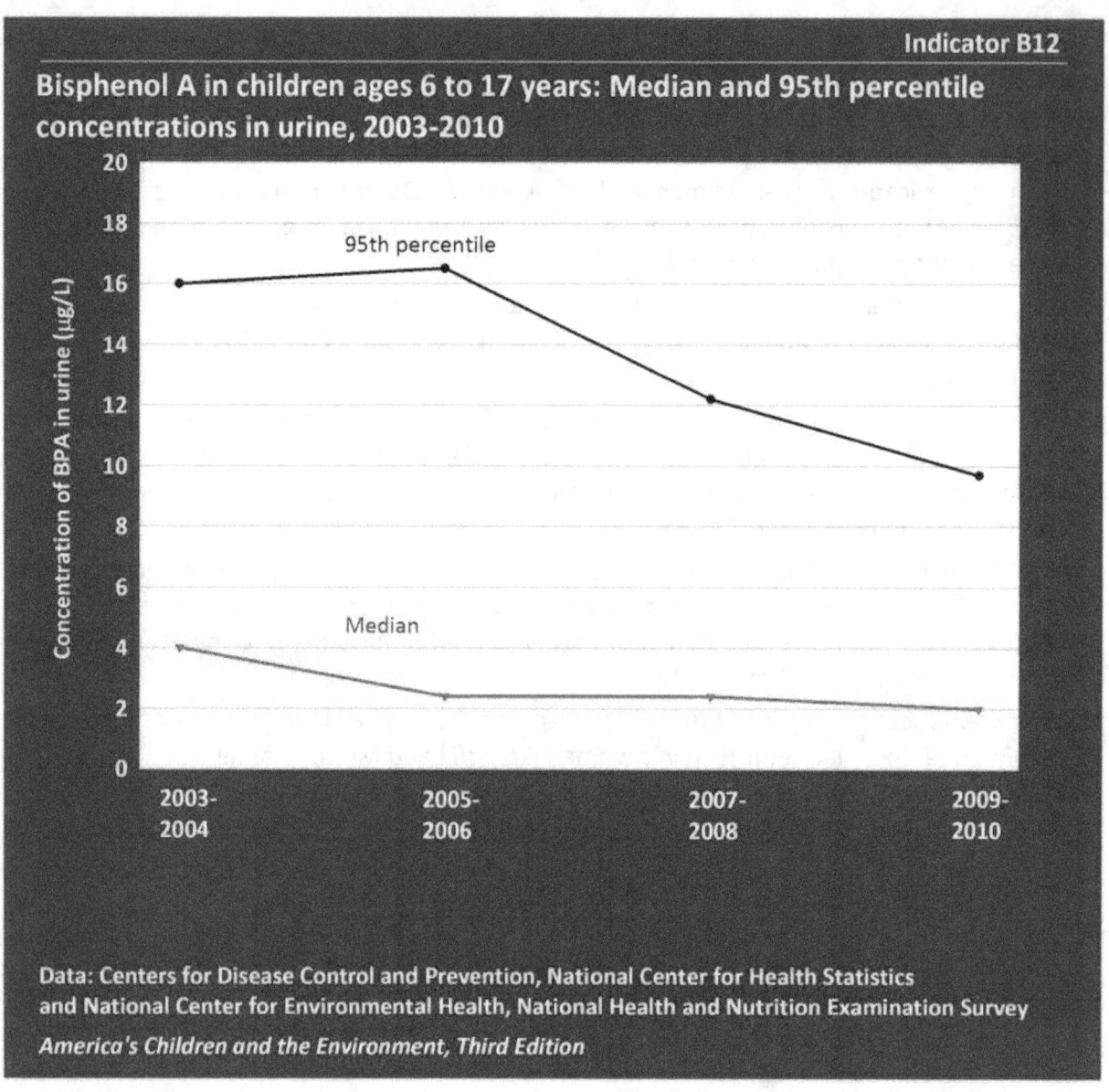

Indicator B12

Bisphenol A in children ages 6 to 17 years: Median and 95th percentile concentrations in urine, 2003-2010

Data: Centers for Disease Control and Prevention, National Center for Health Statistics and National Center for Environmental Health, National Health and Nutrition Examination Survey

America's Children and the Environment, Third Edition

Data characterization
- Data for this indicator are obtained from an ongoing continuous survey conducted by the National Center for Health Statistics.
- Survey data are representative of the U.S. civilian noninstitutionalized population.
- BPA is measured in urine samples obtained from individual survey participants.

- Among children ages 6 to 17 years, the median concentration of BPA in urine of children ages 6 to 17 years decreased from 4 µg/L in 2003–2004 to 2 µg/L in 2009–2010. The concentration of BPA in urine at the 95th percentile decreased from 16 µg/L in 2003–2004 to 10 µg/L in 2009–2010. These decreasing trends were statistically significant.

- Between 2003–2004 and 2009–2010, the concentrations of BPA in the 95th percentile ranged from 4 to 7 times the median levels for children ages 6 to 17 years.

- In 2007–2010, median concentrations of BPA in urine of Black non-Hispanic children ages 6 to 17 years were higher than in White non-Hispanic children, Mexican-American children, and children of "All Other Races/Ethnicities." These differences were statistically significant. (See Table B12a.)

- BPA concentrations at the 95th percentile were similar for Black non-Hispanic, White-non Hispanic, and Mexican-American children ages 6 to 17 years in 2007–2010. (See Table B12b.)

- In 2007–2010, BPA concentrations were similar for age groups 6 to 10 years, 11 to 15 years, and 16 to 17 years, both at the median and at the 95th percentile. (See Table B12c.)

Perchlorate

Perchlorate is a naturally occurring and man-made chemical that is used to manufacture fireworks, explosives, flares, and rocket fuel.[1,2] It is found naturally in groundwater and soils throughout many regions in the United States and other arid regions of the world.[3,4] Perchlorate is presumed to migrate into groundwater during the process of irrigation,[5] and has also been found in groundwater supplies near military and industrial facilities where perchlorate was used.[6] Perchlorate has been detected in surface water; dairy products; and in some food crops, including lettuce, spinach, grapes, carrots, tomatoes, and other fruits and vegetables, produced in the United States and internationally.[5,7-12] Perchlorate has been detected in some fertilizers produced in Chile; however, fertilizers appear to be a negligible source of perchlorate in the United States.[1,13-17] The numerous sources of perchlorate located across the United States result in widespread exposures of perchlorate to the U.S. population.[3,4]

Perchlorate has been detected in human breast milk, urine, blood, amniotic fluid, and saliva.[18-23] A national study representative of the U.S. population ages 6 years and older found perchlorate in the urine of 100% of the more than 5,000 people sampled; children had higher median urinary levels compared with those of adults, including women of child-bearing age.[3,24] Infants are exposed to perchlorate through both breast milk and formula, but those who are fed breast milk may have higher exposures to perchlorate compared with those who are fed cow- or soy-based formula.[25] When comparing perchlorate doses (daily intakes per kilogram of body weight, estimated from urine samples), infants less than 2 months of age experience higher perchlorate doses compared with older infants, and estimated doses of perchlorate in infants are more than twice as high as estimated doses for adults.[3,25,26] A study conducted in China found that blood samples of infants less than 1 year of age have higher mean perchlorate values than blood samples of both older children and adults.[27,28]

Children might be directly exposed to perchlorate through perchlorate-contaminated water and foods containing perchlorate. Surveys conducted by the U.S. Food and Drug Administration have detected varying levels of perchlorate in many foods that may be consumed by both women and young children. The surveys, conducted in 22 states, tested 27 different types of food products and found the highest levels of perchlorate in spinach and tomatoes.[12] Some infant formulas have also been found to contain perchlorate, and the perchlorate content of the formula is increased if it is prepared with perchlorate-contaminated water.[29-32] However, computer modeling studies have concluded that exposure to perchlorate from food consumption is much greater than exposure from drinking water in the United States.[33,34] These modeled predictions are consistent with empirical studies that attribute the majority of perchlorate intake dose in U.S. residents to food consumption.[8,35-37]

Exposure to high doses of perchlorate has been shown to block the uptake of iodide into the thyroid gland.[38,39] Exposure to perchlorate and other thyroid-disrupting chemicals is of particular concern for women of child-bearing age, because thyroid hormones are important for growth and development of the central nervous system in fetuses and infants.[1,40-42] The transfer of iodide from blood into the thyroid gland is an essential step in the synthesis of thyroid hormones that regulate how the body uses energy; influence bone growth; and influence the development of the brain, reproductive, and cardiovascular systems.[43] When this transfer of iodide into the thyroid gland is blocked, the thyroid may not have enough iodide to make thyroid hormones. Reduction in a woman's thyroid hormone levels during the first and second trimester puts the fetus at risk for impaired physical and mental development, with the severity of the impairment depending upon the degree of hormone deficiency.[40,41] Moderate deficits in maternal thyroid hormone levels during early pregnancy have been linked to reduced childhood IQ scores and other neurodevelopmental effects, as well as unsuccessful or complicated pregnancies.[44] Prenatal and newborn hypothyroidism (low thyroid hormone levels) is a risk factor for intellectual disability (mental retardation) and other forms of impaired neurodevelopment.[45] In 2005–2008, approximately 38% of women ages 15 to 44 years in the United States had insufficient iodine intake,[46] potentially increasing the risk for effects on fetal development from exposure to perchlorate.[1]

Associations between perchlorate exposure and thyroid hormones have been based on both epidemiological and animal-based studies. Animal studies have shown that exposure to high doses of perchlorate result in decreased thyroid hormone levels and physical alterations to the thyroid gland,[1] and have also found that these effects of perchlorate can be enhanced with exposure to other chemicals that block uptake of iodide.[47] In 2005, the National Research Council (NRC) concluded that the available epidemiological evidence concerning non-medical exposure to perchlorate did not indicate an association with thyroid disorders in adults or infants, and was inadequate for assessing the potential for adverse associations between prenatal perchlorate exposure and adverse neurodevelopmental outcomes in children.[1] The NRC also indicated that there was a lack of studies to evaluate potential effects of prenatal perchlorate exposures in infants and children, particularly in vulnerable populations.[1]

Some further epidemiological research has been conducted since the NRC report was completed. A study of urinary perchlorate and thyroid hormone levels in more than 11,000 U.S. females ages 12 years and older in 2001–2002 found that increasing levels of perchlorate in urine were associated with decreased thyroid hormone levels.[26] Further analysis of this data set found that tobacco smoke and perchlorate may interact to affect thyroid function at commonly occurring perchlorate levels.[48] In contrast, a study of first-trimester pregnant women identified as iodine-deficient, and a long-term exposure study of women in early pregnancy and late pregnancy in Chile, found that exposure to low levels of perchlorate did not result in decreased levels of thyroid hormones.[49,50]

Other studies have evaluated relationships between drinking water perchlorate levels and thyroid hormone levels in newborns. A study of California infants born in 1998 reported that babies born to mothers in communities with higher drinking water perchlorate levels were more likely to have elevated levels of thyroid stimulating hormone, which is an indication of reduced thyroid hormone levels.[51] An earlier study of the same population and other studies have not found associations between drinking water perchlorate levels and neonatal thyroid hormone function.[50,52-56]

In January 2009, EPA issued an interim health advisory level to help state and local officials manage local perchlorate contamination issues in a health-protective manner, in advance of a final EPA regulatory determination.[2,57] In February 2011, EPA decided to develop a federal drinking water standard for perchlorate, based on the concern for effects on thyroid hormones and the development and growth of fetuses, infants, and children.[2,58] The process for developing the standard will include receiving input from key stakeholders as well as submitting any formal rule to a public comment process. California and Massachusetts have both set their own standards for perchlorate in drinking water.[59] No standards exist for perchlorate in food.

The indicator that follows uses the best nationally representative data currently available on urine perchlorate levels over time for women of child-bearing age. Indicator B13 presents median and 95th percentile urinary perchlorate levels for women ages 16 to 49 years.

About the Indicators: Indicator B13 presents concentrations of perchlorate in urine of U.S. women ages 16 to 49 years. The data are from a national survey that collects urine specimens from a representative sample of the population every two years, and then measures the concentration of perchlorate in the urine. The indicator presents concentrations of perchlorate in urine over time. The focus on women of child-bearing age is based on concern for potential adverse effects in children born to women who have been exposed to perchlorate.

NHANES

The National Health and Nutrition Examination Survey (NHANES) provides nationally representative biomonitoring data for perchlorate. NHANES is designed to assess the health and nutritional status of the civilian noninstitutionalized U.S. population and is conducted by the National Center for Health Statistics, part of the Centers for Disease Control and Prevention (CDC). Interviews and physical examinations are conducted with approximately 10,000 people in each two-year cycle. CDC's National Center for Environmental Health measures concentrations of environmental chemicals in blood and urine samples collected from NHANES participants. Summaries of the measured values for more than 200 chemicals are provided in the *Fourth National Report on Human Exposure to Environmental Chemicals.*[24]

Perchlorate

Indicator B13 presents urinary levels of perchlorate in women of child-bearing age. Perchlorate passes quickly through the body unchanged and is excreted in urine, with an elimination half-life on the order of hours.[3] Therefore, perchlorate measured in humans is indicative of recent exposures. All values are reported as micrograms of perchlorate per liter of urine (µg/L).

Concentrations of perchlorate in urine have been measured in a representative subset of NHANES participants ages 6 years and older in 2001–2002 and 2003–2004, and in all NHANES participants ages 6 years and older in 2005–2006 and 2007–2008.[19]

For 2007-2008, NHANES collected perchlorate biomonitoring data for 7,629 individuals ages 6 years and older, including 1,608 women ages 16 to 49 years. Perchlorate was detected in 100% of the individuals sampled in NHANES 2007-2008. The median and 95th percentile of urinary perchlorate levels for all NHANES participants in 2007-2008 were 4 µg/L and 17 µg/L, respectively. The widespread detection of perchlorate, combined with the fact that perchlorate has a short half-life, indicates that perchlorate exposure is widespread and relatively continuous.

Individual Variability in Urinary Measurements

NHANES data for perchlorate are based on measurements made using a single urine sample for each person surveyed. Due to normal changes in an individual's urinary output throughout the day, this variability in urinary volume, among other factors related to the measurement of chemicals that do not accumulate in the body,[60] may mask differences between individuals in levels of perchlorate. Since perchlorate does not appear to accumulate in bodily tissues, the distribution of NHANES urinary perchlorate levels may over estimate high-end exposures (e.g., at the 95th percentile) as a result of collecting one-time urine samples.[34,61,62] Many studies account for differences in hydration levels by reporting the chemical concentration per gram of creatinine. Creatinine is a byproduct of muscle metabolism that is excreted in urine at a relatively constant rate, independent of the volume of urine, and can in some circumstances partially account for the measurement variability due to changes in urinary output.[63] However, urinary creatinine concentrations differ significantly among different demographic groups, and are strongly associated with an individual's muscle mass, age, sex, diet, health status (specifically renal function), body mass index, and pregnancy status.[64,65] Thus, this indicator presents the unadjusted perchlorate concentrations so that any observed differences in concentrations between demographic groups are not due to differences in creatinine excretion rates. These unadjusted urinary levels from a single sample may either over- or underestimate urinary levels for a sampled individual. However, for a representative group, it can be expected that a median value based on single samples taken throughout the day will provide a good approximation of the median for that group. Furthermore, due to the large number of subjects surveyed, we expect that differences in the concentrations of perchlorate that might be attributed to the volume of the urine sample would average out within and across the various comparison groups.

Birth Rate Adjustment

Indicator B13 uses measurements of perchlorate in urine of women ages 16 to 49 years to reflect the potential distribution of perchlorate exposures to women who are pregnant or may become pregnant. However, women of different ages have a different likelihood of giving birth. For example, in 2003–2004, women aged 27 years had a 12% annual probability of giving birth, and women aged 37 years had a 4% annual probability of giving birth.[66] A birth rate-adjusted distribution of women's perchlorate levels is used in calculating this indicator,[i] meaning that the data are weighted using the age-specific probability of a woman giving birth.[67]

---

[i] There may be multiple ways to implement an adjustment to the data that accounts for birth rates by age. The National Center for Health Statistics has not fully evaluated the method used in ACE, or any other method intended to accomplish the same purpose, and has not used any such method in its publications. NCHS and EPA are working together to further evaluate the birth rate adjustment method used in ACE and alternative methods.

Data Presented in the Indicator

Indicator B13 presents median and 95th percentile concentrations of perchlorate in urine over time for women ages 16 to 49 years, using NHANES data from 2001–2008.

Additional information showing how the median and 95th percentile levels of perchlorate in urine vary by race/ethnicity and family income for women ages 16 to 49 years is presented in supplemental data tables for these indicators. Data tables also display information on the median and 95th percentile levels of perchlorate in urine for children ages 6 to 17 years, including how levels vary by race/ethnicity, family income, and age.

NHANES does not collect urine samples from children less than 6 years of age, and thus cannot assess the exposure of infants, who may be exposed to unhealthy levels of perchlorate due to the presence of perchlorate in breast milk and some infant formula.[18,20,21,31,32]

Please see the Introduction to the Biomonitoring section for an explanation of the terms "median" and "95th percentile," a description of the race/ethnicity and income groups used in the ACE3 biomonitoring indicators, and information on the statistical significance testing applied to these indicators.

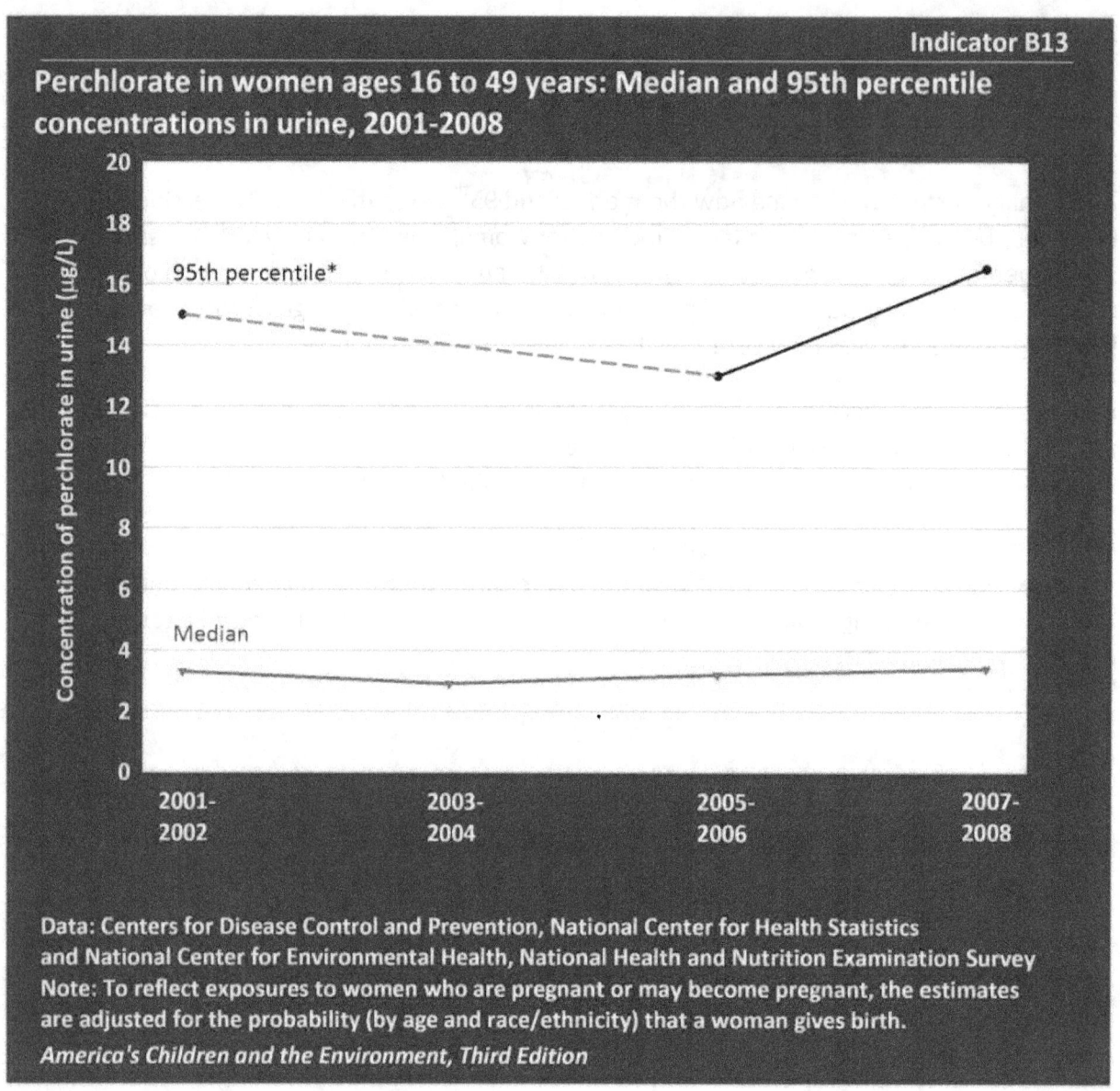

Indicator B13

Perchlorate in women ages 16 to 49 years: Median and 95th percentile concentrations in urine, 2001-2008

Data: Centers for Disease Control and Prevention, National Center for Health Statistics and National Center for Environmental Health, National Health and Nutrition Examination Survey
Note: To reflect exposures to women who are pregnant or may become pregnant, the estimates are adjusted for the probability (by age and race/ethnicity) that a woman gives birth.
America's Children and the Environment, Third Edition

*The 95th percentile concentration for 2003-2004 is not reported because it has large uncertainty: the relative standard error, RSE, is 40% or greater (RSE = standard error divided by the estimate).

Data characterization
- Data for this indicator are obtained from an ongoing continuous survey conducted by the National Center for Health Statistics.
- Survey data are representative of the U.S. civilian noninstitutionalized population.
- Perchlorate is measured in urine samples obtained from individual survey participants.

- From 2001–2002 to 2007–2008, the median level of perchlorate in urine among women ages 16 to 49 years was 3 µg/L with little variation over time. Over the same period, the 95th percentile varied between 13 and 17 µg/L.

- In 2005–2008, there was little variation in median or 95th percentile perchlorate levels by race/ethnicity or income among women ages 16 to 49 years. (See Tables B13a and B13b.)
- From 2001–2002 to 2007–2008, the median level of perchlorate among children ages 6 to 17 years was 5 µg/L with little variation over time. The 95th percentile perchlorate level among children increased from 15 µg/L in 2001–2002 to 19 µg/L in 2007–2008. (See Table B13c.)
  - The increasing trend in children's 95th percentile perchlorate levels was statistically significant.
- The median perchlorate level among children ages 6 to 17 years was about 42% higher than the level found in women of childbearing age in 2005–2008, while the 95th percentile level among children ages 6 to 17 years was about 19% higher than in women of childbearing age. (See Tables B13 and B13c.)
- Differences in urinary perchlorate levels by race/ethnicity and income among children ages 6 to 17 years were relatively limited. (See Tables B13d and B13e.)
- There were minimal differences in urinary perchlorate levels by age group among children ages 6 to 17 years. (See Table B13f.)

Health

Introduction

Why is EPA tracking children's health outcomes in *America's Children and the Environment?*

The central goal of efforts to protect children's environmental health is the reduction of disease, disability, and mortality. Many different factors contribute to children's health, including nutrition, prenatal and childhood exposure to toxins in the environment, genetics, socioeconomic status, access to medical care, and exercise. Data on children's health outcomes can provide important information about changes over time and differences between demographic groups. In particular, monitoring children's health outcomes for which causes are unknown or not well established can help stimulate hypotheses, some of which may point to environmental factors, which then can be examined rigorously in future studies.

What health outcomes are included in *America's Children and the Environment, Third Edition (ACE3)?*

Health outcomes were selected for ACE3 based on: (1) magnitude of prevalence and/or trend in prevalence, and severity of health outcome; (2) research findings that indicate environmental contaminants or characteristics may be contributing factors; and (3) the availability of nationally representative data suitable for constructing an indicator. EPA obtained input from its Children's Health Protection Advisory Committee to assist in selecting topics from among the many diseases and health disorders that affect children. The ACE3 Health indicators address the following topics:

- Respiratory diseases
- Childhood cancer
- Neurodevelopmental disorders
- Obesity
- Adverse birth outcomes

What data sources were used to develop the Health indicators?

Data for all of the selected health outcomes, with the exception of childhood cancer, were based on surveys and registries conducted/maintained by the National Center for Health Statistics (NCHS). These include the National Health Interview Survey (NHIS), National Hospital Ambulatory Medical Care Survey (NHAMCS), National Hospital Discharge Survey (NHDS), National Health and Nutrition Examination Survey (NHANES), and the National Vital Statistics System (NVSS). Data on childhood cancer were obtained from the National Cancer Institute's Surveillance, Epidemiology, and End Results (SEER) Program.

NHIS and NHANES collect health information from a probability sample of the U.S. civilian, noninstitutionalized population, and survey data are weighted to yield national estimates describing this population. NHAMCS and NHDS collect patient visit information from a sample of hospitals, and the survey data are weighted to estimate the rates of respiratory-related emergency room visits and hospital admissions in the United States. NVSS is a registry that captures virtually all births that occur in the United States, and thus does not rely on sampling. The SEER program has attributes of both a survey and a registry. It is based on a collection of registries located across the United States that record all tumors that occur in specific geographical regions. The registry information is then used to estimate the occurrence of cancer, including childhood cancer, for the entire country.

What can we learn from the Health indicators?

The indicators presented in this report focus on health outcome data collected over multiple years, which allow determination of whether the reported prevalence or rate of each outcome is increasing, decreasing, or not changing over time. An additional focus is whether particular groups (defined by race/ethnicity and income) within the population are disproportionately affected by a given health outcome. Such trends and comparisons can generate hypotheses and help identify opportunities for future action.

The topic text provided before the indicators reviews the scientific evidence regarding environmental factors and other factors contributing to the disease or disorder, providing context that informs the interpretation of the indicators. For some of the selected health outcomes, the scientific evidence suggests that environmental contaminants may play a role in the development of the disease or disorder. For other health outcomes, available evidence is less clear as to whether environmental contaminants are involved. The inclusion of the selected health outcomes in this report does not imply that environmental contaminants or other environmental factors definitely play a role in the selected health outcomes. It can be very difficult to develop conclusive evidence that environmental factors cause or contribute to the incidence of childhood health effects, and research is ongoing. Where available, we rely on authoritative reviews of the literature and report their conclusions regarding the strength of the evidence for a causal role of specific environmental factors in the development of childhood diseases and disorders. When such reviews are unavailable, we summarize important findings from individual studies that address the potential role of environmental factors in contributing to an effect.

Furthermore, the inclusion of the selected health outcomes in this report does not imply that environmental factors, in cases where they do play a role, are the only cause of the disease or disorder. Most often, health outcomes are a result of multiple causes that may include genetics, nutrition, and socioeconomic factors, as well as prenatal and childhood exposure to environmental contaminants, and other environmental factors. The various factors may also interact, such as a genetic predisposition that makes a person more susceptible to the effects of an environmental exposure.

In some instances, the indicators show that the prevalence of a health outcome is increasing while important environmental exposures are decreasing. Although this could suggest that the environmental exposures addressed in the indicators are unrelated to the health outcomes being measured, it could also result from a lag between environmental improvements and changes in related health outcomes, or changes in other important environmental exposures that are not currently measured by the indicators in the report. The Health indicators are therefore not intended as a basis for concluding that an environmental factor is or is not related to a particular children's health outcome.

What information is provided for each Health topic?

An introduction section explains the relevance of the topic to children's environmental health, including a description of the health outcomes and a discussion of evidence indicating or suggesting that environmental agents may play a role in contributing to the outcomes. The introduction is followed by a description of the indicators, including a summary of the data available and brief information on how each indicator was calculated. Two to four indicators, each a graphical presentation of the available data, are included for each topic. All Health topics include an indicator that presents a time series. Some of the topics also include indicators that show a comparison of the most current health outcome data by race/ethnicity and income level. Beneath each figure are explanatory bullet points describing dataset characteristics and key findings presented in the figure, along with key data from any supplemental data tables. References are provided for each topic at the end of the report.

Data tables are provided in Appendix A. The tables include all indicator values depicted in the indicator figures, along with additional data of interest not shown in the figures. Metadata describing the data sources are provided in Appendix B. Documents providing details of how the indicators were calculated are available on the ACE website (www.epa.gov/ace).

Many of the topics presented in the Health indicators are addressed in Healthy People 2020, which provides science-based, 10-year national objectives for improving the health of all Americans. Appendix C provides examples of the alignment of the Health topics presented in ACE3 with objectives in Healthy People 2020.

What race/ethnicity groups are used in reporting indicator values?

For each topic in the Health section, indicator values are provided for defined race/ethnicity groups—either in the indicator figures or in the data tables. The race/ethnicity groups vary to some extent across the indicators, depending on the extent to which the data support reporting indicator values for specific race/ethnicity groups. Where possible, the Health indicators provide data for the following races/ethnicities:

- White non-Hispanic
- Black non-Hispanic

- Hispanic[i]
- Asian or Pacific Islander
- American Indian or Alaska Native

Specific races/ethnicities for which the data support stratified group analysis are described in the text preceding each Health indicator.

What income groups are used in reporting indicator values?

For indicators presenting the prevalence of asthma, neurodevelopmental disorders, and obesity, indicator values are presented for income groups defined on the basis of the federal poverty level. Poverty level is defined by the federal government, and is based on income thresholds that vary by family size and composition. In 2010, for example, the poverty threshold was $22,113 for a household with two adults and two related children.[1] These Health indicators (in figures and/or data tables) provide data separately for individuals in families with incomes below poverty level, and those in families with incomes at or above poverty level.

Further detail is provided in the data tables by dividing those above poverty level into two groups: 100–200% of poverty level, and greater than 200% of poverty level. The category of incomes between the poverty level and twice the poverty level (sometimes referred to as "near poor") represents households that have relatively low incomes but are not below the officially defined poverty level, and is frequently used by NCHS in its reporting of health data.

For indicators of respiratory emergency room visits and hospitalization, childhood cancer, and adverse birth outcomes, no income group comparisons can be provided because income data are not collected for these data sets.

How were the indicators calculated and presented?

Data files: The indicators were calculated from publicly available data files obtained from the NCHS and SEER websites. The files include various information such as survey responses (NHIS), diagnosis codes (NHAMCS and NHDS), type of cancer (SEER), gestational age and birth weight (NVSS), and body measurements (NHANES). Depending on the data set, the files may also include information on age, sex, race/ethnicity, and income level (that is, family income above or below poverty level). For the survey data, each individual observation also has a sample weight that is used in calculating population statistics; the weight equals the number of people in the U.S. population represented by the particular observation.

---

[i] For the Obesity indicators, values are provided for "Mexican-American" ethnicity rather than "Hispanic" ethnicity because in all years up to 2006, NHANES was designed to provide statistically reliable estimates for Mexican-Americans rather than all Hispanics. NHANES now oversamples Hispanics instead of Mexican-Americans, beginning with NHANES 2007–2008.
Please see http://www.cdc.gov/nchs/nhanes/nhanes2007-2008/sampling_0708.htm/.

Population age groups: The age groups covered by the indicators differ among the Health topics. The indicators for respiratory diseases used data for children ages 17 years and younger. The indicators for childhood cancer used data for children ages 19 years and younger. The indicators for neurodevelopmental disorders used data for children ages 5 to 17 years. The indicators for adverse birth outcomes used data ascertained at birth. The indicators for obesity used data for children ages 2 to 17 years.

Calculated prevalence or rate of occurrence for each health outcome: Depending on the nature of the available data, some of the indicators present prevalence data while others express the occurrence of health outcomes as a rate. The main difference between these two measures is that prevalence presents data occurring at one point in time. These prevalence measures are proportions, such as the percentage of children who currently have asthma or the percentage of children classified as obese. Rates, on the other hand, express the number of events, such as emergency room visits, hospital admissions, new cancer cases, or cancer deaths, that occur over a definite time period (one year for all of the ACE3 indicators), per the population at risk for the event. The population at risk is either all children or all births, depending on the indicator.

Statistical considerations in presenting and characterizing the indicators: Statistical analysis has been applied to the ACE3 Health indicators to evaluate trends over time in indicator values (for example, percentage of children with asthma) or differences in indicator values between demographic groups. These analyses use a 5% significance level, meaning that a conclusion of statistical significance is made only when there is no more than a 5% probability that the observed trend or difference occurred by chance ($p \leq 0.05$).

The statistical analysis of trends over time for an ACE3 Health indicator is dependent on how the indicator values vary over time, the number of survey years included in the analysis, the number of observations in each survey year, and various aspects of the survey design. The evaluation of trends over time incorporates annual data from each year within the time period reported. A finding of statistical significance for differences in indicator values between demographic groups depends on the magnitude of the difference, the number of observations in each group, and various aspects of the survey design. For example, if the prevalence of a health effect is different between two groups, the statistical test is more likely to detect a difference when data have been obtained from a larger number of people in those groups. It should be noted that when statistical testing is conducted for differences among multiple demographic groups (for example, considering both race/ethnicity and income level), the large number of comparisons involved increases the probability that some differences identified as statistically significant may actually have occurred by chance.

A finding of statistical significance is useful for determining that an observed trend or difference was unlikely to have occurred by chance. However, a determination of statistical significance by itself does not convey information about the magnitude of the increase, decrease, or difference in indicator values. Furthermore, a lack of statistical significance means only that occurrence by chance cannot be ruled out. Thus, a conclusion about statistical significance is only part of the information that should be considered when determining the public health implications of trends or differences in indicator values.

In some cases, calculated indicator values have substantial uncertainty. Uncertainty in these estimates is assessed by looking at the relative standard error (RSE), a measure of how large the variability of the estimate is in relation to the estimate (RSE = standard error divided by the estimate).[ii] The estimate should be interpreted with caution if the RSE is at least 30%; a notation is provided for such estimates in the indicator figures and tables. If the RSE is greater than 40%, the estimate is considered to have very large uncertainty and is not reported.[iii]

[ii] Standard errors for all Health indicator values are provided in a file available on the ACE website (www.epa.gov/ace).
[iii] For respiratory emergency room and hospital visits (Indicator H3), an estimate is also considered to have very large uncertainty and is not reported if it is based on fewer than 30 sampled visits. For obesity (Indicators H10 and H11), values are not reported if the RSE cannot be reliably estimated.

Respiratory Diseases

Respiratory diseases and illness, such as asthma, bronchitis, pneumonia, allergic rhinitis, and sinusitis, can greatly impair a child's ability to function and are an important cause of missed school days and limitations of activities. Symptoms associated with both mild and more severe manifestations of these respiratory conditions, such as cough, wheeze, congestion, chest pain, shortness of breath, respiratory distress, and death in the most extreme cases, are responsible for substantial morbidity and a large cost burden to families and society.

Outdoor and indoor air pollution can adversely affect children's respiratory health.[1-7] Studies have shown that air pollution can exacerbate existing respiratory conditions such as asthma and upper airway allergies.[1,8-10] Increasing evidence suggests that exposure to certain air pollutants may contribute to the onset of asthma in children, although studies relating to the exacerbation of pre-existing asthma are more prevalent because they are easier to conduct.[11-13] Air pollution also increases a child's risk of developing respiratory infections, most likely by causing inflammation and/or impaired immune response.[14-16]

EPA sets health-based National Ambient Air Quality Standards for six air pollutants.[17] These pollutants, referred to as criteria air pollutants, are particulate matter (PM), ground-level ozone, nitrogen oxides, sulfur oxides, carbon monoxide (CO), and lead. Four of these pollutants have extensive evidence linking them to respiratory diseases in children (PM, ground-level ozone, nitrogen oxides, and sulfur oxides). The evidence for respiratory effects is weaker for CO, and lead has not been linked to adverse respiratory outcomes.

PM is associated with significant respiratory problems in children, including aggravated asthma; exacerbation of allergic symptoms; reduced growth of lung function; and increased hospital admissions, emergency room visits, and doctor visits for respiratory diseases, especially in children with lung diseases such as asthma.[6] Particulate air pollution has also been associated with respiratory-related infant mortality, even at relatively low PM levels that are commonly experienced in the United States.[18,19]

Short-term exposure to ground-level ozone can cause a variety of respiratory health effects, including airway inflammation; reduced lung function; increased susceptibility to respiratory infection; and respiratory symptoms such as cough, wheezing, chest pain, and shortness of breath.[3,20,21] Ozone exposure can decrease the capacity to perform exercise and has been associated with the aggravation of respiratory illnesses such as asthma and bronchitis, leading to increased use of medication, absences from school, doctor and emergency department visits, and hospital admissions.[3] Studies have also found that long-term ozone exposure may contribute to the development of asthma, especially among children with certain genetic susceptibilities and children who frequently exercise outdoors.[22-24]

Nitrogen dioxide (NO_2) is an odorless gas that can irritate the eyes, nose, and throat, and can cause shortness of breath. EPA has concluded that exposure to NO_2 can lead to increased respiratory illnesses and symptoms, more severe asthma symptoms, and an increase in the number of emergency department visits and hospital admissions for respiratory causes, especially asthma.[4]

Short-term exposures of persons with asthma to elevated levels of sulfur dioxide (SO_2) while exercising at a moderate level may result in breathing difficulties, accompanied by symptoms such as wheezing, chest tightness, or shortness of breath. Studies also provide consistent evidence of an association between short-term SO_2 exposure and increased respiratory symptoms in children, especially those with asthma or chronic respiratory symptoms. Short-term exposures to SO_2 have also been associated with respiratory-related emergency department visits and hospital admissions, particularly for children.[5]

Exposure to CO reduces the capacity of the blood to carry oxygen, thereby decreasing the supply of oxygen to tissues and organs such as the heart. Research suggests correlations between CO exposure and the exacerbation of asthma; however, CO levels are highly correlated with other combustion-related pollutants, especially in locations near roads. Few analyses clearly distinguish the contributions of CO from those of the larger traffic-related air pollutant mixture, thus it is uncertain whether the observed health effects are truly attributable to CO or whether they are due to other co-occurring air pollutants.[7,10,25]

In addition to the criteria air pollutants, EPA regulates 187 hazardous air pollutants (HAPs) that are known or suspected to cause serious health effects or adverse environmental effects. For many of these pollutants, information on health effects is scarce. HAPs that may be of particular concern for the induction and exacerbation of asthma include acrolein, formaldehyde, nickel, and chromium.[26] Acrolein has been identified as a HAP of particular concern for possible respiratory effects at levels commonly found in outdoor air in the United States.[27-29] Acrolein can cause respiratory irritation in individuals who do not have asthma.[30]

Pollution from traffic-related sources, a mix of criteria air pollutants and HAPs, appears to pose particular threats to a child's respiratory system. Many studies have found a correlation between proximity to traffic (or to traffic-related pollutants) and occurrence of new asthma cases or exacerbation of existing asthma and other respiratory symptoms, including reduced growth of lung function during childhood.[11-13,31-36] A report by the Health Effects Institute concluded that living close to busy roads appears to be an independent risk factor for the onset of childhood asthma. The same report also concluded that the evidence was "sufficient" to infer a causal association between exposure to traffic-related pollution and exacerbations of asthma in children.[37] Some studies have suggested that traffic-related pollutants may contribute to the development of allergic disease, either by affecting the immune response directly or by increasing the concentration or biological activity of the allergens themselves.[38-40]

Children can also be exposed to air pollution inside homes, schools, and other buildings. Indoor air pollutants from biological sources such as mold; dust mites; pet dander (skin flakes); and droppings and body parts from cockroaches, rodents, and other pests or insects, can lead to allergic reactions, exacerbate existing asthma, and have been associated with the development of respiratory symptoms.[1,41,42] Furthermore, the Institute of Medicine concluded that exposure to dust mites can cause asthma in susceptible children, and exposure to cockroaches may cause asthma in young children.[1]

PM and NO$_2$, discussed previously as outdoor air pollutants, also pollute indoor air when they are emitted from gas stoves, gas or oil furnaces, fireplaces, wood stoves, and kerosene or gas space heaters. Indoor concentrations of these combustion byproducts can reach very high levels in developing countries where solid fuels are used extensively for cooking and home heating, but may also affect the respiratory health of children in developed countries, especially during the winter when use of fireplaces and space heaters is more common.[43] Environmental tobacco smoke (ETS), also known as secondhand smoke, is an air pollutant mixture that includes particles and NO$_2$ as well as thousands of other chemicals. The Surgeon General has concluded that exposure to ETS causes sudden infant death syndrome (SIDS), acute lower respiratory infection, ear problems, and more severe asthma in children. Smoking by parents causes respiratory symptoms and slows lung growth in their children.[2]

A number of air pollutants emitted indoors by a variety of household items such as building materials and home furnishings, recently dry-cleaned clothes, cleaning supplies, and room deodorizers, have been associated with respiratory symptoms and may play a role in the exacerbation or development of childhood asthma.[44,45] A recent systematic review of seven studies concluded that there is a significant association between exposure to formaldehyde—a chemical released from particle board, insulation, carpet, and furniture—and self-reported or diagnosed asthma in children.[46]

Air pollutants can enter the bloodstream of pregnant women and cross the placenta to reach the developing fetus; thus the period of fetal development may be a window of special vulnerability for respiratory effects of some air pollutants. Studies indicate that prenatal exposure to ETS may increase the risk of developing asthma during childhood and/or lead to impaired lung function, especially among children with asthma.[2,47-50] Studies have also found that prenatal exposure to polycyclic aromatic hydrocarbons (hazardous air pollutants found in diesel exhaust, ETS, and smoke from burning organic materials) is associated with childhood respiratory illnesses and the development of asthma, particularly when in combination with prenatal or postnatal exposure to ETS.[51-53] Limited studies of prenatal exposure to criteria air pollutants have found that exposure to PM, CO, and oxides of nitrogen and sulfur may increase the risk of developing asthma as well as worsen respiratory outcomes among those children who do develop asthma.[11,54,55] However, it is difficult to distinguish the effects of prenatal and early childhood exposure because exposure to air pollutants is often very similar during both periods.

Asthma

Asthma is a chronic inflammatory disease of the airways. When children with asthma are exposed to an asthma trigger, the airway walls become inflamed and secrete more mucus, and the muscles around the airways tighten. This exaggerates the normal airway constriction that occurs on exhalation, trapping air in the lungs and compromising normal oxygen exchange.

These physiological changes can result in wheezing, coughing, difficulty in breathing, chest tightness, and pain.

Asthma is one of the most common chronic diseases among children: in the year 2009, it affected 7.1 million (or about 10% of) children in the United States.[56] It is costly in both human and monetary terms: estimated national costs of asthma in 2007 were $56 billion.[57] The percentage of children with asthma increased substantially from 1980–1996 and remains high.[58] Researchers do not completely understand why children develop asthma or why the prevalence has increased.

Asthma is a complex disease with many factors, including genetic factors and environmental factors, that interact to influence its development and severity. The percentage of children reported to have current asthma differs by age, family history of asthma and allergies, racial and ethnic group, and family income. Children of color and children of lower-income families are more likely to be diagnosed with asthma. Because minority populations are more likely to be of low socioeconomic status, it is difficult to establish whether racial/ethnic group is an independent risk factor for the development of asthma. While some research has suggested that variations in asthma prevalence among racial groups can be explained by socioeconomic factors,[59,60] another study suggested that the difference persists even after accounting for socioeconomic factors.[61] Other researchers have proposed that the greater prevalence of asthma among Black children can be explained by their disproportionate presence in urban environments.[62]

Children living in poverty are more likely to have poorly maintained housing, which can present risk factors for asthma development and exacerbation. The Institute of Medicine concluded that exposure to dust mites causes asthma in susceptible children, and that cockroaches may cause asthma in young children.[1] Research suggests that lower-income children are more likely to live in homes with high levels of cockroach allergens and homes where someone smokes regularly.[63-66] A nationally representative survey of allergens in U.S. housing reported higher levels of dust mite allergen in bedding from lower-income families.[67] Household mouse allergen was also found at higher concentrations in low-income homes, mobile homes, and older homes.[68] In addition, total dust weight itself has been found to contribute to respiratory symptoms, including asthma and wheeze. Households with lower income, older homes, household pets, a smoker in the house, and less frequent cleaning are more likely to have higher dust weight levels.[69] Furthermore, children living in poverty may also face barriers to medical care, have less access to routine medical care and instructions for asthma management, or may be less likely to use asthma control medications.[70-76] These factors may increase asthma morbidity, as evidenced by increased asthma symptoms among those diagnosed with the disease.

Asthma indicators provide data on the percentage of children who have asthma as well as health outcomes for children with asthma. Indicators H1 and H2 focus on the prevalence of asthma among children. Indicator H1 provides the best nationally representative data available on prevalence of asthma over time among children ages 0 to 17 years. It provides two measures of asthma prevalence by year, from 1997–2010: current asthma prevalence and asthma attack prevalence. While the former measure reports on the percentage of children who have asthma each year, the latter measure presents data on children who had asthma attacks in the past year, and thus represents outcomes for children with asthma by identifying the proportion of children with ongoing or uncontrolled symptoms. Indicator H2 provides the best nationally representative data available to compare the prevalence of current asthma among children 0 to 17 years by race/ethnicity and family income for the years 2007–2010.[i]

Emergency Room Visits and Hospitalizations for Respiratory Diseases

Children who visit emergency rooms or are hospitalized for respiratory diseases (including asthma and upper and lower respiratory infections such as bronchiolitis and pneumonia) usually represent the most severe cases of respiratory disease. Although only a fraction of children with respiratory diseases are admitted to the hospital, asthma is the third leading cause of hospitalization for children in the United States and bronchiolitis is the leading cause of acute illness and hospitalization in infants.[77,78]

Emergency room visits and hospital admissions for respiratory diseases can be related to a number of factors. These factors include exposure to asthma triggers, lack of access to primary health care, lack of or inadequate insurance, inadequate instructions for asthma management, or inadequate compliance with given instructions.[79-83] Changes in emergency room visits and hospital admissions over time may also reflect changes in medical practices, asthma therapy, and access to and use of care.[84,85]

For children with existing respiratory conditions, exposure to air pollution from indoor and outdoor sources can trigger the onset of symptoms and lead to difficulty in breathing, increased use of medication, school absenteeism, visits to the doctor's office, and respiratory-related hospitalizations and trips to the emergency room.[3-6]

Studies have suggested that exacerbation of asthma from exposure to air pollution can be more severe among people with low income compared with other populations,[86,87] and that the gap between Black and White children in both hospitalizations and deaths from asthma appears to be growing.[88-90] The asthma death rate among Black non-Hispanic children with asthma was 4.9 times higher than the rate for White non-Hispanic children with asthma in 2004–2005.[88] Asthma is the leading cause of emergency room visits, hospitalizations, and missed school days in New York City's poorest neighborhoods.[91] In Maryland, the rate of children's emergency room visits for asthma is twice as high for Baltimore City (an area with a relatively high percentage of lower income and Black children) than for any other jurisdiction.[92]

---

[i] State-specific asthma information can be found in the CDC report, *The State of Childhood Asthma, United States, 1980–2005*, located at http://www.cdc.gov/nchs/data/ad/ad381.pdf.

Indicator H3 provides the best nationally representative data available on the frequency with which children experienced asthma or respiratory symptoms resulting in an emergency room visit or hospitalization for the years 1996–2008. This indicator highlights the most severe cases of respiratory illness among children ages 0 to 17 years. Indicator H3 includes further information on health outcomes for children with asthma, in addition to the asthma attack prevalence information in Indicator H1, by reporting on trends in children's hospitalizations and emergency room visits due to asthma.

Indicator H1: Percentage of children ages 0 to 17 years with asthma, 1997–2010

Indicator H2: Percentage of children ages 0 to 17 years reported to have current asthma, by race/ethnicity and family income, 2007–2010

About the Indicators: Indicators H1 and H2 present the percentage of children ages 0 to 17 years with asthma. The data are from a national survey that collects health information from a representative sample of the population each year. Indicator H1 shows how children's asthma rates have changed over time. Indicator H2 shows how children's asthma rates vary by race/ethnicity and family income level.

National Health Interview Survey

The National Health Interview Survey (NHIS) provides nationally representative data on the prevalence of childhood asthma in the United States each year. NHIS is a large-scale household interview survey of a representative sample of the civilian noninstitutionalized U.S. population, conducted by the National Center for Health Statistics (NCHS). The interviews are conducted in person at the participants' homes. From 1997–2005, interviews were conducted for approximately 12,000–14,000 children annually. From 2006–2008, interviews were conducted for approximately 9,000–10,000 children per year. In 2009 and 2010, interviews were conducted for approximately 11,000 children per year.

With a major survey redesign implemented in 1997, the measurement of childhood asthma prevalence in NHIS was changed to reporting the percentage of children ever diagnosed with asthma (lifetime asthma prevalence) and children ever diagnosed with asthma that also had an asthma attack in the previous 12 months (asthma attack prevalence). The data are obtained by asking a parent or other knowledgeable household adult questions regarding the child's health status. NHIS asks "Has a doctor or other health professional ever told you that your child has asthma?" If the answer is YES to this question, NHIS then asks (1) "Does your child still have asthma?" and (2) "during the past 12 months, has your child had an episode of asthma or an asthma attack?" The question "Does your child still have asthma?" was introduced in 2001 and identifies children who were previously diagnosed with asthma and who currently have asthma (current asthma prevalence). Some children may have asthma when they are young and experience fewer symptoms as they get older, or their asthma may be well controlled through medication and by avoiding triggers of asthma attacks. In such cases, children may currently have asthma but may not have experienced any attacks in the previous year.

Data Presented in the Indicators

Indicator H1 presents two different measures of asthma prevalence using data from the NHIS: current asthma and asthma attack prevalence. Indicator H1 provides the annual estimates of asthma prevalence for all children 0 to 17 years of age for the years 1997–2010. Indicator H2 reports on the percentage of children ages 0 to 17 years reported to have current asthma, by race/ethnicity and family income, in 2007–2010. NHIS is also the source of data for this indicator. The 2007, 2008, 2009 and 2010 data are combined for this indicator in order to increase the statistical reliability of the estimates for each race/ethnicity and income group.

For Indicator H2, five race/ethnicity groups are presented: White non-Hispanic, Black non-Hispanic, Asian non-Hispanic, Hispanic, and "All Other Races." The "All Others Races" category includes all other races not specified, together with those individuals who report more than one race. The limits of the sample design and sample size often prevent statistically reliable estimates for smaller race/ethnicity groups. The data are also tabulated for three income groups: all incomes, below the poverty level, and greater than or equal to the poverty level. These prevalence data are based on a survey respondent reporting that asthma has been diagnosed by a health care provider. Accuracy of responses and access to health care providers may vary among population groups.[93,94]

In addition to the data shown in Indicator H1, a supplemental table shows data for the percentage of children who had asthma in the past 12 months (asthma period prevalence), for the years 1980–1996. Estimates for asthma period prevalence are not directly comparable to any of the three prevalence estimates collected since 1997 because of changes in the NHIS survey questions. The data table for Indicator H2 shows the prevalence of current asthma for an expanded set of race/ethnicity categories, including Mexican-American and Puerto Rican. A supplemental data table shows the prevalence of current asthma by age and sex for the years 2007-2010.

Please see the Introduction to the Health section for discussion of statistical significance testing applied to these indicators.

Other Estimates of Asthma Prevalence

In addition to NHIS, other NCHS surveys provide data on asthma prevalence. A telephone-based survey conducted in 2007 by NCHS along with state and local governments found that 11% of high school students currently had asthma.[95] The 2007 National Survey of Children's Health (NSCH) found that nationwide 9.0% of children ages 0 to 17 years currently had asthma, which is very similar to the estimate from NHIS for 2007. The 2007 NSCH also provides information at the state level: South Dakota has the lowest asthma rates, with only 5.2% of children currently having asthma. The District of Columbia has the highest asthma rates, with 14.4% of children currently having asthma.[96]

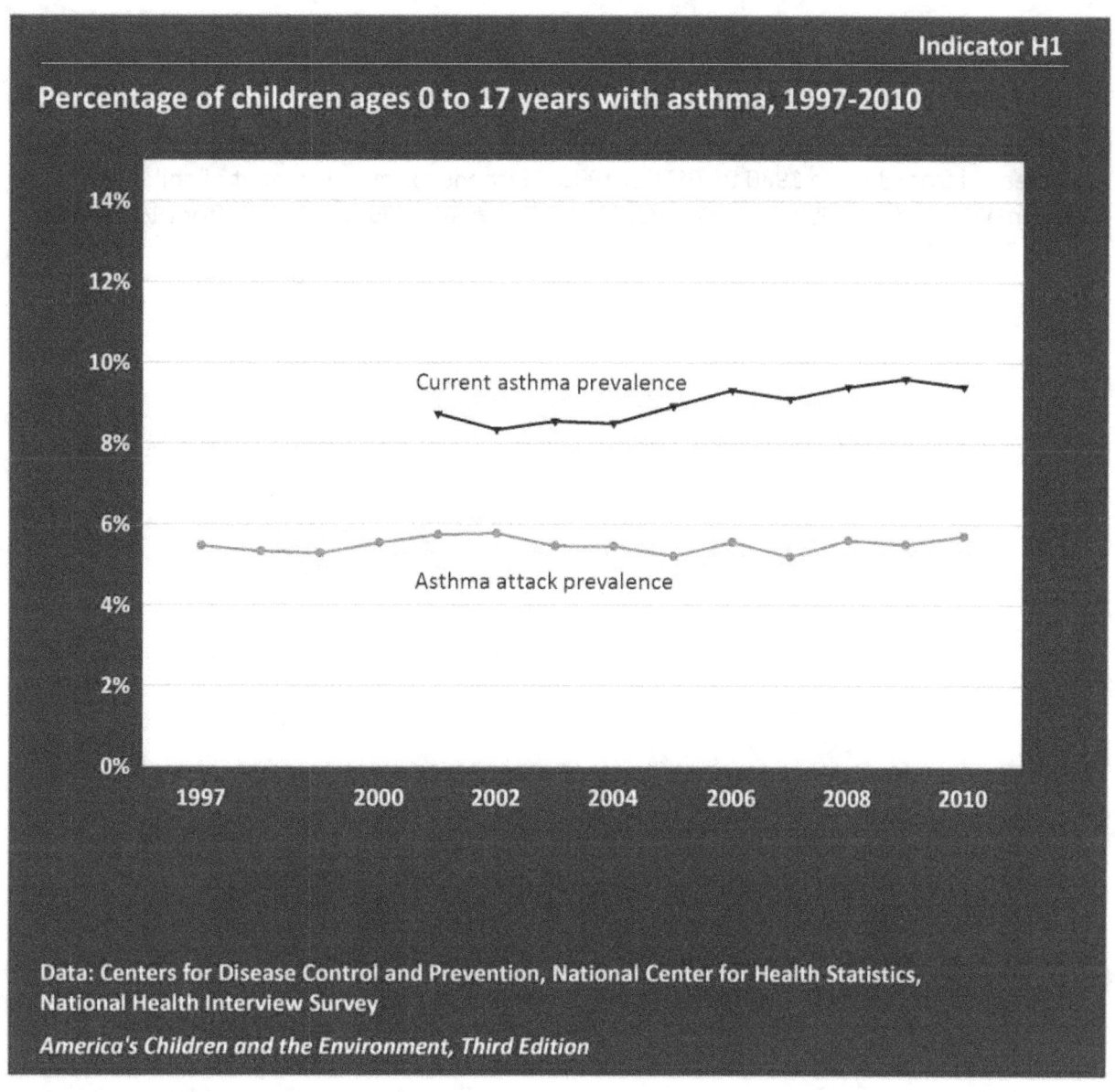

Indicator H1

Percentage of children ages 0 to 17 years with asthma, 1997-2010

Current asthma prevalence

Asthma attack prevalence

Data: Centers for Disease Control and Prevention, National Center for Health Statistics, National Health Interview Survey

America's Children and the Environment, Third Edition

Data characterization

- Data for this indicator are obtained from an ongoing annual survey conducted by the National Center for Health Statistics.
- Survey data are representative of the U.S. civilian noninstitutionalized population.
- A parent or other knowledgeable adult in each sampled household is asked questions regarding the child's health status, including if they have ever been told the child has asthma, if the child has had an asthma attack in the past year, and if the child currently has asthma.

■ The proportion of children reported to currently have asthma increased from 8.7% in 2001 to 9.4% in 2010.

■ In 2010, 5.7% of all children were reported to have had one or more asthma attacks in the previous 12 months. There was little change in this rate between 1997 and 2010.

- In 2001, 61.7% of children with current asthma had one or more asthma attacks in the previous 12 months, and by 2010 this figure had declined to 58.3%.[ii] The decreasing trend from 2001 to 2010 was statistically significant. (See Table H1c.)

- Between 1980 and 1995 the percentage of children who had asthma in the past 12 months increased from 3.6% in 1980 to 7.5% in 1995. Methods for measurement of childhood asthma changed in 1997, so earlier data cannot be compared to the data from 1997–2010. (See Table H1b.)

[ii] See indicator H3 for further information on outcomes for children with asthma.

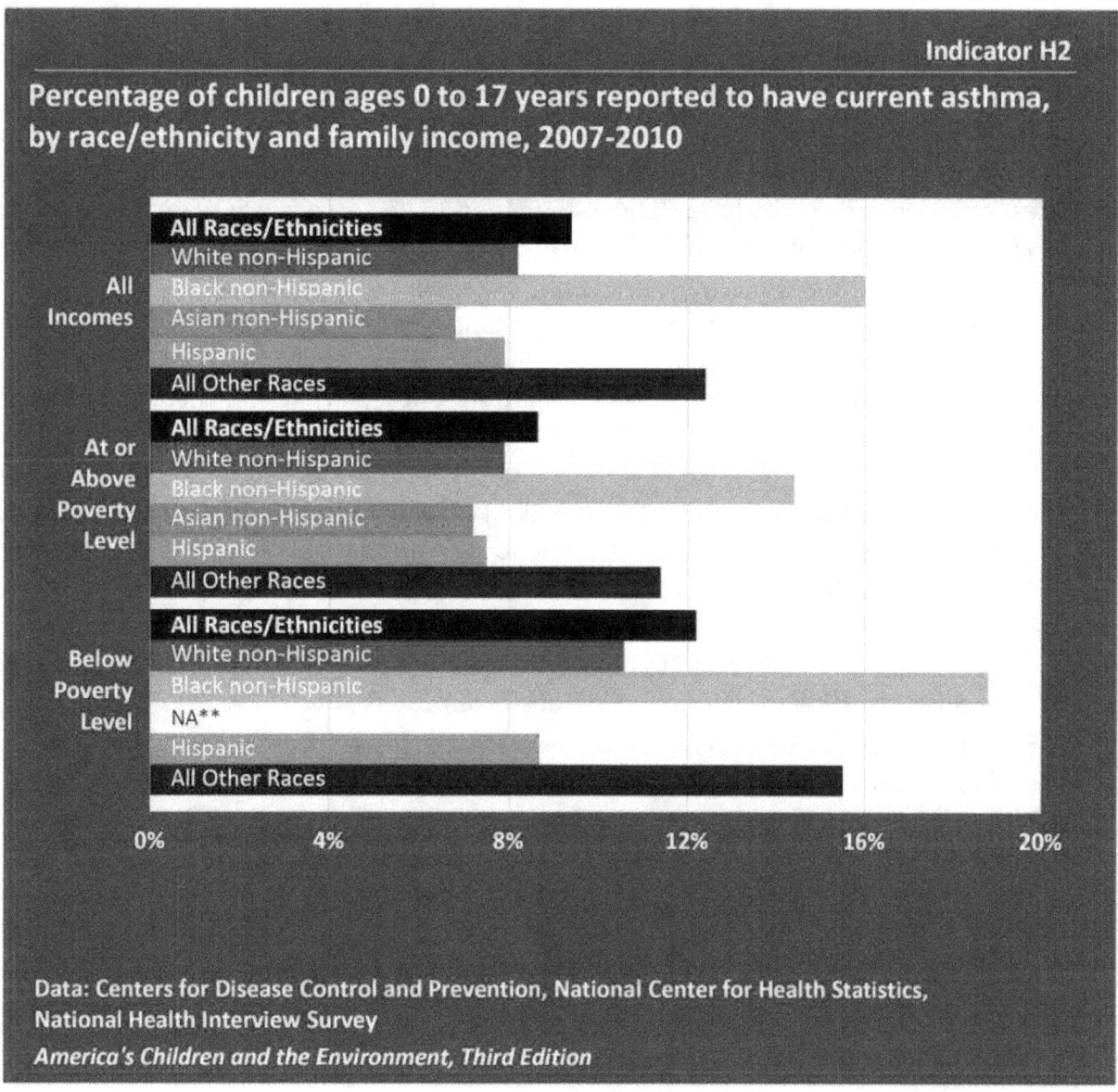

Indicator H2

Percentage of children ages 0 to 17 years reported to have current asthma, by race/ethnicity and family income, 2007-2010

Data: Centers for Disease Control and Prevention, National Center for Health Statistics, National Health Interview Survey

America's Children and the Environment, Third Edition

** Not available. The estimate is not reported because it has large uncertainty: the relative standard error, RSE, is 40% or greater (RSE = standard error divided by the estimate).

Data characterization
- Data for this indicator are obtained from an ongoing annual survey conducted by the National Center for Health Statistics.
- Survey data are representative of the U.S. civilian noninstitutionalized population.
- A parent or other knowledgeable adult in each sampled household is asked questions regarding the child's health status, including if they have ever been told the child has asthma, if the child has had an asthma attack in the past year, and if the child currently has asthma.

- In 2007–2010, 9.4% of all children were reported to currently have asthma.

- Among children living in families with incomes below the poverty level, 12.2% were reported to currently have asthma. Among children living in families with incomes at the poverty level and higher, 8.7% were reported to currently have asthma. This difference was statistically significant.

- In 2007–2010, the percentages of Black non-Hispanic children and children of "All Other Races" reported to currently have asthma, 16.0% and 12.4% respectively, were greater than for White non-Hispanic children (8.2%), Hispanic children (7.9%), and Asian non-Hispanic children (6.8%).

  - The differences in current asthma prevalence among Black non-Hispanic or "All Other Races" children, compared with current asthma prevalence among Hispanic, White non-Hispanic, or Asian non-Hispanic children, were statistically significant. These differences by race/ethnicity also hold true when considering only children below poverty level and only children at or above poverty level.

- Among Hispanic children, about 1 in 4 Puerto Rican children (23.3%) living in families with incomes below the poverty level were reported to currently have asthma. The rate of reported current asthma for Mexican-American children living in families with incomes below the poverty level is 6.6%. This difference was statistically significant. (See Table H2.)

- Among boys, 10.7% were reported to have current asthma compared with 8.0% of girls. This difference was statistically significant. (See Table H2a.)

- Among children ages 0 to 5 years, 7.1% were reported to have current asthma compared with 10.0% of children ages 6 to 10 years and 11.0% of children ages 11 to 17 years. The difference in current asthma by age group was statistically significant. (See Table H2a.)

Indicator H3: Children's emergency room visits and hospitalizations for asthma and other respiratory causes, ages 0 to 17 years, 1996–2008

About the Indicator: Indicator H3 presents information about the number of children's emergency room visits and hospitalizations for asthma and other respiratory causes. The data are from two national surveys that collect information from hospitals each year. Indicator H3 shows how the rates of children's emergency room visits and hospitalizations for respiratory causes have changed over time.

National Hospital Ambulatory Medical Care Survey and National Hospital Discharge Survey

The National Hospital Ambulatory Medical Care Survey (NHAMCS) and the National Hospital Discharge Survey (NHDS), conducted by the National Center for Health Statistics of the Centers for Disease Control and Prevention, provide national data on emergency room visits and hospitalizations. The NHAMCS has collected data for physician diagnoses for visits to hospital emergency rooms and outpatient departments beginning in the year 1992, while the NHDS reports physician diagnoses for discharges from hospitals beginning in the year 1965. The diagnoses in both surveys include asthma and a number of other respiratory conditions. Both surveys exclude federal and military hospitals and report patient demographic information.

Data Presented in the Indicators

Indicator H3 uses data from NHAMCS and NHDS to display emergency room visits and hospitalizations for asthma and other respiratory conditions including bronchitis, pneumonia, and influenza. The top line in each graph represents the total number of children's emergency room visits or hospitalizations for asthma and all other respiratory causes, followed by lines for asthma and for all respiratory causes other than asthma. Indicator H3 presents annual survey results from 1996–2008. 1996 was selected as the initial year for the indicator because not all of the needed hospitalization data for earlier years are available online. The indicator provides data through 2008 because it is the most recent year for which data from both NHAMCS and NHDS are available.

In addition to the data shown in the Indicator H3 graph, supplemental tables show the annual average rates of children's emergency room visits and hospitalizations for asthma and all other respiratory causes, asthma, and all respiratory causes other than asthma (composed of the following subcategories: upper respiratory conditions, pneumonia or influenza, and other lower respiratory conditions besides asthma) by age and race/ethnicity for the years 2005–2008. For emergency room visits, five race/ethnicity groups are presented: White non-Hispanic, Black non-Hispanic, American Indian/Alaska Native non-Hispanic, Asian and Pacific Islander non-Hispanic, and Hispanic. For hospitalizations, race only is reported; the two groups presented are

White and Black. The supplemental tables do not include income data, since neither of these surveys includes the patient's income or family income.

Please see the Introduction to the Health section for discussion of statistical significance testing applied to these indicators.

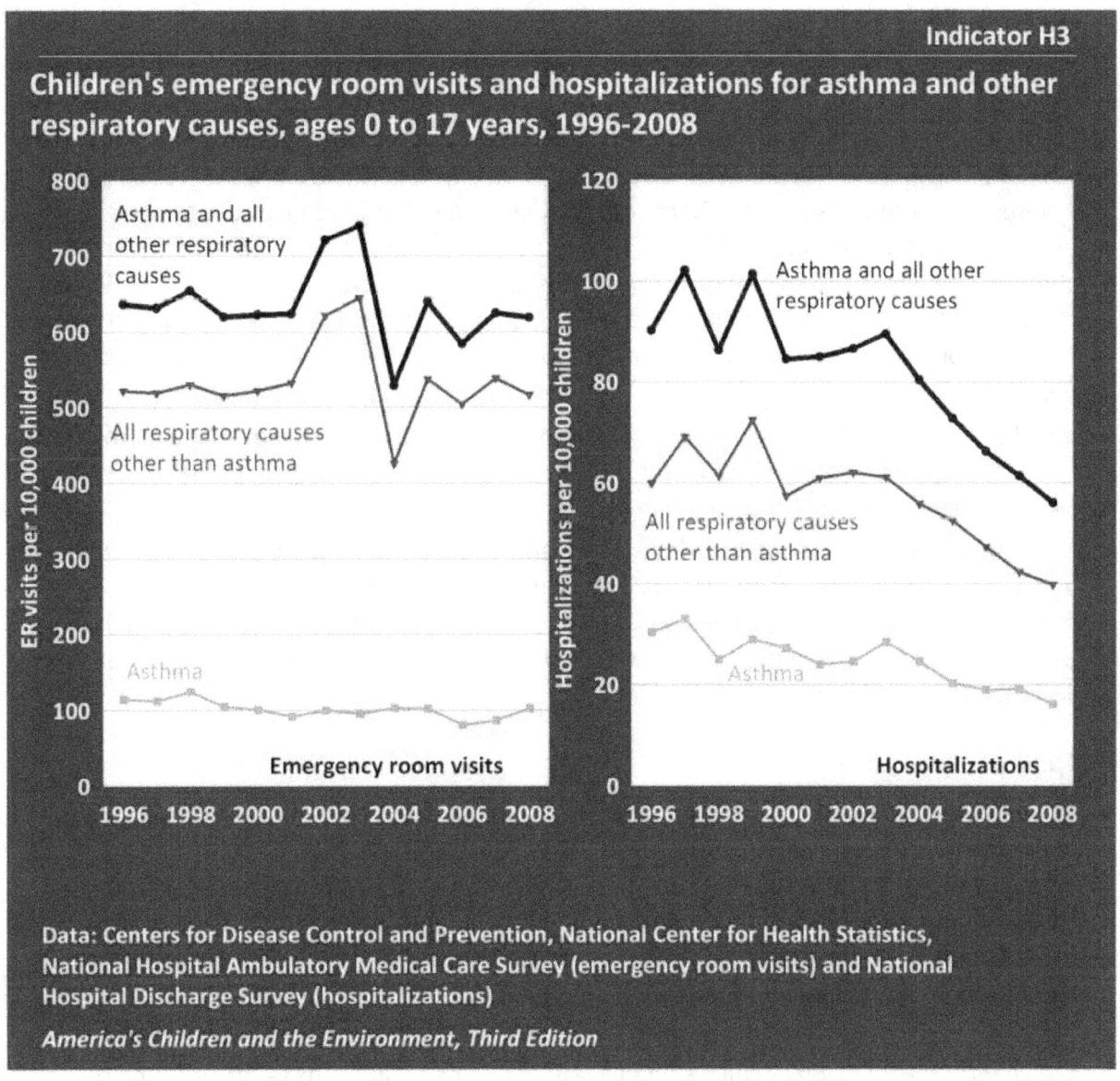

Indicator H3

Children's emergency room visits and hospitalizations for asthma and other respiratory causes, ages 0 to 17 years, 1996-2008

Data: Centers for Disease Control and Prevention, National Center for Health Statistics, National Hospital Ambulatory Medical Care Survey (emergency room visits) and National Hospital Discharge Survey (hospitalizations)

America's Children and the Environment, Third Edition

Data characterization
- Data for this indicator are obtained from two ongoing annual surveys conducted by the National Center for Health Statistics.
- Survey data are representative of U.S. population visits to emergency rooms and stays at non-federal hospitals.
- The surveys collect data on physician diagnoses of patients in sampled hospitals, including diagnoses of asthma and other respiratory conditions.

Emergency Room Visits

- In 2008, the rate of emergency room visits for asthma and all other respiratory causes was 619 visits per 10,000 children. The rate of emergency room visits for asthma alone was 103 visits per 10,000 children, and the rate for all respiratory causes other than asthma was 517 visits per 10,000 children.

- The rate of emergency room visits for asthma decreased from 114 visits per 10,000 children in 1996 to 103 visits per 10,000 children in 2008. This decreasing trend was statistically significant.

- Children's emergency room visits for asthma and all other respiratory causes vary widely by race/ethnicity. For the years 2005–2008, Black non-Hispanic children had a rate of 1,240 emergency room visits per 10,000 children, while Hispanic children had a rate of 672 emergency room visits per 10,000 children, American Indian/Alaska Native non-Hispanic children had a rate of 536 emergency room visits per 10,000 children, White non-Hispanic children had a rate of 487 emergency room visits per 10,000 children, and Asian and Pacific Islander non-Hispanic children had a rate of 371 emergency room visits per 10,000 children. (See Table H3a.)

  - The difference in rates of emergency room visits between Black non-Hispanic children and emergency room visits for each of the other race/ethnicity groups was statistically significant.

- Children's emergency room visits for asthma and all other respiratory causes vary widely by age. For the years 2005–2008, infants less than 12 months of age had a rate of 2,142 emergency room visits per 10,000 children, while children 16 to 17 years of age had a rate of 338 emergency room visits per 10,000 children. The differences between age groups were statistically significant. (See Table H3b.)

Hospitalizations

- Between 1996 and 2008, hospitalizations for asthma and for all other respiratory causes decreased from 90 hospitalizations per 10,000 children to 56 hospitalizations per 10,000 children. Between 1996 and 2008, hospitalizations for asthma alone decreased from 30 per 10,000 children to 16 per 10,000 children, and hospitalizations for all other respiratory causes decreased from 60 per 10,000 children to 40 per 10,000 children. These decreasing trends were statistically significant.

- Children's hospitalizations for asthma and all other respiratory causes vary widely by race. For the years 2005–2008, Black children had a rate of 84 hospitalizations for asthma and other respiratory causes per 10,000 children, while White children had a rate of 52 hospitalizations per 10,000 children. This difference was statistically significant. (See Table H3c.)

- Children's hospitalizations for asthma and all other respiratory causes vary widely by age. For the years 2005–2008, infants less than 12 months of age had a rate of 396 hospitalizations per 10,000 children, while children 16 to 17 years of age had a rate of 13 hospitalizations per 10,000 children. The differences between age groups were statistically significant. (See Table H3d.)

Childhood Cancer

Cancer is not a single disease, but includes a variety of malignancies in which abnormal cells divide in an uncontrolled manner. These cancer cells can invade nearby tissues and can migrate by way of the blood or lymph systems to other parts of the body.[1] The most common childhood cancers are leukemias (cancers of the white blood cells) and cancers of the brain or central nervous system, which together account for more than half of new childhood cancer cases.[2]

Cancer in childhood is rare compared with cancer in adults, but still causes more deaths than any factor, other than injuries, among children from infancy to age 15 years.[2] The annual incidence of childhood cancer has increased slightly over the last 30 years; however, mortality has declined significantly for many cancers due largely to improvements in treatment.[2,3] Part of the increase in incidence may be explained by better diagnostic imaging or changing classification of tumors, specifically brain tumors.[4] However, the President's Cancer Panel recently concluded that the causes of the increased incidence of childhood cancers are not fully understood, and cannot be explained solely by the introduction of better diagnostic techniques. The Panel also concluded that genetics cannot account for this rapid change. The proportion of this increase caused by environmental factors has not yet been determined.[5]

The causes of cancer in children are poorly understood, though in general it is thought that different forms of cancer have different causes. According to scientists at the National Cancer Institute, established risk factors for the development of childhood cancer include family history, specific genetic syndromes (such as Down syndrome), high levels of radiation, and certain pharmaceutical agents used in chemotherapy.[4,6] A number of studies suggest that environmental contaminants may play a role in the development of childhood cancers. The majority of these studies have focused on pesticides and solvents, such as benzene. According to the President's Cancer Panel, "the true burden of environmentally induced cancer has been grossly underestimated."[5]

The development of cancer, or carcinogenesis, is a multistep process leading to the uncontrolled growth and division of cells. This process can begin with an inherited genetic mutation or DNA damage initiated by an exogenous agent, such as exposure to a carcinogenic chemical or radiation. Additionally, many external influences, such as environmental exposures or nutrition, can alter gene expression without changing the DNA sequence.[7] These alterations, referred to as epigenetic changes, can promote alterations in the expression of genes important for controlling cell growth and division.[8,9] Because the initiation of carcinogenesis is a multistep process, multiple factors are thought to contribute to the development of cancer.[9] Newer research suggests that childhood cancer may be caused by a combination of genetic predisposition and environmental exposure.[10-16]

Different types of cancer affect children at different ages. This pattern may reflect the different types of exposures and windows of vulnerability experienced by children as they grow older, and the time between the initiation of cancer and its clinical presentation. Children can be affected by exposures that occur during different developmental stages, such as during infancy and early childhood. Scientific evidence suggests that early childhood cancers may be related to exposure in the womb, or even to parents' exposures prior to conception.[17-21] Furthermore, recent studies suggest that susceptibility to some cancers that arise later in adulthood also may be determined while in the womb.[7]

Leukemia is the most common form of cancer in children. According to the Centers for Disease Control and Prevention, adults and children who undergo chemotherapy and radiation therapy for cancer treatment; take immune suppressing drugs; or have certain genetic conditions, such as Down syndrome; are at a higher risk of developing acute leukemia.[22] Multiple review articles have concluded that ionizing radiation from sources such as x-rays is associated with an increased risk of leukemia.[23-25] CT scans are also an increasing source of ionizing radiation exposure to children,[26] and may be associated with an increased risk of childhood leukemia.[27] Further, studies have consistently shown an approximately 40% increased risk of childhood leukemia after maternal exposure to ionizing radiation during pregnancy.[18,23-25] These confirmed risk factors, however, explain less than 10% of the incidence of childhood leukemia, meaning that the cause is unknown in at least 90% of leukemia cases.[18]

Associations between proximity to extremely low frequency electromagnetic radiation, such as radiation from electrical power lines, and childhood leukemia have been investigated for many years.[5] Some studies suggest an effect on cancer risk, while others do not.[28,29] At this time, a variety of national and international organizations have concluded that the link between exposure to extremely low frequency electromagnetic fields and cancer is controversial or weak.[4,5] Radon is a naturally occurring radioactive element that has been associated with lung cancer; some studies have also found an association between childhood leukemia and radon while other studies have not.[4,30-32] A recent study also reported an association between naturally occurring gamma radiation and childhood leukemia.[33]

Pesticides, solvents, hazardous air pollutants, motor vehicle exhaust, and environmental tobacco smoke have been studied for a potential role in childhood leukemia. Numerous studies have examined the link between parents' (parental), prenatal, and childhood exposures to pesticides and childhood leukemia, and several meta-analyses of these studies have found associations between pesticide exposure and childhood leukemia in both residential and occupational settings.[20,34-46] Recent literature has also suggested an association between childhood exposures to multiple hazardous air pollutants and leukemia.[47-49] A study exploring the relationship between childhood leukemia and hazardous air pollutants (HAPs) in outdoor air found an increased risk for childhood leukemia in census tracts with the highest concentrations of a group of 25 potentially carcinogenic HAPs, including several solvents.[48] Several other studies have found associations between leukemia and surrogate measures of exposure to motor vehicle exhaust, including residential proximity to traffic and gas stations.[18,50-53] However, other studies conducted in California and Denmark did not find an association between these proxy measures of motor vehicle exhaust and childhood leukemia,[54-57] and review studies have concluded that the overall evidence for a relationship is inconclusive.[18,58]

According to the U.S. Surgeon General, there is also suggestive evidence that prenatal and postnatal exposure to environmental tobacco smoke can lead to leukemia in children.[59]

Cancers of the nervous system, including brain tumors, are the second most common form of cancer in children. Known risk factors for childhood brain tumors include radiation therapy and certain genetic syndromes, although these factors explain only a small portion of cases.[6] Some studies have also reported an association between prenatal exposure to ionizing radiation and brain tumors while a few smaller studies have not.[25,60,61] Other research reports that head CT scans may be associated with an increased risk of brain tumors in children.[27] Research also suggests that parental, prenatal, and childhood exposure to pesticides may lead to brain tumors in children.[43,45,46] There is suggestive evidence linking prenatal and postnatal exposure to environmental tobacco smoke and childhood brain tumors, according to the U.S. Surgeon General.[59] Many studies have examined whether there is an association between cellular phone use and brain cancer. Some of these studies have found an association between cellular phone use and some types of brain cancer, while other studies have found no association.[62-69] Because the use of cellular phones by children has only recently become more common, no long-term epidemiological studies of cancer related to cellular phone use by children are available.[5]

Lymphomas, which affect a child's lymph system, are another common form of childhood cancer. The cause of most cases of childhood lymphoma is unknown, but it is clear that children with compromised immune systems are at a greater risk of developing lymphomas.[6] Extensive review studies have found suggestive associations between parental, prenatal, and childhood exposure to pesticides and childhood lymphomas.[43,46] The U.S. Surgeon General has concluded that there is also suggestive evidence linking prenatal and postnatal exposure to environmental tobacco smoke and childhood lymphomas.[59]

Other childhood cancers that have been associated with environmental exposures include thyroid cancer, Wilms' tumor (a type of kidney cancer), Ewing's sarcoma (a cancer of the bone or soft tissue), and melanoma. Some research has reported an increased risk of thyroid cancer in childhood or early adulthood from exposure to ionizing radiation.[70-72] Much of the evidence for this association comes from studies of individuals in areas with high ionizing radiation exposure due to the Chernobyl accident in eastern Europe. While the only known causal factors for Wilms' tumor and Ewing's sarcoma are certain birth defects and genetic conditions, there is limited research indicating that exposure to pesticides may also be a causal factor in the development of Wilms' tumor and Ewing's sarcoma in children.[36,46,73] Although childhood melanoma is rare, the incidence of melanoma is increasing in children, especially in adolescents. Environmental factors associated with melanoma include sunburns, especially in childhood, and increased exposure to ultraviolet (UV) radiation.[74-76] Depletion of the ozone layer causes more ultraviolet radiation to reach the earth's surface. Even though the use of ozone depleting compounds has been largely phased out and the ozone layer will eventually be restored, higher levels of ultraviolet radiation reaching the earth's surface will persist for many years to come.[77,78] Finally, the increased rates of melanoma in adolescent girls and young women may reflect increased UV exposure from sunbathing or from the widespread practice of indoor tanning.[79,80]

The two indicators that follow provide the best nationally representative data available on cancer incidence and mortality among U.S. children over time. Indicator H4 presents cancer incidence and mortality for children ages 0 to 19 years for the period 1992–2009. Indicator H5 presents cancer incidence, by cancer type, for children ages 0 to 19 years for the period 1992–2006. Changes in childhood cancer mortality are most likely reflective of changes in treatment options, rather than environmental exposures. However, showing childhood cancer mortality rates in conjunction with childhood cancer incidence rates highlights the magnitude and severity of childhood cancer and indicates the proportion of children that survive.

Indicator H4 provides an indication of broad trends in childhood cancer over time, while Indicator H5 provides more detailed information about the incidence of specific types of cancer in children.

Indicator H4: Cancer incidence and mortality for children ages 0 to 19 years, 1992–2009

Indicator H5: Cancer incidence for children ages 0 to 19 years by type, 1992–2006

About the Indicators: Indicators H4 and H5 present information about the number of new childhood cancer cases and the number of deaths caused by childhood cancer. The childhood cancer case data come from a program that collects information from tumor registries located in specific geographic regions around the country each year. The childhood cancer death data come from a national database of vital statistics that collects data on numbers and causes of all deaths each year. Indicator H4 shows how the rates of all new childhood cancers and all childhood cancer deaths have changed over time, and Indicator H5 shows how the rates of specific types of childhood cancers have changed over time.

SEER

The National Cancer Institute's Surveillance, Epidemiology, and End Results (SEER) Program has provided data on cancer incidence, survival, and prevalence since 1973. SEER obtains its cancer case data from tumor registries in various locations throughout the United States and its cancer mortality data from a national database of vital statistics that collects data on numbers and causes of all deaths each year. Each of the tumor registries collects information for all tumors within a specified geographic region. The sample population covered by the SEER tumor registries is comparable to the general U.S. population in terms of poverty and education. However, the population covered by the SEER tumor registries tends to be more urban and has a higher proportion of foreign-born persons compared with the general U.S. population.[81]

Since its initiation in 1973, the SEER program has expanded to include a greater number of tumor registries. Currently, the SEER program includes data from 18 tumor registries, but complete data from all 18 registries are only available beginning with the year 2000. SEER data are available from 13 different tumor registries that provide data starting in 1992, and represent geographic areas containing 13.8% of the total U.S. population.[82] The registries include the Alaska Native, Atlanta, Connecticut, Detroit, Hawaii, Iowa, Los Angeles, New Mexico, Rural Georgia, San Francisco-Oakland, San Jose-Monterey, Seattle-Puget Sound, and Utah tumor registries.

Data Presented in the Indicators

Childhood cancer incidence refers to the number of new childhood cancer cases reported for a specified period of time. Childhood cancer incidence is shown in Indicator H4 and Indicator H5 as the number of childhood cancer cases reported per million children for one year. The incidence rate is age-adjusted, meaning that each year's incidence calculation uses the age distribution of children from the year 2000. For example, 25.3% of all U.S. children were between the ages of 5 and 9 years in 2000, and this percentage is assumed to be the same for each year from 1992 to 2009. This age adjustment ensures that differences in cancer rates over time are not simply due to changes in the age composition of the population. Indicator H4 also shows childhood cancer mortality as the number of deaths per million children for each year.

SEER reports the incidence data by single year of age, but reports mortality data in five age groups for children under the age of 20: under 1 year, 1–4, 5–9, 10–14, and 15–19 years. For this reason, both indicators use data for all children 0 to 19 years of age, in contrast to the other indicators in this report that define children as younger than age 18 years.

Trends in the total incidence of childhood cancer, as shown by Indicator H4, are useful for assessing the overall burden of cancer among children. However, broad trends mask changes in the frequency of specific types of cancers that often have patterns that diverge from the overall trend. Moreover, environmental factors may be more likely to contribute to some childhood cancers than to others. Indicator H5 shows trends in incidence for specific types of childhood cancers.

Some types of childhood cancers are very rare, and as such the yearly incidence is particularly low and variable. Due to this fact, Indicator H5 shows the incidence of individual childhood cancers in groupings of three years. Each bar in the graph represents the annual number of cases of that specific cancer diagnosed per million children, calculated as the average number of cases per year divided by the average population of children (in millions) per year for each three-year period.

The SEER cancer incidence data for the 13 longer-established registries, instead of all 18, were used to develop the H4 and H5 indicators because this allowed for more comprehensive trend analysis while still covering a substantial portion of the population. Indicator H4 begins with the earliest available SEER13 incidence data from 1992 and ends with 2009. Childhood cancer mortality data for 1992 to 2009 are also used for indicator H4. Indicator H5 presents data for the series of three-year periods beginning in 1992 and ending in 2006. In addition to the data shown in the Indicator H4 graph, supplemental tables show childhood cancer incidence and mortality by race/ethnicity and sex, as well as childhood cancer incidence by age group. These data tables use data from the three most current years shown in Indicator H4, which are 2007–2009. Combining three years of data allows for more statistically reliable estimates by race/ethnicity, sex, and age group. Five race/ethnicity groups are used in the supplemental tables for Indicator H4: White non-Hispanic, Black non-Hispanic, American Indian/Alaska Native non-Hispanic, Asian or Pacific Islander non-Hispanic, and Hispanic. In addition to the data shown in the Indicator H5 graph, a supplemental table shows childhood cancer incidence by cancer type and age group.

Please see the Introduction to the Health section for discussion of statistical significance testing applied to these indicators.

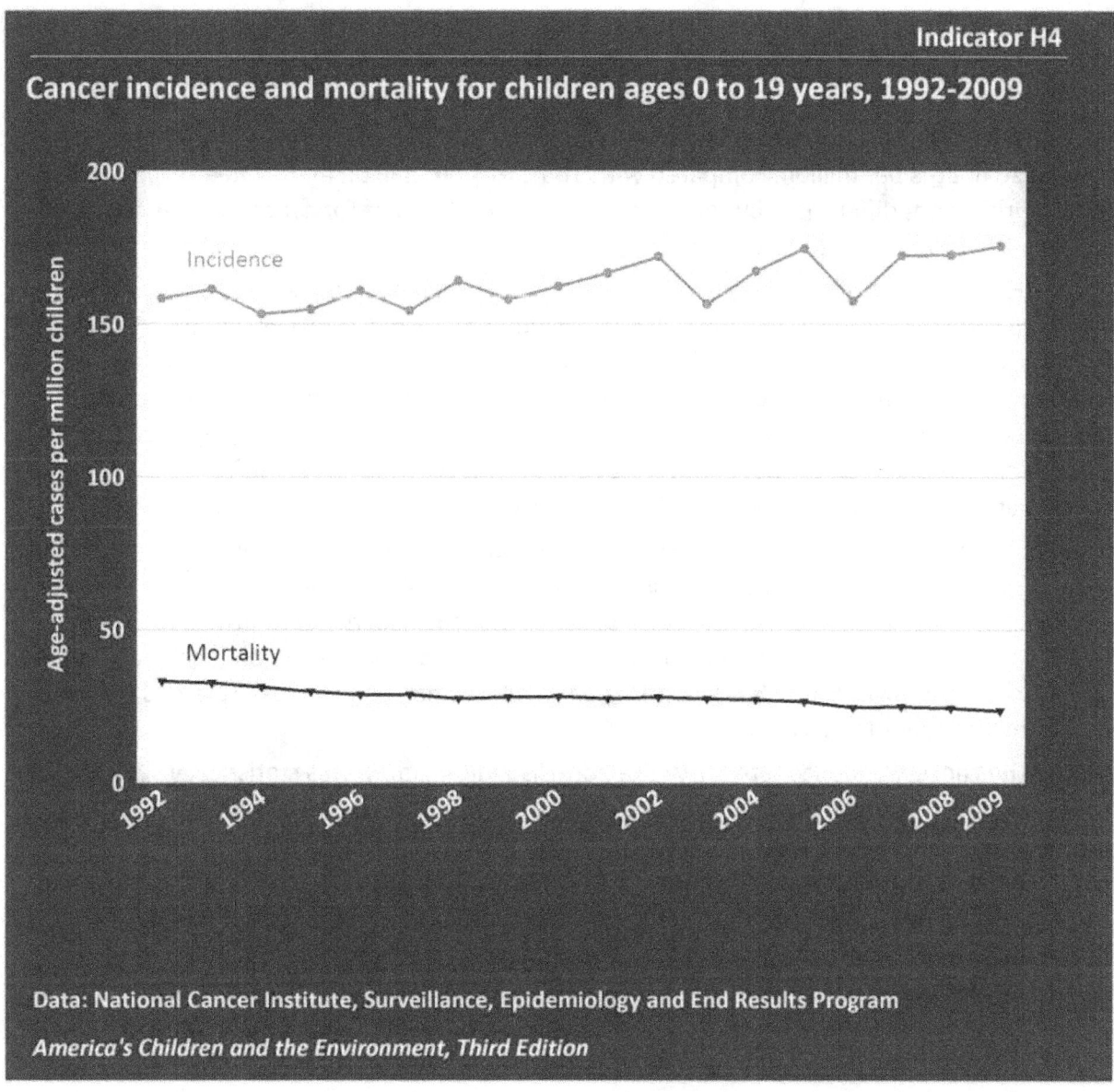

Data: National Cancer Institute, Surveillance, Epidemiology and End Results Program

America's Children and the Environment, Third Edition

Data characterization
- Cancer incidence data for this indicator are obtained from a database of 13 regional tumor registries located throughout the country, maintained by the National Cancer Institute.
- The population covered by the 13 registries is comparable to the general U.S. population regarding poverty and education, but is more urban and has more foreign-born persons.
- Cancer mortality data for this indicator are obtained from a database of all death certificates in the United States; cause of death is recorded on the death certificates.

- The age-adjusted annual incidence of cancer ranged from 153 to 161 cases per million children between 1992 and 1994 and from 172 to 175 cases per million children between 2007 and 2009. This increasing trend from 1992–2009 was statistically significant.

- Childhood cancer mortality decreased from 33 deaths per million children in 1992 to 24 deaths per million children in 2009, a statistically significant decreasing trend.

- Childhood cancer incidence and mortality rates were generally higher for boys than for girls. In 2007–2009, rates of cancer incidence and mortality for boys were 183 cases per million and 26 deaths per million, compared with 163 cases per million and 22 deaths per million for girls. These differences by sex were statistically significant for cancer incidence (after adjustment for age and race/ethnicity) and cancer mortality. (See Tables H4a and H4b.)

- In 2007–2009, the difference in cancer incidence between boys and girls was not consistent for all races/ethnicities. No statistically significant difference in cancer incidence by sex was seen among Black non-Hispanic children or Asian or Pacific Islander non-Hispanic children. Among American Indian and Alaska Native non-Hispanic children, cancer incidence was greater for girls than for boys, although this difference was not statistically significant. Cancer incidence was greater for boys than for girls and statistically significant (after adjustment for age) among White non-Hispanic children and Hispanic children. (See Table H4a.)

- In 2007–2009, childhood cancer incidence was highest among White non-Hispanic children at 188 cases per million. Hispanic children had an incidence rate of 169 cases per million, Asian and Pacific Islander non-Hispanic children had an incidence rate of 152 cases per million, American Indian and Alaska Native non-Hispanic children had an incidence rate of 137 cases per million, and Black non-Hispanic children had an incidence rate of 133 cases per million. (See Table H4a.)

  - The cancer incidence rate for White non-Hispanic children was statistically significantly higher than the rates of each of the other race/ethnicity categories after accounting for differences by age and sex. The cancer incidence rate for Black non-Hispanic children was also statistically significantly lower than the rates for Hispanic children and Asian and Pacific Islander non-Hispanic children after adjustment for differences by age and sex. The cancer incidence rate for Asian and Pacific Islander non-Hispanic was also statistically significantly lower than the rate for Hispanic children after adjustment for differences by age and sex. The remaining differences between race/ethnicity groups were not statistically significant.

- Childhood cancer incidence rates vary by age. In 2007–2009, children under 5 and those of ages 15 to 19 years experienced the highest incidence rates of cancer at approximately 208 and 232 cases per million, respectively. Children ages 5 to 9 years and 10 to 14 years had lower incidence rates at 117 and 139 cases per million, respectively. These differences among age groups were statistically significant. (See Table H4c.)

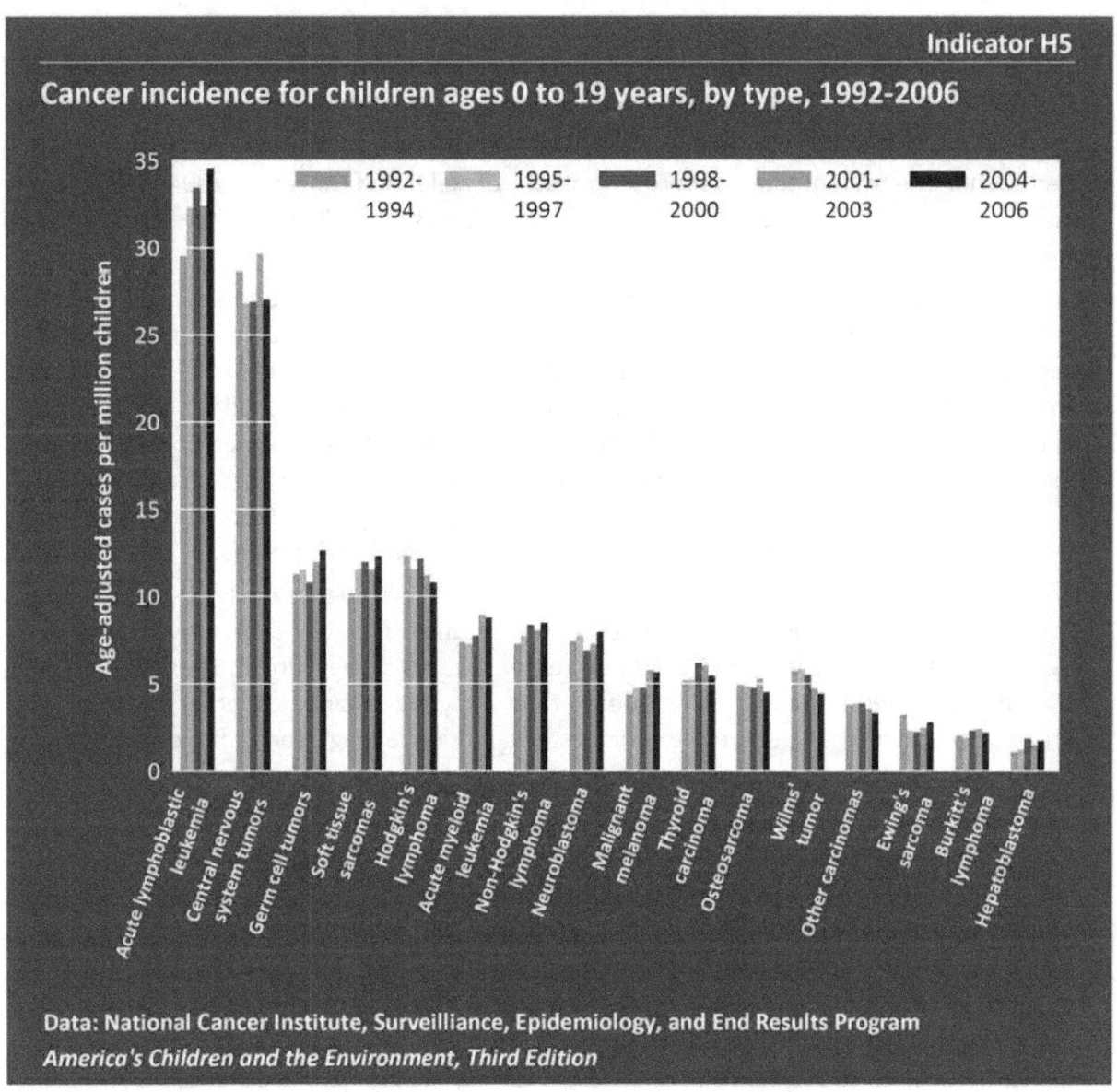

Data: National Cancer Institute, Surveilliance, Epidemiology, and End Results Program
America's Children and the Environment, Third Edition

Data characterization

- Data for this indicator are obtained from a database of 13 regional tumor registries located throughout the country, maintained by the National Cancer Institute.
- The population covered by the 13 registries is comparable to the general U.S. population regarding poverty and education, but is more urban and has more foreign-born persons.

■ Leukemia, which includes acute lymphoblastic leukemia and acute myeloid leukemia, was the most common cancer diagnosis for children from 2004–2006, representing 28% of total cancer cases. Incidence of acute lymphoblastic (lymphocytic) leukemia was 30 cases per million in 1992–1994 and 35 cases per million in 2004–2006. The rate of acute myeloid (myelogenous) leukemia was 7 cases per million in 1992–1994 and 9 cases per million in 2004–2006.

- The increasing trend for incidence of acute lymphoblastic leukemia was statistically significant after accounting for differences by age, sex, and race/ethnicity. The trend for acute myeloid leukemia was not statistically significant.

- Central nervous system tumors represented 18% of childhood cancers in 2004–2006. The incidence of central nervous system tumors was 27 cases per million in 2004–2006, with no statistically significant trend for 1992–2006.

- Lymphomas, which include Hodgkin's lymphoma, non-Hodgkin's lymphoma, and Burkitt's lymphoma, represented 14% of childhood cancers in 2004–2006. Incidence of Hodgkin's lymphoma was 12 cases per million in 1992–1994 and 11 per million in 2004–2006. There were approximately 7 cases of non-Hodgkin's lymphoma per million children in 1992–1994 and 9 per million in 2004–2006. Incidence of Burkitt's lymphoma remained constant from 1992–2006 (2 cases per million children). The increasing trend in the incidence rate of non-Hodgkin's lymphoma was statistically significant, while there was no statistically significant trend in the incidence rate of Hodgkin's lymphoma or Burkitt's lymphoma.

- Between the years 1992 and 2006, increasing trends in the incidence of soft tissue sarcomas, malignant melanomas, and hepatoblastomas were statistically significant, as was the decreasing trend in the incidence of Wilms' tumor (tumors of the kidney). There was no statistically significant trend in the incidence rate of thyroid carcinomas, other and unspecified carcinomas, germ cell tumors, osteosarcomas, Ewing's sarcomas, or neuroblastomas.

  - The increasing trend in the incidence rate of hepatoblastomas was statistically significant after accounting for differences by age, sex, and race/ethnicity.

- Different types of cancer affect children at different ages. The incidence of neuroblastomas and Wilms' tumor (tumors of the kidney) was highest for young children (ages 0 to 4 years). Leukemias occur in all age groups, but the incidence is highest among 0- to 4-year-olds. The incidence of Hodgkin's and non-Hodgkin's lymphomas, thyroid carcinomas, malignant melanomas, other and unspecified carcinomas, germ cell tumors, and osteosarcomas was higher in those 15 to 19 years old. Differences among age groups were statistically significant for each of these cancer types. (See Table H5a.)

Neurodevelopmental Disorders

Neurodevelopmental disorders are disabilities associated primarily with the functioning of the neurological system and brain. Examples of neurodevelopmental disorders in children include attention-deficit/hyperactivity disorder (ADHD), autism, learning disabilities, intellectual disability (also known as mental retardation), conduct disorders, cerebral palsy, and impairments in vision and hearing. Children with neurodevelopmental disorders can experience difficulties with language and speech, motor skills, behavior, memory, learning, or other neurological functions. While the symptoms and behaviors of neurodevelopmental disabilities often change or evolve as a child grows older, some disabilities are permanent. Diagnosis and treatment of these disorders can be difficult; treatment often involves a combination of professional therapy, pharmaceuticals, and home- and school-based programs.

Based on parental responses to survey questions, approximately 15% of children in the United States ages 3 to 17 years were affected by neurodevelopmental disorders, including ADHD, learning disabilities, intellectual disability, cerebral palsy, autism, seizures, stuttering or stammering, moderate to profound hearing loss, blindness, and other developmental delays, in 2006–2008.[1] Among these conditions, ADHD and learning disabilities had the greatest prevalence. Many children affected by neurodevelopmental disorders have more than one of these conditions: for example, about 4% of U.S. children have both ADHD and a learning disability.[2] Some researchers have stated that the prevalence of certain neurodevelopmental disorders, specifically autism and ADHD, has been increasing over the last four decades.[3-7] Long-term trends in these conditions are difficult to detect with certainty, due to a lack of data to track prevalence over many years as well as changes in awareness and diagnostic criteria. However, some detailed reviews of historical data have concluded that the actual prevalence of autism seems to be rising.[4,8-10] Surveys of educators and pediatricians have reported a rise in the number of children seen in classrooms and exam rooms with behavioral and learning disorders.[11-13]

Genetics can play an important role in many neurodevelopmental disorders, and some cases of certain conditions such as intellectual disability are associated with specific genes. However, most neurodevelopmental disorders have complex and multiple contributors rather than any one clear cause. These disorders likely result from a combination of genetic, biological, psychosocial and environmental risk factors. A broad range of environmental risk factors may affect neurodevelopment, including (but not limited to) maternal use of alcohol, tobacco, or illicit drugs during pregnancy; lower socioeconomic status; preterm birth; low birthweight; the physical environment; and prenatal or childhood exposure to certain environmental contaminants.[14-21]

Lead, methylmercury, and PCBs are widespread environmental contaminants associated with adverse effects on a child's developing brain and nervous system in multiple studies. The National Toxicology Program (NTP) has concluded that childhood lead exposure is associated with reduced cognitive function, including lower intelligence quotient (IQ) and reduced academic achievement.[22] The NTP has also concluded that childhood lead exposure is associated with attention-related behavioral problems (including inattention, hyperactivity, and diagnosed attention-deficit/hyperactivity disorder) and increased incidence of problem behaviors (including delinquent, criminal, or antisocial behavior).[22]

EPA has determined that methylmercury is known to have neurotoxic and developmental effects in humans.[23] Extreme cases of such effects were seen in people prenatally exposed during two high-dose mercury poisoning events in Japan and Iraq, who experienced severe adverse health effects such as cerebral palsy, mental retardation, deafness, and blindness.[24-26] Prospective cohort studies have been conducted in island populations where frequent fish consumption leads to methylmercury exposure in pregnant women at levels much lower than in the poisoning incidents but much greater than those typically observed in the United States. Results from such studies in New Zealand and the Faroe Islands suggest that increased prenatal mercury exposure due to maternal fish consumption was associated with adverse effects on intelligence and decreased functioning in the areas of language, attention, and memory.[26-32] These associations were not seen in initial results reported from a similar study in the Seychelles Islands.[33] However, further studies in the Seychelles found associations between prenatal mercury exposure and some neurodevelopmental deficits after researchers had accounted for the developmental benefits of fish consumption.[34-36] More recent studies conducted in the United States have found associations between neurodevelopmental effects and blood mercury levels within the range typical for U.S. women, after accounting for the beneficial effects of fish consumption during pregnancy.[32,37,38]

Several studies of children who were prenatally exposed to elevated levels of polychlorinated biphenyls (PCBs) have suggested linkages between these contaminants and neurodevelopmental effects, including lowered intelligence and behavioral deficits such as inattention and impulsive behavior.[39-44] Studies have also reported associations between PCB exposure and deficits in learning and memory.[39,45] Most of these studies found that the effects are associated with exposure in the womb resulting from the mother having eaten food contaminated with PCBs,[46-51] although some studies have reported relationships between adverse effects and PCB exposure during infancy and childhood.[45,51-53] Although there is some inconsistency in the epidemiological literature, several reviews of the literature have found that the overall evidence supports a concern for effects of PCBs on children's neurological development.[52,54-58] The Agency for Toxic Substances and Disease Registry has determined that "Substantial data suggest that PCBs play a role in neurobehavioral alterations observed in newborns and young children of women with PCB burdens near background levels."[59] In addition, adverse effects on intelligence and behavior have been found in children of women who were highly exposed to mixtures of PCBs, chlorinated dibenzofurans, and other pollutants prior to conception.[60-63]

A wide variety of other environmental chemicals have been identified as potential concerns for childhood neurological development, but have not been as well studied for these effects as lead, mercury, and PCBs. Concerns for these additional chemicals are based on both laboratory animal studies and human epidemiological research; in most cases, the epidemiological studies are relatively new and the literature is just beginning to develop. Among the chemicals being studied for potential effects on childhood neurological development are organophosphate pesticides, polybrominated diphenyl ether flame retardants (PBDEs), phthalates, bisphenol A (BPA), polycyclic aromatic hydrocarbons (PAHs), arsenic, and perchlorate. Exposure to all of these chemicals is widespread in the United States for both children and adults.[64]

Organophosphate pesticides can interfere with the proper function of the nervous system when exposure is sufficiently high.[65] Many children may have low capacity to detoxify organophosphate pesticides through age 7 years.[66] In addition, recent studies have reported an association between prenatal organophosphate exposure and childhood ADHD in a U.S. community with relatively high exposures to organophosphate pesticides,[67] as well as with exposures found within the general U.S. population.[68] Other recent studies have described associations between prenatal organophosphate pesticide exposures and a variety of neurodevelopmental deficits in childhood, including reduced IQ, perceptual reasoning, and memory.[69-71]

Studies of certain PBDEs have found adverse effects on behavior, learning, and memory in laboratory animals.[72-74] A recent epidemiological study in New York City reported significant associations between children's prenatal exposure to PBDEs and reduced performance on IQ tests and other tests of neurological development in 6-year-old children.[75] Another study in the Netherlands reported significant associations between children's prenatal exposure to PBDEs and reduced performance on some neurodevelopmental tests in 5- and 6-year-old children, while associations with improved performance were observed for other tests.[76]

Two studies of a group of New York City children ages 4 to 9 years reported associations between prenatal exposure to certain phthalates and behavioral deficits, including effects on attention, conduct, and social behaviors.[77,78] Some of the behavioral deficits observed in these studies are similar to those commonly displayed in children with ADHD and conduct disorder. Studies conducted in South Korea of children ages 8 to 11 years reported that children with higher levels of certain phthalate metabolites in their urine were more inattentive and hyperactive, displayed more symptoms of ADHD, and had lower IQ compared with those who had lower levels.[79,80] The exposure levels in these studies are comparable to typical exposures in the U.S. population.

In 2008, the NTP concluded that there is "some concern" for effects of early-life (including prenatal) BPA exposure on brain development and behavior, based on findings of animal studies conducted at relatively low doses.[81] An epidemiological study conducted in Ohio reported an association between prenatal exposure to BPA and effects on children's behavior (increased hyperactivity and aggression) at age 2 years.[82] Another study of prenatal BPA exposure in New York City reported no association between prenatal BPA exposure and social behavior deficits in testing conducted at ages 7 to 9 years.[78]

A series of recent studies conducted in New York City has reported that children of women who were exposed to increased levels of polycyclic aromatic hydrocarbons (PAHs, produced when gasoline and other materials are burned) during pregnancy are more likely to have experienced adverse effects on neurological development (for example, reduced IQ and behavioral problems).[83,84]

Early-life exposure to arsenic has been associated with measures of reduced cognitive function, including lower scores on tests that measure neurobehavioral and intellectual development, in four studies conducted in Asia; however there are some inconsistencies in the findings of these studies.[85] These findings are from countries where arsenic levels in drinking water are generally much higher than in the United States due to high levels of naturally occurring arsenic in groundwater.[86]

Perchlorate is a naturally occurring and man-made chemical that has been found in drinking water[87] and foods[88,89] in the United States. Exposure to elevated levels of perchlorate inhibits iodide uptake into the thyroid gland, thus possibly disrupting the function of the thyroid and potentially leading to a reduction in the production of thyroid hormone.[90,91] Moderate deficits in maternal thyroid hormone levels during early pregnancy have been linked to reduced childhood IQ scores and other neurodevelopmental effects.[92-94]

Interactions of environmental contaminants and other environmental factors may combine to increase the risk of neurodevelopmental disorders. For example, exposure to lead may have stronger effects on neurodevelopment among children with lower socioeconomic status.[21,95]

A child's brain and nervous system are vulnerable to adverse impacts from pollutants because they go through a long developmental process beginning shortly after conception and continuing through adolescence.[96,97] This complex developmental process requires the precise coordination of cell growth and movement, and may be disrupted by even short-term exposures to environmental contaminants if they occur at critical stages of development. This disruption can lead to neurodevelopmental deficits that may have an effect on the child's achievements and behavior even when they do not result in a diagnosable disorder.

Attention-Deficit/Hyperactivity Disorder (ADHD)

Attention-deficit/hyperactivity disorder (ADHD) is a disruptive behavior disorder characterized by symptoms of inattention and/or hyperactivity-impulsivity, occurring in several settings and more frequently and severely than is typical for other individuals in the same stage of development.[98] ADHD can make family and peer relationships difficult, diminish academic performance, and reduce vocational achievement.

As the medical profession has developed a greater understanding of ADHD through the years, the name of this condition has changed. The American Psychiatric Association adopted the name "attention deficit disorder" in the early 1980s and revised it to "attention-deficit/hyperactivity disorder" in 1987.[99] Many children with ADHD have a mix of inattention and hyperactivity/impulsivity behaviors, while some may display primarily hyperactive behavior traits, and others display primarily inattentive traits. It is possible for an individual's primary symptoms of ADHD to change over time.[20] Children with ADHD frequently have other disorders, with parents reporting that about half of children with ADHD have a learning disability and about one in four have a conduct disorder.[2,100]

Other disorders, including anxiety disorders, depression, and learning disabilities, can be expressed with signs and symptoms that resemble those of ADHD. A diagnosis of ADHD requires a certain amount of judgment on the part of a doctor, similar to diagnosis of other mental disorders. Despite the variability among children diagnosed with the disorder and the challenges involved in diagnosis, ADHD has good clinical validity, meaning that impaired children share similarities, exhibit symptoms, respond to treatment, and are recognized with general consistency across clinicians.[20]

A great deal of research on ADHD has focused on aspects of brain functioning that are related to the behaviors associated with ADHD. Although this research is not definitive, it has found that children with ADHD generally have trouble with certain skills involved in problem-solving (referred to collectively as executive function). These skills include working memory (keeping information in mind while briefly doing something else), planning (organizing a sequence of activities to complete a task), response inhibition (suppressing immediate responses when they are inappropriate), and cognitive flexibility (changing an approach when a situation changes). Children with ADHD also generally have problems in maintaining sustained attention to a task (referred to as vigilance), and/or maintaining readiness to respond to new information (referred to as alertness).[20,101,102]

While uncertainties remain, findings to date indicate that ADHD is caused by combinations of genetic and environmental factors. [20,103-106] Much of the research on environmental factors has focused on the fetal environment. Maternal smoking during pregnancy has been associated with increased risk of ADHD in the child in numerous studies, however, this continues to be an active area of research as scientists consider whether other factors related to smoking (e.g., genetic factors, maternal mental health, stress, alcohol use, and low birth weight) may be responsible for associations attributed to smoking.[17,19,107] Findings regarding ADHD and maternal consumption of alcohol during pregnancy are considered more limited and inconsistent.[19,20] Preterm birth and low birth weight have also been found to increase the likelihood that a child will have ADHD.[16,18,20] Psychosocial adversity (representing factors such as low socioeconomic status and in-home conflict) in childhood may also play a role in ADHD.[108]

The potential role of environmental contaminants in contributing to ADHD, either alone or in conjunction with certain genetic susceptibilities or other environmental factors, is becoming better understood as a growing number of studies look explicitly at the relationship between ADHD and exposures to environmental contaminants.

Among environmental contaminants known or suspected to be developmental neurotoxicants, lead has the most extensive evidence of a potential contribution to ADHD. A number of recent epidemiological studies (all published since 2006, with data gathered beginning in 1999 or more recently) conducted in the United States and Asia have reported relationships between increased levels of lead in a child's blood and increased likelihood of ADHD.[55,109-115] In most of these studies, blood lead levels were comparable to levels observed currently in the United States. The potential contribution of childhood lead exposure to the risk of ADHD may be amplified in children of women who smoked cigarettes during pregnancy.[110] In addition, several studies have reported relationships between blood lead levels and the aspects of brain functioning that are most affected in children with ADHD, including sustained attention, alertness, and problem-solving skills (executive functions, specifically cognitive flexibility, working memory, planning, and response inhibition).[22,44,55,116-119] Similar results have been observed in laboratory animal studies.[55,96,120-122] The NTP has concluded that childhood lead exposure is "associated with increased diagnosis of attention-related behavioral problems."[22]

Although no studies evaluating a potential association between PCBs and ADHD itself have been published, a study in Massachusetts reported a relationship between levels of PCBs measured in cord blood and increased ADHD-like behaviors observed by teachers in children at ages 7 to 11 years. PCB levels in this study were generally lower than those measured in other epidemiological studies of PCBs and childhood neurological development.[40] Other research findings also suggest that PCBs may play a role in contributing to ADHD. Several studies in U.S. and European populations, most having elevated exposure to PCBs through the diet, have found generally consistent associations with aspects of brain function that are most affected in children with ADHD, including alertness and problem-solving skills (executive functions, specifically response inhibition, working memory, cognitive flexibility, and planning).[54,55] Studies in laboratory animals have similar findings regarding the mental functions affected by PCB exposure.[55,96]

Studies of other environmental chemicals reporting associations with ADHD or related outcomes have been published in recent years, but findings tend to be much more limited than for lead and PCBs. Findings for phthalates and organophosphate pesticides were noted above. In addition, three studies have reported associations between ADHD or impulsivity and concentrations of certain perfluorinated chemicals measured in the blood of children.[123-125] Studies of mercury have produced generally mixed findings of associations with ADHD or related symptoms and mental functions.[29,111,118,126-128]

Learning Disability

Learning disability (or learning disorder) is a general term for a neurological disorder that affects the way in which a child's brain can receive, process, retain, and respond to information. A child with a learning disability may have trouble learning and using certain skills, including reading, writing, listening, speaking, reasoning, and doing math, although learning disabilities vary from child to child. Children with learning disabilities usually have average or above-average intelligence, but there are differences in the way their brains process information.[129]

As with many other neurodevelopmental disorders, the causes of learning disabilities are not well understood. Often learning disabilities run in the family, suggesting that heredity may play a role in their development. Problems during pregnancy and birth, such as drug or alcohol use during pregnancy, low birth weight, lack of oxygen, or premature or prolonged labor, may also lead to learning disabilities.[130]

As is the case with other neurodevelopmental outcomes, there are generally many more studies of lead exposure that are relevant to learning disabilities than for other environmental contaminants. Several studies have found associations between lead exposure and learning disabilities or reduced classroom performance that are independent of IQ.[119,120,131-133] Exposures to lead have been associated with impaired memory and difficulties or impairments in rule learning, following directions, planning, verbal abilities, speech processing, and classroom performance in children.[22,119,131,134-137] Other findings that may indicate contributions from environmental contaminants to learning disabilities include a study that found associations of both maternal smoking during pregnancy and childhood exposure to environmental tobacco smoke with parent report of a child with a learning disability diagnosis;[138] associations of prenatal mercury exposure with dysfunctions in children's language abilities and memory,[29,30] and associations of prenatal PCB exposure with poorer concentration and memory deficits compared with unexposed children.[39,45]

Autism Spectrum Disorders

Autism spectrum disorders (ASDs) are a group of developmental disabilities defined by significant social, communication, and behavioral impairments. The term "spectrum disorders" refers to the fact that although people with ASDs share some common symptoms, ASDs affect different people in different ways, with some experiencing very mild symptoms and others experiencing severe symptoms. ASDs encompass autistic disorder and the generally less severe forms, Asperger's syndrome and pervasive developmental disorder-not otherwise specified (PDD-NOS). Children with ASDs may lack interest in other people, have trouble showing or talking about feelings, and avoid or resist physical contact. A range of communication problems are seen in children with ASDs: some speak very well, while many children with an ASD do not speak at all. Another hallmark characteristic of ASDs is the demonstration of restrictive or repetitive interests or behaviors, such as lining up toys, flapping hands, rocking his or her body, or spinning in circles.[139]

To date, no single risk factor sufficient to cause ASD has been identified; rather each case is likely to be caused by the combination of multiple genetic and environmental risk factors.[140-142] Several ASD research findings and hypotheses may imply an important role for environmental contaminants. First, there has been a sharp upward trend in reported prevalence that cannot be fully explained by factors such as younger ages at diagnosis, migration patterns, changes in diagnostic criteria, inclusion of milder cases, or increased parental age.[8,9,143-146] Also, the neurological signaling systems that are impaired in children with ASDs can be affected by certain environmental chemicals. For example, several pesticides are known to interfere with acetylcholine (Ach) and γ-aminobutyric acid (GABA) neurotransmission, chemical messenger systems that have been altered in certain subsets of autistic individuals.[147] Some studies have reported associations between certain pharmaceuticals taken by pregnant women and increased incidence of autism, which may suggest that there are biological pathways by which other chemical exposures during pregnancy could increase the risk of autism.[148]

Furthermore, some of the identified genetic risk factors for autism are *de novo* mutations, meaning that the genetic defect is not present in either of the parents' genes, yet can be found in the genes of the child when a new genetic mutation forms in a parent's germ cells (egg or sperm), potentially from exposure to contaminants.[140,142,149,150] Many environmental contaminants have been identified as agents capable of causing mutations in DNA, by leading to oxidative DNA damage and by inhibiting the body's normal ability to repair DNA damage.[151] Some children with autism have been shown to display markers of increased oxidative stress, which may strengthen this line of reasoning.[152-154] Many studies have linked increasing paternal and maternal age with increased risk of ASDs.[144,146,155-157] The role of parental age in increased autism risk might be explained by evidence that shows advanced parental age can contribute significantly to the frequency of *de novo* mutations in a parent's germ cells.[151,158,159] Advanced parental age signifies a longer period of time when environmental exposures may act on germ cells and cause DNA damage and *de novo* mutations. Finally, a recent study concluded that the role of genetic factors in ASDs has been overestimated, and that environmental factors play a greater role than genetic factors in contributing to autism.[141] This study did not evaluate the role of any particular environmental factors, and in this context "environmental factors" are defined broadly to include any influence that is not genetic.

Studies, limited in number and often limited in research design, have examined the possible role that certain environmental contaminants may play in the development of ASDs. A number of these studies have focused on mercury exposures. Earlier studies reported higher levels of mercury in the blood, baby teeth, and urine of children with ASDs compared with control children;[160-162] however, another more recent study reported no difference in the blood mercury levels of children with autism and typically developing children.[163] Proximity to industrial and power plant sources of environmental mercury was reported to be associated with increased autism prevalence in a study conducted in Texas.[164]

Thimerosal is a mercury-containing preservative that is used in some vaccines to prevent contamination and growth of harmful bacteria in vaccine vials. Since 2001, thimerosal has not been used in routinely administered childhood vaccines, with the exception of some influenza vaccines.[165] The Institute of Medicine has rejected the hypothesis of a causal relationship between thimerosal-containing vaccines and autism.[166]

Some studies have also considered air pollutants as possible contributors to autism. A study conducted in the San Francisco Bay Area reported an association between the amount of certain airborne pollutants at a child's place of birth (mercury, cadmium, nickel, trichloroethylene, and vinyl chloride) and the risk for autism, but a similar study in North Carolina and West Virginia did not find such a relationship.[167,168] Another study in California reported that mothers who lived near a freeway at the time of delivery were more likely to have children diagnosed with autism, suggesting that exposure to traffic-related air pollutants may play a role in contributing to ASDs.[169]

Finally, a study in Sweden reported an increased risk of ASDs in children born to families living in homes with polyvinyl chloride (PVC) flooring, which is a source of certain phthalates in indoor environments.[170]

Intellectual Disability (Mental Retardation)

The most commonly used definitions of intellectual disability (also referred to as mental retardation) emphasize subaverage intellectual functioning before the age of 18, usually defined as an IQ less than 70 and impairments in life skills such as communication, self-care, home living, and social or interpersonal skills. Different severity categories, ranging from mild to severe retardation, are defined on the basis of IQ scores.[171,172]

"Intellectual disability" is used as the preferred term for this condition in the disabilities sector, but the term "mental retardation" continues to be used in the contexts of law and public policy when designating eligibility for state and federal programs.[171]

Researchers have identified some causes of intellectual disability, including genetic disorders, traumatic injuries, and prenatal events such as maternal infection or exposure to alcohol.[172,173] However, the causes of intellectual disability are unknown in 30–50% of all cases.[173] The causes are more frequently identified for cases of severe retardation (IQ less than 50), whereas the cause of mild retardation (IQ between 50 and 70) is unknown in more than 75% of cases.[174,175] Exposures to environmental contaminants could be a contributing factor to the cases of mild retardation where the cause is unknown. Exposure to high levels of lead and mercury have been associated with intellectual disability.[23,176-178] Furthermore, lead, mercury, and PCBs all have been found to have adverse effects on intelligence and cognitive functioning in children,[22,26,43,52,179] and recent studies have reported associations of a number of other environmental contaminants with childhood IQ deficits, including organophosphate pesticides,[69-71] PBDEs,[75] phthalates,[79] and PAHs.[83,180] Exposure to environmental contaminants that reduce IQ has the potential to increase the proportion of the population with IQ less than 70, thus increasing the incidence of intellectual disability in an exposed population.[181-183]

Indicators in this Section

The four indicators that follow provide the best nationally representative data available on the prevalence of neurodevelopmental disorders among U.S. children over time. The indicators present the number of children ages 5 to 17 years reported to have ever been diagnosed with ADHD (Indicator H6), learning disabilities (Indicator H7), autism (Indicator H8), and intellectual disability (Indicator H9). These four conditions are examples of neurodevelopmental disorders that may be influenced by exposures to environmental contaminants. Intellectual disability and learning disabilities are disorders in which a child's cognitive or intellectual development is affected, and ADHD is a disorder in which a child's behavioral development is affected. Autism spectrum disorders are disorders in which a child's behavior, communication, and social skills are affected.

Indicator H6: Percentage of children ages 5 to 17 years reported to have attention-deficit/hyperactivity disorder, by sex, 1997–2010

Indicator H7: Percentage of children ages 5 to 17 years reported to have a learning disability, by sex, 1997–2010

Indicator H8: Percentage of children ages 5 to 17 years reported to have autism, 1997–2010

Indicator H9: Percentage of children ages 5 to 17 years reported to have intellectual disability (mental retardation), 1997–2010

About the Indicators: Indicators H6, H7, H8, and H9 present information about the number of children who are reported to have ever been diagnosed with four different neurodevelopmental disorders: attention-deficit/hyperactivity disorder (ADHD), learning disabilities, autism, and intellectual disability. The data come from a national survey that collects health information from a representative sample of the population each year. The four indicators show how the prevalence of children's neurodevelopmental disorders has changed over time, and, when possible, how the prevalence differs between boys and girls.

National Health Interview Survey

The National Health Interview Survey (NHIS) provides nationally representative data on the prevalence of ADHD, learning disabilities, autism, and intellectual disability (mental retardation) in the United States each year. NHIS is a large-scale household interview survey of a representative sample of the civilian noninstitutionalized U.S. population, conducted by the National Center for Health Statistics (NCHS). The interviews are conducted in person at the participants' homes. From 1997–2005, interviews were conducted for approximately 12,000–14,000 children annually. From 2006–2008, interviews were conducted for approximately 9,000–10,000 children per year. In 2009 and 2010, interviews were conducted for approximately 11,000 children per year. The data are obtained by asking a parent or other knowledgeable household adult questions regarding the child's health status. NHIS asks "Has a doctor or health professional ever told you that <child's name> had Attention Deficit/Hyperactivity Disorder (ADHD) or Attention Deficit Disorder (ADD)? Autism? Mental Retardation?" Another question on the NHIS survey asks "Has a representative from a school or a health professional ever told you that <child's name> had a learning disability?"

Data Presented in the Indicators

The following indicators display the prevalence of ADHD, learning disabilities, autism, and intellectual disability among U.S. children, for the years 1997–2010. Diagnosing neurodevelopmental disorders in young children can be difficult: many affected children may not receive a diagnosis until they enter preschool or kindergarten. For this reason, the indicators here show children ages 5 to 17 years. Where data are sufficiently reliable, the indicators provide separate prevalence estimates for boys and girls.

Although the NHIS provides national-level data on the prevalence of neurodevelopmental disorders over a span of many years, NHIS data could underestimate the prevalence of neurodevelopmental disorders. Reasons for underestimation may include late identification of affected children and the exclusion of institutionalized children from the NHIS survey population. A diagnosis of a neurodevelopmental disorder depends not only on the presence of particular symptoms and behaviors in a child, but on concerns being raised by a parent or teacher about the child's behavior, as well as the child's access to a doctor and the accuracy of the doctor's diagnosis. Further, the NHIS relies on parents reporting that their child has been diagnosed with a neurodevelopmental disorder, and the accuracy of parental responses could be affected by cultural and other factors.

Long-term trends in these conditions are difficult to detect with certainty due to a lack of data to track prevalence over many years, as well as changes in awareness and diagnostic criteria, which could explain at least part of the observed increasing trends.[184-186] The NHIS questions also do not assess whether a child currently has a disorder; instead, they provide data on whether a child has ever been diagnosed with a disorder, regardless of their current status.

Survey responses for learning disabilities may be more uncertain than for the other three disorders presented. Whereas survey respondents are asked whether the child has been diagnosed with ADHD, autism, or intellectual disability (mental retardation) by a health professional, for learning disabilities an affirmative response may also include a school representative. It is possible that some parents may respond "yes" to the question regarding learning disabilities based on informal comments made at school, rather than a formal evaluation to determine whether the child has any specific learning disability; similarly, they may give a "yes" answer for children with diagnosed disorders that are not learning disabilities. For example, parents of children with intellectual disability might also respond "yes" to the learning disability question, thinking that any learning problems may apply, even though intellectual disability and learning disabilities are distinct conditions.[2]

Because autism is the only autism spectrum disorder (ASD) referred to in the survey, it is not clear how parents of children with other ASDs, i.e., Asperger's syndrome and PDD-NOS, may have responded. The estimates shown by Indicator H8 could represent underestimates of ASD prevalence if parents of children with Asperger's syndrome and PDD-NOS did not answer yes to the NHIS questions about autism.

In addition to the data shown in the indicator graphs, supplemental tables provide information regarding the prevalence of neurodevelopmental disorders for different age groups and prevalence by race/ethnicity, sex, and family income. These comparisons use the most current four years of data available. The data from four years are combined to increase the statistical reliability of the estimates for each race/ethnicity, sex, and family income group. The tables include prevalence estimates for the following race/ethnicity groups: White non-Hispanic, Black non-Hispanic, Asian non-Hispanic, Hispanic, and "All Other Races." The "All Others Races" category includes all other races not specified, together with those individuals who report more than one race. The limits of the sample design and sample size often prevent statistically reliable estimates for smaller race/ethnicity groups. The data are also tabulated for three income groups: all incomes, income below the poverty level, and greater than or equal to the poverty level.

Please see the Introduction to the Health section for discussion of statistical significance testing applied to these indicators.

Other Estimates of ADHD and Autism Prevalence

In addition to NHIS, other NCHS studies provide data on prevalence of ADHD and ASDs among children. The National Survey of Children's Health (NSCH), conducted in 2003 by NCHS, found that 7.8% of children ages 4 to 17 years had ever been diagnosed with ADHD. The same survey, when conducted again in 2007, found that 9.5% of children ages 4 to 17 years had ever been diagnosed with ADHD.[7] Both estimates are somewhat higher than the ADHD prevalence estimates from the NHIS for those years. The 2007 NSCH also estimates that 7.2% of children ages 4 to 17 years currently have ADHD. The 2007 NSCH also provides information at the state level: North Carolina had the highest rate, with 15.6% of children ages 4 to 17 years having ever been diagnosed with ADHD; the rate was lowest in Nevada, at 5.6%.[7]

In 2002 and 2006, the Centers for Disease Control and Prevention performed thorough data gathering in selected areas to examine the prevalence of ASDs in eight-year-old children. The ASD prevalence estimate for 2002 was 0.66%, or 1 in 152 eight-year-old children, and the estimate for 2006 was 0.9%, or 1 in 110 eight-year-old children.[8,187] The 2007 NSCH also provides an estimate of 1.1% of children ages 3 to 17 years reported to have ASDs, or about 1 in 90.[188]

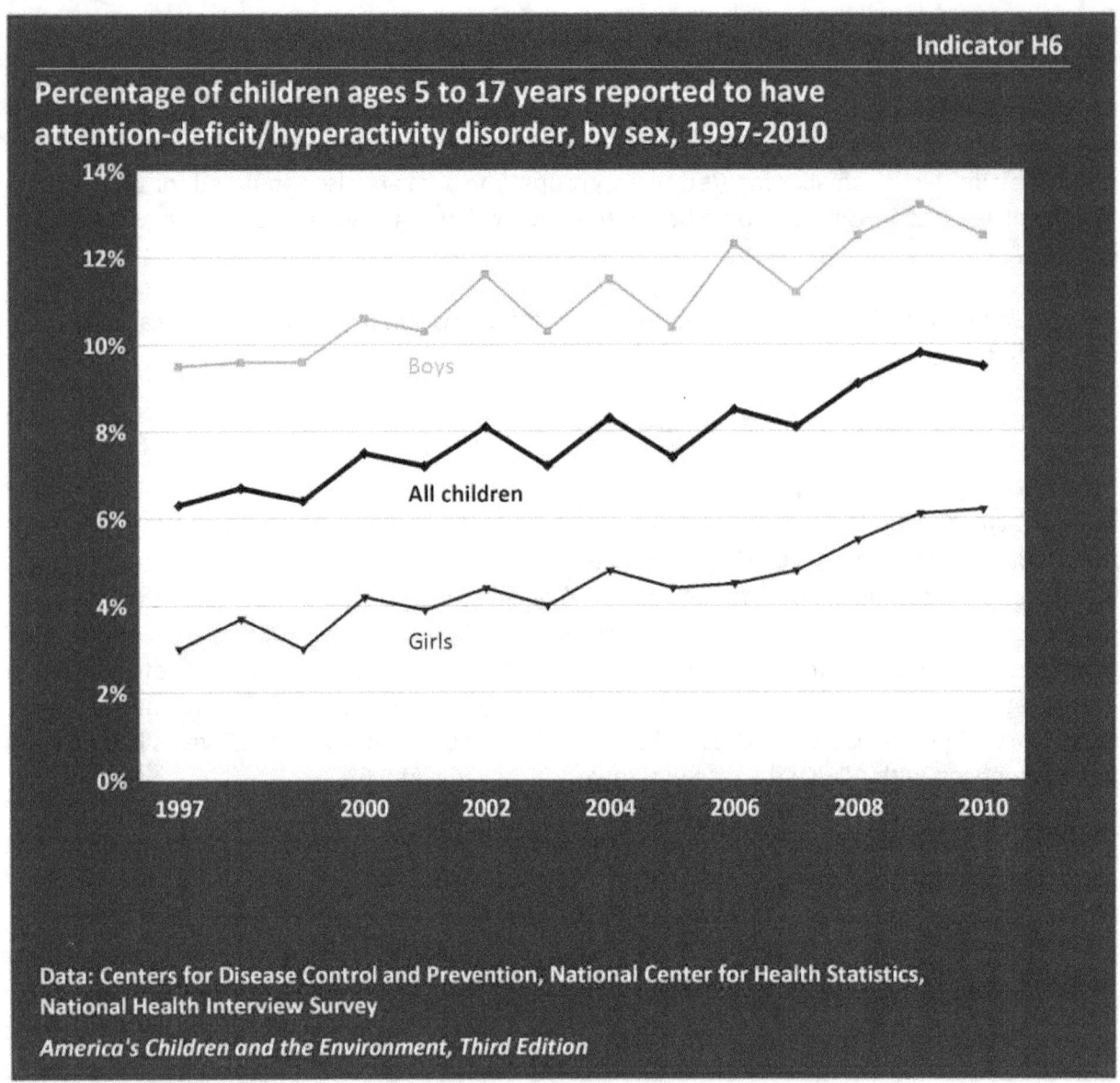

Indicator H6

Percentage of children ages 5 to 17 years reported to have attention-deficit/hyperactivity disorder, by sex, 1997-2010

Data: Centers for Disease Control and Prevention, National Center for Health Statistics, National Health Interview Survey

America's Children and the Environment, Third Edition

Data characterization

- Data for this indicator are obtained from an ongoing annual survey conducted by the National Center for Health Statistics.
- Survey data are representative of the U.S. civilian noninstitutionalized population.
- A parent or other knowledgeable adult in each sampled household is asked questions regarding the child's health status, including if they have ever been told the child has Attention Deficit/Hyperactivity Disorder (ADHD).

- From 1997 to 2010, the proportion of children ages 5 to 17 years reported to have ever been diagnosed with attention-deficit/hyperactivity disorder (ADHD) increased from 6.3% to 9.5%.

- The increasing trend was statistically significant for children overall, and for both boys and girls considered separately.
- For the years 2007–2010, the percentage of boys reported to have ADHD (12.4%) was higher than the rate for girls (5.7%). This difference was statistically significant. (See Table H6a.)
- In 2007–2010, 11.6% of children of "All Other Races," 10.7% of White non-Hispanic children, 10.2% of Black non-Hispanic children, 4.8% of Hispanic children, and 1.7% of Asian non-Hispanic children were reported to have ADHD. (See Table H6b.)
  - These differences were statistically significant, after accounting for the influence of other demographic differences (i.e., differences in age, sex, and family income), with two exceptions: there was no statistically significant difference between children of "All Other Races" and White non-Hispanic children, or between children of "All Other Races" and Black non-Hispanic children.
- In 2007–2010, 11.3% of children from families living below the poverty level were reported to have ADHD compared with 8.6% of children from families living at or above the poverty level. This difference was statistically significant. (See Table H6b.)

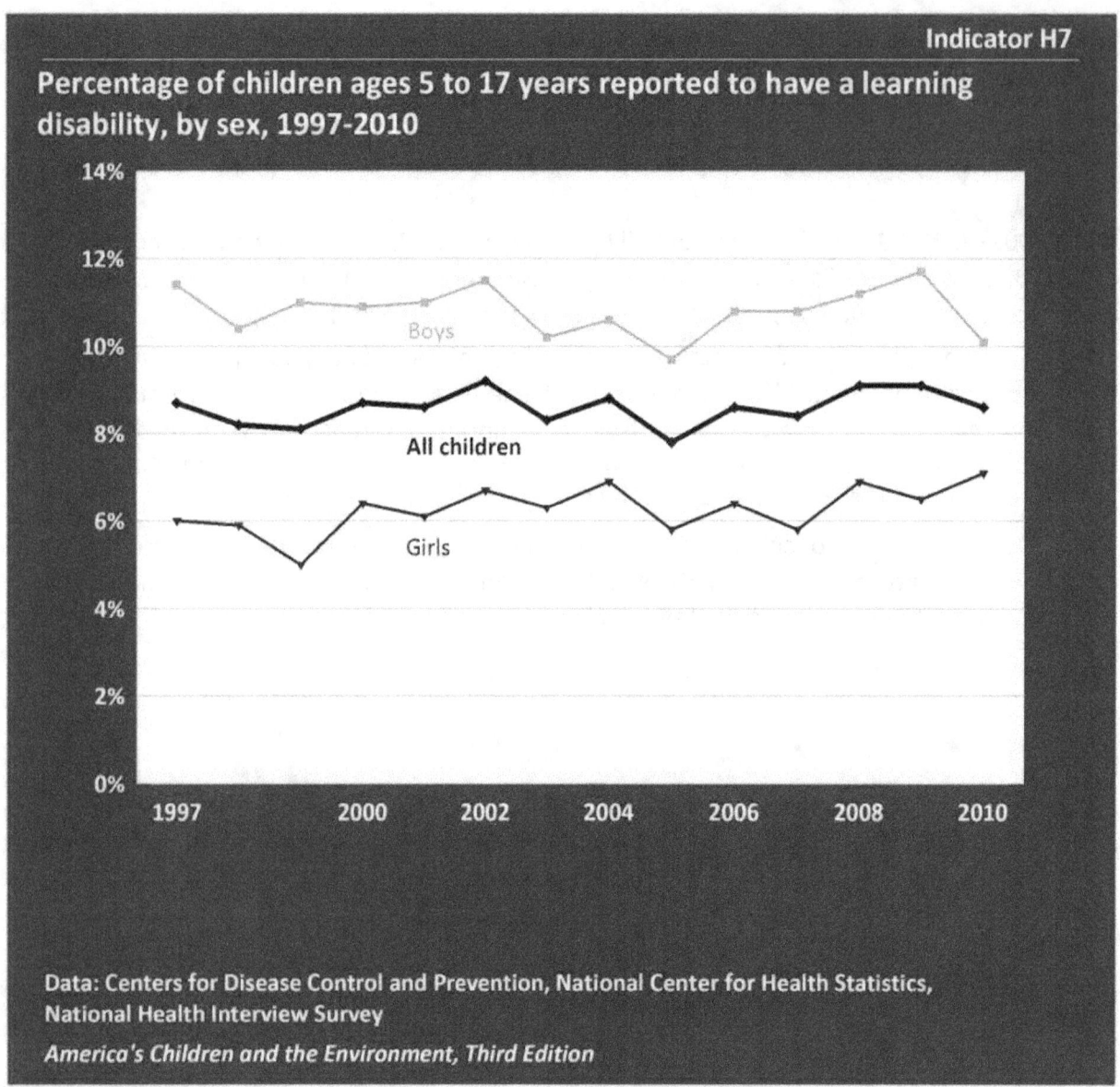

Indicator H7

Percentage of children ages 5 to 17 years reported to have a learning disability, by sex, 1997-2010

Data: Centers for Disease Control and Prevention, National Center for Health Statistics, National Health Interview Survey

America's Children and the Environment, Third Edition

Data characterization

- Data for this indicator are obtained from an ongoing annual survey conducted by the National Center for Health Statistics.
- Survey data are representative of the U.S. civilian noninstitutionalized population.
- A parent or other knowledgeable adult in each sampled household is asked questions regarding the child's health status, including if they have ever been told the child has a learning disability.

- In 2010, 8.6% of children ages 5 to 17 years had ever been diagnosed with a learning disability. There was little change in this percentage between 1997 and 2010.

- For the years 2007–2010, the percentage of boys reported to have a learning disability (10.9%) was higher than for girls (6.6%). This difference was statistically significant. (See Table H7a.)

- The reported prevalence of learning disability varies by race and ethnicity. The highest percentages of learning disability are reported for children of "All Other Races" (11.2%), Black non-Hispanic children (10.2%), and White non-Hispanic children (9.3%). By comparison, 7.2% of Hispanic children are reported to have a learning disability, and Asian non-Hispanic children have the lowest prevalence of learning disability, at 2.7%. (See Table H7b.)

  - The prevalence of learning disability reported for Hispanic children and for Asian non-Hispanic children were lower than for the remaining race/ethnicity groups, and these differences were statistically significant. The difference in prevalence between Hispanic and Asian non-Hispanic children was also statistically significant.

- For the years 2007–2010, the percentage of children reported to have a learning disability was higher for children living below the poverty level (12.6%) compared with those living at or above the poverty level (7.9%), a statistically significant difference. (See Table H7b.)

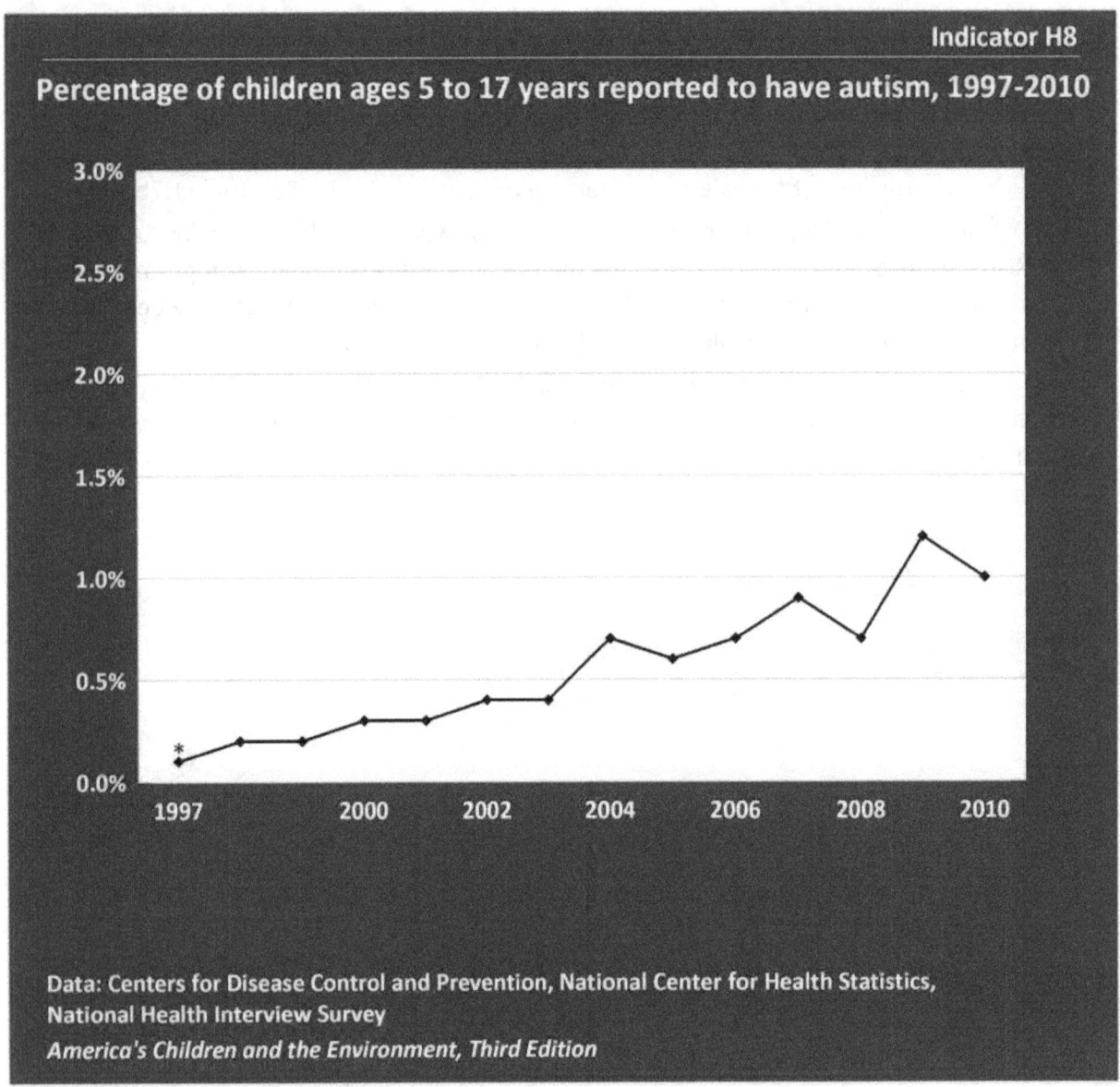

Percentage of children ages 5 to 17 years reported to have autism, 1997-2010

Data: Centers for Disease Control and Prevention, National Center for Health Statistics,
National Health Interview Survey

America's Children and the Environment, Third Edition

* The estimate should be interpreted with caution because the standard error of the estimate is relatively large: the relative standard error, RSE, is at least 30% but is less than 40% (RSE = standard error divided by the estimate).

Data characterization
- Data for this indicator are obtained from an ongoing annual survey conducted by the National Center for Health Statistics.
- Survey data are representative of the U.S. civilian noninstitutionalized population.
- A parent or other knowledgeable adult in each sampled household is asked questions regarding the child's health status, including if they have ever been told the child has autism.

- The percentage of children ages 5 to 17 years reported to have ever been diagnosed with autism rose from 0.1% in 1997 to 1.0% in 2010. This increasing trend was statistically significant.

- For the years 2007–2010, the rate of reported autism was more than three times higher in boys than in girls, 1.5% and 0.4%, respectively. This difference was statistically significant. (See Table H8a.)

- The reported prevalence of autism varies by race/ethnicity. The highest prevalence of autism is for children of "All Other Races" (1.7%) and White non-Hispanic children (1.1%). Autism prevalence was lower among Asian non-Hispanic children (0.8%), Black non-Hispanic children (0.7%), and Hispanic children (0.6%). (See Table H8b.)

  - The prevalence of autism for both White non-Hispanic children and children of "All Other Races" was statistically significantly different from the prevalence for both Black non-Hispanic children and Hispanic children.

- For the years 2007–2010, the prevalence of autism was similar for children living below the poverty level and those living at or above the poverty level. (See Table H8b.)

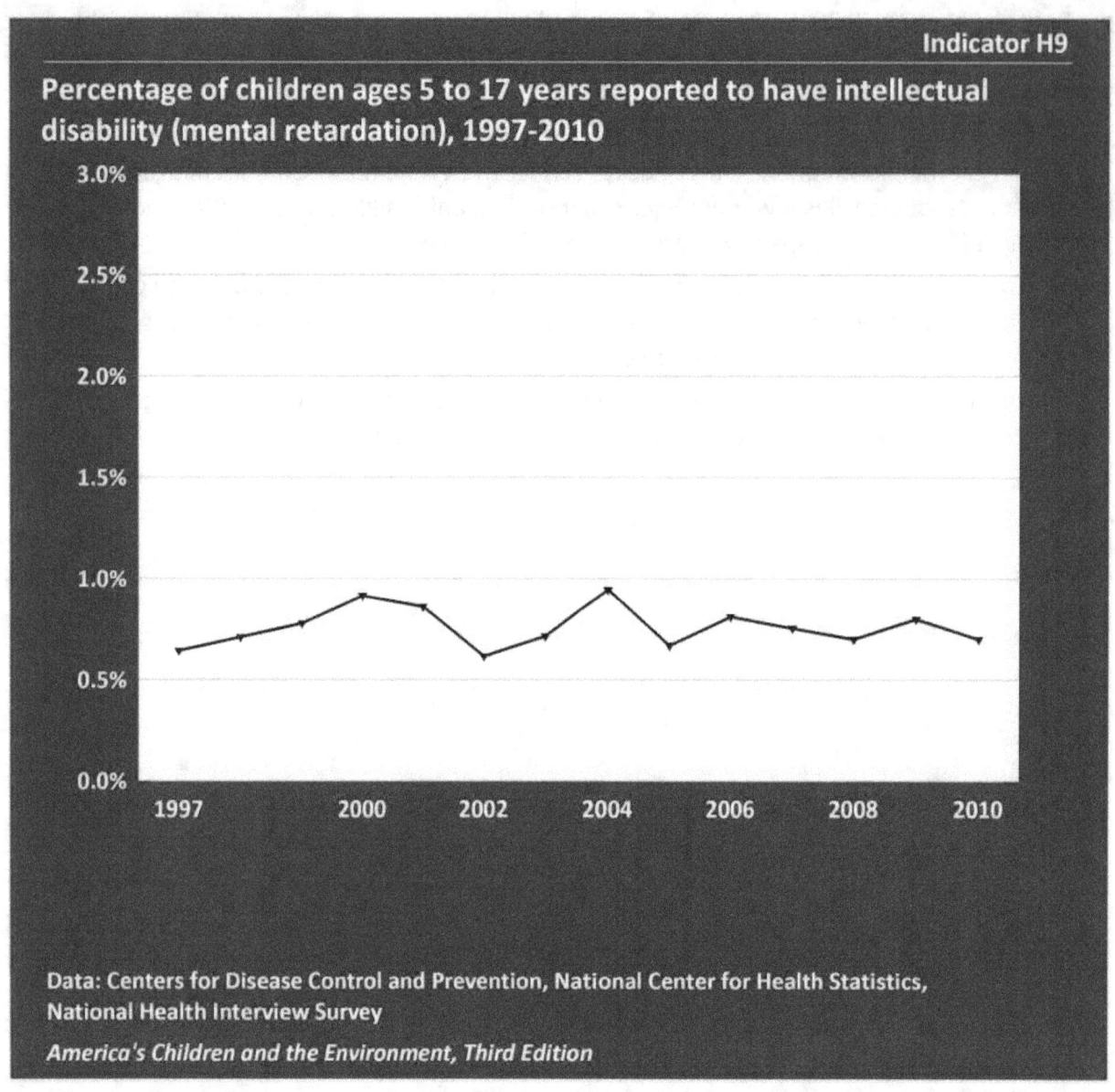

Indicator H9

Percentage of children ages 5 to 17 years reported to have intellectual disability (mental retardation), 1997-2010

Data: Centers for Disease Control and Prevention, National Center for Health Statistics, National Health Interview Survey

America's Children and the Environment, Third Edition

Data characterization

- Data for this indicator are obtained from an ongoing annual survey conducted by the National Center for Health Statistics.
- Survey data are representative of the U.S. civilian noninstitutionalized population.
- A parent or other knowledgeable adult in each sampled household is asked questions regarding the child's health status, including if they have ever been told the child has mental retardation.

- In 2010, 0.7% of children ages 5 to 17 years were reported to have ever been diagnosed with intellectual disability (mental retardation). There was little change in this percentage between 1997 and 2010.

- In 2007–2010, the percentage of boys reported to have intellectual disability (0.9%) was higher than for girls (0.6%). This difference was statistically significant. (See Table H9a.)

- In 2007–2010, there was little difference by race/ethnicity in the reported prevalence of intellectual disability. (See Table H9b.)

- In 2007–2010, 1.2% of children from families with incomes below the poverty level were reported to have intellectual disability, compared with 0.7% of children from families at or above the poverty level, a statistically significant difference. (See Table H9b.)

Obesity | Health

Obesity

Obesity is the term used to indicate the high range of weight for an individual of given height that is associated with adverse health effects.[1] Definitions of overweight and obesity for adults are based on set cutoff points directly related to an individual's body mass index (BMI, weight in kilograms divided by the square of height in meters). Essential to this definition is that a high degree of body weight be associated with a large amount of body fat. The BMI is correlated to body fat, but BMI varies with age and sex in children more than it does in adults. Thus the designation of a child or adolescent (ages 2 to 19 years) as either overweight or obese is based on comparing his or her BMI to a sex- and age-specific reference population (the CDC growth charts). Children and adolescents between the 85th and 94th percentiles of BMI-for-age are considered overweight; those greater than or equal to the 95th percentile are considered obese. The percentiles used to identify children as overweight or obese are fixed, and based on data collected from 1963–1980 (or, for children ages 2 to 6 years, data from 1963–1994).[1-3]

The prevalence of excessive body weight in the United States population has been increasing for several decades, though it has stabilized over the last several years.[4-7] BMI is the most common screening measure used to determine whether an individual may be overweight or obese. The BMI does not measure body fat directly, but is used as a surrogate measure since it correlates with direct measures of body fat, especially at high BMI levels, and is inexpensive and easy to obtain in a clinical setting. The significance of a child being overweight is complicated by the BMI's inability to distinguish between differences in mass due to muscle or due to the unhealthy accumulation of fatty tissue. A recent study found that less than half of "overweight" children had excess body fat, and that there are differences among race/ethnicity groups in the amount of body fat for a given BMI in children.[8] Among children with an elevated BMI, some may have excess body fat, and others may be incorrectly identified as overweight because they have a higher amount of mass attributed to nonfatty tissue. Despite the limitations imposed by measuring the BMI, a rise in the prevalence of overweight children is cause for concern, since overweight children are more likely to become overweight or obese adults.[9-11]

Obesity has rapidly become a serious public health concern in the United States, and is associated with several adverse health effects in childhood and later in life, including cardiovascular disease risk factors (which includes hypertension and altered lipid levels),[12-22] cancer,[15,23,24] psychological stress,[25-28] asthma,[29-31] and diabetes.[32-37] Some studies have found a relationship between obesity and early onset of puberty and early menarche in girls,[38-40] although other research has found differences in the timing of puberty even after controlling for BMI in the population.[41] As such, the extent to which the obesity epidemic may contribute to early puberty is unclear.

An emerging body of research suggests there may be common biological mechanisms underlying a cluster of adverse health effects (obesity, hypertension, altered lipid levels, and other metabolic abnormalities) referred to as metabolic syndrome. While the clinical utility of a diagnosis of metabolic syndrome is debated in the medical literature,[42-44] the term describes an area of active research, and prospective data demonstrate the relevance of metabolic syndrome in obese children for both type 2 diabetes[45] and cardiovascular disease.[46] Metabolic syndrome has been identified in obese children and adolescents, and studies suggest a developmental origin of the condition.[47-49] The consideration of obesity and metabolic effects as a group is supported by findings in laboratory animals, where early-life exposure to certain organophosphate pesticides can disrupt adult lipid metabolism, induce weight gain, and cause other metabolic responses that mimic those seen in diabetes and obesity.[50-52] Given these relationships, obesity and other health conditions related to metabolism are discussed below.

Obesity is due primarily to an imbalance between caloric intake and activity. Increased caloric intake and reduced physical activity are likely the major drivers of obesity in children. Researchers are also investigating whether exposures to certain environmental chemical exposures may play a contributing role in childhood obesity.[53,54] These chemicals, which are referred to as obesogens, are thought to be capable of disrupting the human body's regulation of metabolism and the accumulation of fatty tissue.[55] Studies have also reported associations between exposure to certain chemicals and diabetes in adults. Diabetes (Type 2) results from the body's inability to regulate blood sugar levels with insulin in response to dietary intake, and is positively associated with the increasing rates of obesity seen in the U.S. population.[56] Excess body weight is a risk factor for Type 2 diabetes. In the past, Type 2 diabetes has been diagnosed almost exclusively in adult populations, but it is now being diagnosed in youth—although with low prevalence (0.25%).[56-59] However, the clinical designation of prediabetes (elevated blood glucose levels that do not meet the diagnostic criteria for diabetes) is prevalent in obese youth.[35]

While the possible contribution of chemical exposures to obesity is not clear, a number of animal and cellular studies provide some evidence that environmental chemical exposures may contribute to obesity and diabetes. Studies finding associations between chemical exposures and obesity in children are limited. A recent study reported that prenatal exposure to high levels of hexachlorobenzene was associated with increased BMI and weight in children at 6.5 years.[60] Another recent study in Belgium, at relatively high exposure levels within the general population, reported an association between prenatal exposure to DDE (the primary metabolite of the pesticide DDT) and BMI, as well as an association between exposure to polychlorinated biphenyls (PCBs) and increased BMI during early childhood.[61] In adults, associations have been reported between diabetes and both PCBs and dioxins at levels of exposure seen in the U.S. population.[62,63] A study of adult occupational exposures to organochlorine and organophosphate pesticides reported an increased risk of diabetes in exposed workers.[64] However, other studies have reported no association between these exposures and markers of obesity or diabetes.[65-68] Several animal and cellular studies suggest that endocrine-disrupting chemicals (including bisphenol A, diethylstilbestrol, and tributyltin) may contribute to increased weight and diabetes.[69-73] After reviewing these findings, scientists at a National Toxicology Program-sponsored workshop concluded that existing research provides evidence of plausibility (varying from "suggestive" to "strong" evidence) that several environmental chemicals could contribute to obesity and/or diabetes.[74] For example, scientists concluded that the available data support the biological plausibility that exposure to a number of classes of pesticides may affect risk factors for obesity and diabetes. The National Institutes of Health Strategic Plan for Obesity Research and the White House Task Force on Childhood Obesity Report to the President also acknowledge a potential relationship between environmental exposures and obesity and cite the need for further research.[75,76]

Research has also considered a role for air pollution in childhood obesity and diabetes. In one recent study, adult mice fed a high-fat diet and exposed to concentrated particulate air pollution ($PM_{2.5}$) experienced an increase in blood glucose levels and insulin resistance, which are precursors of diabetes.[77] Other studies in animals and children have reported that obesity may result in greater susceptibility to the adverse effects of airborne pollutants such as $PM_{2.5}$ and ozone, including airway inflammation, cardiovascular effects, and increased deposition of particles in the lungs.[30,78,79] Air pollution may contribute to childhood obesity by limiting the number of days when air quality is appropriate for outdoor recreational activity, particularly in children with pre-existing respiratory conditions such as wheeze and asthma.[80] Animal studies further suggest that diet-induced obesity may increase susceptibility to the effects of environmental toxicants such as PCBs, dioxins, and acrylamide.[81-83]

Other environmental factors are thought to contribute to the increasing rates of overweight and obesity seen in the U.S. population. The term "built environment" is used to describe the physical elements of the environment for a population.[84,85] Multiple reviews of the literature have concluded that several properties of the built environment, including the extent of urban sprawl, housing density, access to food outlets, and access to recreational facilities, may be associated with overweight and obesity and/or levels of physical activity in children.[84-90] The relationship between characteristics of the built environment and obesity is likely more significant in children than adults, because children are less able to leave their local environment without the help of an adult.[91,92] Built environments that promote exercise through the inclusion of nearby recreational areas and walkable communities, and those that provide healthy eating options through reducing the number of fast food restaurants while providing access to fresh produce, are thought to reduce the frequency of obesity in children.[84,85,93] "Green" environments that contain a greater number of natural environments and features such as parks, trees, and nature trails, may contribute to increased levels of physical activity in children that can reduce rates of obesity.[94]

Socioeconomically disadvantaged populations are more likely to be located in built environments with characteristics that promote lifestyles that increase rates of obesity in children.[95-97] However, a child living in a suburban community with a higher socioeconomic status may spend greater amounts of time commuting in a car rather than walking, which may also contribute to a sedentary lifestyle that promotes obesity.[98,99] Factors contributing to the prevalence of obesity may differ among environments. Previous research had suggested that differences in obesity rates in rural or urban environments were small.[100,101] However, other recent studies have identified a higher prevalence of obesity in rural compared to urban environments.[87,102] The complex interplay of behavioral, environmental, and physiological factors and the disparities in pediatric obesity observed in the population add to the difficulty in identifying effective interventions.

The following indicators present the best nationally representative data on obesity in the U.S. child population. The first indicator shows the prevalence of obesity among children ages 2 to 17 years from 1976–2008. The second indicator presents the current prevalence of obesity by race/ethnicity and family income, using data from 2005–2008. Together these indicators highlight basic trends and current status in prevalence of childhood obesity in the United States.

Indicator H10: Percentage of children ages 2 to 17 years who were obese, 1976–2008

Indicator H11: Percentage of children ages 2 to 17 years who were obese, by race/ethnicity and family income, 2005–2008

About the Indicators: Indicators H10 and H11 present the prevalence of obesity in U.S. children ages 2 to 17 years. The data are from a national survey that measures weight and height in a representative sample of the U.S. population every two years. Indicator H10 shows the trend in obesity prevalence from 1976–2008. Indicator H11 presents comparisons of current obesity rates in children of different race/ethnicities and income levels, using data for 2005–2008.

NHANES

The National Health and Nutrition Examination Survey (NHANES) provides data on childhood obesity in the United States. NHANES is a nationally representative survey of the health and nutritional status of the civilian noninstitutionalized U.S. population, conducted by the National Center for Health Statistics. Interviews and physical examinations are conducted with approximately 10,000 people in each two-year year survey cycle. Height and weight are measured for survey participants of all ages.

Obesity and BMI

Determination of obesity in children is based on the calculation of body mass index (BMI), which is correlated with body fat.[103] First, the BMI is calculated by dividing an individual's weight in kilograms by the square of his or her height in meters. For children and teenagers in the United States, the BMI number is then compared with an age- and sex-specific reference population based on the 2000 CDC growth charts. These charts are based on national data collected from 1963–1994 for children 2 through 6 years of age and from 1963–1980 for children ages 7 years and older.[2] These growth charts apply to all racial and ethnic groups, and were obtained from nationally representative surveys. Children and teenagers with BMIs at or above the 95th percentile on the growth charts are classified as obese.[1,3]

Data Presented in the Indicators

Indicator H10 presents the percentage of children ages 2 to 17 years who were obese from NHANES surveys conducted from 1976 through 2008. Indicator H11 presents the current prevalence of childhood obesity by race/ethnicity and family income using the 2005–2006 and 2007–2008 surveys combined. The data from two NHANES cycles are combined to increase the statistical reliability of the estimates for each race/ethnicity and income group, and to reduce any possible influence of geographic variability that may occur in two-year NHANES data. Four race/ethnicity groups are presented in Indicator H11: White non-Hispanic, Black non-Hispanic, Mexican-American, and "All Other Races/Ethnicities." The "All Other Races/Ethnicities" category includes all other races and ethnicities not specified, together with those individuals who report more than one race. The limits of the sample design and sample size often prevent statistically reliable estimates for smaller race/ethnicity groups. The data are also tabulated across three income categories: all incomes, below the poverty level, and greater than or equal to the poverty level.

Please see the Introduction to the Health section for discussion of statistical significance testing applied to these indicators.

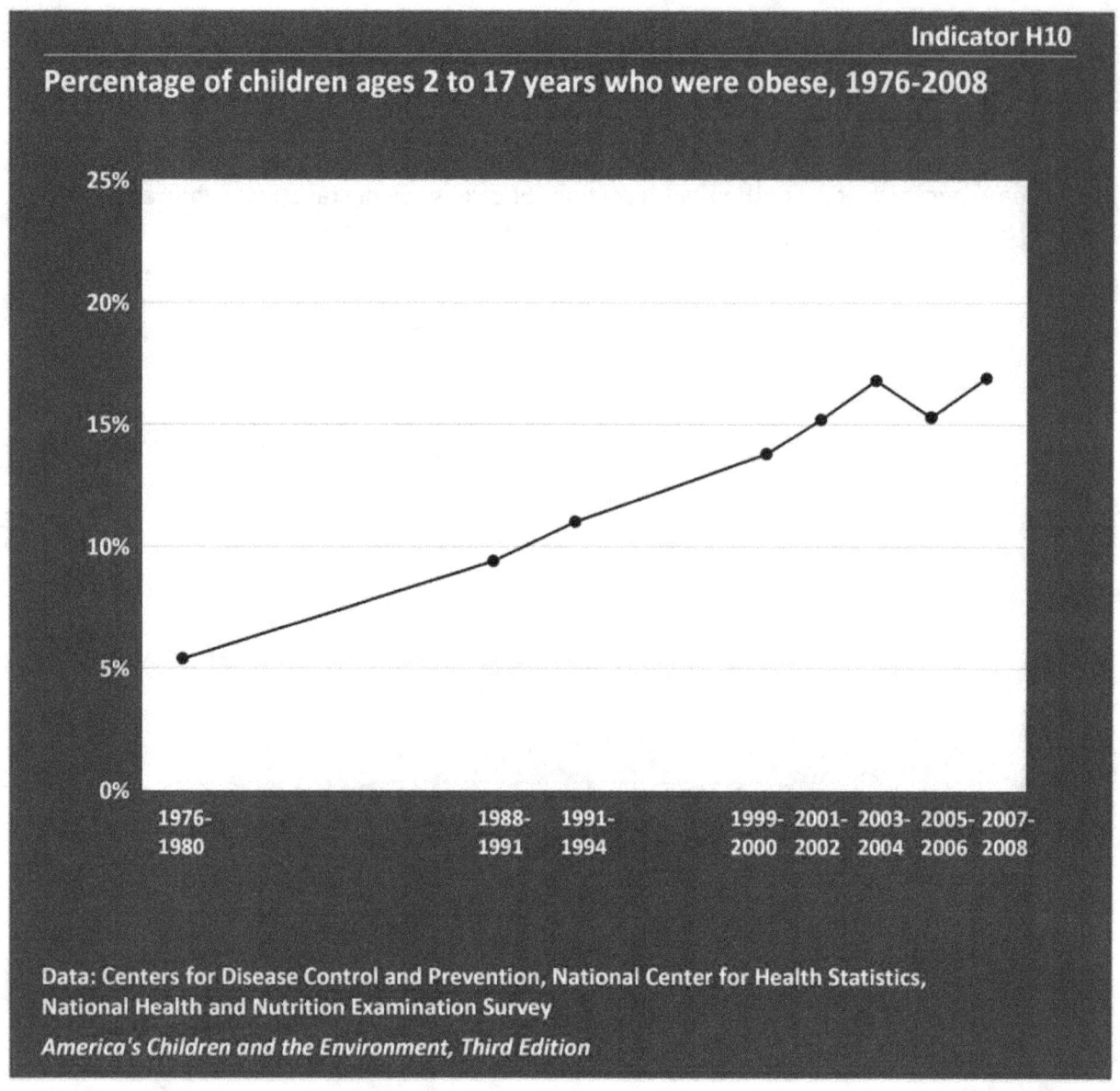

Indicator H10

Percentage of children ages 2 to 17 years who were obese, 1976-2008

Data: Centers for Disease Control and Prevention, National Center for Health Statistics,
National Health and Nutrition Examination Survey

America's Children and the Environment, Third Edition

Data characterization

- Data for this indicator are obtained from an ongoing continuous survey conducted by the National Center for Health Statistics.
- Survey data are representative of the U.S. civilian noninstitutionalized population.
- Height and weight are measured in individual survey participants.

- Between 1976–1980 and 2007–2008, the percentage of children identified as obese showed an increasing trend. In 1976–1980, 5% of children ages 2 to 17 years were obese. This percentage reached a high of 17% in 2007–2008. Between 1999–2000 and 2007–2008, the percentage of children identified as obese remained between 15% and 17%.

  - From 1976–2008, the increasing trend in prevalence of obese children was statistically significant for children overall, and for children of each race/ethnicity (See Table H10) and age group (Table H10a). From 1999–2008, the trends were not statistically significant for each of these groups.

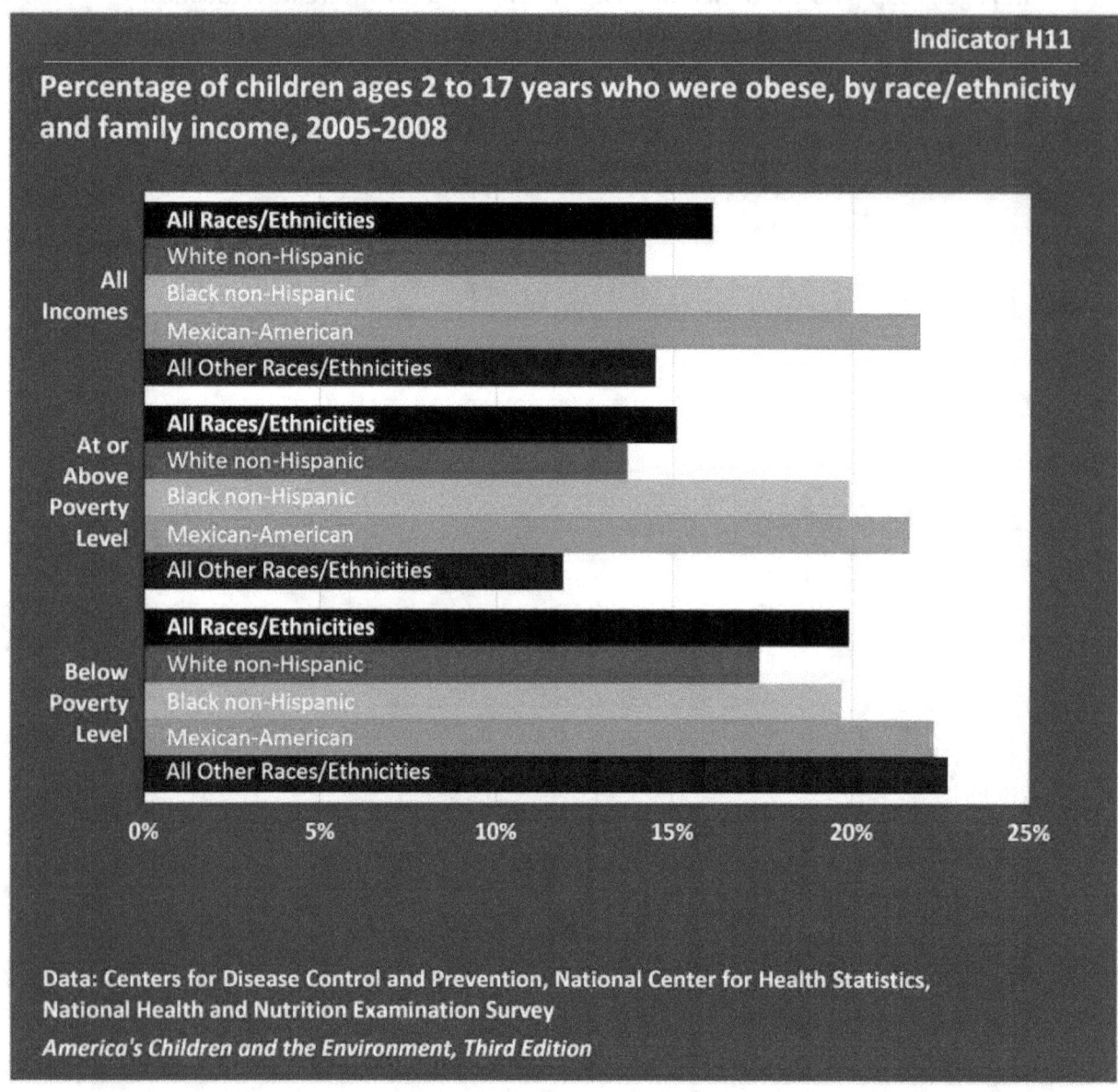

Indicator H11

Percentage of children ages 2 to 17 years who were obese, by race/ethnicity and family income, 2005-2008

Data: Centers for Disease Control and Prevention, National Center for Health Statistics, National Health and Nutrition Examination Survey

America's Children and the Environment, Third Edition

Data characterization

- Data for this indicator are obtained from an ongoing continuous survey conducted by the National Center for Health Statistics.
- Survey data are representative of the U.S. civilian noninstitutionalized population.
- Height and weight are measured in individual survey participants.

- In 2005–2008, 16% of children ages 2 to 17 years were classified as obese.

- In 2005–2008, a higher percentage of Mexican-American and Black non-Hispanic children were obese at 22% and 20%, respectively, compared with 14% of White non-Hispanic children and 14% of children of "All Other Races/Ethnicities."

- The greater prevalence of obesity among Mexican-American and Black non-Hispanic children, compared with the lower prevalence among White non-Hispanic children and children of "All Other Races/Ethnicities," was statistically significant.

- Among children overall, the prevalence of obesity was greater in children with family incomes below poverty level than in those above poverty level. However, when accounting for differences by race/ethnicity as well as poverty status, children of "All Other Races/Ethnicities" were the only group to have a statistically significant association between low family income and higher prevalence of obesity.

Adverse Birth Outcomes

The period of gestation is a crucial determinant of an infant's health and survival for years to come. Two measures that may be used to understand the quality of an infant's gestation are 1) length of gestation (pregnancy length) and 2) birth weight. Normal term pregnancies last between 37 and 41 completed weeks, allowing for more complete development of an infant's organs and systems.[1] Preterm birth is defined as a live birth before 37 completed weeks of gestation.[1] Birth weight is determined by two factors: length of gestation and fetal growth (the rate at which an infant develops and increases in size). Low birth weight infants are defined as weighing less than 2,500 grams (about 5 pounds, 8 ounces).[2] Infants may be born with a low birth weight because they were born early, because their growth while in utero has been restricted, or both. Because they have had sufficient time to develop, infants born at term with low birth weight are usually considered growth restricted. Because birth weight alone does not always indicate whether an infant's fetal growth has been restricted, other measurements such as birth length, head circumference, and abdominal circumference are also used.

Other adverse birth outcomes that are not discussed here include post-term birth, high birth weight, neonatal mortality, and birth defects, a specific group of adverse birth outcomes that include structural and functional abnormalities.

Preterm and low birth weight infants are at greater risk for mortality and a variety of health and developmental problems. As a result, the birth of a preterm or low birth weight infant can have significant emotional and economic effects on the infant's family.[3] Conditions related to preterm birth and low birth weight are the second leading cause of infant death in the United States (after birth defects).[4] The infant mortality rate for low birth weight infants is about 25 times that of the infant mortality rate for normal weight babies. Likewise, the infant mortality rate for late preterm babies (34–36 weeks of gestation) is about three times the infant mortality rate for term babies, and the infant mortality rate for very preterm babies (less than 32 weeks of gestation) is 75 times that of term babies.[4] Preterm infants may experience complications such as acute respiratory, gastrointestinal, immunologic, and central nervous system problems. Longer-term effects of preterm birth, including motor, cognitive, visual, hearing, behavioral, social-emotional, health, and growth problems, may not become apparent for years and may persist throughout a child's life into adulthood. It is important to recognize that not all infants born before 37 completed weeks have the same risk of adverse health outcomes. As gestational age decreases, the risk of morbidity and mortality increases greatly. Also, recent research suggests that even early term births, those at 37 or 38 weeks, are at increased risk of respiratory and other adverse neonatal outcomes.[5-7]

Because many of the effects of low birth weight are due to being born immature and unprepared for life outside the womb, morbidities associated with low birth weight often overlap with those of preterm birth. Low birth weight infants are more likely to have underdeveloped lungs and breathing problems; heart problems (which can lead to heart failure); immature and improperly functioning livers; too many or too few red blood cells (polycythemia or anemia); inadequate body fat, leading to trouble maintaining a normal body temperature; feeding problems; and increased risk of infection.[2] Furthermore, the process of growth restriction may exert its own negative effects aside from often producing low birth weight infants. Data suggest that fetuses with a declining growth rate may make adaptations, such as preserving brain growth, in order to survive adverse intrauterine conditions. Such adaptations can have physiological costs, and may have effects on fetal brain development, cardiac and renal function, and adult health.[8] The theory of fetal origins of adult disease postulates that certain types of chemical, nutritional, or stress-related exposures in utero can alter the programming of fetal cells in ways that are not apparent at birth, but are predictive of disease risk later in life. Birth weight and measures of growth restriction are used as proxies for these changes and have been associated with diseases in adulthood, including cardiovascular disease, obesity, metabolic disorders, and cancer.[9]

For many years, the rates of both preterm birth and low birth weight have been increasing;[10] however, starting in 2006 this pattern seems to be partially reversing as the rate of preterm birth is now declining. A number of factors may contribute to increasing rates of preterm birth and low birth weight, including increases in maternal age, rates of multiple births (e.g., twins, triplets), use of early Cesarean sections and labor inductions, changes in neonatal technology, and use of assisted reproductive technologies (e.g., *in vitro* fertilization).[3] Multiple births run a higher risk of preterm birth and low birth weight, and the rates of multiple births have increased in recent decades. The rate of twin births increased 70% from 1980–2004, but has been essentially stable since that time. The rate of triplet and higher-order births increased 400% from 1980 to 1998, but since that time has been trending downward.[11] Advances in medical technology that allow for resuscitation of infants born at increasingly early gestational ages may also contribute to the increase in percentage of births that are preterm, since many of those infants would not have survived previously and thus would have be characterized as fetal deaths. Other factors linked to preterm birth and low birth weight include birth defects; chronic maternal health problems (e.g., high blood pressure); maternal use of tobacco, alcohol, and illicit drugs; maternal and fetal infections; placental problems; inadequate maternal weight gain; and socioeconomic factors (e.g., low income and poor education).[12-16]

Rates of low birth weight and preterm birth can vary greatly by maternal race/ethnicity. Black women have consistently had higher rates of preterm and low birth weight babies.[17] While it has been suggested that race is a proxy for differences in socioeconomic status (SES), most studies that have controlled for differences in SES continue to find persistent birth outcomes differences between Black and White women.[17-20] Similarly, studies that have adjusted for other risk factors, such as risky behavior during pregnancy and use of prenatal care, have found these persistent Black-White differences in birth outcomes as well.[4,21,22]

While maternal characteristics and obstetric practices play an important role in preterm birth and low birth weight, other factors—including environmental contaminants—may also contribute to adverse birth outcomes.[23] A growing number of studies have examined the possible role that exposure to environmental contaminants may play in the causation of preterm birth and low birth weight. The evidence is particularly strong for environmental tobacco smoke (ETS) and lead. The Surgeon General has determined that exposure of pregnant women to ETS causes a small reduction in mean birth weight, and that the evidence is suggestive (but not sufficient to infer causation) of a relationship between maternal exposure to environmental tobacco smoke during pregnancy and preterm delivery.[24] The National Toxicology Program has concluded that maternal exposure to lead is known to cause reduced fetal growth, and that there is limited evidence of an association with preterm birth.[25]

In recent years, the potential effects of common air pollutants on adverse birth outcomes have received more attention. A number of large epidemiological studies (many with 10,000+ participants) from several countries have identified potential links between elevated levels of exposure to particulate matter (PM), sulfur dioxide (SO_2), nitrogen dioxide (NO_2), and carbon monoxide (CO) exposure and outcomes such as decreased fetal growth, low birth weight, and preterm birth.[26-40] Several of these studies have identified such links to adverse birth outcomes even in regions with relatively low ambient air pollution levels.[27,29,33,36] In such epidemiological studies, researchers make an effort, when data are available, to adjust for other factors that may also lead to an increased risk of low birth weight or preterm birth, such as mother's age, smoking status, race, and income.[41] Articles reviewing the findings from these studies have generally concluded that these air pollutants likely have an adverse effect on birth outcomes, although methodological inconsistencies across studies have made definitive conclusions difficult.[42-45] In addition, studies have reported associations between elevated levels of exposure to airborne polycyclic aromatic hydrocarbons (PAHs), generated largely by fossil fuel combustion, and reduced birth weight and fetal growth restriction, especially when in combination with ETS exposure.[46-49] Other studies have reported associations between living in proximity to traffic during pregnancy and increased risk of preterm birth and low birth weight, although an extensive review study concluded that there is inadequate and insufficient evidence to infer a causal relationship.[50-54]

In addition to air pollutants, several other environmental chemicals have been studied for possible roles in contributing to adverse birth outcomes. A handful of studies with typical population-level exposure levels have reported associations between prenatal exposure to some phthalates and preterm birth, shorter gestational length, and low birth weight; however, one study reported phthalate exposure to be associated with longer gestational length and increased risk of delivery by Cesarean section.[55-59]

A limited number of studies suggest that prenatal exposure to another class of chemicals, polychlorinated biphenyls (PCBs), may lead to preterm birth and low birth weight or otherwise restrict fetal growth.[60-63] One study examining women from the Danish National Birth Cohort reported that elevated exposure to PCBs from fatty fish consumption was associated with lower birth weight. The study found that infants born to highly exposed women weighed, on average, about 5.5 ounces less than infants born to women with relatively low PCB exposure.[64] Another study looked at a historical cohort of women who were pregnant prior to the 1979 ban of PCBs, and did not observe any relation between levels of PCB exposure and low birth weight or shorter pregnancy length.[65] Some human health studies have reported associations between prenatal exposure to perfluorinated compounds (PFCs)—particularly perfluorooctane sulfonic acid (PFOS) and perfluorooctanoic acid (PFOA)—and a range of adverse birth outcomes, such as low birth weight, decreased head circumference, reduced birth length, and smaller abdominal circumference.[66-70] However, there are inconsistencies in the results of these studies, and two other studies did not find an association between prenatal PFC exposure and birth weight.[71,72] The participants in all of these studies had PFC blood serum levels comparable to levels in the general population. Studies of disinfection byproducts in drinking water as possible causes of adverse birth outcomes are also conflicting, with recent evidence indicating that there may be no effect on preterm birth.[73-75] Studies of arsenic in drinking water and birth outcomes have produced similarly mixed results.[76-78] For the following environmental contaminants, there is some evidence from animal studies and a limited number of studies in humans of possible associations with adverse birth outcomes, particularly reduced fetal growth: benzene,[79] herbicides,[80] bisphenol A (BPA),[81] dioxins and dioxin-like chemicals,[82] and manganese.[83]

This section presents two indicators of adverse birth outcomes: Indicator H12 presents the rate of preterm birth, and Indicator H13 presents the rate of term low birth weight. These two indicators were chosen because for each there is a wealth of quality data available.

Indicator H12: Percentage of babies born preterm, by race/ethnicity, 1993–2008

Indicator H13: Percentage of babies born at term with low birth weight, by race/ethnicity, 1993–2008

About the Indicator: Indicator H12 shows the percentage of babies born preterm, and Indicator H13 shows the percentage of babies who are born at term with low birth weight. Both graphs show separate lines for the different race/ethnicity groups. The data come from a national data system that collects data from birth certificates for virtually every baby born in the United States each year. Indicators H12 and H13 show the change in preterm and term low birth weight over time.

The National Vital Statistics System

The National Vital Statistics System (NVSS), operated by the National Center for Health Statistics (NCHS), provides national data on gestational ages and birth weights. The NVSS data are provided through contracts between the NCHS and vital registration systems operated in each state, which are legally responsible for the registration of vital events including births, deaths, marriages, divorces, and fetal deaths. The collection and publication of this information is mandated by federal law. Together NCHS and the states have developed standard forms and procedures to use for the data collection. The NVSS captures virtually all of the births occurring in the United States. The most current NVSS data available are for 2008.

Birth certificates provide information on characteristics of both the infant and his/her parents, including the weight of the infant and the length of gestation. Length of gestation is recorded in completed weeks, so for example a pregnancy of 36 weeks and 6 days would be recorded as 36 weeks, and would therefore be considered preterm.[3] Pregnancy duration is most often estimated from the date of a woman's last menstrual period. Many factors, including age, levels of physical activity, and body mass, can cause variation in menstrual cycle timing, making this method of estimating gestational length subject to some error.[3] NVSS data report pregnancy duration based on a clinical estimation, often determined using ultrasound, if information on last menstrual period is unavailable or is inconsistent with the reported birth weight. Because ultrasound measurements tend to give lower gestational age estimates than last menstrual period,[3] the slight increase in use of ultrasound data in recent years could contribute to any increase in the rate of preterm birth.

Data Presented in the Indicators

Indicator H12 displays the trend in the percentage of preterm births for all births (singletons, as well as multiples), with a separate line for each maternal race/ethnicity group and a single line for all maternal races and ethnicities combined for the years 1993–2008.

Indicator H13 displays the trend in the percentage of low birth weight births at term among all births (singletons, as well as multiples), with a separate line for each maternal race/ethnicity group and a single line for all maternal races and ethnicities combined for the years 1993–2008. Presentation of low birth weight data for only term births (babies with a gestational age of 37 completed weeks or more) is intended to identify trends in growth restriction separate from trends in gestational duration. This indicator does not include all infants with low birth weight, nor does it include all infants who are growth-restricted; therefore, it is designed as a surveillance tool and not as a way to identify a group of infants that are particularly at risk for adverse health effects.

Five maternal race/ethnicity groups are presented in these indicators: White non-Hispanic, Black non-Hispanic, Hispanic, American Indian/Alaska Native non-Hispanic, and Asian Pacific Islander non-Hispanic. Prior to the year 1993, not all states recorded Hispanic origin on birth certificates; for this reason, both Indicator H12 and H13 begin with data from 1993. Birth certificates do not include information on family or maternal income, so it is not possible to examine differences or trends by income level.

The indicator graphs show data for all births, singletons and multiples combined. The rates for singletons and multiples are provided in supplemental data tables. Additional supplemental tables highlight differences in rates of preterm birth and term low birth weight by age of the mother.

Please see the Introduction to the Health section for discussion of statistical significance testing applied to these indicators. The NVSS records virtually all births in the United States— approximately 4 million per year. Because of this very large sample size, differences in birth outcomes that appear to be small in magnitude may be found to be statistically significant. Extensive research has been conducted with NVSS data to assess the presence of statistically significant trends and demographic differences, including analyses with much more detail than the one conducted here.[23,84,85]

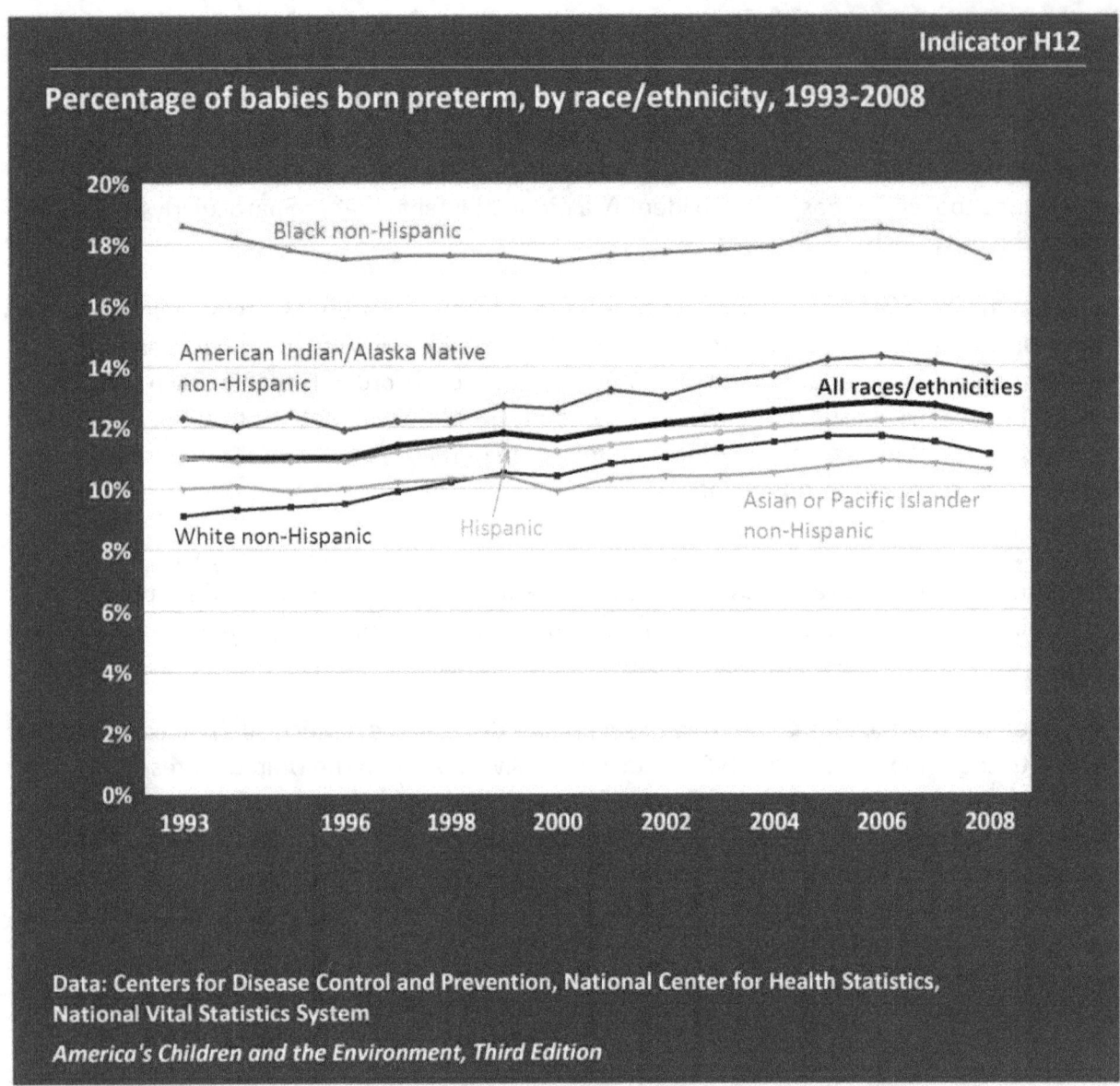

Percentage of babies born preterm, by race/ethnicity, 1993-2008

Black non-Hispanic

American Indian/Alaska Native
non-Hispanic

All races/ethnicities

White non-Hispanic   Hispanic

Asian or Pacific Islander
non-Hispanic

Data: Centers for Disease Control and Prevention, National Center for Health Statistics,
National Vital Statistics System

America's Children and the Environment, Third Edition

Data characterization
- Data from this indicator are obtained from a database maintained by the National Center for Health Statistics.
- The database collects information from birth certificates for virtually all births in the United States.
- Length of gestation is recorded on each birth certificate.

- Between 1993 and 2008, the rate of preterm birth showed an increasing trend, ranging from 11.0% in 1993 to its highest value of 12.8% in 2006. This increasing trend was statistically significant.

- In 2008, Black non-Hispanic women had the highest rate of preterm birth, compared with women of other races/ethnicities. More than 1 in 6 infants born to Black non-Hispanic women were born prematurely in that year.

- The difference between the rate of preterm birth for Black non-Hispanic women and the rates for the other race/ethnicity groups was statistically significant.
- Between 1993 and 2008, the preterm birth rate showed an increasing trend for each race/ethnicity group except Black non-Hispanic women. The preterm birth rate for Black non-Hispanic women stayed relatively constant, ranging between 17% and 19%.
  - The increasing trend in the rate of preterm birth was statistically significant for White non-Hispanic, Hispanic, American Indian/Alaska Native non-Hispanic, and Asian or Pacific Islander non-Hispanic women.
- The preterm birth rate varies depending on the age of the mother. Women ages 20 to 39 years have the lowest rate of preterm birth, compared with women under 20 years and women 40 years and older. The rates of preterm birth for women ages 20 to 39 years and women 40 years and older showed an increasing trend between 1993 and 2008; however, the increase for women ages 20 to 39 years was smaller. (See Table H12a.)
  - The differences between the preterm birth rates for the different age groups were statistically significant. The increasing trends in the rate of preterm birth for women ages 20 to 39 years and women 40 years and older were statistically significant as well.
- Twins, triplets, and other higher-order multiple birth babies are more than 5 times as likely to be born preterm compared with singleton babies (60.4% vs. 10.6% in 2008). The preterm birth rates for both singletons and multiples showed an increasing trend from 1993 to 2008; however, the increase for multiples was larger than for singletons. (See Table H12b.)
  - The increasing trend for both singleton and multiple births was statistically significant.

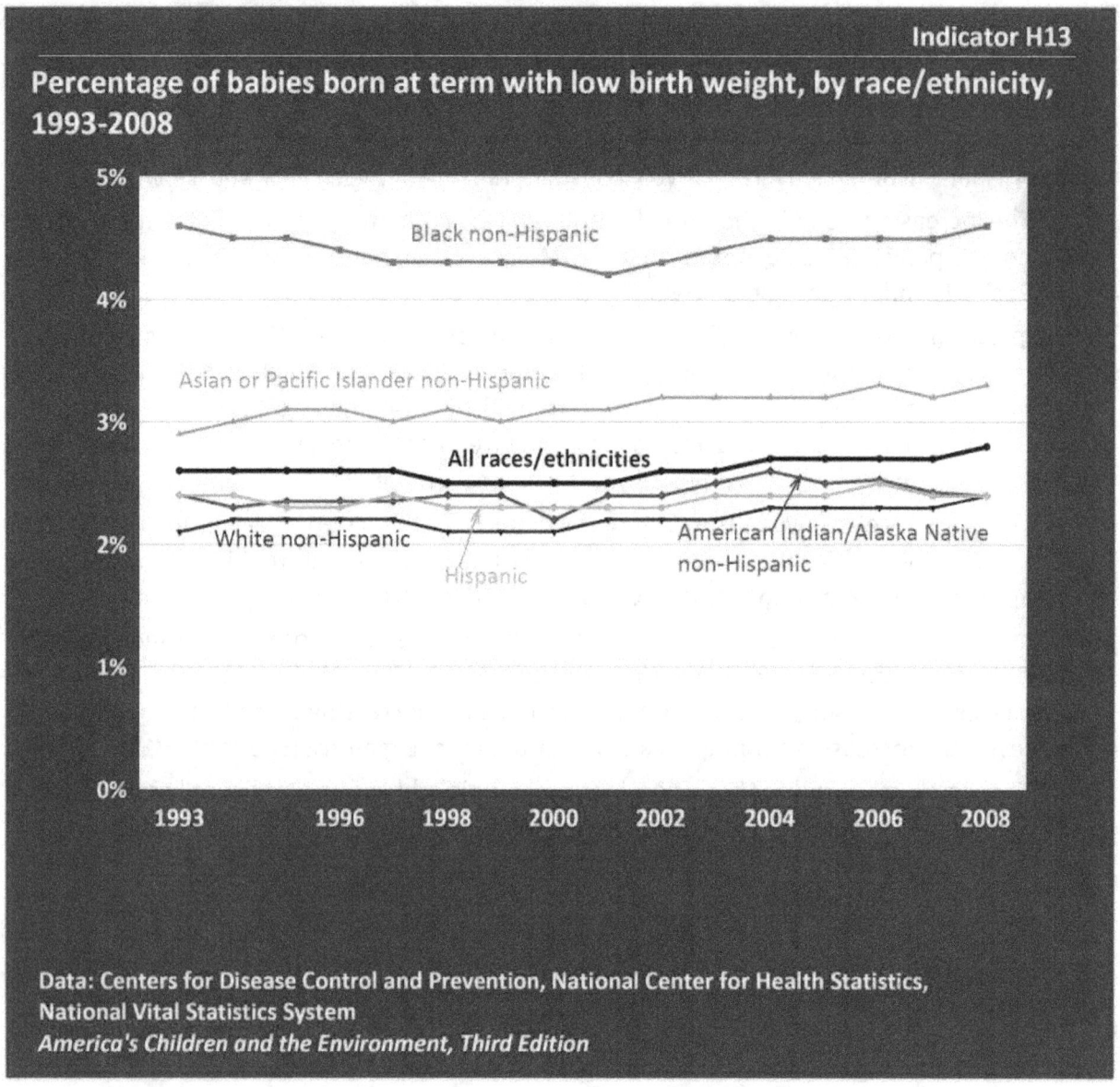

Data: Centers for Disease Control and Prevention, National Center for Health Statistics, National Vital Statistics System
America's Children and the Environment, Third Edition

Data characterization
- Data from this indicator are obtained from a database maintained by the National Center for Health Statistics.
- The database collects information from birth certificates for virtually all births in the United States.
- Birth weight and length of gestation are recorded on each birth certificate.

■ Between 1993 and 2008, the rate of term low birth weight for all races/ethnicities stayed relatively constant, ranging between 2.5% and 2.8%. The rates of term low birth weight for infants born to White non-Hispanic mothers and Asian or Pacific Islander non-Hispanic mothers showed increasing trends between 1993 and 2008, while the rates of term low birth weight for infants born to mothers of the other race/ethnicity groups stayed relatively constant.

- The rate of term low birth weight varies by race/ethnicity. In 2008, the rate was highest for infants born to Black non-Hispanic mothers, and next highest for infants born to Asian or Pacific Islander non-Hispanic mothers. The rate of term low birth weight is lowest for infants born to White non-Hispanic mothers, Hispanic mothers, and American Indian/Alaska Native non-Hispanic mothers.

    - The rate of term low birth weight for Black non-Hispanic women was statistically significantly higher than for all other race/ethnicity groups. The rate of term low birth weight for Asian or Pacific Islander non-Hispanic women was significantly lower than for Black non-Hispanic women but significantly higher than the other race/ethnicity groups.

- Term low birth weight rates vary by the age of the mother. In 2008, women ages 20 to 39 years had the lowest rate of term low birth weight infants, while women under 20 years had the highest rate of term low birth weight infants. These differences were statistically significant. (See Table H13a.)

- Between 1993 and 2008, the rate of term low birth weight for women 40 years and older showed an increasing trend, ranging from 2.9% to 3.4%. This increasing trend was statistically significant. (See Table H13a.)

- Twins, triplets, and other higher-order multiple birth babies are more than 5 times as likely to be born at term with low birth weight compared with singleton babies (12.6% vs. 2.4% in 2008). The rate of term low birth weight for singleton and multiple babies stayed relatively constant over the period of 1993–2008. (See Table H13b.)

Supplementary Topics

Introduction

The three main sections of *America's Children and the Environment, Third Edition (ACE3)* present environment and contaminant, biomonitoring, and health indicators derived from data sources of national interest, updated on a regular basis. Some topics of interest for children's environmental health do not have suitable national data sources. For some of these topics, ACE3 Supplementary Topics measures have been developed using data sets from a single state, or produced by one-time studies that have not been repeated. The data presentations in this section are referred to as "measures" rather than "indicators" because they are lacking in at least one key characteristic desired for ACE3 indicators.

Measures have been prepared for two Supplementary Topics in ACE3:

- Birth Defects
- Contaminants in Schools and Child Care Facilities

The birth defects topic includes a measure summarizing data from the Texas Birth Defects Registry. The contaminants in schools and child care facilities topic includes three measures of conditions in educational environments. Contaminants in child care facilities are represented by measures drawn from EPA's Children's Total Exposure to Persistent Pesticides and Other Persistent Organic Pollutants (CTEPP) Study, conducted in North Carolina and Ohio, and the First National Environmental Health Survey of Child Care Centers, a federal government study of a nationally representative sample of child care facilities. Contaminants in schools are represented by a measure calculated using a database on pesticide application in schools from the California Department of Pesticide Regulation.

For each Supplementary Topic, an introduction section explains the relevance of the topic to children's environmental health. The introduction section is followed by a description of the measures, including a summary of the data and information on how each measure was calculated. The measures are then presented in graphical form. Beneath each figure are explanatory bullet points describing dataset characteristics and key findings from the data presented in the figure, along with key data from any supplemental data tables. References are provided for each topic at the end of the report.

Data tables are provided in Appendix A. The tables include all values depicted in the Supplementary Topics figures, along with additional data of interest not shown in the figures. Metadata describing the data sources are provided in Appendix B. Documents providing details of how the measures were calculated are available on the ACE website (www.epa.gov/ace).[i]

---

[i] Detailed methods documents are not provided for Measures S2 and S3 (contaminants in child care facilities) because all values were taken directly from published sources.

The topics presented in this section are addressed in Healthy People 2020, which provides science-based, 10-year national objectives for improving the health of all Americans. Appendix C provides examples of the alignment of the ACE3 Supplementary Topics with objectives in Healthy People 2020.

Birth Defects

The term "birth defects" covers a range of structural and chromosomal abnormalities that occur while the baby is developing in the mother's body.[1,2] A birth defect may affect how the body looks, works, or both. Some birth defects can be detected before birth, others can be detected when the baby is born, and others may not be detected until some time has passed after birth.

Birth defects are the leading cause of infant death in the first year of life, accounting for about 20% of infant deaths in 2005.[3] Infants who do survive with a birth defect often have lifelong disabilities, such as intellectual disability, heart problems, or difficulty in performing everyday activities such as walking.

Some birth defects are inherited. Others have known risk factors that can be avoided such as prenatal exposure of the fetus to certain pharmaceuticals (such as Accutane® or Thalidomide); exposure to alcohol; maternal smoking, and insufficient folate in a woman's diet.[3-5] For example, birth defects resulting from fetal alcohol syndrome are prevented when a woman does not consume alcohol during pregnancy, and reported cases of neural tube defects such as spina bifida and anencephaly have been shown to decrease following mandatory folic acid fortification of cereal grain products.[6,7] About 60–70% of birth defects have unknown causes, but research suggests that some defects could be modified or caused by environmental factors, possibly in conjunction with genetic factors.[3,8-10] Several environmental contaminants cause birth defects when pregnant women are exposed to high concentrations. Mercury poisoning in Minamata, Japan resulted in birth defects such as deafness and blindness.[11] Prenatal exposures to high concentrations of polychlorinated biphenyls (PCBs) and related chemicals have resulted in skin alterations, including chloracne, a potentially serious inflammatory condition.[12] However, any possible relationship between exposures to lower concentrations of these or other environmental contaminants and birth defects is less clear.

A number of epidemiological studies have evaluated the relationship between environmental and occupational exposures to chemicals and birth defects. The majority of studies consider the relationship of birth defects to exposures to specific types of environmental contaminants, including solvents, pesticides, drinking water disinfection byproducts, endocrine disrupting chemicals, and air pollutants. Some studies consider other scenarios in which individuals may have elevated exposures without measuring or estimating exposure to any particular substances. These studies evaluate factors such as occupational category, or residence near a contaminated site or industrial facility.

Several studies have evaluated the relationship between maternal and paternal solvent exposure and birth defects. An extensive review of the literature concluded that the evidence linking neural tube defects to paternal exposures to solvents was suggestive of an association, although not strong enough to draw a conclusion regarding a causal relationship.[10] A meta-analysis that included multiple studies of women's occupational exposure to organic solvents reported an increased risk for birth defects such as heart defects and oral cleft defects in children born to exposed women.[13] In a recent study conducted in Massachusetts, women who were exposed to drinking water contaminated with the solvent tetrachloroethylene around the time of conception were reported to have an increased risk of giving birth to a child with a birth defect.[14]

Multiple studies have suggested an association between maternal and paternal exposure to pesticides (both before and after conception) and increased risk of offspring having or dying from birth defects.[15-31] A subsequent review study that evaluated many of these individual studies together, however, concluded that the data are inadequate at this time to confirm an association between pesticide exposure and the risk of birth defects.[10]

Disinfection byproducts in drinking water have also been linked to birth defects in some epidemiological studies. Disinfection byproducts are formed when organic material found in source water reacts with chemicals (primarily chlorine) used in treatment of drinking water to control microbial contaminants. Some individual epidemiological studies have reported associations between the presence of disinfection byproducts in drinking water and increased risk of birth defects, especially neural tube defects and oral clefts; however, recent articles reviewing the body of literature determined that the evidence is too limited to make conclusions about a possible association between exposure to disinfection byproducts and birth defects.[10,32-35]

Some studies have also reported associations between exposure to endocrine disrupting chemicals and urogenital malformations in newborn boys, such as cryptorchidism (undescended testes) and hypospadias (abnormally placed urinary opening).[19,22,36-44] An analysis of a large national database showed a significant increase in the incidence of congenital penile anomalies, particularly hypospadias, from 1988–2000.[45] According to studies by the Centers for Disease Control and Prevention, the prevalence of hypospadias in the United States has doubled in recent decades.[46] This considerable increase, combined with evidence of an association between endocrine-disrupting contaminants and urogenital birth defects in animal studies, has led to the hypothesis that environmental exposures are a contributing factor.[47] However, a review study recently concluded that there is inadequate evidence at this time of associations between male genital birth defects and exposure to environmental contaminants such as pesticides, PCBs, wood preservatives, and phthalates.[10]

A limited number of studies have investigated the relationship between birth defects and prenatal exposure to air pollution, specifically carbon monoxide, ozone, particulate matter, nitrogen dioxide, and sulfur dioxide.[48-57] Most of these studies have focused on cardiac and oral cleft birth defects. A recent pooled analysis of these studies reported statistically significant associations between nitrogen dioxide, sulfur dioxide and particulate matter and certain cardiac birth defects.[58] No statistically significant associations were found between any of the pollutants and oral cleft defects.

Since the discovery of extensive environmental contamination in the Love Canal community in New York State in the 1970s, there has been increased awareness that contaminated sites can be associated with negative birth outcomes, including birth defects.[59,60] Multiple epidemiological studies conducted over the last 25 years have found possible associations between residence near contaminated sites and an increased risk of birth defects, particularly neural tube defects and congenital heart defects.[38,61-64] Studies have also reported associations between residence near hazardous waste sites or active industrial facilities and chromosomal birth defects.[65,66] The majority of these studies use maternal proximity to sites of interest in order to classify exposure and do not distinguish between specific types of contaminant exposures; however, a few studies have reported associations between birth defects and sites that emit heavy metals or solvents.[64,65] Some studies have suggested that the greatest impact may be for mothers residing within a half mile of a contaminated site.[61,67] Studies comparing Superfund sites undergoing assessment or remediation to active industrial facilities reporting toxic chemical releases reported no association between birth defect rates and proximity to Superfund sites, but did report significant associations with proximity to the active industrial sites.[65,68] A recent study of birth defect records for children born to mothers living with proximity to any of 154 Superfund cleanup sites reported an overall reduced incidence of birth defects.[69]

The process of fetal development is intensely complicated, requiring the precise coordination of cell division, growth, and movement. During the process of fetal development there are critical periods of susceptibility or vulnerability, at which point exposure to environmental contaminants may be especially damaging.[70] For example, two air pollution epidemiology studies found that the first two months of gestation are a particularly vulnerable period, during which exposure to air pollutants may cause birth defects of the heart and oral clefts.[52,56] Similarly, studies hypothesizing a role for pesticide exposure in birth defects have reported that conception during the spring is a risk factor for birth defects.[25,29,71] Agricultural use of certain pesticides is at its highest during spring, potentially leading to increased exposures that could contribute to the observed seasonal pattern in the incidence of birth defects.[25,29,71] These types of studies are useful for generating hypotheses for future research investigating the relationship between environmental exposures and the development of birth defects.

There is currently no unified national monitoring system for birth defects. Information on prevalence of birth defects comes from birth certificates and from state birth defects monitoring systems. Many birth defects can be observed shortly after delivery and are recorded on birth certificates. A national-scale indicator could be constructed using birth certificate data, but would miss any birth defect that is not immediately recognized and recorded at birth. Comparisons of birth defects recorded on birth certificates and birth defect registries have indicated that typically, less than half of birth defects are recorded on birth certificates.[72,73] Most states have some type of birth defects monitoring program, although the type of tracking varies widely among the states. As of 2008, 45 states had some type of existing birth defects monitoring program.[74] A small portion of these states have the most complete type of tracking system, which includes actively researching medical records for birth defects and following children through at least the first year of life. The remaining states have some type of monitoring program, but do not have all the aspects of a complete surveillance system. The National Birth Defects Prevention Network has pooled data from several state registries to derive prevalence estimates for a subset of 21 selected birth defects for the years 1999–2001 and 2004–2006.[75]

The Texas monitoring program, which has monitored birth defects since 1995, is considered one of the most complete in the nation.[76] Data from the Texas registry for several categories of birth defects are presented in this section, as an example.

Measure S1: Birth defects in Texas, 1999–2007

About the Measure: Measure S1 presents information about the number of infants born with birth defects in Texas. The data come from a registry of birth defects for the state of Texas, which compiles data on any birth defects identified in the first year after each child is born. The Texas Registry staff routinely review medical records at all hospitals and birthing centers where babies are delivered or treated to identify birth defects. Measure S1 shows how the rates of different types of birth defects have changed over time. The rates of birth defects in Texas are not necessarily representative of those In other states.

The Texas Birth Defects Registry

The Texas Birth Defects Epidemiology and Surveillance Branch of the Texas Department of State Health Services provides information on birth defects in the state of Texas. The Texas program began monitoring the Houston/Galveston and South Texas areas in 1995 and expanded so that beginning in 1999, it covered the entire state. The Texas monitoring program covers approximately 380,000 births each year, which represents almost 10% of all births in the United States. In addition to live births, the Texas monitoring program also covers birth defects occurring in a fetal death or pregnancy termination. The Texas monitoring program reports a wide array of birth defects.

Although most states have a birth defects monitoring program in place, the comprehensiveness of these programs varies. Texas's birth defects monitoring program is one of the most complete in the nation, using high-quality active surveillance methods to examine a wide range of birth defects throughout a child's first year of life.[76] Specifically, the Texas Registry staff employ robust approaches to collecting, verifying, and ascertaining cases of birth defects such as routinely visiting all hospitals and birthing centers where babies are delivered or treated to individually review logs, discharge lists, and medical records.[77] As a result, a joint review by the Trust for America's Health and the National Birth Defects Prevention Network of the birth defects tracking activities in all 50 states assigned the Texas Registry their highest grade ranking, based on a number of criteria such as the ability to carry out tracking and the resources devoted to the task.[76] Although the Texas Registry data are of high quality, the rates and types of birth defects in Texas are not necessarily representative of those in other states.

Comparing the Texas Birth Defects Registry with Other Data Sources

To examine whether the rate of birth defects in Texas is similar to the rate for the country as a whole, it is useful to compare birth defect rates from birth certificates. Birth certificates record only those birth defects apparent at birth, and do not represent defects that become apparent after some time. Most states report birth defects on birth certificates using the standard birth certificate format recommended by the National Center for Health Statistics. The birth certificate reported rates of birth defects for Texas are generally similar to the nationwide rates.[78]

Comparing the Texas Birth Defects Registry data to the birth certificate data for Texas reveals that the active surveillance strategies detect a far greater number of birth defects than can be detected at an infant's birth. For specific birth defects that could be directly compared, the Texas monitoring program typically detects two to three times the number of birth defects reported on birth certificates, demonstrating the importance of tracking birth defects that are not observed at the time of delivery.[77,78] Texas birth certificates list potential birth defects for clinicians to choose from when recording the details of an infant's birth. An analysis by the Texas Birth Defects Registry found that birth certificates identify these listed birth defects only 15% of the time that they occur. Furthermore, of those birth defects listed on Texas birth certificates, the most obvious birth defects, such as spina bifida and cleft palate, are only identified 36-42% of the time.[73]

As mentioned previously, there is currently no unified national monitoring system for birth defects. However, CDC, in collaboration with the National Birth Defects Prevention Network, pools data from states with active and passive monitoring programs to estimate national prevalence rates for several selected birth defects. The pooled data set currently accounts for about 30% of births nationwide.[75]

Data Presented in the Measure

Measure S1 displays the number of birth defects per 10,000 live births for the state of Texas. Measure S1 shows data for 1999–2007 and groups birth defects by structural categories. A supplemental data table for this measure provides information showing how birth defect rates vary by race/ethnicity.[i]

Trends in the rates of birth defects may be influenced by differences in clinical practice. For example, increasing trends in the prevalence of some birth defects could represent more accurate recording of birth defects and/or better diagnosis of subtle defects due to the use of more sensitive examinations and technology.[79-82] Trends for specific birth defects may also be masked when grouping birth defects by structural categories. For example, anencephaly is included in the structural category of central nervous system defects. Incidence of central nervous system birth defects overall in Texas increased from 1999–2007, but the incidence of anencephaly defects specifically appear to be decreasing in the same years.[83]

Statistical Testing

Statistical analysis has been applied to Measure S1 to evaluate trends over time or differences between demographic groups in the prevalence of birth defects. These analyses use a 5% significance level, meaning that a conclusion of statistical significance is made only when there is no more than a 5% probability that the observed trend or difference occurred by chance ($p \leq 0.05$). The statistical analysis of trends over time is dependent on how the values in the

---

[i] 95% confidence intervals for the birth defects rates are provided in a file available on the ACE website (www.epa.gov/ace).

measure vary over time as well as on the number of time periods. For example, the statistical test is more likely to detect a trend when data have been obtained over a longer period. A finding of statistical significance for differences between demographic groups depends on the magnitude of the difference and the number of observations in each group. It should be noted that conducting statistical testing for multiple categories of birth defects increases the probability that some trends or differences identified as statistically significant may actually have occurred by chance.

A finding of statistical significance is useful for determining that an observed trend or difference was unlikely to have occurred by chance. However, a determination of statistical significance by itself does not convey information about the magnitude of the increase, decrease, or difference. Furthermore, a lack of statistical significance means only that occurrence by chance cannot be ruled out. Thus a conclusion about statistical significance is only part of the information that should be considered when determining the public health implications of trends or differences in the prevalence of birth defects.

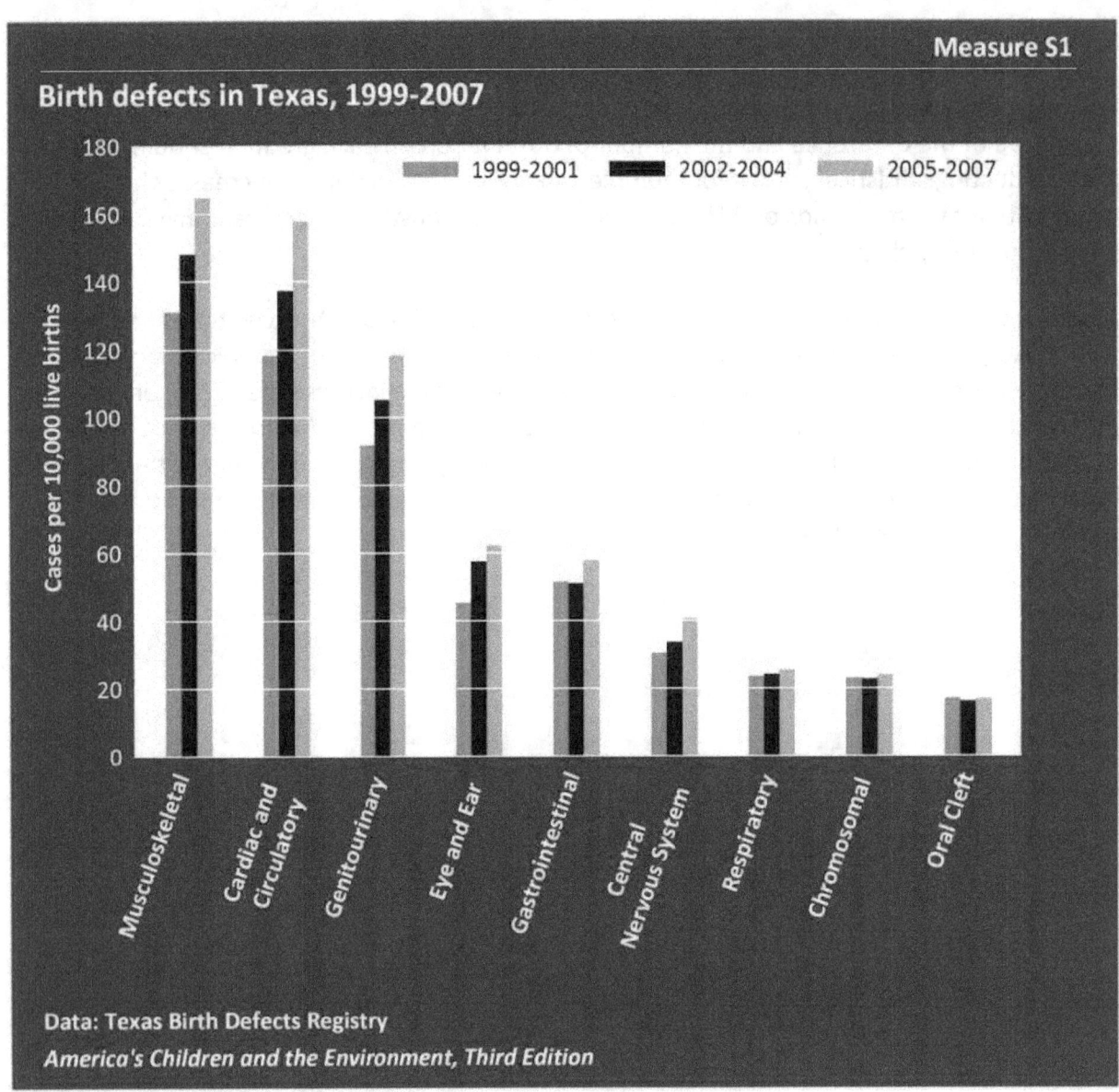

Measure S1

Birth defects in Texas, 1999-2007

Cases per 10,000 live births

Legend: 1999-2001 | 2002-2004 | 2005-2007

Categories: Musculoskeletal, Cardiac and Circulatory, Genitourinary, Eye and Ear, Gastrointestinal, Central Nervous System, Respiratory, Chromosomal, Oral Cleft

Data: Texas Birth Defects Registry

America's Children and the Environment, Third Edition

Data characterization

– Data for this measure are obtained from the Texas Birth Defects Registry.
– The Registry employs robust surveillance methods to monitor all births in Texas and identify cases of birth defects.
– The Registry represents almost 10% of all births in the United States, but the rates and types of birth defects in Texas are not necessarily representative of those in other states.

■ Musculoskeletal defects are the most common type of birth defect in Texas, with 165 cases per 10,000 live births for the years 2005–2007. The second most common type of birth defect in Texas is cardiac and circulatory, with 158 cases per 10,000 live births for the years 2005–2007.

- The rates for all categories of birth defects in Texas have increased or remained stable for the period of 1999–2007. Some of the biggest increases were seen for musculoskeletal defects, cardiac and circulatory defects, genitourinary defects, eye and ear defects, and central nervous system defects.

  - The increases were statistically significant for musculoskeletal defects, cardiac and circulatory defects, genitourinary defects, eye and ear defects, gastrointestinal defects, and central nervous system defects.

- The prevalence of birth defects varies by race/ethnicity for most of the anatomical categories examined. Compared with White non-Hispanics, Black non-Hispanics had lower rates of musculoskeletal, genitourinary, eye and ear, gastrointestinal, chromosomal, and oral cleft birth defects, and these differences were statistically significant. There were no statistically significant differences between Black non-Hispanics and White non-Hispanics in rates of cardiac and circulatory, central nervous system, and respiratory birth defects. (See Table S1a.)

- Compared with White non-Hispanics, Hispanics had higher rates of cardiac and circulatory, eye and ear, and respiratory defects, whereas rates of musculoskeletal and genitourinary birth defects were lower. These differences were statistically significant. There were no statistically significant differences between Hispanics and White non-Hispanics in rates of gastrointestinal, central nervous system, chromosomal, and oral cleft defects. (See Table S1a.)

Contaminants in Schools and Child Care Facilities

The indoor and outdoor environmental quality of schools and child care facilities plays an important role in affecting children's health and academic performance. Depending on the type of facility and its particular characteristics (i.e., age, usage, and maintenance), children may be exposed to contaminants from a variety of indoor and outdoor sources. Potential indoor exposure sources include building materials and furnishings (such as paint, treated wood, furniture, carpet, and fabrics), products used for building maintenance (such as cleaning products and pesticides), and products used for hobbies, science projects, and arts and crafts projects or within the learning environment (such as paint, markers, and correction fluid). Potential outdoor exposure sources include air pollution from nearby traffic and industry. In addition to these specific exposures, children may also experience unsatisfactory environmental conditions such as inadequate lighting, ventilation, indoor air quality, or noise control.[1] These exposures potentially impact the comfort and health of students, which may adversely affect their academic performance and increase their risk of both short- and long-term health problems.[2-4]

These potential exposures are of particular concern because children generally spend most of their active, awake time at schools and child care facilities. Children are especially sensitive to contamination, for several reasons. First, children are biologically more vulnerable than adults since their bodies are still growing and developing.[5-7] Second, children's intake of air and food is proportionally greater than that of adults. For example, relative to body weight, a child may breathe up to twice as much air as adults do; this increases their sensitivity to indoor air pollutants.[8] In particular for younger children, the inhalation and ingestion of contaminated dust is a major route of exposure due to their frequent and extensive contact with floors, carpets, and other surfaces where dust gathers, such as windowsills, as well as their high rate of hand-to-mouth activity.[8] Lastly, children have many years of future life in which to develop disease associated with exposure.[7]

School and child care environments share many characteristics influencing children's exposure to indoor environmental contaminants, such as the sources and types of potential environmental contaminants. Both environments also tend to house a large number of occupants in a small confined space, so that without proper ventilation a large number of children can be at risk for potential exposure to indoor contaminants.[9] However, there are also a number of important differences between the two. Children in child care facilities are generally much younger than those in schools, sometimes as young as a few weeks old. The behaviors of very young children (e.g., crawling, hand-to-mouth activity) increase their exposure to contaminants in dust, on surfaces, or in toys and other objects.[6,10] Younger children may also spend more time in child care facilities, some as many as 10 hours per day, 5 days a week.[11,12] Also, compared with schools, child care facilities can be located in a much wider variety of settings, including office buildings, individual homes, and religious buildings. As a result, the indoor and outdoor environments can differ widely between child care facilities and may not be directly under the control of those running the child care itself. Furthermore, child care facilities are more often operated independently, while schools are frequently part of a school district with centralized facilities management. This has important implications for strategies to address environmental issues in these facilities.

Building upkeep characteristics are extremely important, because the design, construction, and current condition of school and child care center facilities can contribute to children's exposure to environmental contaminants.[13-17] Age, level of deterioration, and ventilation efficiency are key characteristics that determine a building's indoor environmental quality. Many substances are released into the indoor environment as a result of deterioration of the building from old age, poor maintenance, or through improperly managed removal and renovation processes.[15,18]

Children may be exposed to a variety of contaminants in school and child care settings, such as lead, asbestos, polychlorinated biphenyls (PCBs), pesticides, brominated flame retardants, phthalates, and perfluorinated chemicals. Exposure to indoor contaminants can occur through multiple routes, such as dermal (through the skin), inhalation, and direct and indirect ingestion. These types of indoor environmental contaminants have been associated with a variety of adverse health outcomes, as well as outcomes related to educational performance for which impaired health is a suspected cause.[19,20] These adverse health effects may be short-term (headache, dizziness, nausea, allergy attacks, or respiratory problems) or longer-term and more serious (asthma, neurodevelopmental effects, or cancer).[21] Children exposed to indoor air pollution also miss more days of school due to illness.[14,22] A child's overall academic performance can suffer as a result of such an illness or absence.[23] For example, exposure to indoor air pollutants has been associated with decreased concentration and poor testing outcomes.[24-26]

There is evidence that many schools and child care facilities in the United States have significant and serious problems with indoor environmental contaminants,[27] and certain groups of children are especially susceptible to such exposures.[28] Children with allergies, asthma, and other respiratory problems are especially susceptible to the effects of indoor air pollution. Asthma attacks and allergies are often triggered by indoor allergens (pollen, dust, cockroaches), as well as by mold.[29]

Lead

Lead is a pervasive and serious environmental health threat for children in the United States.[30,31] The most common sources of lead exposure in schools and child care environments are lead-based paint, lead dust, and lead-contaminated soil in outdoor play areas.[32] This is a particular concern for young children, due to their frequent and extensive contact with floors, carpets, window areas, and other surfaces where dust gathers, as well as their frequent hand-to-mouth activity.[33] A nationally representative sample of licensed child care facilities in 2001 estimated that approximately 14% of these facilities in the United States have significant lead-based paint hazards. Most of these are facilities in older buildings: 26% of facilities located in a building built before 1960 were found to have lead-based paint hazards, compared with 4% in newer buildings. [11]

Additional sources of lead may include lead in drinking water, lead-contaminated products (such as toys, books, and jewelry), and outdoor air from nearby industry.[33-35] The ingestion and inhalation of lead-contaminated dust are the primary pathways of childhood exposure to lead.[36] The National Toxicology Program has concluded that childhood lead exposure is associated with reduced cognitive function, reduced academic achievement, and increased attention-related behavioral problems.[30] Studies have reported associations of childhood exposure to lead with behavioral problems such as attention-deficit/hyperactivity disorder,[37-44] increased likelihood of school absenteeism and of dropping out of high school,[45] increased risks of juvenile delinquency and antisocial behaviors,[46-49] higher total arrest rates, and arrest rates for violent crimes in early adulthood.[50,51]

Polychlorinated Biphenyls (PCBs)

PCBs are a family of industrial chemicals used primarily as cooling or insulating fluids for electrical equipment or as additives to paints, plastics, and rubber products.[20] While the manufacture of PCBs was banned in 1979, PCBs continue to be present in products and materials produced before the ban. Many schools in the United States have lighting systems containing PCBs. When contained in the lighting systems, PCBs pose very little health risk or environmental hazard.[52] However, lighting systems degrade as they age, increasing the risk of PCB leaks or even fires, which pose health and environmental hazards. In December 2010, EPA issued guidance recommending that schools take steps to reduce potential exposures to PCBs from these types of older lighting fixtures.[53] PCBs are also found in caulk and paint used in building structures before 1980,[54,55] which may mobilize into the surroundings from removal efforts, natural weathering, or deterioration over time, and contribute significantly to PCB levels in indoor air and dust in schools.[56,57] Although there is some inconsistency in the epidemiological literature, several reviews of the literature have concluded that the overall evidence supports a concern for adverse effects of PCBs on children's neurological development.[58-62]

Asbestos

Asbestos is a naturally occurring mineral fiber that has been used in building materials as an insulator and fire retardant.[63] The production and use of building materials containing asbestos is currently limited by law in the United States,[64] but many older schools and other buildings may have asbestos-containing materials that were previously allowed in construction. The Asbestos Hazard Emergency Response Act provides rules for the management of asbestos in schools.[65] Under this law, some asbestos-containing products are removed when found, but most often it is recommended that they are "managed-in-place"—i.e., maintaining and managing the contaminated material to reduce potential exposure. Properly managed asbestos that has not been disturbed poses little health risk to students. However, if asbestos-containing materials are disturbed or begin to deteriorate, they can release hazardous fibers into the air and water. Long-term exposure to these fibers can lead to lung cancer, asbestosis (lung scarring), or mesothelioma (cancer of the lung cavity lining).[66,67] These diseases require a long time to develop following exposure, putting children at greater risk of disease development later in life.

Other Indoor Contaminants

Cleaning products and maintenance activities in schools and child care facilities are significant sources of exposure to chemical contaminants. Many conventional cleaning supplies contain harmful chemicals that have been associated with various health effects, including asthma and cancer.[68] Additionally, maintenance activities, from routine cleaning to renovation, can cause dust and particulate matter to become airborne, leading to increased opportunity for inhalation and ingestion of contaminated particles.[69]

Children also may be exposed to a variety of other hazardous chemicals in these environments, such as glues, paints, and other art supplies; mercury from older thermometers; a range of chemicals in chemistry labs; lead acid in batteries and other automotive and trade shop supplies; formaldehyde in pressed wood furniture, flooring, carpets, curtains, and cleaning products; volatile organic compounds (VOCs) in paints, aerosol sprays and fresheners, cleaning supplies, and building materials and furnishings; and the wide variety of toxic chemicals found in environmental tobacco smoke.[70] These and other chemicals commonly found in indoor air have been associated with a range of short-term effects, such as eye, lung, and skin irritation; headaches; nausea; fatigue; and a range of long-term health effects, from chronic lung irritation to cancer, depending on the specific chemical.

In addition to these direct sources of potential exposure, inefficient or malfunctioning heating, ventilation, and air conditioning (HVAC) systems may increase the risks of adverse health effects or even become an additional source for indoor contaminant exposures. First, failing to provide sufficient circulation and filtration of the indoor air mixed with fresh outdoor air can lead to an accumulation of existing air pollutants to dangerous levels.[9,71] This includes increased levels of the chemical contaminants already discussed, as well as other environmental contaminants such as particulate matter and allergens such as cockroach allergen, rodent dander, or pollen.[72] Second, failing to adequately control moisture and temperature levels can trigger the growth of dust mites and mold, which thrive in damp, warm environments.[19,72] Exposure to these are known to cause asthma or trigger asthma attacks.[73,74] Inefficient HVAC capabilities are of particular concern in temporary classroom structures, such as trailers and portable classrooms, which have been associated with poor indoor air quality due to a combination of inadequate ventilation along with use of toxic building materials. A state-wide survey of permanent and portable classrooms in California found that, on average, portable classrooms had worse indoor air quality than permanent ones did, including less efficient or improperly functioning HVAC systems; higher levels of indoor air formaldehyde, particulate matter, polycyclic aromatic hydrocarbons (PAHs), and humidity; and temperatures above and below thermal comfort standards during warm and cool seasons, respectively.[75]

School Siting

School siting (selecting a site, or location, for a new school) is a complex process that often requires assessment of several considerations, such as whether to renovate an old school or to build a new one, cost of land and location preparation, and the availability of infrastructure including roads and utilities. EPA has recently developed voluntary guidelines for school siting as a way to support states, tribes, communities, local officials, and the public in understanding and appropriately considering environmental and public health factors when making school siting decisions. These siting guidelines address issues such as the special vulnerabilities of children to hazardous substances or pollution exposures, modes of transportation available to students and staff, the efficient use of energy, and the potential use of the school as an emergency shelter.[17]

School locations may have underlying causes of potential exposure, such as site contamination, neighborhood emission sources, or indoor air quality problems.[17] Radon, a naturally occurring gas, can seep into buildings from soil. A nationwide survey of radon levels in schools estimates that nearly one in five schools has at least one schoolroom with a short-term radon level above the level at which EPA recommends that schools take action.[76] Additionally, children attending schools near highways or industrial sources may be exposed to various air pollutants such as ozone, particulate matter, carbon monoxide, VOCs, and lead. These potential exposures may pose either short-term or long-term health risks to children who utilize school facilities.[17]

Pesticides in Schools and Child Care Facilities

Pesticides are used in the indoor and outdoor environment to prevent, destroy, repel, or otherwise control pests such as rodents, insects, unwanted plants, and microbials (such as bacteria). They can be sold in many different forms, such as sprays, powders, crystals, or balls, and thus their application inside or outside of schools and child care facilities may lead to several potential routes of exposure for children. For example, application of pesticides in the indoor environment has been shown to contaminate untreated surfaces, including kitchen counters and toys,[77-83] indoor air,[77-79,83-87] and dust.[84,88-92]

Once applied, pesticide residues may take anywhere from a few hours to several months or years to completely break down (degrade). Pesticide residues in the indoor environment are less exposed to factors, such as sunlight, that enable their degradation, and therefore are more persistent than those pesticide residues in the outdoor environment.[82,93,94] This persistence means that pesticide exposures can remain a potential concern for a long period of time, even if the area is no longer being treated. For example, an assessment of pesticide residues in dust of inner city homes found a high prevalence of the pesticide chlorpyrifos two to three years after its indoor use was banned.[90] DDT also continues to be measured in indoor dust several decades after its use was banned in the United States.[91,92,95,96] Furthermore, the persistence of pesticides in the environment after application creates not only an opportunity for children to be exposed directly to the residues, but also the potential for residue migration, leading to contamination of untreated areas.[82,97] As a result, exposures may occur long after application and through a variety of routes such as inhalation and indirect ingestion of dust.[77]

Outdoor pesticide applications on school property, as well as on nearby agricultural fields, lawns, or house perimeters, may contaminate nearby schools and child care facilities.[77] Several studies demonstrate increased levels of pesticides in indoor air[82,98] and dust[95,98] following pesticide applications in an adjacent outdoor area. This often occurs when outdoor air contaminated with pesticide residues mixes with the indoor air (through natural drifting into the building or being brought in through HVAC systems), or residue particles are tracked in on the shoes and clothing of people entering the building.[80,82,95,98,79]

Few studies have evaluated pesticide exposures in the school environment. Some states have conducted studies of pesticide occurrence in their schools. A comprehensive survey of public K–12 classrooms was conducted by the state of California between October 2001 and February 2002.[99] The California study found residues of both available and restricted-use pesticides in all floor dust samples, and concluded that pesticides enter classrooms either during application or by being tracked in on clothing or shoes from outdoors. Pesticides detected in more than 80% of the samples include cis- and trans-permethrin, chlorpyrifos, and piperonyl butoxide. The First National Environmental Health Survey of Child Care Centers evaluated potential pesticide exposures in child care facilities, and reported that 75% of licensed child care facilities had at least one pesticide application in the past year.[97] The study detected numerous organophosphate and pyrethroid pesticides in indoor floor wipe samples. Chlorpyrifos, diazinon, and permethrin were detected in more than 67% of the tested centers.[97]

Several studies have reported associations between exposure to pesticides in early life and adverse health effects such as cancer and neurodevelopmental disorders. Childhood leukemia in particular has been associated with childhood exposures to pesticides.[100-104] Permethrin and resmethrin, which both belong to the commonly used class of pesticides known as pyrethroids, were recently classified by EPA as "likely to be carcinogenic to humans."[104] Childhood exposures to organophosphate pesticides have been associated with various adverse neurodevelopmental effects.[105-107] Exposure to herbicides and/or other pesticides in the first year of life has been associated with higher risk of asthma.[108]

The short- and long-term health effects of exposure to pesticides in the school environment are largely unknown, due to a lack of data. Between 1993 and 1996, there were 2,300 pesticide-related exposures reported to poison control centers that involved individuals at schools, resulting in 329 people seen in health care facilities, 15 hospitalized, and 4 treated in intensive care units.[109] Data on the long-term effects of pesticide exposure in schools are not available.[109]

Currently, there is no federal law on pesticide use in the school environment. However, at least 35 states have adopted laws on pesticide use in schools.[110] The state laws are generally focused on the adoption of certain types of practices that eliminate or minimize the use of hazardous pesticides: adoption of Integrated Pest Management (IPM) programs, prohibiting when and where pesticides can be applied, requiring signs before and after indoor and outdoor pesticide application, requiring prior written notification to parents and staff for pesticide use, and establishing restricted buffer zones to address chemicals drifting into school yards and buildings. Strategies such as restrictions on the use of pesticides and adoption of IPM have been shown to be effective at reducing human exposure.[87,111,112]

There is no national system for compiling data on the amount of pesticides used in schools.[109] Some states require reporting on pesticide use in schools. The state of Louisiana requires schools to submit a written record of "restricted use" pesticides used annually.[113] In the state of New York, commercial applicators are required by a 1996 law to report the amount of each specific pesticide used and the location where it was applied. Also, six states—Arizona, California, Connecticut, Massachusetts, New Hampshire, and New Mexico—require commercial applicators to report the amount of specific pesticides used.[109]

Measures in This Section

Data on school or child care environmental exposures are not systematically collected. Over the years, there have been few national and state-specific surveys or assessments to acquire information on environmental hazards in educational facilities. The following two measures provide data on the use or presence of pesticides and other chemicals of concern indoors in schools and child care facilities. Measures S2 and S3 present data on detectable levels of pesticides and other contaminants in a regional and national sample of child care centers. Measure S4 presents data on the amount of pesticides applied in schools in California.

Measure S2: Percentage of environmental and personal media samples with detectable pesticides in child care facilities, 2001

Measure S3: Percentage of environmental and personal media samples with detectable industrial chemicals in child care facilities, 2001

About the Measures: Measures S2 and S3 present information about the types of contaminants that were detected in child care facilities. The data come from two different studies. One study collected information from selected child care facilities in Ohio and North Carolina, while the other study collected information from child care facilities throughout the United States. The measures show how frequently different contaminants were detected in various media samples (e.g., indoor air, dust) taken at the testing locations.

CTEPP Study and the First National Environmental Health Survey of Child Care Centers

Measures S2 and S3 present data on the relative potential exposures of children to a variety of pesticides and other contaminants found in child care centers. The measures are based on data from two different federal studies: the Children's Total Exposure to Persistent Pesticides and Other Persistent Organic Pollutants (CTEPP) Study and the First National Environmental Health Survey of Child Care Centers. Data shown in these measures were obtained directly from these sources:

Tulve, N.S., P.A. Jones, M.G. Nishioka, R.C. Fortmann, C.W. Croghan, J.Y. Zhou, A. Fraser, C. Cave, and W. Friedman. 2006. Pesticide Measurements from the First National Environmental Health Survey of Child Care Centers Using a Multi-Residue GC/MS Analysis Method. *Environmental Science and Technology* 40(20) 6269-6274.

Morgan, M.K., L.S. Sheldon, C.W. Croghan, J.C. Chuang, R.A. Lordo, N.K. Wilson, C. Lyu, M. Brinkman, N. Morse, Y.L. Chou, C. Hamilton, J.K. Finegold, K. Hand, and S.M. Gordon. 2004. A Pilot Study of Children's Total Exposure to Persistent Pesticides and Other Persistent Organic Pollutants (CTEPP), Appendix I and Appendix J. Research Triangle Park, NC: U.S. Environmental Protection Agency. http://www.epa.gov/heasd/ctepp/.

The CTEPP study investigated the potential exposures of 257 preschool children, ages 1.5 to 5 years, and their primary adult child care providers to more than 50 anthropogenic chemicals, including pesticides, PAHs, PCBs, phthalates, and phenols. This regional study was conducted by EPA at 29 child care centers in North Carolina and Ohio in 2000–2001. Environmental (indoor and outdoor air, carpet house dust, and soil) and personal (hand wipe, solid and liquid food, drinking water, and urine) samples were collected for each child in the study at home and at the child care center over a 48-hour period.[114]

The First National Environmental Health Survey of Child Care Centers was conducted by the U.S. Department of Housing and Urban Development, the Consumer Product Safety Commission, and EPA in 2001. Indoor and outdoor environmental media samples (surface wipes and soil samples) from a nationally representative sample of 168 child care centers were tested for lead, allergens, and pesticides. No personal samples were collected.

Data Presented in the Measures

Measure S2 presents the percentage of environmental and personal media samples (indoor air, hand wipe, dust, and floor wipe samples) taken from selected regional and national child care facilities with detectable pesticides. Measure S3 presents the percentage of environmental and personal media samples (indoor air, hand wipe, and dust samples) taken from selected regional child care facilities with detectable industrial chemicals. The "Regional Data" in the first graph and all data in the second graph are derived from the CTEPP study, and reflect the percentage of media samples with detectable pesticides and chemical residues. The chemicals were each measured in 42–43 indoor air and dust samples collected from child care centers in Ohio and North Carolina, and the chemicals were measured in hand wipe samples collected from 60–61 children attending those child care centers. The "National Data" in the first graph are derived from The First National Environmental Health Survey of Child Care Centers, and reflect the percentage of 168 floor wipe samples with detectable chemical residues. The level that is detectable is determined by the capabilities of the sampling and testing equipment used in a study; therefore, it cannot be completely ruled out that contaminants are present at lower levels in samples classified as being below the detection limit. Both measures are based on whether the contaminant is detected or not detected, and thus provide an indication of potential for exposure, but they do not provide data on concentrations of the chemicals or levels of exposure.

The "indoor air" category reflects children's potential exposure to airborne chemicals through inhalation. The "hand wipes" category is based on sampling for the presence of chemicals on children's hands. Due to children's high levels of hand-to-mouth activity, hand wipe data indicate potential exposure via ingestion.[8] The "dust" category captures contaminants that accumulate in dust on various indoor surfaces, and reflects potential inhalation exposure to contaminants if dust is resuspended in the air, as well as indirect ingestion if dust contaminates items that children put in their mouths, such as food, toys, and their hands.

The specific pesticides shown in Measure S2 are pentachlorophenol, an organochlorine pesticide that has been used in the past in some paints, and in industrial and agricultural practices, but which is now limited to use in wood railroad ties and utility poles; chlorpyrifos, an organophosphate insecticide used previously indoors against cockroaches, fleas, and termites, and currently used on farms to control pests on animals and crops and in warehouses, factories, and food processing plants; cis-permethrin, a synthetic pyrethroid used to kill and repel domestic insects; and diazinon, an organophosphate pesticide with current agricultural uses and previous residential uses.

The industrial chemicals shown in Measure S3 are PCB-52, polycyclic aromatic hydrocarbons (PAHs, represented in the measure with data for the PAH benzo[b]fluoranthene), dibutyl phthalate, and bisphenol A. While the manufacture of PCBs was banned in 1979, PCBs continue to be present in electrical equipment and some building materials, such as caulk, produced before the ban. Several PCBs were measured in the CTEPP study; data for PCB-52 are displayed in the graph because it is one of the PCBs most frequently detected in the study, and thus gives an indication of potential for exposure to PCBs in general. Benzo[b]fluoranthene is one of several PAHs measured in the CTEPP study. Mixtures of PAHs are produced when carbon-based fuels are burned. Data for benzo[b]fluoranthene are displayed in the graph because it is one of the PAHs most frequently detected in the study, and thus gives an indication of potential for exposure to PAHs in general. Dibutyl phthalate is a chemical commonly used in adhesives, plastics, and personal care products. Bisphenol A is a high-volume industrial chemical used in the production of epoxy resins and polycarbonate plastics. Polycarbonate plastics may be encountered in many products, notably food and drink containers, while epoxy resins are frequently used as inner liners of metallic food and drink containers to prevent corrosion.

Many of these pesticides and industrial chemicals are no longer available or have highly restricted uses. Manufacture of PCBs and PCB-containing equipment and materials was banned in 1979, though equipment and materials manufactured with PCBs prior to the ban remain in use. Pentachlorophenol has not been used other than as a wood preservative since 1987. Indoor application of chlorpyrifos, and any use at schools, was restricted beginning in 2001. All indoor uses of diazinon were banned in 2001.

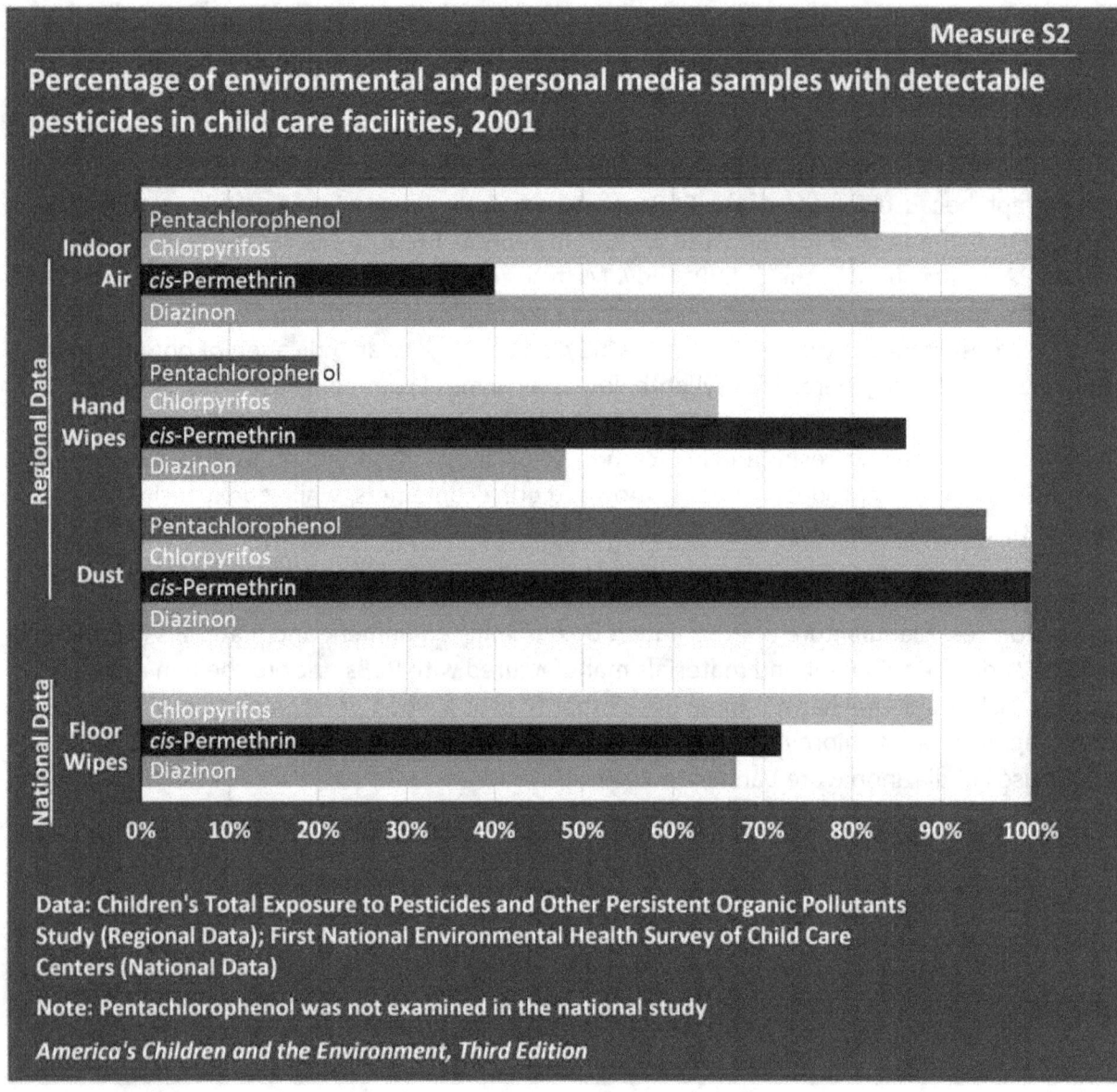

Measure S2

Percentage of environmental and personal media samples with detectable pesticides in child care facilities, 2001

Data: Children's Total Exposure to Pesticides and Other Persistent Organic Pollutants Study (Regional Data); First National Environmental Health Survey of Child Care Centers (National Data)

Note: Pentachlorophenol was not examined in the national study

America's Children and the Environment, Third Edition

Data characterization
- National data for this measure were obtained from a federal government study of a nationally representative sample of 168 child care centers. Pesticides were measured in environmental samples collected from the child care centers.
- Regional data for this measure were obtained from an EPA study of 29 child care centers in Ohio and North Carolina. Pesticides were measured in environmental samples collected from the child care centers and from the hands of children in the centers.

■ Chlorpyrifos, *cis*-permethrin, and diazinon were detected in all of the dust samples collected at Ohio and North Carolina child care centers included in the CTEPP study in 2000-2001. Chlorpyrifos and diazinon were also detected in all of the indoor air samples collected at these child care centers.

- Pesticide residues were detected least often in the hand wipe samples collected at the selected Ohio and North Carolina child care centers, but chlorpyrifos and *cis*-permethrin were detected in more than half of the hand wipe samples.

- The national level floor wipe sampling found chlorpyrifos most frequently, in 89% of samples. *Cis*-permethrin and diazinon were also detected frequently, in 72% and 67% of floor wipe samples, respectively. (Pentachlorophenol was not examined in the national study.)

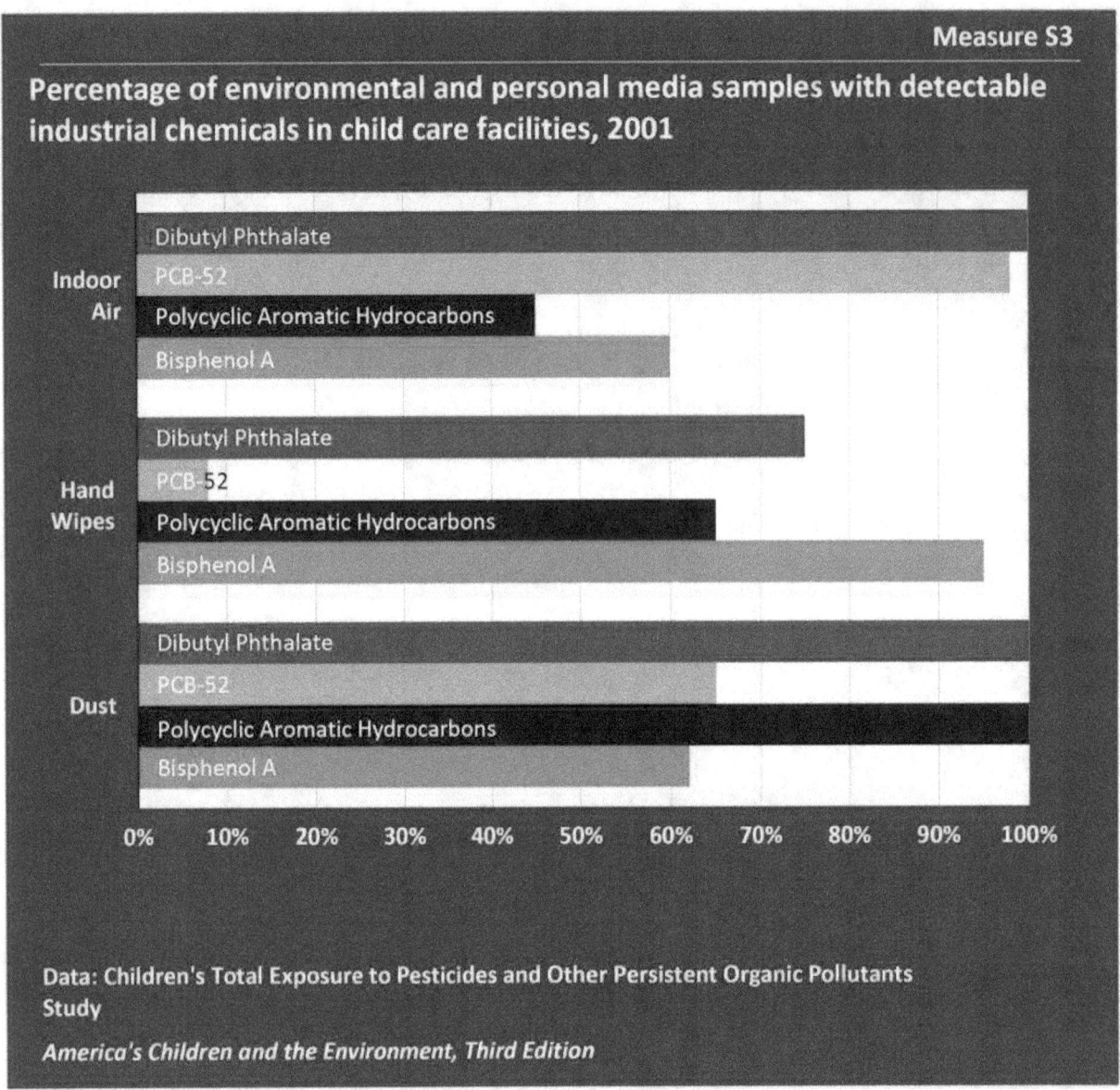

Measure S3

Percentage of environmental and personal media samples with detectable industrial chemicals in child care facilities, 2001

Data: Children's Total Exposure to Pesticides and Other Persistent Organic Pollutants Study

America's Children and the Environment, Third Edition

Data characterization

- Data for this measure were obtained from an EPA study of 29 child care centers in Ohio and North Carolina.
- Chemicals were measured in environmental samples collected from the child care centers and from the hands of children in the centers.

- Of the chemicals shown in this measure, dibutyl phthalate was the most frequently detected in indoor air and dust samples collected at Ohio and North Carolina child care centers included in the CTEPP study in 2000–2001.

- Dibutyl phthalate and PAHs (represented by benzo[b]fluoranthene) were detected in 100% of the dust samples. PCB-52 and bisphenol A were detected in 65% and 62% or dust samples, respectively.

- Dibutyl phthalate, PAHs, and bisphenol A were detected in more than 60% of hand wipe samples, while PCB-52 was detected in less than 10% of these samples.

- Dibutyl phthalate was detected in all of the indoor air samples and PCB-52 was detected in almost all (98%) of the samples. PAHs were detected in slightly less than half of the indoor air samples, while bisphenol A was detected in slightly more than half of the indoor air samples.

> **About the Measure:** Measure S4 presents information about pesticides used inside California schools. The data for this measure come from the California Department of Pesticide Regulation, which collects data on all commercial pesticide application in California schools. The measure shows how the application amounts of different pesticide categories have changed over the years.

California Schools Pesticide Use Reporting Database

The California Department of Pesticide Regulation collects data on all commercial pesticide application in California schools. In the year 2000, California passed the Healthy Schools Act of 2000, which required all public child care facilities and school sites to report pesticide use on school sites by pest control businesses.[115] Schools are required to report pesticide use at least once per year, and all schools are required to maintain records of their reports on-site for four years. The California Healthy Schools Act requires reporting for application of pesticides to the buildings or structures (including attics and crawl spaces), playgrounds, athletic fields, school vehicles, or any other area of school property, indoors and outdoors, visited or used by pupils.[115] The law does not apply to products used as self-contained baits or traps; gels or pastes used as crack-and-crevice treatments; pesticides exempted from regulation by EPA; or antimicrobial pesticides, including sanitizers and disinfectants. All other pesticides must be reported.

Data Presented in the Measure

Measure S4 displays the annual amount (pounds per year) of pesticides used inside California schools and child care facilities by commercial applicators. The measure presents data for the indoor applications of pesticides for all years for which data are available: 2002–2007. Although the measure presents data for schools and child care facilities, nearly all of the data reported are from schools.

The measure presents the amount of pesticides applied in California schools and child care facilities, in pounds per year, with pesticides grouped into seven categories: pyrethrin and pyrethroid insecticides, organophosphate insecticides, other insecticides, herbicides, fumigants, rodenticides, and miscellaneous pesticides. Most use of the "other insecticides" category inside of California schools is accounted for by imidacloprid, which is marketed for indoor termite and cockroach control. Most of the "miscellaneous pesticides" category use inside of schools is accounted for by a borate compound used as a fungicide and insecticide.

Routinely collected pesticide use data can provide helpful information about the types of pesticides used and the extent of such use, including changes over time. However, these data do not indicate the extent of pesticide exposure experienced by children in California schools.

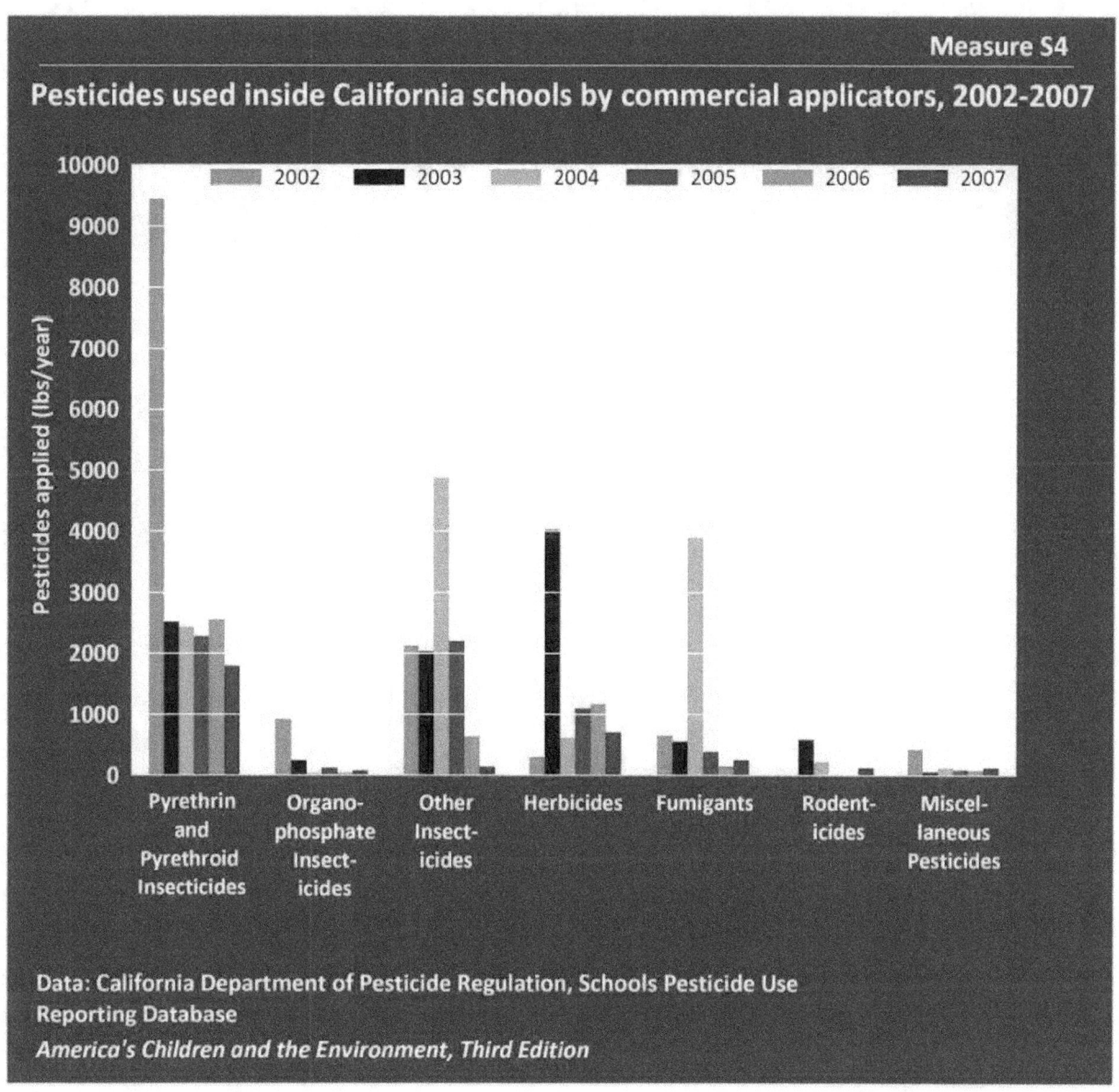

Measure S4

Pesticides used inside California schools by commercial applicators, 2002-2007

Data: California Department of Pesticide Regulation, Schools Pesticide Use Reporting Database
America's Children and the Environment, Third Edition

Data characterization
- Data for this measure are obtained from a reporting database maintained by the California Department of Pesticide Regulation.
- Reporting is required for all pesticide applications by pest control companies at all school and childcare facilities in California.
- Pesticide reports are submitted to the database at least annually and report all pesticide application on any area of school or childcare facility property visited or used by pupils.

- Pyrethrin and pyrethroid insecticides accounted for the greatest volume of pesticide use in California schools overall from 2002 to 2007, although there was greater use of herbicides in 2003, and of the "other" insecticides category and fumigants in 2004.

- The application of pyrethrin and pyrethroid insecticides, and organophosphate insecticides inside California schools has decreased since 2002.

References

About This Report

1. World Health Organization. 2012. *Environmental Health.* WHO Department of Public Health and the Environment. Retrieved August 1, 2012 from http://www.who.int/topics/environmental_health/en/.

2. Executive Office of the President. 1997. Executive Order 1305: Protection of Children from Environmental Health Risks and Safety Risks. *Federal Register* 62 (78).

3. Children's Health Protection Advisory Committee. 2009. *Letter (Dated September 9, 2009) to EPA Administrator Lisa P. Jackson Regarding EPA's America's Children and the Environment Report.* Oakland, CA: Children's Health Protection Advisory Committee. http://yosemite.epa.gov/ochp/ochpweb.nsf/content/ACERecs.htm/$file/ACE%20Recommendation.pdf.

4. Children's Health Protection Advisory Committee. 2009. *Report of the Task Group of the Children's Health Protection Advisory Committee on America's Children and the Environment, Third Edition.* Oakland, CA: Children's Health Protection Advisory Committee. http://yosemite.epa.gov/ochp/ochpweb.nsf/content/ACETask.htm/$file/ACE%20Task%20Group%20Report.pdf.

5. U.S. Census Bureau. 2011. *Poverty Thresholds by Size of Family and Number of Related Children Under 18 Years: 2010.* U.S. Census Bureau. Retrieved August 20, 2011 from http://www.census.gov/hhes/www/cpstables/032011/pov/new35_000.htm.

6. U.S. Census Bureau. 2011. *Income, Poverty and Health Insurance Coverage in the United States: 2010.* Washington, DC: U.S. Government Printing Office. Current Population Reports, P60-239. http://www.census.gov/prod/2011pubs/p60-239.pdf.

7. Executive Office of the President. 1994. Executive Order 12898. Federal Actions to Address Environmental Justice in Minority Populations and Low-Income Populations. *Federal Register* 59 (32).

8. Healthy People. 2011. *About Healthy People.* U.S. Department of Health and Human Services. Retrieved August 20, 2011 from http://www.healthypeople.gov/2020/about/default.aspx.

9. Healthy People. 2011. *Disparities.* U.S. Department of Health of Human Services. Retrieved August 20, 2011 from http://www.healthypeople.gov/2020/about/disparitiesAbout.aspx.

10. National Research Council. 2008. *Phthalates and Cumulative Risk Assessment: The Tasks Ahead.* Washington, DC: National Academies Press. 978-0-309-12841-4. http://www.nap.edu/catalog.php?record_id=12528#toc.

11. Olden, K., N. Freudenberg, J. Dowd, and A.E. Shields. 2011. Discovering how environmental exposures alter genes could lead to new treatments for chronic illnesses. *Health Affairs* 30 (5):833-41.

12. Rappaport, S.M., and M.T. Smith. 2010. Epidemiology: Environment and disease risks. *Science* 330 (6003):460-461.

13. Willett, W.C. 2002. Balancing life-style and genomics research for disease prevention. *Science* 296 (5568):695-698.

14. Selevan, S.G., C.A. Kimmel, and P. Mendola. 2000. Identifying critical windows of exposure for children's health. *Environmental Health Perspectives* 108 Supplement 3:451-5.

15. U.S. Environmental Protection Agency. 2006. *A Framework for Assessing Health Risks of Environmental Exposures to Children.* Washington, DC: U.S. EPA, National Center for Environmental Assessment. http://cfpub.epa.gov/ncea/cfm/recordisplay.cfm?deid=158363.

Environments and Contaminants

Criteria Air Pollutants

1. U.S. Environmental Protection Agency. 2010. *Clean Air Act.* U.S. EPA, Office of Air and Radiation. Retrieved December 28, 2010 from http://www.epa.gov/air/caa/.

2. U.S. Environmental Protection Agency. 2006. *Air Quality Criteria for Ozone and Related Photochemical Oxidants (Final Report).* Washington, DC: U.S. EPA, National Center for Environmental Assessment. EPA. EPA/600/R-05/004aF-cF. http://cfpub.epa.gov/ncea/isa/recordisplay.cfm?deid=149923.

3. U.S. Environmental Protection Agency. 2006. *Air Quality Criteria for Lead (Final Report).* Washington, DC: U.S. EPA, National Center for Environmental Assessment. EPA/600/R-05/144aF-bF. http://cfpub.epa.gov/ncea/isa/recordisplay.cfm?deid=158823.

4. U.S. Environmental Protection Agency. 2008. *Integrated Science Assessment for Oxides of Nitrogen — Health Criteria (Final Report).* Washington, DC: National Center for Environmental Assessment. EPA/600/R-08/071. http://cfpub.epa.gov/ncea/isa/recordisplay.cfm?deid=194645.

5. U.S. Environmental Protection Agency. 2008. *Integrated Science Assessment for Sulfur Oxides — Health Criteria (Final Report).* Washington, DC: U.S. EPA, National Center for Environmental Assessment. EPA/600/R-08/047F. http://cfpub.epa.gov/ncea/isa/recordisplay.cfm?deid=198843.

6. U.S. Environmental Protection Agency. 2009. *Integrated Science Assessment for Particulate Matter (Final Report).* Washington, DC: U.S. EPA, National Center for Environmental Assessment. EPA/600/R-08/139F. http://cfpub.epa.gov/ncea/CFM/recordisplay.cfm?deid=216546.

7. U.S. Environmental Protection Agency. 2010. *Integrated Science Assessment for Carbon Monoxide (Final Report).* Washington, DC: U.S. EPA, National Center for Environmental Assessment. EPA/600/R-09/019F. http://cfpub.epa.gov/ncea/cfm/recordisplay.cfm?deid=218686.

8. U.S. Environmental Protection Agency. 2009. *National Ambient Air Quality Standards (NAAQS).* Retrieved May 21, 2009 from http://www.epa.gov/ttn/naaqs/.

9. Bateson, T.F., and J. Schwartz. 2008. Children's response to air pollutants. *Journal of Toxicology and Environmental Health* 71 (3):238-43.

10. Kampa, M., and E. Castanas. 2008. Human health effects of air pollution. *Environmental Pollution* 151 (2):362-7.

11. Latza, U., S. Gerdes, and X. Baur. 2008. Effects of nitrogen dioxide on human health: Systematic review of experimental and epidemiological studies conducted between 2002 and 2006. *International Journal of Hygiene and Environmental Health.*

12. Salvi, S. 2007. Health effects of ambient air pollution in children. *Paediatric Respiratory Reviews* 8 (4):275-80.

13. Wigle, D.T., T.E. Arbuckle, M. Walker, M.G. Wade, S. Liu, and D. Krewski. 2007. Environmental hazards: Evidence for effects on child health. *Toxicology and Environmental Health Part B: Critical Reviews* 10 (1-2):3-39.

14. Ginsberg, G., B. Foos, R.B. Dzubow, and M. Firestone. 2010. Options for incorporating children's inhaled dose into human health risk assessment. *Inhalation Toxicology* 22 (8):627-47.

15. Ginsberg, G.L., B. Asgharian, J.S. Kimbell, J.S. Ultman, and A.M. Jarabek. 2008. Modeling approaches for estimating the dosimetry of inhaled toxicants in children. *Journal of Toxicology and Environmental Health* 71 (3):166-95.

16. Ginsberg, G.L., B.P. Foos, and M.P. Firestone. 2005. Review and analysis of inhalation dosimetry methods for application to children's risk assessment. *Journal of Toxicology and Environmental Health* 68 (8):573-615.

17. Makri, A., M. Goveia, J. Balbus, and R. Parkin. 2004. Children's susceptibility to chemicals: A review by developmental stage. *Toxicology and Environmental Health Part B: Critical Reviews* 7 (6):417-35.

18. U.S. Environmental Protection Agency. 2008. National Ambient Air Quality Standards for Lead: Final Rule. *Federal Register* 73 (219):66964-67062.

19. U.S. Environmental Protection Agency. 2008. National Ambient Air Quality Standards for Ozone: Final Rule. *Federal Register* 73 (60):16436-16514.

20. U.S. Environmental Protection Agency. 2006. National Ambient Air Quality Standards for Particulate Matter: Final Rule. *Federal Register* 71 (200):61143-61233.

21. Kajekar, R. 2007. Environmental factors and developmental outcomes in the lung. *Pharmacology & Therapeutics* 114 (2):129-45.

22. Islam, T., K. Berhane, R. McConnell, W.J. Gauderman, E. Avol, J.M. Peters, and F.D. Gilliland. 2009. Glutathione-S-transferase (GST) P1, GSTM1, exercise, ozone and asthma incidence in school children. *Thorax* 64 (3):197-202.

23. Islam, T., R. McConnell, W.J. Gauderman, E. Avol, J.M. Peters, and F.D. Gilliland. 2008. Ozone, oxidant defense genes, and risk of asthma during adolescence. *American Journal of Respiratory and Critical Care Medicine* 177 (4):388-95.

Criteria Air Pollutants (continued)

24. McConnell, R., K. Berhane, F. Gilliland, S.J. London, T. Islam, W.J. Gauderman, E. Avol, H.G. Margolis, and J.M. Peters. 2002. Asthma in exercising children exposed to ozone: A cohort study. *Lancet* 359 (9304):386-91.

25. Clark, N.A., P.A. Demers, C.J. Karr, M. Koehoorn, C. Lencar, L. Tamburic, and M. Brauer. 2010. Effect of early life exposure to air pollution on development of childhood asthma. *Environmental Health Perspectives* 118 (2):284-90.

26. Mortimer, K., R. Neugebauer, F. Lurmann, S. Alcorn, J. Balmes, and I. Tager. 2008. Air pollution and pulmonary function in asthmatic children: Effects of prenatal and lifetime exposures. *Epidemiology* 19 (4):550-7; discussion 561-2.

27. Mortimer, K., R. Neugebauer, F. Lurmann, S. Alcorn, J. Balmes, and I. Tager. 2008. Early-lifetime exposure to air pollution and allergic sensitization in children with asthma. *Journal of Asthma* 45 (10):874-81.

28. Gauderman, W.J., E. Avol, F. Gilliland, H. Vora, D. Thomas, K. Berhane, R. McConnell, N. Kuenzli, F. Lurmann, E. Rappaport, et al. 2004. The effect of air pollution on lung development from 10 to 18 years of age. *The New England Journal of Medicine* 351 (11):1057-67.

29. Gehring, U., A.H. Wijga, M. Brauer, P. Fischer, J.C. de Jongste, M. Kerkhof, M. Oldenwening, H.A. Smit, and B. Brunekreef. 2010. Traffic-related air pollution and the development of asthma and allergies during the first 8 years of life. *American Journal of Respiratory and Critical Care Medicine* 181 (6):596-603.

30. Jerrett, M., K. Shankardass, K. Berhane, W.J. Gauderman, N. Kunzli, E. Avol, F. Gilliland, F. Lurmann, J.N. Molitor, J.T. Molitor, et al. 2008. Traffic-related air pollution and asthma onset in children: A prospective cohort study with individual exposure measurement. *Environmental Health Perspectives* 116 (10):1433-8.

31. Karr, C.J., P.A. Demers, M.W. Koehoorn, C.C. Lencar, L. Tamburic, and M. Brauer. 2009. Influence of ambient air pollutant sources on clinical encounters for infant bronchiolitis. *American Journal of Respiratory and Critical Care Medicine* 180 (10):995-1001.

32. McConnell, R., K. Berhane, L. Yao, M. Jerrett, F. Lurmann, F. Gilliland, N. Kunzli, J. Gauderman, E. Avol, D. Thomas, et al. 2006. Traffic, susceptibility, and childhood asthma. *Environmental Health Perspectives* 114 (5):766-72.

33. McConnell, R., T. Islam, K. Shankardass, M. Jerrett, F. Lurmann, F. Gilliland, J. Gauderman, E. Avol, N. Kunzli, L. Yao, et al. 2010. Childhood incident asthma and traffic-related air pollution at home and school. *Environmental Health Perspectives* 118 (7):1021-6.

34. Morgenstern, V., A. Zutavern, J. Cyrys, I. Brockow, U. Gehring, S. Koletzko, C.P. Bauer, D. Reinhardt, H.E. Wichmann, and J. Heinrich. 2007. Respiratory health and individual estimated exposure to traffic-related air pollutants in a cohort of young children. *Occupational and Environmental Medicine* 64 (1):8-16.

35. Salam, M.T., T. Islam, and F.D. Gilliland. 2008. Recent evidence for adverse effects of residential proximity to traffic sources on asthma. *Current Opinion in Pulmonary Medicine* 14 (1):3-8.

36. Health Effects Institute. 2010. *Traffic-Related Air Pollution: A Critical Review of the Literature on Emissions, Exposure, and Health Effects.* Boston, Massachusetts: Health Effects Institute.

37. Bartra, J., J. Mullol, A. del Cuvillo, I. Davila, M. Ferrer, I. Jauregui, J. Montoro, J. Sastre, and A. Valero. 2007. Air pollution and allergens. *Journal of Investigative Allergology and Clinical Immunology* 17 Suppl 2:3-8.

38. Bråbäck, L., and B. Forsberg. 2009. Does traffic exhaust contribute to the development of asthma and allergic sensitization in children: Findings from recent cohort studies. *Environmental Health* 8:17.

39. Krzyzanowski, M., B. Kuna-Dibbert, and J. Schneider, eds. 2005. *Health Effects of Transport-related Air Pollution.* Copenhagen, Denmark: World Health Organization.

40. U.S. Environmental Protection Agency. 2009. *Air Quality Index: A Guide to Air Quality and Your Health.* Research Triangle Park, NC: U.S. EPA, Office of Air Quality Planning and Standards. EPA-456/F-09-002. http://www.epa.gov/airnow/aqi_brochure_08-09.pdf.

41. U.S. Environmental Protection Agency. 2010. *The Green Book Nonattainment Areas for Criteria Pollutants.* U.S. EPA. Retrieved December 28, 2010 from http://www.epa.gov/air/oaqps/greenbk/.

42. U.S. Environmental Protection Agency. 2010. Primary National Ambient Air Quality Standards for Nitrogen Dioxide: Final Rule. *Federal Register* 75 (26):6474-6537.

43. U.S. Environmental Protection Agency. 2010. Primary National Ambient Air Quality Standard for Sulfur Dioxide: Final Rule. *Federal Register* 75 (119):35520-35603.

44. U.S. Environmental Protection Agency. 2010. *Revisions to Lead Ambient Air Monitoring Requirements: Final Rule.* U.S. EPA. Retrieved February 8, 2011 from http://www.gpo.gov/fdsys/pkg/FR-2010-12-27/pdf/2010-32153.pdf.

45. Harnett, W. 2009. *Guidance on SIP Elements Required Under Sections 110(a)(1) and (2) for the 2006 24-Hour Fine Particle (PM_{2.5}) National Ambient Air Quality Standards (NAAQS).* Washington, DC: U.S. EPA, Office of Air Quality Planning and Standards. Memo from William Harnett, Director, Air Quality Policy Division, U.S. EPA Office of Air Quality Planning and Standards, to Regional Air Division Directors, EPA Regions 1-10, September 25, 2009.

Hazardous Air Pollutants

1. U.S. Environmental Protection Agency. 2009. *About Air Toxics*. Retrieved August 6, 2009 from www.epa.gov/ttn/atw/allabout.html.

2. Leikauf, G.D. 2002. Hazardous air pollutants and asthma. *Environmental Health Perspectives* 110 Suppl 4:505-26.

3. McGwin, G., J. Lienert, and J.I. Kennedy. 2010. Formaldehyde exposure and asthma in children: a systematic review. *Environmental Health Perspectives* 118 (3):313-7.

4. McMartin, K.I., M. Chu, E. Kopecky, T.R. Einarson, and G. Koren. 1998. Pregnancy outcome following maternal organic solvent exposure: a meta-analysis of epidemiologic studies. *American Journal of Industrial Medicine* 34 (3):288-92.

5. National Toxicology Program. 2011. *Report on Carcinogens, 12th Edition*. Research Triangle Park, NC: U.S. Department of Health and Human Services, National Toxicology Program. http://ntp.niehs.nih.gov/ntp/roc/twelfth/roc12.pdf.

6. Perera, F.P., Z. Li, R. Whyatt, L. Hoepner, S. Wang, D. Camann, and V. Rauh. 2009. Prenatal airborne polycyclic aromatic hydrocarbon exposure and child IQ at age 5 years. *Pediatrics* 124 (2):e195-202.

7. Perera, F.P., S. Wang, J. Vishnevetsky, B. Zhang, K.J. Cole, D. Tang, V. Rauh, and D.H. Phillips. 2011. PAH/Aromatic DNA Adducts in Cord Blood and Behavior Scores in New York City Children. *Environmental Health Perspectives* 119 (8):1176-81.

8. Schantz, S.L., J.J. Widholm, and D.C. Rice. 2003. Effects of PCB exposure on neuropsychological function in children. *Environmental Health Perspectives* 111 (3):357-576.

9. Wigle, D.T., T.E. Arbuckle, M.C. Turner, A. Berube, Q. Yang, S. Liu, and D. Krewski. 2008. Epidemiologic evidence of relationships between reproductive and child health outcomes and environmental chemical contaminants. *Journal of Toxicology and Environmental Health Part B: Critical Reviews* 11 (5-6):373-517.

10. U.S. Environmental Protection Agency. 2011. *IRIS Summaries: Benzene (CASRN 71-43-2)*. U.S. EPA, National Center for Environmental Assessment. Retrieved June 6, 2011 from http://www.epa.gov/iris/subst/0276.htm.

11. U.S. Environmental Protection Agency. 2011. *IRIS Summaries: Vinyl Chloride (CASRN 75-01-4)*. U.S. EPA, National Center for Environmental Assessment. Retrieved June 6, 2011 from http://www.epa.gov/iris/subst/1001.htm.

12. U.S. Environmental Protection Agency. 2011. *IRIS Summaries: 1,3 Butadiene (CASRN 106-99-0)*. U.S. EPA, National Center for Environmental Assessment. Retrieved June 6, 2011 from http://www.epa.gov/iris/subst/0139.htm.

13. U.S. Environmental Protection Agency. 2011. *IRIS Summaries: Chromium (VI) (CASRN 18540-29-9)*. U.S. EPA, National Center for Environmental Assessment. Retrieved June 6, 2011 from http://www.epa.gov/iris/subst/0144.htm.

14. U.S. Environmental Protection Agency. 2011. *IRIS Summaries: Nickel Subsulfide (CASRN 12035-72-2)*. U.S. EPA, National Center for Environmental Assessment. Retrieved June 6, 2011 from http://www.epa.gov/iris/subst/0273.htm.

15. U.S. Environmental Protection Agency. 2011. *IRIS Summaries: 2,4-/2,6-Toluene diisocyanate mixture (TDI) (CASRN 26471-62-5)*. U.S. EPA, National Center for Environmental Assessment. Retrieved June 6, 2011 from http://www.epa.gov/iris/subst/0503.htm.

16. U.S. Environmental Protection Agency. 2011. *IRIS Summaries: Manganese (CASRN 7439-96-5)*. U.S. EPA, National Center for Environmental Assessment. Retrieved June 6, 2011 from http://www.epa.gov/iris/subst/0373.htm.

17. Jedrychowski, W., A. Galas, A. Pac, E. Flak, D. Camman, V. Rauh, and F. Perera. 2005. Prenatal ambient air exposure to polycyclic aromatic hydrocarbons and the occurrence of respiratory symptoms over the first year of life. *European Journal of Epidemiology* 20 (9):775-82.

18. Miller, R.L., R. Garfinkel, M. Horton, D. Camann, F.P. Perera, R.M. Whyatt, and P.L. Kinney. 2004. Polycyclic aromatic hydrocarbons, environmental tobacco smoke, and respiratory symptoms in an inner-city birth cohort. *Chest* 126 (4):1071-8.

19. Rosa, M.J., K.H. Jung, M.S. Perzanowski, E.A. Kelvin, K.W. Darling, D.E. Camann, S.N. Chillrud, R.M. Whyatt, P.L. Kinney, F.P. Perera, et al. 2011. Prenatal exposure to polycyclic aromatic hydrocarbons, environmental tobacco smoke and asthma. *Respiratory Medicine* 105 (6):869-76.

20. U.S. Environmental Protection Agency. 2011. *IRIS Summaries: Benzo[a]pyrene (BaP) (CASRN 50-32-8)*. U.S. EPA, National Center for Environmental Assessment. Retrieved June 6, 2011 from http://www.epa.gov/iris/subst/0136.htm.

21. U.S. Environmental Protection Agency. 2011. *IRIS Summaries: Acetaldehyde (CASRN 75-07-0)*. U.S. EPA, National Center for Environmental Assessment. Retrieved June 6, 2011 from http://www.epa.gov/iris/subst/0290.htm.

22. U.S. Environmental Protection Agency. *IRIS Summaries: Carbon Tetrachloride (CASRN 56-23-5)*. U.S. EPA, National Center for Environmental Assessment. Retrieved August 9, 2011 from http://www.epa.gov/iris/subst/0020.htm.

23. Cook, R., M. Strum, J.S. Touma, T. Palma, J. Thurman, D. Ensley, and R. Smith. 2007. Inhalation exposure and risk from mobile source air toxics in future years. *Journal of Exposure Science and Environmental Epidemiology* 17 (1):95-105.

24. Woodruff, T.J., D.A. Axelrad, J. Caldwell, R. Morello-Frosch, and A. Rosenbaum. 1998. Public health implications of 1990 air toxics concentrations across the United States. *Environmental Health Perspectives* 106 (5):245-51.

Hazardous Air Pollutants (continued)

25. U.S. Environmental Protection Agency. 2003. *Toxicological Review of Acrolein (CAS No. 107-02-8)*. Washington, DC: U.S. EPA, National Center for Environmental Assessment. EPA/635/R-03/003. http://www.epa.gov/iris/toxreviews/0364tr.pdf.

26. U.S. Environmental Protection Agency. 2011. *National Air Toxics Assessment, 2005*. Office of Air Quality Planning and Standards. Retrieved June 6, 2011 from http://www.epa.gov/ttn/atw/nata2005/.

27. U.S. Environmental Protection Agency. 2010. *Results of the 2005 NATA Model-to-Monitor Comparison, Final Report*. Research Triangle Park, NC: U.S. EPA, Office of Air Quality Planning and Standards. http://www.epa.gov/ttn/atw/nata2005/05pdf/nata2005_model2monitor.pdf.

28. U.S. Environmental Protection Agency. *Health Effects Information Used in Cancer and Noncancer Risk Characterization for the 2005 National-Scale Assessment*. U.S. EPA, Office of Air and Radiation. Retrieved June 6, 2011 from http://www.epa.gov/ttn/atw/nata2005/05pdf/health_effects.pdf.

29. U.S. Environmental Protection Agency. 2011. *An Overview of Methods for EPA's National-Scale Air Toxics Assessment*. Research Triangle Park, NC: U.S. EPA, Office of Air Quality, Planning, and Standards. http://www.epa.gov/ttn/atw/nata2005/05pdf/nata_tmd.pdf.

30. U.S. Environmental Protection Agency. 2011. *Summary of Results for the 2005 National-Scale Assessment*. Washington, DC: U.S. EPA, Office of Air and Radiation http://www.epa.gov/ttn/atw/nata2005/05pdf/sum_results.pdf.

31. Ginsberg, G., B. Foos, R.B. Dzubow, and M. Firestone. 2010. Options for incorporating children's inhaled dose into human health risk assessment. *Inhalation Toxicology* 22 (8):627-47.

32. U.S. Environmental Protection Agency. 2005. *Supplemental Guidance for Assessing Susceptibility from Early-Life Exposure to Carcinogens*. Washington, DC: U.S. EPA, Risk Assessment Forum. EPA/630/R-03/003F. http://www.epa.gov/ttn/atw/childrens_supplement_final.pdf.

33. Brody, J.G., R. Morello-Frosch, A. Zota, P. Brown, C. Perez, and R. Rudel. 2009. Linking exposure assessment science with policy objectives for environmental justice and breast cancer advocacy: The Northern California Household Exposure Study. *American Journal of Public Health* 99 (Suppl 3):S600-9.

34. Long, T., T. Johnson, J. Laurenson, and A. Rosenbaum. 2004. *Memorandum: Development of Penetration and Proximity Microenvironment Factor Distributions for the HAPEM5 in Support of the 1999 National-Scale Air Toxics Assessment (NATA)*. Washington, DC: U.S. EPA. http://www.epa.gov/ttn/fera/hapem5/hapem5_me_factor_memo.pdf.

35. Payne-Sturges, D.C., T.A. Burke, P. Breysse, M. Diener-West, and T.J. Buckley. 2004. Personal exposure meets risk assessment: a comparison of measured and modeled exposures and risks in an urban community. *Environmental Health Perspectives* 112 (5):589-98.

36. Weisel, C.P., J. Zhang, B.J. Turpin, M.T. Morandi, S. Colome, T.H. Stock, D.M. Spektor, L. Korn, A.M. Winer, J. Kwon, et al. 2005. *Relationships of Indoor, Outdoor, and Personal Air (RIOPA): Part I. Collection Methods and Descriptive Analyses*. Boston, MA and Houston, TX: Health Effects Institute and Mickey Leland National Urban Air Toxics Research Center. HEI Research Report 130; NUATRC Research Report 7. http://pubs.healtheffects.org/getfile.php?u=25.

37. California Air Resources Board. 2004. *Children's School Bus Exposure Study*. California Environmental Protection Agency, Air Resources Board. Retrieved October 3, 2011 from http://www.arb.ca.gov/research/schoolbus/schoolbus.htm.

38. Sabin, L.D., E. Behrentz, A.M. Winer, S. Jeong, D.R. Fitz, D.V. Pankratz, S.D. Colome, and S.A. Fruin. 2004. Characterizing the range of children's air pollutant exposure during school bus commutes. *Journal of Exposure Analysis and Environmental Epidemiology* 15 (5):377-387.

39. California Air Resources Board. 2005. *Measuring Inside Vehicle Pollutants*. California Environmental Protection Agency, California Air Resources Board. Retrieved October 3, 2011 from http://www.arb.ca.gov/research/indoor/in-vehsm.htm.

40. National Research Council. 2000. *Toxicological Effects of Methylmercury*. Washington, DC: National Academy Press.

41. U.S. Environmental Protection Agency. 1997. *Mercury Study Report to Congress Volumes I to VII*. Washington DC: U.S. Environmental Protection Agency Office of Air Quality Planning and Standards and Office of Research and Development. EPA-452/R-97-003. http://www.epa.gov/hg/report.htm.

42. U.S. Environmental Protection Agency. 2000. *Deposition of Air Pollutants to the Great Waters: Third Report to Congress*. Washington, DC. http://epa.gov/ttncaaa1/t3/reports/head_2kf.pdf.

43. U.S. Department of Education. 2010. *Average number of hours and percentage of the student school week that public school teachers of first- through fourth-grade, self-contained classrooms spent on each of four subjects, total instruction hours per week on four subjects, total time spent delivering all instruction per week, and average length of student school week: Selected years 1987-88 through 2007-08*. National Center for Education Statistics. Retrieved July 22, 2010 from http://nces.ed.gov/surveys/sass/tables/sass0708_005_t1n.asp.

44. South Coast Air Quality Management District. 2000. *Multiple Air Toxics Exposure Study II*. Retrieved July 22, 2010 from: http://www.aqmd.gov/matesiidf/matestoc.htm.

Indoor Environments

1. U.S. Environmental Protection Agency. 2008. *Child-Specific Exposure Factors Handbook (Final Report)*. Washington, DC: U.S. EPA, National Center for Environmental Assessment. http://cfpub.epa.gov/ncea/cfm/recordisplay.cfm?deid=199243.

2. Gale, R.W., W.L. Cranor, D.A. Alvarez, J.N. Huckins, J.D. Petty, and G.L. Robertson. 2009. Semivolatile organic compounds in residential air along the Arizona-Mexico border. *Environmental Science and Technology* 43 (9):3054-60.

3. Rudel, R.A., D.E. Camann, J.D. Spengler, L.R. Korn, and J.G. Brody. 2003. Phthalates, alkylphenols, pesticides, polybrominated diphenyl ethers, and other endocrine-disrupting compounds in indoor air and dust. *Environmental Science and Technology* 37 (20):4543-53.

4. Weschler, C.J. 2009. Changes in indoor pollutants since the 1950s. *Atmospheric Environment* 43 (1):153-69.

5. Hwang, H.M., E.K. Park, T.M. Young, and B.D. Hammock. 2008. Occurrence of endocrine-disrupting chemicals in indoor dust. *Science of the Total Environment* 404 (1):26-35.

6. Matt, G.E., P.J. Quintana, M.F. Hovell, J.T. Bernert, S. Song, N. Novianti, T. Juarez, J. Floro, C. Gehrman, M. Garcia, et al. 2004. Households contaminated by environmental tobacco smoke: sources of infant exposures. *Tobacco Control* 13 (1):29-37.

7. Stapleton, H.M., J.G. Allen, S.M. Kelly, A. Konstantinov, S. Klosterhaus, D. Watkins, M.D. McClean, and T.F. Webster. 2008. Alternate and new brominated flame retardants detected in U.S. house dust. *Environmental Science and Technology* 42 (18):6910-6.

8. Strynar, M.J., and A.B. Lindstrom. 2008. Perfluorinated compounds in house dust from Ohio and North Carolina, USA. *Environmental Science and Technology* 42 (10):3751-6.

9. Tulve, N.S., P.A. Jones, M.G. Nishioka, R.C. Fortmann, C.W. Croghan, J.Y. Zhou, A. Fraser, C. Cavel, and W. Friedman. 2006. Pesticide measurements from the first national environmental health survey of child care centers using a multi-residue GC/MS analysis method. *Environmental Science and Technology* 40 (20):6269-74.

10. Butte, W. 2004. Sources and impacts of pesticides in indoor environments. *The Handbook of Environmental Chemistry* 4F:89-116.

11. Weschler, C.J., and W.W. Nazaroff. 2008. Semivolatile organic compounds in indoor environments. *Atmospheric Environment* 42 (40):9018-9040.

12. Egeghy, P.P., L.S. Sheldon, D.M. Stout, E.A. Cohen-Hubal, N.S. Tulve, L.J. Melnyk, M.K. Morgan, R.C. Fortmann, D.A. Whitaker, C.W. Croghan, et al. 2007. *Important Exposure Factors for Children: An Analysis of Laboratory and Observational Field Data Characterizing Cumulative Exposure to Pesticides*. Washington, DC: U.S. EPA, Office of Research and Development. http://www.epa.gov/nerl/research/data/exposure-factors.pdf.

13. Stapleton, H.M., S.M. Kelly, J.G. Allen, M.D. McClean, and T.F. Webster. 2008. Measurement of polybrominated diphenyl ethers on hand wipes: estimating exposure from hand-to-mouth contact. *Environmental Science and Technology* 42 (9):3329-34.

14. Adar, S.D., M. Davey, J.R. Sullivan, M. Compher, A. Szpiro, and L.J. Liu. 2008. Predicting Airborne Particle Levels Aboard Washington State School Buses. *Atmospheric Environment* 42 (33):7590-7599.

15. Sabin, L.D., E. Behrentz, A.M. Winer, S. Jeong, D.R. Fitz, D.V. Pankratz, S.D. Colome, and S.A. Fruin. 2004. Characterizing the range of children's air pollutant exposure during school bus commutes. *Journal of Exposure Analysis and Environmental Epidemiology* 15 (5):377-387.

16. Jones, M.R., A. Navas-Acien, J. Yuan, and P.N. Breysse. 2009. Secondhand tobacco smoke concentrations in motor vehicles: a pilot study. *Tobacco Control* 18 (5):399-404.

17. Matt, G.E., P.J. Quintana, M.F. Hovell, D.A. Chatfield, D.S. Ma, R. Romero, and A.M. Uribe. 2008. Residual tobacco smoke pollution in used cars for sale: air, dust, and surfaces. *Nicotine and Tobacco Research* 10 (9):1467-75.

18. Brody, J.G., R. Morello-Frosch, A. Zota, P. Brown, C. Perez, and R. Rudel. 2009. Linking exposure assessment science with policy objectives for environmental justice and breast cancer advocacy: The Northern California Household Exposure Study. *American Journal of Public Health* 99 (Suppl 3):S600-9.

19. U.S. Environmental Protection Agency. 2009. *Integrated Science Assessment for Particulate Matter (Final Report)*. Washington, DC: U.S. EPA, National Center for Environmental Assessment. EPA/600/R-08/139F. http://cfpub.epa.gov/ncea/CFM/recordisplay.cfm?deid=216546.

20. Hunt, A., D.L. Johnson, and D.A. Griffith. 2006. Mass transfer of soil indoors by track-in on footwear. *Science of the Total Environment* 370 (2-3):360-71.

21. Nishioka, M.G., R.G. Lewis, M.C. Brinkman, H.M. Burkholder, C.E. Hines, and J.R. Menkedick. 2001. Distribution of 2,4-D in air and on surfaces inside residences after lawn applications: comparing exposure estimates from various media for young children. *Environmental Health Perspectives* 109 (11):1185-91.

22. Kerger, B.D., C.E. Schmidt, and D.J. Paustenbach. 2000. Assessment of airborne exposure to trihalomethanes from tap water in residential showers and baths. *Risk Analysis* 20 (5):637-51.

23. Nuckols, J.R., D.L. Ashley, C. Lyu, S.M. Gordon, A.F. Hinckley, and P. Singer. 2005. Influence of tap water quality and household water use activities on indoor air and internal dose levels of trihalomethanes. *Environmental Health Perspectives* 113 (7):863-70.

Indoor Environments (continued)

24. Mills, W.B., S. Liu, M.C. Rigby, and D. Brenner. 2007. Time-variable simulation of soil vapor intrusion into a building with a combined crawl space and basement. *Environmental Science and Technology* 41 (14):4993-5001.

25. New Jersey Department of Environmental Protection. 2005. Vapor Intrusion Guidance. Updated in March 2007.

26. U.S. Environmental Protection Agency. 2011. *Radon (Rn)*. U.S. EPA, Office of Radiation and Indoor Air. Retrieved February 10, 2011 from http://www.epa.gov/radon/.

27. Cohn, R.D., S.J.A. Jr., R. Jaramillo, L.H. Reid, and D.C. Zeldin. 2006. National prevalence and exposure risk for cockroach allergen in U.S. households. *Environmental Health Perspectives* 114 (4):522-6.

28. Dales, R., L. Liu, A.J. Wheeler, and N.L. Gilbert. 2008. Quality of indoor residential air and health. *Canadian Medical Association Journal* 179 (2):147-52.

29. Institute of Medicine. 2000. *Clearing the Air: Asthma and Indoor Air Exposure*. Washington, D.C.: National Academy Press. http://books.nap.edu/openbook.php?record_id=9610&page=R1.

30. U.S. Environmental Protection Agency. 2010. *An Introduction to Indoor Air Quality*. U.S. EPA, Indoor Environments Division. Retrieved July 11, 2011 from http://www.epa.gov/iaq/biologic.html.

31. Seltzer, J.M., and M.J. Fedoruk. 2007. Health effects of mold in children. *Pediatric Clinics of North America* 54 (2):309-33, viii-ix.

32. U.S. Department of Health and Human Services. 2006. *The Health Consequences of Involuntary Exposure to Tobacco Smoke: A Report of the Surgeon General* Atlanta, GA: U.S. Department of Health and Human Services, Centers for Disease Control and Prevention, Office on Smoking and Health. http://www.surgeongeneral.gov/library/reports/secondhandsmoke/fullreport.pdf.

33. National Toxicology Program. 2011. *Report on Carcinogens, 12th Edition*. Research Triangle Park, NC: U.S. Department of Health and Human Services, National Toxicology Program. http://ntp.niehs.nih.gov/ntp/roc/twelfth/roc12.pdf.

34. U.S. Environmental Protection Agency. 1992. *Respiratory Health Effects of Passive Smoking: Lung Cancer and Other Disorders*. Washington, D.C.: Office of Research and Development. EPA/600/6-90/006F. http://cfpub.epa.gov/ncea/cfm/ets/etsindex.cfm.

35. Gergen, P.J., J.A. Fowler, K.R. Maurer, W.W. Davis, and M.D. Overpeck. 1998. The burden of environmental tobacco smoke exposure on the respiratory health of children 2 months through 5 years of age in the United States: Third National Health and Nutrition Examination Survey, 1988 to 1994. *Pediatrics* 101 (2):E8.

36. Yousey, Y.K. 2006. Household characteristics, smoking bans, and passive smoke exposure in young children. *Journal of Pediatric Health Care* 20 (2):98-105.

37. King, B.A., M.J. Travers, K.M. Cummings, M.C. Mahoney, and A.J. Hyland. 2010. Secondhand smoke transfer in multiunit housing. *Nicotine and Tobacco Research* 12 (11):1133-41.

38. Wamboldt, F.S., R.C. Balkissoon, A.E. Rankin, S.J. Szefler, S.K. Hammond, R.E. Glasgow, and W.P. Dickinson. 2008. Correlates of household smoking bans in low-income families of children with and without asthma. *Family Process* 47 (1):81-94.

39. Wilson, K.M., J.D. Klein, A.K. Blumkin, M. Gottlieb, and J.P. Winickoff. 2011. Tobacco-smoke exposure in children who live in multiunit housing. *Pediatrics* 127 (1):85-92.

40. Matt, G.E., P.J. Quintana, J.M. Zakarian, A.L. Fortmann, D.A. Chatfield, E. Hoh, A.M. Uribe, and M.F. Hovell. 2011. When smokers move out and non-smokers move in: residential thirdhand smoke pollution and exposure. *Tobacco Control* 20 (1):e1.

41. Singer, B.C., A.T. Hodgson, K.S. Guevarra, E.L. Hawley, and W.W. Nazaroff. 2002. Gas-phase organics in environmental tobacco smoke. 1. Effects of smoking rate, ventilation, and furnishing level on emission factors. *Environmental Science and Technology* 36 (5):846-53.

42. Winickoff, J.P., J. Friebely, S.E. Tanski, C. Sherrod, G.E. Matt, M.F. Hovell, and R.C. McMillen. 2009. Beliefs about the health effects of "thirdhand" smoke and home smoking bans. *Pediatrics* 123 (1):e74-9.

43. Singer, B.C., A.T. Hodgson, and W.W. Nazaroff. 2003. Gas-phase organics in environmental tobacco smoke: 2. Exposure-relevant emission factors and indirect exposures from habitual smoking. *Atmospheric Environment* 37:5551-61.

44. Pirkle, J.L., J.T. Bernert, S.P. Caudill, C.S. Sosnoff, and T.F. Pechacek. 2006. Trends in the Exposure of Nonsmokers in the U.S. Population to Secondhand Smoke: 1988–2002. *Environmental Health Perspectives* 114 (6).

45. Centers for Disease Control and Prevention. 2007. Cigarette smoking among adults—United States, 2006. *Morbidity and Mortality Weekly Report* 56 (44):1157-1161.

46. Centers for Disease Control and Prevention. 2011. Vital Signs: Current cigarette smoking among adults aged ≥ 18 years--United States, 2005-2010. *Morbidity and Mortality Weekly Report* 60 (35):1207-1212.

47. Centers for Disease Control and Prevention. 2007. State-specific prevalence of smoke-free home rules - United States, 1992-2003. *Morbidity and Mortality Weekly Report* 56 (20):501-504.

48. Mackay, D., S. Haw, J.G. Ayres, C. Fischbacher, and J.P. Pell. 2010. Smoke-free legislation and hospitalizations for childhood asthma. *New England Journal of Medicine* 363 (12):1139-45.

Indoor Environments (continued)

49. Rayens, M.K., P.V. Burkhart, M. Zhang, S. Lee, D.K. Moser, D. Mannino, and E.J. Hahn. 2008. Reduction in asthma-related emergency department visits after implementation of a smoke-free law. *Journal of Allergy and Clinical Immunology* 122 (3):537-41 e3.

50. King, K., M. Martynenko, M.H. Bergman, Y.-H. Liu, J.P. Winickoff, and M. Weitzman. 2009. Family composition and children's exposure to adult smokers in their homes. *Pediatrics* 123 (4):559-64.

51. Jacobs, D.E., R.P. Clickner, J.Y. Zhou, S.M. Viet, D.A. Marker, J.W. Rogers, D.C. Zeldin, P. Broene, and W. Friedman. 2002. The prevalence of lead-based paint hazards in U.S. housing. *Environmental Health Perspectives* 110 (10):A599-606.

52. Laidlaw, M.A.S., and G.M. Filippelli. 2008. Resuspension of urban soils as a persistent source of lead poisoning in children: A review and new directions. *Applied Geochemistry* 23 (8):2021-2039.

53. Dixon, S.L., J.M. Gaitens, D.E. Jacobs, W. Strauss, J. Nagaraja, T. Pivetz, J.W. Wilson, and P.J. Ashley. 2009. Exposure of U.S. children to residential dust lead, 1999-2004: II. The contribution of lead-contaminated dust to children's blood lead levels. *Environmental Health Perspectives* 117 (3):468-74.

54. Lanphear, B.P., R. Hornung, M. Ho, C.R. Howard, S. Eberly, and K. Knauf. 2002. Environmental lead exposure during early childhood. *The Journal of Pediatrics* 140 (1):40-7.

55. Lanphear, B.P., R. Hornung, J. Khoury, K. Yolton, P. Baghurst, D.C. Bellinger, R.L. Canfield, K.N. Dietrich, R. Bornschein, T. Greene, et al. 2005. Low-level environmental lead exposure and children's intellectual function: an international pooled analysis. *Environmental Health Perspectives* 113 (7):894-9.

56. Lanphear, B.P., T.D. Matte, J. Rogers, R.P. Clickner, B. Dietz, R.L. Bornschein, P. Succop, K.R. Mahaffey, S. Dixon, W. Galke, et al. 1998. The contribution of lead-contaminated house dust and residential soil to children's blood lead levels. A pooled analysis of 12 epidemiologic studies. *Environmental Research* 79 (1):51-68.

57. Lanphear, B.P., M. Weitzman, N.L. Winter, S. Eberly, B. Yakir, M. Tanner, M. Emond, and T.D. Matte. 1996. Lead-contaminated house dust and urban children's blood lead levels. *American Journal of Public Health* 86 (10):1416-21.

58. U.S. Environmental Protection Agency. 2006. *Air Quality Criteria for Lead. Volume I of II*. Washington, DC: United States Environmental Protection Agency. EPA/600/R-5/144aF.

59. Levin, R., M.J. Brown, M.E. Kashtock, D.E. Jacobs, E.A. Whelan, J. Rodman, M.R. Schock, A. Padilla, and T. Sinks. 2008. Lead exposures in U.S. children, 2008: implications for prevention. *Environmental Health Perspectives* 116 (10):1285-93.

60. U.S. Environmental Protection Agency. 2010. *Lead in Paint, Dust, and Soil: Renovation, Repair, and Painting*. U.S. EPA, Office of Pollution Prevention and Toxics. Retrieved October 4, 2010 from http://www.epa.gov/lead/pubs/renovation.htm.

61. U.S. Department of Housing and Urban Development. 1999. Lead-Safe Housing Rule, 24 CFR Part 35.

62. U.S. Environmental Protection Agency, U.S. Consumer Product Safety Commission, and U.S. Department of Housing and Urban Development. 2003. *Protect Your Family from Lead in Your Home*. Washington, DC: U.S. EPA, U.S. CPSC, U.S. HUD. EPA747-K-99-001. http://www.hud.gov/offices/lead/library/lead/pyf_eng.pdf.

63. U.S. Environmental Protection Agency. 2008. *40 CFR Part 745, Final Rule; Lead; Renovation, Repair, and Painting Program*. Washington, DC: U.S. EPA. EPA-HQ-OPPT-2005-0049. http://www.epa.gov/fedrgstr/EPA-TOX/2008/April/Day-22/t8141.htm.

64. National Toxicology Program. 2012. *NTP Monograph on Health Effects of Low-Level Lead*. Research Triangle Park, NC: National Institute of Environmental Health Sciences, National Toxicology Program. http://ntp.niehs.nih.gov/go/36443.

65. Bellinger, D., J. Sloman, A. Leviton, M. Rabinowitz, H.L. Needleman, and C. Waternaux. 1991. Low-level lead exposure and children's cognitive function in the preschool years. *Pediatrics* 87 (2):219-27.

66. Canfield, R.L., C.R. Henderson, Jr., D.A. Cory-Slechta, C. Cox, T.A. Jusko, and B.P. Lanphear. 2003. Intellectual impairment in children with blood lead concentrations below 10 microg per deciliter. *New England Journal of Medicine* 348 (16):1517-26.

67. Jusko, T.A., C.R. Henderson, B.P. Lanphear, D.A. Cory-Slechta, P.J. Parsons, and R.L. Canfield. 2008. Blood lead concentrations < 10 microg/dL and child intelligence at 6 years of age. *Environmental Health Perspectives* 116 (2):243-8.

68. Lanphear, B.P., K. Dietrich, P. Auinger, and C. Cox. 2000. Cognitive deficits associated with blood lead concentrations <10 microg/dL in US children and adolescents. *Public Health Reports* 115 (6):521-9.

69. Schnaas, L., S.J. Rothenberg, M.F. Flores, S. Martinez, C. Hernandez, E. Osorio, S.R. Velasco, and E. Perroni. 2006. Reduced intellectual development in children with prenatal lead exposure. *Environmental Health Perspectives* 114 (5):791-7.

70. Surkan, P.J., A. Zhang, F. Trachtenberg, D.B. Daniel, S. McKinlay, and D.C. Bellinger. 2007. Neuropsychological function in children with blood lead levels <10 microg/dL. *Neurotoxicology* 28 (6):1170-7.

71. Centers for Disease Control and Prevention. 1997. *Screening Young Children for Lead Poisoning: Guidance for State and Local Public Health Officials*. Atlanta, GA.

Indoor Environments (continued)

72. Centers for Disease Control and Prevention. 2002. *Managing Elevated Blood Lead Levels Among Young Children: Recommendations from the Advisory Committee on Childhood Lead Poisoning Prevention*. Atlanta, GA.

73. Advisory Committee on Childhood Lead Poisoning Prevention. 2012. *Low Level Lead Exposure Harms Children: A Renewed Call for Primary Prevention*. Atlanta, GA: Centers for Disease Control and Prevention. http://www.cdc.gov/nceh/lead/ACCLPP/Final_Document_030712.pdf.

74. Centers for Disease Control and Prevention. 2012. *CDC Response to Advisory Committee on Childhood Lead Poisoning Prevention Recommendations in Low Level Lead Exposure Harms Children: A Renewed Call for Primary Prevention*. Atlanta, GA: Centers for Disease Control and Prevention. http://www.cdc.gov/nceh/lead/acclpp/cdc_response_lead_exposure_recs.pdf.

75. Centers for Disease Control and Prevention. 1991. *Preventing Lead Poisoning in Young Children*. Atlanta, GA.

76. U.S. Environmental Protection Agency. 2010. *Section 21 Petitions Filed with EPA Since September 2007: Lead Dust Hazard Standard and Definition of Lead-based Paint*. U.S. EPA. Retrieved February 9, 2011 from http://www.epa.gov/oppt/chemtest/pubs/petitions.html#petition5.

77. Gaitens, J.M., S.L. Dixon, D.E. Jacobs, J. Nagaraja, W. Strauss, J.W. Wilson, and P.J. Ashley. 2009. Exposure of U.S. children to residential dust lead, 1999-2004: I. Housing and demographic factors. *Environmental Health Perspectives* 117 (3):461-7.

78. Jones, R.L., D.M. Homa, P.A. Meyer, D.J. Brody, K.L. Caldwell, J.L. Pirkle, and M.J. Brown. 2009. Trends in blood lead levels and blood lead testing among U.S. children aged 1 to 5 years, 1988–2004. *Pediatrics* 123 (3):e376-e385.

79. Kim, D.Y., F. Staley, G. Curtis, and S. Buchanan. 2002. Relation between housing age, housing value, and childhood blood lead levels in children in Jefferson County, Ky. *American Journal of Public Health* 92 (5):769-72.

80. Mannino, D.M., J.E. Moorman, B. Kingsley, D. Rose, and J. Repace. 2001. Health effects related to environmental tobacco smoke exposure in children in the United States: data from the Third National Health and Nutrition Examination Survey. *Archives of Pediatrics and Adolescent Medicine* 155 (1):36-41.

81. Sexton, K., J.L. Adgate, T.R. Church, S.S. Hecht, G. Ramachandran, I.A. Greaves, A.L. Fredrickson, A.D. Ryan, S.G. Carmella, and M.S. Geisser. 2004. Children's exposure to environmental tobacco smoke: using diverse exposure metrics to document ethnic/racial differences. *Environmental Health Perspectives* 112 (3):392-7.

82. Braun, J.M., T.E. Froehlich, J.L. Daniels, K.N. Dietrich, R. Hornung, P. Auinger, and B.P. Lanphear. 2008. Association of environmental toxicants and conduct disorder in U.S. children: NHANES 2001-2004. *Environmental Health Perspectives* 116 (7):956-62.

83. DeLorenze, G.N., M. Kharrazi, F.L. Kaufman, B. Eskenazi, and J.T. Bernert. 2002. Exposure to environmental tobacco smoke in pregnant women: the association between self-report and serum cotinine. *Environmental Research* 90 (1):21-32.

84. Kalkbrenner, A.E., R.W. Hornung, J.T. Bernert, S.K. Hammond, J.M. Braun, and B.P. Lanphear. 2010. Determinants of serum cotinine and hair cotinine as biomarkers of childhood secondhand smoke exposure. *Journal of Exposure Science and Environmental Epidemiology* 20 (7):615-24.

85. U.S. Department of Health and Human Services. 2000. *Healthy People 2010. 2nd ed. With Understanding and Improving Health and Objectives for Improving Health. 2 vols*. Washington, DC: U.S. Government Printing Office, November 2000.

86. U.S. Environmental Protection Agency. 2001. *40 CFR Part 745, Final Rule; Lead; Identification of Dangerous Levels of Lead*. http://www.epa.gov/fedrgstr/EPA-TOX/2001/January/Day-05/t84.pdf.

87. U.S. Department of Housing and Urban Development. 2001. *National Survey of Lead and Allergens in Housing, Final Report, Volume I: Analysis of Lead Hazards*. Washington, DC: HUD, Office of Lead Hazard Control. http://www.nmic.org/nyccelp/documents/HUD_NSLAH_Vol1.pdf.

Drinking Water Contaminants

1. Kumar, A., and I. Xagoraraki. 2010. Pharmaceuticals, personal care products and endocrine-disrupting chemicals in U.S. surface and finished drinking waters: a proposed ranking system. *Science of the Total Environment* 408 (23):5972-89.

2. U.S. Environmental Protection Agency. 2008. *National Primary Drinking Water Regulations*. U.S. EPA, Office of Water. Retrieved November 25, 2010 from http://water.epa.gov/drink/contaminants/index.cfm.

3. U.S. Geological Survey. 2010. *Source Water-Quality Assessment (SWQA) Program*. Retrieved November 25, 2010 from http://water.usgs.gov/nawqa/swqa/.

4. U.S. Environmental Protection Agency. 2010. *Drinking Water Glossary: Surface Water*. U.S. EPA, Office of Water. Retrieved November 25, 2010 from http://owpubauthor.epa.gov/aboutow/ogwdw/glossary.cfm#slink.

5. U.S. Environmental Protection Agency. 2010. *Drinking Water Glossary: Ground Water*. U.S. EPA, Office of Water. Retrieved November 25, 2010 from http://water.epa.gov/aboutow/ogwdw/glossary.cfm#glink.

6. Winter, T.C., J.W. Harvey, O.W. Franke, and W.M. Alley. 1998. *Ground Water and Surface Water: A Single Resource*. Reston, VA: U.S. Geological Survey. http://pubs.usgs.gov/circ/circ1139/pdf/circ1139.pdf.

Drinking Water Contaminants (continued)

7. U.S. Environmental Protection Agency. 1992. *Secondary Drinking Water Regulations: Guidance for Nuisance Chemicals*. U.S. EPA, Office of Water. Retrieved November 26, 2010 from http://water.epa.gov/drink/contaminants/secondarystandards.cfm.

8. Dietert, R.R. 2009. Developmental immunotoxicology: focus on health risks. *Chemical Research in Toxicology* 22 (1):17-23.

9. Garcia, A.M., S.A. Fadel, S. Cao, and M. Sarzotti. 2000. T cell immunity in neonates. *Immunologic Research* 22 (2-3):177-90.

10. Nwachuku, N., and C.P. Gerba. 2004. Microbial risk assessment: don't forget the children. *Current Opinion in Microbiology* 7 (3):206-9.

11. Thompson, S.C. 1994. Giardia lamblia in children and the child care setting: a review of the literature. *Journal of Paediatrics and Child Health* 30 (3):202-9.

12. Woodruff, T., L. Zeise, D. Axelrad, K.Z. Guyton, S. Janssen, M. Miller, G. Miller, J. Schwartz, G. Alexeeff, H. Anderson, et al. 2008. Moving Upstream: A workshop on evaluating adverse upstream endpoints for improved risk assessment and decision making. *Environmental Health Perspectives* 116 (11):1568–1575.

13. Yoder, J.S., and M.J. Beach. 2010. Cryptosporidium surveillance and risk factors in the United States. *Experimental Parasitology* 124 (1):31-9.

14. Centers for Disease Control and Prevention. 2010. Giardiasis Surveillance---United States, 2006--2008. *Morbidity and Mortality Weekly Report* 59 (SS06):15-25.

15. Edwards, M., S. Triantafyllidou, and D. Best. 2009. Elevated blood lead in young children due to lead-contaminated drinking water: Washington, DC, 2001-2004. *Environmental Science and Technology* 43 (5):1618-1623.

16. Levin, R., M.J. Brown, M.E. Kashtock, D.E. Jacobs, E.A. Whelan, J. Rodman, M.R. Schock, A. Padilla, and T. Sinks. 2008. Lead exposures in U.S. children, 2008: implications for prevention. *Environmental Health Perspectives* 116 (10):1285-93.

17. Miranda, M.L., D. Kim, A.P. Hull, C.J. Paul, and M.A. Galeano. 2007. Changes in blood lead levels associated with use of chloramines in water treatment systems. *Environmental Health Perspectives* 115 (2):221-5.

18. National Toxicology Program. 2012. *NTP Monograph on Health Effects of Low-Level Lead*. Research Triangle Park, NC: National Institute of Environmental Health Sciences, National Toxicology Program. http://ntp.niehs.nih.gov/go/36443.

19. Hunter, W.J. 2008. Remediation of Drinking Water for Rural Populations. In *Nitrogen in the Environment: Sources, Problems, and Management, Second Edition*, edited by J. L. Hatfield and R. F. Follett. Boston, MA: Academic Press/Elsevier.

20. U.S. Environmental Protection Agency. 2009. *Consumer Factsheet on: Nitrates/Nitrites*. U.S. EPA, Office of Water. Retrieved November 25, 2010 from http://www.epa.gov/safewater/pdfs/factsheets/ioc/nitrates.pdf.

21. Gupta, S.K., R.C. Gupta, A.K. Seth, A.B. Gupta, J.K. Bassin, and A. Gupta. 2000. Methaemoglobinaemia in areas with high nitrate concentration in drinking water. *National Medical Journal of India* 13 (2):58-61.

22. Knobeloch, L., B. Salna, A. Hogan, J. Postle, and H. Anderson. 2000. Blue babies and nitrate-contaminated well water. *Environmental Health Perspectives* 108 (7):675-8.

23. U.S. Environmental Protection Agency. 2010. *Basic Information About Nitrate in Drinking Water*. U.S. EPA, Office of Water. Retrieved November 25, 2010 from http://water.epa.gov/drink/contaminants/basicinformation/nitrate.cfm.

24. Gatseva, P.D., and M.D. Argirova. 2008. High-nitrate levels in drinking water may be a risk factor for thyroid dysfunction in children and pregnant women living in rural Bulgarian areas. *International Journal of Hygiene and Environmental Health* 211 (5-6):555-9.

25. Tajtakova, M., Z. Semanova, Z. Tomkova, E. Szokeova, J. Majoros, Z. Radikova, E. Sebokova, I. Klimes, and P. Langer. 2006. Increased thyroid volume and frequency of thyroid disorders signs in schoolchildren from nitrate polluted area. *Chemosphere* 62 (4):559-64.

26. Haddow, J.E., G.E. Palomaki, W.C. Allan, J.R. Williams, G.J. Knight, J. Gagnon, C.E. O'Heir, M.L. Mitchell, R.J. Hermos, S.E. Waisbren, et al. 1999. Maternal thyroid deficiency during pregnancy and subsequent neuropsychological development of the child. *New England Journal of Medicine* 341 (8):549-55.

27. Pop, V.J., E.P. Brouwers, H.L. Vader, T. Vulsma, A.L. van Baar, and J.J. de Vijlder. 2003. Maternal hypothyroxinaemia during early pregnancy and subsequent child development: a 3-year follow-up study. *Clinical Endocrinology* 59 (3):282-8.

28. Morreale de Escobar, G., M.J. Obregon, and F. Escobar del Rey. 2000. Is neuropsychological development related to maternal hypothyroidism or to maternal hypothyroxinemia? *The Journal of Clinical Endocrinology and Metabolism* 85 (11):3975-87.

29. U.S. Environmental Protection Agency. 2008. *Arsenic in Drinking Water*. U.S. EPA, Office of Water. Retrieved November 25, 2010 from http://water.epa.gov/lawsregs/rulesregs/sdwa/arsenic/index.cfm.

30. Sambu, S., and R. Wilson. 2008. Arsenic in food and water--a brief history. *Toxicology and Industrial Health* 24 (4):217-26.

31. Schuhmacher-Wolz, U., H.H. Dieter, D. Klein, and K. Schneider. 2009. Oral exposure to inorganic arsenic: evaluation of its carcinogenic and non-carcinogenic effects. *Critical Reviews in Toxicology* 39 (4):271-98.

Drinking Water Contaminants (continued)

32. Smith, A.H., G. Marshall, Y. Yuan, C. Ferreccio, J. Liaw, O. von Ehrenstein, C. Steinmaus, M.N. Bates, and S. Selvin. 2006. Increased mortality from lung cancer and bronchiectasis in young adults after exposure to arsenic in utero and in early childhood. *Environmental Health Perspectives* 114 (8):1293-6.

33. Smith, A.H., and C.M. Steinmaus. 2009. Health effects of arsenic and chromium in drinking water: recent human findings. *Annual Review of Public Health* 30:107-22.

34. U.S. Environmental Protection Agency. 2004. *Drinking Water Treatment*. Washington, DC: U.S. EPA, Office of Water. http://water.epa.gov/lawsregs/guidance/sdwa/upload/2009_08_28_sdwa_fs_30ann_treatment_web.pdf.

35. Burkholder, J., B. Libra, P. Weyer, S. Heathcote, D. Kolpin, P.S. Thorne, and M. Wichman. 2007. Impacts of waste from concentrated animal feeding operations on water quality. *Environmental Health Perspectives* 115 (2):308-12.

36. Centers for Disease Control and Prevention. 2009. *Water Treatment*. CDC Healthy Water Site. Retrieved November 25, 2010 from http://www.cdc.gov/healthywater/drinking/public/water_treatment.html.

37. U.S. Environmental Protection Agency. 2008. *Drinking Water Contaminants: Disinfection Byproducts*. U.S. EPA, Office of Water. Retrieved November 25, 2010 from http://water.epa.gov/drink/contaminants/#Byproducts.

38. Bove, F., Y. Shim, and P. Zeitz. 2002. Drinking water contaminants and adverse pregnancy outcomes: a review. *Environmental Health Perspectives* 110 (Suppl 1):61-74.

39. Colman, J., G.E. Rice, J.M. Wright, E.S. Hunter, 3rd, L.K. Teuschler, J.C. Lipscomb, R.C. Hertzberg, J.E. Simmons, M. Fransen, M. Osier, et al. 2011. Identification of developmentally toxic drinking water disinfection byproducts and evaluation of data relevant to mode of action. *Toxicology and Applied Pharmacology* 254 (2):100-26.

40. Federal Register. 2006. *National Primary Drinking Water Regulations: Stage 2 Disinfectants and Disinfection Byproducts Rule*. Washington, DC: U.S. EPA. January 4, 2006. http://www.epa.gov/fedrgstr/EPA-WATER/2006/January/Day-04/w03.pdf.

41. Villanueva, C.M., K.P. Cantor, S. Cordier, J.J. Jaakkola, W.D. King, C.F. Lynch, S. Porru, and M. Kogevinas. 2004. Disinfection byproducts and bladder cancer: a pooled analysis. *Epidemiology* 15 (3):357-67.

42. Hwang, B.F., J.J. Jaakkola, and H.R. Guo. 2008. Water disinfection by-products and the risk of specific birth defects: a population-based cross-sectional study in Taiwan. *Environmental Health* 7:23.

43. Hwang, B.F., P. Magnus, and J.J. Jaakkola. 2002. Risk of specific birth defects in relation to chlorination and the amount of natural organic matter in the water supply. *American Journal of Epidemiology* 156 (4):374-82.

44. Nieuwenhuijsen, M.J., D. Martinez, J. Grellier, J. Bennett, N. Best, N. Iszatt, M. Vrijheid, and M.B. Toledano. 2009. Chlorination disinfection by-products in drinking water and congenital anomalies: review and meta-analyses. *Environmental Health Perspectives* 117 (10):1486-93.

45. Wigle, D.T., T.E. Arbuckle, M.C. Turner, A. Berube, Q. Yang, S. Liu, and D. Krewski. 2008. Epidemiologic evidence of relationships between reproductive and child health outcomes and environmental chemical contaminants. *Journal of Toxicology and Environmental Health Part B: Critical Reviews* 11 (5-6):373-517.

46. U.S. Environmental Protection Agency. 2004. *Pesticides Industry Sales and Usage: 2000-2001 Market Estimates*. Washington, DC: U.S. EPA, Office of Pesticide Programs. http://www.epa.gov/opp00001/pestsales/01pestsales/market_estimates2001.pdf.

47. U.S. Environmental Protection Agency. 2010. *Basic Information about Regulated Drinking Water Contaminants*. U.S. EPA, Office of Water. Retrieved November 26, 2010 from http://water.epa.gov/drink/contaminants/basicinformation/index.cfm.

48. U.S. Environmental Protection Agency. 2011. *Source Water Assessment*. U.S. EPA, Office of Water. Retrieved June 21, 2011 from http://water.epa.gov/infrastructure/drinkingwater/sourcewater/protection/sourcewaterassessments.cfm.

49. Chevrier, C., G. Limon, C. Monfort, F. Rouget, R. Garlantezec, C. Petit, G. Durand, and S. Cordier. 2011. Urinary biomarkers of prenatal atrazine exposure and adverse birth outcomes in the PELAGIE birth cohort. *Environmental Health Perspectives* 119 (7):1034-41.

50. Munger, R., P. Isacson, S. Hu, T. Burns, J. Hanson, C.F. Lynch, K. Cherryholmes, P. Van Dorpe, and W.J. Hausler, Jr. 1997. Intrauterine growth retardation in Iowa communities with herbicide-contaminated drinking water supplies. *Environmental Health Perspectives* 105 (3):308-14.

51. Ochoa-Acuna, H., J. Frankenberger, L. Hahn, and C. Carbajo. 2009. Drinking-water herbicide exposure in Indiana and prevalence of small-for-gestational-age and preterm delivery. *Environmental Health Perspectives* 117 (10):1619-24.

52. Villanueva, C.M., G. Durand, M.B. Coutte, C. Chevrier, and S. Cordier. 2005. Atrazine in municipal drinking water and risk of low birth weight, preterm delivery, and small-for-gestational-age status. *Occupational and Environmental Medicine* 62 (6):400-5.

53. U.S. Geological Survey. 2010. *Glyphosate Herbicide Found in Many Midwestern Streams, Antibiotics Not Common*. U.S.G.S., Toxic Substances Hydrology Program. Retrieved November 25, 2010 from http://toxics.usgs.gov/highlights/glyphosate02.html.

54. U.S. Environmental Protection Agency. 1990. *IRIS Summaries: Glyphosate (CASRN 1071-83-6)*. U.S. EPA, National Center for Environmental Assessment. Retrieved November 25, 2010 from http://www.epa.gov/ncea/iris/subst/0057.htm#studoral.

Drinking Water Contaminants (continued)

55. Williams, G.M., R. Kroes, and I.C. Munro. 2000. Safety evaluation and risk assessment of the herbicide Roundup and its active ingredient, glyphosate, for humans. *Regulatory Toxicology and Pharmacology* 31 (2 Pt 1):117-65.

56. Dallegrave, E., F.D. Mantese, R.T. Oliveira, A.J. Andrade, P.R. Dalsenter, and A. Langeloh. 2007. Pre- and postnatal toxicity of the commercial glyphosate formulation in Wistar rats. *Archives of Toxicology* 81 (9):665-73.

57. Romano, R.M., M.A. Romano, M.M. Bernardi, P.V. Furtado, and C.A. Oliveira. 2010. Prepubertal exposure to commercial formulation of the herbicide glyphosate alters testosterone levels and testicular morphology. *Archives of Toxicology* 84 (4):309-17.

58. Sanin, L.H., G. Carrasquilla, K.R. Solomon, D.C. Cole, and E.J. Marshall. 2009. Regional differences in time to pregnancy among fertile women from five Colombian regions with different use of glyphosate. *Journal of Toxicology and Environmental Health* 72 (15-16):949-60.

59. National Research Council. 2006. *Assessing the Human Health Risks of Trichloroethylene: Key Scientific Issues.* Washington, DC: National Academies Press. http://www.nap.edu/catalog.php?record_id=11707.

60. National Research Council. 2010. *Review of the Environmental Protection Agency's Draft IRIS Assessment of Tetrachloroethylene.* Washington, DC: National Academies Press. http://www.nap.edu/catalog.php?record_id=12863.

61. U.S. Environmental Protection Agency. 2011. *Toxicological Review of Trichloroethylene (CAS No. 79-01-6).* Washington, DC: U.S. EPA, National Center for Environmental Assessment. EPA/635/R-09/011F. http://www.epa.gov/iris/toxreviews/0199tr/0199tr.pdf.

62. Aschengrau, A., J.M. Weinberg, P.A. Janulewicz, L.G. Gallagher, M.R. Winter, V.M. Vieira, T.F. Webster, and D.M. Ozonoff. 2009. Prenatal exposure to tetrachloroethylene-contaminated drinking water and the risk of congenital anomalies: a retrospective cohort study. *Environmental Health* 8:44.

63. Sonnenfeld, N., I. Hertz-Picciotto, and W.E. Kaye. 2001. Tetrachloroethylene in drinking water and birth outcomes at the US Marine Corps Base at Camp Lejeune, North Carolina. *American Journal of Epidemiology* 154 (10):902-8.

64. Aschengrau, A., J. Weinberg, S. Rogers, L. Gallagher, M. Winter, V. Vieira, T. Webster, and D. Ozonoff. 2008. Prenatal exposure to tetrachloroethylene-contaminated drinking water and the risk of adverse birth outcomes. *Environmental Health Perspectives* 116 (6):814-20.

65. Aschengrau, A., J.M. Weinberg, L.G. Gallagher, M.R. Winter, V.M. Vieira, T.F. Webster, and D.M. Ozonoff. 2009. Exposure to Tetrachloroethylene-Contaminated Drinking Water and the Risk of Pregnancy Loss. *Water Quality, Exposure, and Health* 1 (1):23-34.

66. U.S. Environmental Protection Agency. 2010. *Pharmaceuticals and Personal Care Products: Frequent Questions.* U.S. EPA, Office of Research and Development. Retrieved November 25, 2010 from http://www.epa.gov/ppcp/faq.html.

67. Focazio, M.J., D.W. Kolpin, K.K. Barnes, E.T. Furlong, M.T. Meyer, S.D. Zaugg, L.B. Barber, and M.E. Thurman. 2008. A national reconnaissance for pharmaceuticals and other organic wastewater contaminants in the United States--II) untreated drinking water sources. *Science of the Total Environment* 402 (2-3):201-16.

68. U.S. Environmental Protection Agency. 2004. *Drinking Water Health Advisory for Manganese.* Washington, DC: U.S. EPA, Office of Water. EPA-822-R-04-003. http://www.epa.gov/ogwdw000/ccl/pdfs/reg_determine1/support_cc1_magnese_dwreport.pdf.

69. Bouchard, M., F. Laforest, L. Vandelac, D. Bellinger, and D. Mergler. 2007. Hair manganese and hyperactive behaviors: pilot study of school-age children exposed through tap water. *Environmental Health Perspectives* 115 (1):122-7.

70. Bouchard, M.F., S. Sauve, B. Barbeau, M. Legrand, M.E. Brodeur, T. Bouffard, E. Limoges, D.C. Bellinger, and D. Mergler. 2011. Intellectual impairment in school-age children exposed to manganese from drinking water. *Environmental Health Perspectives* 119 (1):138-43.

71. Khan, K., P. Factor-Litvak, G.A. Wasserman, X. Liu, E. Ahmed, F. Parvez, V. Slavkovich, D. Levy, J. Mey, A. van Geen, et al. 2011. Manganese exposure from drinking water and children's classroom behavior in Bangladesh. *Environmental Health Perspectives* 119 (10):1501-6.

72. Takser, L., D. Mergler, G. Hellier, J. Sahuquillo, and G. Huel. 2003. Manganese, monoamine metabolite levels at birth, and child psychomotor development. *Neurotoxicology* 24 (4-5):667-74.

73. Wasserman, G.A., X. Liu, F. Parvez, H. Ahsan, D. Levy, P. Factor-Litvak, J. Kline, A. van Geen, V. Slavkovich, N.J. Lolacono, et al. 2006. Water manganese exposure and children's intellectual function in Araihazar, Bangladesh. *Environmental Health Perspectives* 114 (1):124-9.

74. Ericson, J.E., F.M. Crinella, K.A. Clarke-Stewart, V.D. Allhusen, T. Chan, and R.T. Robertson. 2007. Prenatal manganese levels linked to childhood behavioral disinhibition. *Neurotoxicology and Teratology* 29 (2):181-7.

75. Wright, R.O., C. Amarasiriwardena, A.D. Woolf, R. Jim, and D.C. Bellinger. 2006. Neuropsychological correlates of hair arsenic, manganese, and cadmium levels in school-age children residing near a hazardous waste site. *Neurotoxicology* 27 (2):210-6.

76. California Department of Public Health. *History of Perchlorate in California Drinking Water.* California Department of Public Health, Drinking Water Program. Retrieved November 25, 2010 from http://www.cdph.ca.gov/certlic/drinkingwater/Pages/Perchloratehistory.aspx.

Drinking Water Contaminants (continued)

77. Rao, B., T.A. Anderson, G.J. Orris, K.A. Rainwater, S. Rajagopalan, R.M. Sandvig, B.R. Scanlon, D.A. Stonestrom, M.A. Walvoord, and W.A. Jackson. 2007. Widespread natural perchlorate in unsaturated zones of the southwest United States. *Environmental Science and Technology* 41 (13):4522-8.

78. U.S. Environmental Protection Agency. 2011. *Perchlorate*. U.S. EPA, Office of Water. Retrieved February 11, 2011 from http://water.epa.gov/drink/contaminants/unregulated/perchlorate.cfm.

79. Blount, B.C., and L. Valentin-Blasini. 2007. Biomonitoring as a method for assessing exposure to perchlorate. *Thyroid* 17 (9):837-41.

80. Centers for Disease Control and Prevention. 2009. *Perchlorate in Baby Formula Fact Sheet*. CDC. Retrieved August 13, 2009 from http://cdc.gov/nceh/features/perchlorate_factsheet.htm.

81. Pearce, E.N., A.M. Leung, B.C. Blount, H.R. Bazrafshan, X. He, S. Pino, L. Valentin-Blasini, and L.E. Braverman. 2007. Breast milk iodine and perchlorate concentrations in lactating Boston-area women. *Journal of Clinical Endocrinology & Metabolism* 92 (5):1673-7.

82. Schier, J.G., A.F. Wolkin, L. Valentin-Blasini, M.G. Belson, S.M. Kieszak, C.S. Rubin, and B.C. Blount. 2010. Perchlorate exposure from infant formula and comparisons with the perchlorate reference dose. *Journal of Exposure Science and Environmental Epidemiology* 20 (3):281-7.

83. Greer, M.A., G. Goodman, R.C. Pleus, and S.E. Greer. 2002. Health effects assessment for environmental perchlorate contamination: the dose response for inhibition of thyroidal radioiodine uptake in humans. *Environmental Health Perspectives* 110 (9):927-37.

84. National Research Council. 2005. *Health Implications of Perchlorate Ingestion*. Washington, DC: National Academies Press. http://www.nap.edu/catalog.php?record_id=11202.

85. U.S. Environmental Protection Agency. 2008. *Interim Drinking Water Health Advisory for Perchlorate*. Washington, DC: U.S. EPA, Office of Water. EPA 822-R-08-025. http://www.epa.gov/ogwdw/contaminants/unregulated/pdfs/healthadvisory_perchlorate_interim.pdf.

86. U.S. Environmental Protection Agency. 2008. *2006 Drinking Water Data Reliability Analysis and Action Plan*. Washington, DC. EPA 816-R-07-010. http://www.epa.gov/ogwdw/databases/pdfs/report_data_datareliability_2006.pdf.

87. U.S. Environmental Protection Agency. 1989. Drinking Water; National Primary Drinking Water Regulations; Total Coliforms (Including Fecal Coliforms and E. Coli) Final Rule. *Federal Register* 54 (124):27544-68.

88. U.S. Environmental Protection Agency. 2010. *Total Coliform Rule: Basic Information*. U.S. EPA, Office of Water. Retrieved November 25, 2010 from http://water.epa.gov/lawsregs/rulesregs/sdwa/tcr/basicinformation.cfm.

89. U.S. Environmental Protection Agency. 2010. *Microbial and Disinfection Byproducts Rules: Microbials and Disinfection Byproducts*. U.S. EPA, Office of Water. Retrieved November 25, 2010 from http://water.epa.gov/lawsregs/rulesregs/sdwa/mdbp/index.cfm.

90. U.S. Environmental Protection Agency. 2001. National Primary Drinking Water Regulations; Arsenic and Clarifications to Compliance and New Source Contaminants Monitoring. *Federal Register* 66 (14):6975-7066.

91. U.S. Environmental Protection Agency. 2006. National Primary Drinking Water Regulations: Stage 2 Disinfectants and Disinfection Byproducts Rule. *Federal Register* 71 (2):387-493.

92. U.S. Environmental Protection Agency. 2009. *Factoids: Drinking Water and Ground Water Statistics for 2009*. Washington, DC: U.S. EPA, Office of Water. EPA 816-K-09-004. http://www.epa.gov/safewater/databases/pdfs/data_factoids_2009.pdf.

93. U.S. Environmental Protection Agency. 2005. *Water Health Series: Bottled Water Basics*. Washington, DC: U.S. EPA, Office of Water. 816-K-05-003. http://www.epa.gov/safewater/faq/pdfs/fs_healthseries_bottlewater.pdf.

94. U.S. Environmental Protection Agency. 2010. *Public Drinking Water Systems Programs*. U.S. EPA, Office of Water. Retrieved November 25, 2010 from http://water.epa.gov/infrastructure/drinkingwater/pws/index.cfm.

95. U.S. Food and Drug Administration. 2011. *FDA Regulates the Safety of Bottled Water Beverages Including Flavored Water and Nutrient-Added Water Beverages*. U.S. FDA. Retrieved June 21, 2011 from http://www.fda.gov/Food/ResourcesForYou/Consumers/ucm046894.htm.

96. U.S. Environmental Protection Agency. 2010. *Private Drinking Water Wells*. U.S. EPA, Office of Water. Retrieved January 10, 2011 from http://water.epa.gov/drink/info/well/index.cfm.

97. U.S. Geological Survey. 2004. Estimated Use of Water in the United States in 2000. In *USGS Circular 1268*. Denver, CO.

98. Focazio, M.J., D. Tipton, S. Dunkle Shapiro, and L.H. Geiger. 2006. The Chemical Quality of Self-Supplied Domestic Well Water in the United States. *Ground Water Monitoring and Remediation* 26 (3):92-104.

99. DeSimone, L.A. 2009. *Quality of Water from Domestic Wells in Principal Aquifers of the United States, 1991–2004: U.S. Geological Survey Scientific Investigations Report*. Reston, VA: U.S.G.S. SIR 2008–5227. http://pubs.usgs.gov/sir/2008/5227.

100. U.S. Environmental Protection Agency. 2009. *Potential Environmental Impacts of Animal Feeding Operations*. U.S. EPA, Ag Center. Retrieved January 10, 2011 from http://www.epa.gov/agriculture/ag101/impacts.html.

Chemicals in Food

1. National Research Council. 1993. *Pesticides in the Diets of Infants and Children*. Washington, DC: National Academy Press. http://www.nap.edu/catalog/2126.html?se_side.

2. Scallan, E., P.M. Griffin, F.J. Angulo, R.V. Tauxe, and R.M. Hoekstra. 2011. Foodborne illness acquired in the United States--unspecified agents. *Emerging Infectious Diseases* 17 (1):16-22.

3. Centers for Disease Control and Prevention. 2011. Vital Signs: Incidence and trends of infection with pathogens transmitted commonly through food--foodborne diseases active surveillance network, 10 U.S. sites, 1996-2010. *Morbidity and Mortality Weekly Report* 60 (22):749-755.

4. Guimaraes, J.R.D., J. Ikingura, and H. Akagi. 2000. Methyl mercury production and distribution in river water-sediment systems investigated through radiochemical techniques. *Water, Air, and Soil Pollution* 124 (1-2):113-124.

5. Chen, C.Y., R.S. Stemberger, B. Klaue, J.D. Blum, P.C. Pickhardt, and C.L. Folt. 2000. Accumulation of heavy metals in food web components across a gradient of lakes. *Limnology and Oceanography* 45 (7):1525-1536.

6. Dietz, R., F. Riget, M. Cleemann, A. Aarkrog, P. Johansen, and J.C. Hansen. 2000. Comparison of contaminants from different trophic levels and ecosystems. *Science of the Total Environment* 245 (1-3):221-231.

7. Gilmour, C.C., and G.S. Riedel. 2000. A survey of size-specific mercury concentrations in game fish from Maryland fresh and estuarine waters. *Archives of Environmental Contamination and Toxicology* 39 (1):53-59.

8. Mason, R.P., J.R. Reinfelder, and F.M.M. Morel. 1995. Bioaccumulation of mercury and methylmercury. *Water, Air, and Soil Pollution* 80:915-921.

9. Mahaffey, K.R. 2004. Fish and shellfish as dietary sources of methylmercury and the omega-3 fatty acids, eicosahexaenoic acid and docosahexaenoic acid: risks and benefits. *Environmental Research* 95 (3):414-28.

10. U.S. Environmental Protection Agency. 1997. *Mercury Study Report to Congress Volumes I to VII*. Washington DC: U.S. Environmental Protection Agency Office of Air Quality Planning and Standards and Office of Research and Development. EPA-452/R-97-003. http://www.epa.gov/hg/report.htm.

11. Karagas, M.R., A.L. Choi, E. Oken, M. Horvat, R. Schoeny, E. Kamai, W. Cowell, P. Grandjean, and S. Korrick. 2012. Evidence on the human health effects of low-level methylmercury exposure. *Environmental Health Perspectives* 120 (6):799-806.

12. Lederman, S.A., R.L. Jones, K.L. Caldwell, V. Rauh, S.E. Sheets, D. Tang, S. Viswanathan, M. Becker, J.L. Stein, R.Y. Wang, et al. 2008. Relation between cord blood mercury levels and early child development in a World Trade Center cohort. *Environmental Health Perspectives* 116 (8):1085-91.

13. Lynch, M.L., L.S. Huang, C. Cox, J.J. Strain, G.J. Myers, M.P. Bonham, C.F. Shamlaye, A. Stokes-Riner, J.M. Wallace, E.M. Duffy, et al. 2011. Varying coefficient function models to explore interactions between maternal nutritional status and prenatal methylmercury toxicity in the Seychelles Child Development Nutrition Study. *Environmental Research* 111 (1):75-80.

14. National Research Council. 2000. *Toxicological Effects of Methylmercury*. Washington, DC: National Academy Press.

15. Oken, E., J.S. Radesky, R.O. Wright, D.C. Bellinger, C.J. Amarasiriwardena, K.P. Kleinman, H. Hu, and M.W. Gillman. 2008. Maternal fish intake during pregnancy, blood mercury levels, and child cognition at age 3 years in a US cohort. *American Journal of Epidemiology* 167 (10):1171-81.

16. Institute of Medicine. 2006. *Seafood Choices: Balancing Benefits and Risks*. Washington, DC: Committee on Nutrient Relationships in Seafood: Selections to Balance Benefits and Risks. Food and Nutrition Board. Institute of Medicine. http://iom.edu/Reports/2006/Seafood-Choices-Balancing-Benefits-and-Risks.aspx.

17. U.S. Environmental Protection Agency, and U.S. Food and Drug Administration. 2004. *What You Need to Know About Mercury in Fish and Shellfish. Advice for Women who Might Become Pregnant, Women who are Pregnant, Nursing Mothers and Children*. Washington, DC: U.S. Environmental Protection Agency and U.S. Food and Drug Administration. EPA-823-F-04-009. http://www.epa.gov/waterscience/fish/files/MethylmercuryBrochure.pdf.

18. Ginsberg, G.L., and B.F. Toal. 2009. Quantitative approach for incorporating methylmercury risks and omega-3 fatty acid benefits in developing species-specific fish consumption advice. *Environmental Health Perspectives* 117 (2):267-75.

19. U.S. Department of Agriculture, and U.S. Department of Health and Human Services. 2010. *Dietary Guidelines for Americans, 2010*. Washington, DC: U.S. Government Printing Office. http://www.cnpp.usda.gov/Publications/DietaryGuidelines/2010/PolicyDoc/PolicyDoc.pdf.

20. Agency for Toxic Substances and Disease Registry (ATSDR). 2000. *Toxicological Profile for Polychlorinated Biphenyls (PCBs)*. Atlanta, GA: U.S. Department of Health and Human Services, Public Health Service. http://www.atsdr.cdc.gov/toxprofiles/tp.asp?id=142&tid=26.

21. Choi, A.L., J.I. Levy, D.W. Dockery, L.M. Ryan, P.E. Tolbert, L.M. Altshul, and S.A. Korrick. 2006. Does living near a Superfund site contribute to higher polychlorinated biphenyl (PCB) exposure? *Environmental Health Perspectives* 114 (7):1092-8.

Chemicals in Food (continued)

22. U.S. Department of Agriculture. 2009. *DIOXIN 08 Survey: Dioxin and Dioxin-Like Compounds in the U.S. Domestic Meat and Poultry Supply*. Washington, DC: USDA Food Safety and Inspection Service, Office of Public Health Science, Risk Assessment Division. http://www.fsis.usda.gov/PDF/Dioxin_Report_1009.pdf.

23. Axelrad, D.A., S. Goodman, and T.J. Woodruff. 2009. PCB body burdens in US women of childbearing age 2001-2002: An evaluation of alternate summary metrics of NHANES data. *Environmental Research* 109 (4):368-78.

24. Patterson, D.G., Jr., L.Y. Wong, W.E. Turner, S.P. Caudill, E.S. Dipietro, P.C. McClure, T.P. Cash, J.D. Osterloh, J.L. Pirkle, E.J. Sampson, et al. 2009. Levels in the U.S. population of those persistent organic pollutants (2003-2004) included in the Stockholm Convention or in other long range transboundary air pollution agreements. *Environmental Science & Technology* 43 (4):1211-8.

25. Hickey, J.P., S.A. Batterman, and S.M. Chernyak. 2006. Trends of chlorinated organic contaminants in great lakes trout and walleye from 1970 to 1998. *Archives of Environmental Contamination and Toxicology* 50 (1):97-110.

26. Schecter, A., O. Papke, K.C. Tung, J. Joseph, T.R. Harris, and J. Dahlgren. 2005. Polybrominated diphenyl ether flame retardants in the U.S. population: current levels, temporal trends, and comparison with dioxins, dibenzofurans, and polychlorinated biphenyls. *Journal of Occupational and Environmental Medicine* 47 (3):199-211.

27. Sjodin, A., R.S. Jones, J.F. Focant, C. Lapeza, R.Y. Wang, E.E. McGahee, 3rd, Y. Zhang, W.E. Turner, B. Slazyk, L.L. Needham, et al. 2004. Retrospective time-trend study of polybrominated diphenyl ether and polybrominated and polychlorinated biphenyl levels in human serum from the United States. *Environmental Health Perspectives* 112 (6):654-8.

28. Sun, P., I. Basu, P. Blanchard, K.A. Brice, and R.A. Hites. 2007. Temporal and spatial trends of atmospheric polychlorinated biphenyl concentrations near the Great Lakes. *Environmental Science and Technology* 41 (4):1131-6.

29. Schantz, S.L., J.C. Gardiner, D.M. Gasior, R.J. McCaffrey, A.M. Sweeney, and H.E.B. Humphrey. 2004. Much ado about something: the weight of evidence for PCB effects on neuropsychological function. *Psychology in the Schools* 41 (6):669-679.

30. Schantz, S.L., J.J. Widholm, and D.C. Rice. 2003. Effects of PCB exposure on neuropsychological function in children. *Environmental Health Perspectives* 111 (3):357-376.

31. Selgrade, M.K. 2007. Immunotoxicity: the risk is real. *Toxicological Sciences* 100 (2):328-32.

32. Boucher, O., G. Muckle, and C.H. Bastien. 2009. Prenatal exposure to polychlorinated biphenyls: a neuropsychologic analysis. *Environmental Health Perspectives* 117 (1):7-16.

33. Ribas-Fito, N., M. Sala, M. Kogevinas, and J. Sunyer. 2001. Polychlorinated biphenyls (PCBs) and neurological development in children: a systematic review. *Journal of Epidemiology and Community Health* 55 (8):537-46.

34. Wigle, D.T., T.E. Arbuckle, M.C. Turner, A. Berube, Q. Yang, S. Liu, and D. Krewski. 2008. Epidemiologic evidence of relationships between reproductive and child health outcomes and environmental chemical contaminants. *Journal of Toxicology and Environmental Health B Critical Reviews* 11 (5-6):373-517.

35. Jacobson, J.L., S.W. Jacobson, and H.E. Humphrey. 1990. Effects of exposure to PCBs and related compounds on growth and activity in children. *Neurotoxicology and Teratology* 12 (4):319-26.

36. Vreugdenhil, H.J., P.G. Mulder, H.H. Emmen, and N. Weisglas-Kuperus. 2004. Effects of perinatal exposure to PCBs on neuropsychological functions in the Rotterdam cohort at 9 years of age. *Neuropsychology* 18 (1):185-93.

37. Walkowiak, J., J.A. Wiener, A. Fastabend, B. Heinzow, U. Kramer, E. Schmidt, H.J. Steingruber, S. Wundram, and G. Winneke. 2001. Environmental exposure to polychlorinated biphenyls and quality of the home environment: effects on psychodevelopment in early childhood. *Lancet* 358 (9293):1602-7.

38. Institute of Medicine. 2003. *Dioxins and Dioxin-like Compounds in the Food Supply*. Washington, DC: National Academy Press. http://books.nap.edu/openbook.php?record_id=10763&page-R1.

39. U.S. Environmental Protection Agency. 2010. *DecaBDE Phase-out Initiative*. U.S. Environmental Protection Agency. Retrieved February 26, 2010 from http://www.epa.gov/oppt/existingchemicals/pubs/actionplans/deccadbe.html.

40. Costa, L.G., and G. Giordano. 2007. Developmental neurotoxicity of polybrominated diphenyl ether (PBDE) flame retardants. *Neurotoxicology* 28 (6):1047-1067.

41. Frederiksen, M., K. Vorkamp, M. Thomsen, and L.E. Knudsen. 2009. Human internal and external exposure to PBDEs--a review of levels and sources. *International Journal of Hygiene and Environmental Health* 212 (2):109-34.

42. Huwe, J.K., and G.L. Larsen. 2005. Polychlorinated dioxins, furans, and biphenyls, and polybrominated diphenyl ethers in a U.S. meat market basket and estimates of dietary intake. *Environmental Science and Technology* 39 (15):5606-11.

43. Rose, M., D.H. Bennett, A. Bergman, B. Fangstrom, I.N. Pessah, and I. Hertz-Picciotto. 2010. PBDEs in 2-5 year-old children from California and associations with diet and indoor environment. *Environmental Science and Technology* 44 (7):2648-53.

44. Schecter, A., J. Colacino, K. Patel, K. Kannan, S.H. Yun, D. Haffner, T.R. Harris, and L. Birnbaum. 2010. Polybrominated diphenyl ether levels in foodstuffs collected from three locations from the United States. *Toxicology and Applied Pharmacology* 243 (2):217-24.

Chemicals in Food (continued)

45. Schecter, A., D. Haffner, J. Colacino, K. Patel, O. Papke, M. Opel, and L. Birnbaum. 2010. Polybrominated diphenyl ethers (PBDEs) and hexabromocyclodecane (HBCD) in composite U.S. food samples. *Environmental Health Perspectives* 118 (3):357-62.

46. Schecter, A., O. Papke, T.R. Harris, K.C. Tung, A. Musumba, J. Olson, and L. Birnbaum. 2006. Polybrominated diphenyl ether (PBDE) levels in an expanded market basket survey of U.S. food and estimated PBDE dietary intake by age and sex. *Environmental Health Perspectives* 114 (10):1515-20.

47. Wu, N., T. Herrmann, O. Paepke, J. Tickner, R. Hale, L.E. Harvey, M. La Guardia, M.D. McClean, and T.F. Webster. 2007. Human exposure to PBDEs: associations of PBDE body burdens with food consumption and house dust concentrations. *Environmental Science & Technology* 41 (5):1584-9.

48. Yogui, G.T., and J.L. Sericano. 2009. Polybrominated diphenyl ether flame retardants in the U.S. marine environment: a review. *Environment International* 35 (3):655-66.

49. Fraser, A.J., T.F. Webster, and M.D. McClean. 2009. Diet contributes significantly to the body burden of PBDEs in the general U.S. population. *Environmental Health Perspectives* 117 (10):1520-5.

50. Johnson-Restrepo, B., and K. Kannan. 2009. An assessment of sources and pathways of human exposure to polybrominated diphenyl ethers in the United States. *Chemosphere* 76 (4):542-8.

51. Lorber, M. 2008. Exposure of Americans to polybrominated diphenyl ethers. *Journal of Exposure Science and Environmental Epidemiology* 18 (1):2-19.

52. Stapleton, H.M., S.M. Kelly, J.G. Allen, M.D. McClean, and T.F. Webster. 2008. Measurement of polybrominated diphenyl ethers on hand wipes: estimating exposure from hand-to-mouth contact. *Environmental Science and Technology* 42 (9):3329-34.

53. U.S. Environmental Protection Agency. 2010. *An Exposure Assessment of Polybrominated Diphenyl Ethers.* Washington, DC: U.S. EPA, National Center for Environmental Assessment. EPA/600/R-08/086F. http://cfpub.epa.gov/ncea/cfm/recordisplay.cfm?deid=210404.

54. Wei, H., M. Turyk, S. Cali, S. Dorevitch, S. Erdal, and A. Li. 2009. Particle size fractionation and human exposure of polybrominated diphenyl ethers in indoor dust from Chicago. *Journal of Environmental Science and Health, Part A: Toxic/Hazardous Substances and Environmental Engineering* 44 (13):1353-61.

55. Birnbaum, L.S., and D.F. Staskal. 2004. Brominated flame retardants: cause for concern? *Environmental Health Perspectives* 112 (1):9-17.

56. Branchi, I., F. Capone, E. Alleva, and L.G. Costa. 2003. Polybrominated diphenyl ethers: neurobehavioral effects following developmental exposure. *Neurotoxicology* 24 (3):449-62.

57. Costa, L.G., G. Giordano, S. Tagliaferri, A. Caglieri, and A. Mutti. 2008. Polybrominated diphenyl ether (PBDE) flame retardants: Environmental contamination, human body burden and potential adverse health effects. *Acta Biomedica* 79 (3):172-183.

58. Herbstman, J.B., A. Sjodin, M. Kurzon, S.A. Lederman, R.S. Jones, V. Rauh, L.L. Needham, D. Tang, M. Niedzwiecki, R.Y. Wang, et al. 2010. Prenatal exposure to PBDEs and neurodevelopment. *Environmental Health Perspectives* 118 (5):712-9.

59. McDonald, T.A. 2005. Polybrominated diphenylether levels among United States residents: daily intake and risk of harm to the developing brain and reproductive organs. *Integrated Environmental Assessment and Management* 1 (4):343-54.

60. Le, H.H., E.M. Carlson, J.P. Chua, and S.M. Belcher. 2008. Bisphenol A is released from polycarbonate drinking bottles and mimics the neurotoxic actions of estrogen in developing cerebellar neurons. *Toxicology Letters* 176 (2):149-56.

61. National Toxicology Program. 2008. *NTP-CERHR Monograph on the Potential Human Reproductive and Developmental Effects of Bisphenol A.* Research Triangle Park, NC: National Institute of Environmental Health Sciences, National Toxicology Program. http://ntp.niehs.nih.gov/ntp/ohat/bisphenol/bisphenol.pdf.

62. Vandenberg, L.N., R. Hauser, M. Marcus, N. Olea, and W.V. Welshons. 2007. Human exposure to bisphenol A (BPA). *Reproductive Toxicology* 24 (2):139-77.

63. Diamanti-Kandarakis, E., J.P. Bourguignon, L.C. Giudice, R. Hauser, G.S. Prins, A.M. Soto, R.T. Zoeller, and A.C. Gore. 2009. Endocrine-disrupting chemicals: an Endocrine Society scientific statement. *Endocrine Reviews* 30 (4):293-342.

64. vom Saal, F.S., B.T. Akingbemi, S.M. Belcher, L.S. Birnbaum, D.A. Crain, M. Eriksen, F. Farabollini, L.J. Guillette, Jr., R. Hauser, J.J. Heindel, et al. 2007. Chapel Hill bisphenol A expert panel consensus statement: integration of mechanisms, effects in animals and potential to impact human health at current levels of exposure. *Reproductive Toxicology* 24 (2):131-8.

65. Kavlock, R.J., G.P. Daston, C. DeRosa, P. Fenner-Crisp, L.E. Gray, S. Kaattari, G. Lucier, M. Luster, M.J. Mac, C. Maczka, et al. 1996. Research needs for the risk assessment of health and environmental effects of endocrine disruptors: a report of the U.S. EPA-sponsored workshop. *Environmental Health Perspectives* 104 (Suppl 4):715-40.

66. Howdeshell, K.L., J. Furr, C.R. Lambright, V.S. Wilson, B.C. Ryan, and L.E. Gray, Jr. 2008. Gestational and lactational exposure to ethinyl estradiol, but not bisphenol A, decreases androgen-dependent reproductive organ weights and epididymal sperm abundance in the male long evans hooded rat. *Toxicological Sciences* 102 (2):371-82.

Chemicals in Food (continued)

67. Palanza, P.L., K.L. Howdeshell, S. Parmigiani, and F.S. vom Saal. 2002. Exposure to a low dose of bisphenol A during fetal life or in adulthood alters maternal behavior in mice. *Environmental Health Perspectives* 110 (Suppl 3):415-22.

68. Sharpe, R.M. 2010. Is it time to end concerns over the estrogenic effects of Bisphenol A? *Toxicological Sciences* 114 (1):1-4.

69. Vandenberg, L.N., M.V. Maffini, C. Sonnenschein, B.S. Rubin, and A.M. Soto. 2009. Bisphenol-A and the great divide: a review of controversies in the field of endocrine disruption. *Endocrine Reviews* 30 (1):75-95.

70. U.S. Food and Drug Administration. 2012. *BPA*. USFDA. Retrieved July 20, 2012 from http://www.fda.gov/Food/FoodIngredientsPackaging/ucm166145.htm.

71. Agency for Toxic Substances and Disease Registry (ATSDR). 1995. *Toxicological profile for diethyl phthalate*. Atlanta, GA: U.S. Department of Health and Human Services, Public Health Service.

72. Agency for Toxic Substances and Disease Registry (ATSDR). 1997. *Toxicological profile for di-n-octylphthalate (DNOP)*. Atlanta, GA: U.S. Department of Health and Human Services, Public Health Service.

73. Agency for Toxic Substances and Disease Registry (ATSDR). 2001. *Toxicological profile for Di-n-butyl Phthalate. Update*. Atlanta, GA: U.S. Department of Health and Human Services, Public Health Service. .

74. Agency for Toxic Substances and Disease Registry (ATSDR). 2002. *Toxicological profile for Di(2-ethylhexyl)phthalate (DEHP)*. Atlanta, GA: U.S. Department of Health and Human Services, Public Health Service.

75. Mortensen, G.K., K.M. Main, A.M. Andersson, H. Leffers, and N.E. Skakkebaek. 2005. Determination of phthalate monoesters in human milk, consumer milk, and infant formula by tandem mass spectrometry (LC-MS-MS). *Analytical and Bioanalytical Chemistry* 382 (4):1084-92.

76. Andrade, A.J., S.W. Grande, C.E. Talsness, K. Grote, A. Golombiewski, A. Sterner-Kock, and I. Chahoud. 2006. A dose-response study following in utero and lactational exposure to di-(2-ethylhexyl) phthalate (DEHP): effects on androgenic status, developmental landmarks and testicular histology in male offspring rats. *Toxicology* 225 (1):64-74.

77. Barlow, N.J., B.S. McIntyre, and P.M. Foster. 2004. Male reproductive tract lesions at 6, 12, and 18 months of age following in utero exposure to di(n-butyl) phthalate. *Toxicologic Pathology* 32 (1):79-90.

78. Christiansen, S., M. Scholze, M. Axelstad, J. Boberg, A. Kortenkamp, and U. Hass. 2008. Combined exposure to anti-androgens causes markedly increased frequencies of hypospadias in the rat. *International Journal of Andrology* 31 (2):241-8.

79. Gray, L.E., Jr., J. Ostby, J. Furr, M. Price, D.N. Veeramachaneni, and L. Parks. 2000. Perinatal exposure to the phthalates DEHP, BBP, and DINP, but not DEP, DMP, or DOTP, alters sexual differentiation of the male rat. *Toxicological Sciences* 58 (2):350-65.

80. Howdeshell, K.L., V.S. Wilson, J. Furr, C.R. Lambright, C.V. Rider, C.R. Blystone, A.K. Hotchkiss, and L.E. Gray, Jr. 2008. A mixture of five phthalate esters inhibits fetal testicular testosterone production in the sprague-dawley rat in a cumulative, dose-additive manner. *Toxicological Sciences* 105 (1):153-65.

81. Lehmann, K.P., S. Phillips, M. Sar, P.M. Foster, and K.W. Gaido. 2004. Dose-dependent alterations in gene expression and testosterone synthesis in the fetal testes of male rats exposed to di (n-butyl) phthalate. *Toxicological Sciences* 81 (1):60-8.

82. Metzdorff, S.B., M. Dalgaard, S. Christiansen, M. Axelstad, U. Hass, M.K. Kiersgaard, M. Scholze, A. Kortenkamp, and A.M. Vinggaard. 2007. Dysgenesis and histological changes of genitals and perturbations of gene expression in male rats after in utero exposure to antiandrogen mixtures. *Toxicological Sciences* 98 (1):87-98.

83. Mylchreest, E., D.G. Wallace, R.C. Cattley, and P.M. Foster. 2000. Dose-dependent alterations in androgen-regulated male reproductive development in rats exposed to Di(n-butyl) phthalate during late gestation. *Toxicological Sciences* 55 (1):143-51.

84. National Research Council. 2008. *Phthalates and Cumulative Risk Assessment: The Tasks Ahead*. Washington, DC: The National Academies Press. http://www.nap.edu/catalog.php?record_id=12528.

85. Sharpe, R.M. 2008. "Additional" effects of phthalate mixtures on fetal testosterone production. *Toxicological Sciences* 105 (1):1-4.

86. Main, K.M., G.K. Mortensen, M.M. Kaleva, K.A. Boisen, I.N. Damgaard, M. Chellakooty, I.M. Schmidt, A.M. Suomi, H.E. Virtanen, D.V. Petersen, et al. 2006. Human breast milk contamination with phthalates and alterations of endogenous reproductive hormones in infants three months of age. *Environmental Health Perspectives* 114 (2):270-6.

87. Nassar, N., P. Abeywardana, A. Barker, and C. Bower. 2009. Parental occupational exposure to potential endocrine disrupting chemicals and risk of hypospadias in infants. *Occupational and Environmental Medicine* 67 (9):585-9.

88. Swan, S.H. 2008. Environmental phthalate exposure in relation to reproductive outcomes and other health endpoints in humans. *Environmental Research* 108 (2):177-84.

89. Swan, S.H., K.M. Main, F. Liu, S.L. Stewart, R.L. Kruse, A.M. Calafat, C.S. Mao, J.B. Redmon, C.L. Ternand, S. Sullivan, et al. 2005. Decrease in anogenital distance among male infants with prenatal phthalate exposure. *Environmental Health Perspectives* 113 (8):1056-61.

90. Agency for Toxic Substances and Disease Registry (ATSDR). 2009. *Toxicological profile for Perfluoroalkyls. (Draft for Public Comment)*. Atlanta, GA: U.S. Department of Health and Human Services, Public Health Service. http://www.atsdr.cdc.gov/toxprofiles/tp200.pdf.

Chemicals in Food (continued)

91. Calafat, A.M., L.Y. Wong, Z. Kuklenyik, J.A. Reidy, and L.L. Needham. 2007. Polyfluoroalkyl chemicals in the U.S. population: data from the National Health and Nutrition Examination Survey (NHANES) 2003-2004 and comparisons with NHANES 1999-2000. *Environmental Health Perspectives* 115 (11):1596-602.

92. Egeghy, P.P., and M. Lorber. 2010. An assessment of the exposure of Americans to perfluorooctane sulfonate: A comparison of estimated intake with values inferred from NHANES data. *Journal of Exposure Science and Environmental Epidemiology* Epub Date 2010/02/11.

93. Trudel, D., L. Horowitz, M. Wormuth, M. Scheringer, I.T. Cousins, and K. Hungerbuhler. 2008. Estimating consumer exposure to PFOS and PFOA. *Risk Analysis* 28 (2):251-69.

94. Begley, T.H., K. White, P. Honigfort, M.L. Twaroski, R. Neches, and R.A. Walker. 2005. Perfluorochemicals: potential sources of and migration from food packaging. *Food Additives and Contaminants* 22 (10):1023-31.

95. Tittlemier, S.A., K. Pepper, C. Seymour, J. Moisey, R. Bronson, X.L. Cao, and R.W. Dabeka. 2007. Dietary exposure of Canadians to perfluorinated carboxylates and perfluorooctane sulfonate via consumption of meat, fish, fast foods, and food items prepared in their packaging. *Journal of Agricultural and Food Chemistry* 55 (8):3203-10.

96. Ericson, I., R. Marti-Cid, M. Nadal, B. Van Bavel, G. Lindstrom, and J.L. Domingo. 2008. Human exposure to perfluorinated chemicals through the diet: intake of perfluorinated compounds in foods from the Catalan (Spain) market. *Journal of Agricultural and Food Chemistry* 56 (5):1787-94.

97. Schecter, A., J. Colacino, D. Haffner, K. Patel, M. Opel, O. Papke, and L. Birnbaum. 2010. Perfluorinated compounds, polychlorinated biphenyl, and organochlorine pesticide contamination in composite food samples from Dallas, Texas. *Environmental Health Perspectives* 118:796-802.

98. Era, S., K.H. Harada, M. Toyoshima, K. Inoue, M. Minata, N. Saito, T. Takigawa, K. Shiota, and A. Koizumi. 2009. Cleft palate caused by perfluorooctane sulfonate is caused mainly by extrinsic factors. *Toxicology* 256 (1-2):42-7.

99. Lau, C., J.R. Thibodeaux, R.G. Hanson, J.M. Rogers, B.E. Grey, M.E. Stanton, J.L. Butenhoff, and L.A. Stevenson. 2003. Exposure to perfluorooctane sulfonate during pregnancy in rat and mouse. II: postnatal evaluation. *Toxicological Sciences* 74 (2):382-92.

100. Apelberg, B.J., F.R. Witter, J.B. Herbstman, A.M. Calafat, R.U. Halden, L.L. Needham, and L.R. Goldman. 2007. Cord serum concentrations of perfluorooctane sulfonate (PFOS) and perfluorooctanoate (PFOA) in relation to weight and size at birth. *Environmental Health Perspectives* 115 (11):1670-6.

101. Fei, C., J.K. McLaughlin, R.E. Tarone, and J. Olsen. 2007. Perfluorinated chemicals and fetal growth: a study within the Danish National Birth Cohort. *Environmental Health Perspectives* 115 (11):1677-82.

102. Fei, C., J.K. McLaughlin, R.E. Tarone, and J. Olsen. 2008. Fetal growth indicators and perfluorinated chemicals: a study in the Danish National Birth Cohort. *American Journal of Epidemiology* 168 (1):66-72.

103. Washino, N., Y. Saijo, S. Sasaki, S. Kato, S. Ban, K. Konishi, R. Ito, A. Nakata, Y. Iwasaki, K. Saito, et al. 2009. Correlations between prenatal exposure to perfluorinated chemicals and reduced fetal growth. *Environmental Health Perspectives* 117 (4):660-7.

104. Hamm, M.P., N.M. Cherry, E. Chan, J.W. Martin, and I. Burstyn. 2010. Maternal exposure to perfluorinated acids and fetal growth. *Journal of Exposure Science and Environmental Epidemiology* 20:589-597.

105. Monroy, R., K. Morrison, K. Teo, S. Atkinson, C. Kubwabo, B. Stewart, and W.G. Foster. 2008. Serum levels of perfluoroalkyl compounds in human maternal and umbilical cord blood samples. *Environmental Research* 108 (1):56-62.

106. Centers for Disease Control and Prevention. *Perchlorate in Baby Formula Fact Sheet*. Retrieved August 13, 2009 from http://cdc.gov/nceh/features/perchlorate_factsheet.htm.

107. National Research Council. 2005. *Health Implications of Perchlorate Ingestion*. Washington, DC: National Academy Press. http://www.nap.edu/catalog.php?record_id=11202.

108. Sanchez, C.A., L.M. Barraj, B.C. Blount, C.G. Scrafford, L. Valentin-Blasini, K.M. Smith, and R.I. Krieger. 2009. Perchlorate exposure from food crops produced in the lower Colorado River region. *Journal of Exposure Science and Environmental Epidemiology* 19 (4):359-68.

109. U.S. Environmental Protection Agency. *Perchlorate*. Retrieved August 13, 2009 from http://www.epa.gov/safewater/contaminants/unregulated/perchlorate.html.

110. Dasgupta, P.K., A.B. Kirk, J.V. Dyke, and S. Ohira. 2008. Intake of iodine and perchlorate and excretion in human milk. *Environmental Science & Technology* 42 (21):8115-21.

111. Kirk, A.B., J.V. Dyke, C.F. Martin, and P.K. Dasgupta. 2007. Temporal patterns in perchlorate, thiocyanate, and iodide excretion in human milk. *Environmental Health Perspectives* 115 (2):182-6.

112. Kirk, A.B., P.K. Martinelango, K. Tian, A. Dutta, E.E. Smith, and P.K. Dasgupta. 2005. Perchlorate and iodide in dairy and breast milk. *Environmental Science & Technology* 39 (7):2011-7.

Chemicals in Food (continued)

113. Murray, C.W., S.K. Egan, H. Kim, N. Beru, and P.M. Bolger. 2008. US Food and Drug Administration's Total Diet Study: dietary intake of perchlorate and iodine. *Journal of Exposure Science and Environmental Epidemiology* 18 (6):571-80.

114. Sanchez, C.A., K.S. Crump, R.I. Krieger, N.R. Khandaker, and J.P. Gibbs. 2005. Perchlorate and nitrate in leafy vegetables of North America. *Environmental Science & Technology* 39 (24):9391-7.

115. U.S. Food and Drug Administration. 2007. *2004-2005 Exploratory Survey Data on Perchlorate in Food.* U.S. FDA. Retrieved January 18, 2012 from http://www.fda.gov/Food/FoodSafety/FoodContaminantsAdulteration/ChemicalContaminants/Perchlorate/ucm077685.htm.

116. Blount, B.C., and L. Valentin-Blasini. 2007. Biomonitoring as a method for assessing exposure to perchlorate. *Thyroid* 17 (9):837-41.

117. Pearce, E.N., A.M. Leung, B.C. Blount, H.R. Bazrafshan, X. He, S. Pino, L. Valentin-Blasini, and L.E. Braverman. 2007. Breast milk iodine and perchlorate concentrations in lactating Boston-area women. *The Journal of Clinical Endocrinology and Metabolism* 92 (5):1673-7.

118. Schier, J.G., A.F. Wolkin, L. Valentin-Blasini, M.G. Belson, S.M. Kieszak, C.S. Rubin, and B.C. Blount. 2009. Perchlorate exposure from infant formula and comparisons with the perchlorate reference dose. *Journal of Exposure Science and Environmental Epidemiology* 20 (3):281-7.

119. Greer, M.A., G. Goodman, R.C. Pleus, and S.E. Greer. 2002. Health effects assessment for environmental perchlorate contamination: the dose response for inhibition of thyroidal radioiodine uptake in humans. *Environmental Health Perspectives* 110 (9):927-37.

120. U.S. Food and Drug Administration. *Perchlorate Questions and Answers.* Retrieved August 13, 2009 from http://www.fda.gov/Food/FoodSafety/FoodContaminantsAdulteration/ChemicalContaminants/Perchlorate/ucm077572.htm#effects.

121. Morreale de Escobar, G., M.J. Obregon, and F. Escobar del Rey. 2000. Is Neuropsychological Development Related to Maternal Hypothyroidism or to Maternal Hypothyroxinemia? *The Journal of Clinical Endocrinology and Metabolism* 85 (11):3975-87.

122. Eskenazi, B., A. Bradman, and R. Castorina. 1999. Exposures of children to organophosphate pesticides and their potential adverse health effects. *Environmental Health Perspectives* 107 (Suppl. 3):409-19.

123. Huen, K., Harley, K., Brooks, J., Hubbard, A., Bradman, A., Eskenazi, B., Holland, N. 2009. Developmental changes in PON1 enzyme activity in young children and effects of PON1 polymorphisms. *Environmental Health Perspectives* 117 (10):1632-8.

124. Marks, A.R., K. Harley, A. Bradman, K. Kogut, D.B. Barr, C. Johnson, N. Calderon, and B. Eskenazi. 2010. Organophosphate pesticide exposure and attention in young Mexican-American children: the CHAMACOS study. *Environmental Health Perspectives* 118 (12):1768-74.

125. Bouchard, M.F., D.C. Bellinger, R.O. Wright, and M.G. Weisskopf. 2010. Attention-deficit/hyperactivity disorder and urinary metabolites of organophosphate pesticides. *Pediatrics* 125 (6):e1270-7.

126. Bouchard, M.F., J. Chevrier, K.G. Harley, K. Kogut, M. Vedar, N. Calderon, C. Trujillo, C. Johnson, A. Bradman, D.B. Barr, et al. 2011. Prenatal exposure to organophosphate pesticides and IQ in 7-year old children. *Environmental Health Perspectives* 119 (8):1189-95.

127. Engel, S.M., J. Wetmur, J. Chen, C. Zhu, D.B. Barr, R.L. Canfield, and M.S. Wolff. 2011. Prenatal exposure to organophosphates, Paraoxonase 1, and cognitive development in childhood. *Environmental Health Perspectives* 119 (8):1182-8.

128. Rauh, V., S. Arunajadai, M. Horton, F. Perera, L. Hoepner, D.B. Barr, and R. Whyatt. 2011. 7-year neurodevelopmental scores and prenatal exposure to Chlorpyrifos, a common agricultural pesticide. *Environmental Health Perspectives* 119 (8):1196-201.

129. U.S. Environmental Protection Agency. 2002. *Interim Reregistration Eligibility Decision for Chlorpyrifos.* Washington, DC: U.S. EPA, Office of Prevention, Pesticides, and Toxic Substances. EPA 738-R-01-007. http://www.epa.gov/oppsrrd1/REDs/chlorpyrifos_ired.pdf.

130. U.S. Environmental Protection Agency. 2008. *Azinphos-Methyl (AZM) Registration Review Status.* Washington, DC: U.S. EPA, Office of Pesticide Programs. EPA-HQ-OPP-2005-0061. http://www.epa.gov/oppsrrd1/registration_review/azm/azm-status.pdf.

131. U.S. Environmental Protection Agency. 2006. *Interim Reregistration Eligibility Decision for Methyl Parathion.* Washington, DC: U.S. EPA, Office of Pesticide Programs. EPA-HQ-OPP-2006-0618. http://www.epa.gov/oppsrrd1/REDs/methylparathion_ired.pdf.

132. U.S. Environmental Protection Agency. 2010. *Food Quality Protection Act (FQPA) of 1996.* U.S. EPA, Office of Pesticide Programs. Retrieved December 28, 2010 from http://www.epa.gov/pesticides/regulating/laws/fqpa/.

133. U.S. Department of Agriculture. 2010. *Pesticide Data Program.* U.S. Department of Agriculture, Agricultural Marketing Service. Retrieved December 28, 2010 from http://www.ams.usda.gov/AMSv1.0/pdp.

134. Lu, C., F.J. Schenck, M.A. Pearson, and J.W. Wong. 2010. Assessing children's dietary pesticide exposure - direct measurement of pesticide residues in 24-hour duplicate food samples. *Environmental Health Perspectives* 118 (11):1625-30.

135. U.S. Food and Drug Administration. 2010. *FDA Pesticide Program Residue Monitoring: 1993-2008.* U.S. FDA. Retrieved January 18, 2012 from http://www.fda.gov/Food/FoodSafety/FoodContaminantsAdulteration/Pesticides/ResidueMonitoringReports/default.htm.

136. U.S. Food and Drug Administration. 2012. *Total Diet Study.* USFDA. Retrieved July 20, 2012 from http://www.fda.gov/Food/FoodSafety/FoodContaminantsAdulteration/TotalDietStudy/default.htm.

Chemicals in Food (continued)

137. U.S. Department of Agriculture. 2011. *Pesticide Data Program Annual Summary, Calendar Year 2009*. Washington, DC: USDA Marketing and Regulatory Programs, Agricultural Marketing Service. http://www.ams.usda.gov/AMSv1.0/getfile?dDocName=STELPRDC5091055.

138. U.S. Environmental Protection Agency. 2002. Endocrine Disruptor Screening Program, proposed chemical selection approach for initial round of screening; request for comment. *Federal Register* 67 (250):79611-29.

Contaminated Lands

1. U.S. Environmental Protection Agency. 2012. *Protecting and Restoring Land. OSWER FY11 Accomplishment Report*. Washington, DC: U.S. EPA, Office of Solid Waste and Emergency Response. http://www.epa.gov/aboutepa/oswer_accomplishment_report_2011.pdf.

2. U.S. Environmental Protection Agency. 2006. *Interim Guidance for OSWER Cross-Program Revitalization Measures*. Washington, DC: EPA, Office of Solid Waste and Emergency Response. http://www.epa.gov/landrecycling/download/cprmguidance_10_20_06covermemo.pdf.

3. U.S. Environmental Protection Agency. 2009. *OSWER Cross-Program Revitalization Measures*. Washington, DC: EPA, Office of Solid Waste and Emergency Response. http://www.epa.gov/landrecycling/download/cprm_report_031709.pdf.

4. Agency for Toxic Substances and Disease Registry. 1995. *Public Health Assessments - RSR Corporation*. Atlanta, GA: Agency for Toxic Substances and Disease Registry. http://www.atsdr.cdc.gov/hac/pha/pha.asp?docid=134&pg=0.

5. Agency for Toxic Substances and Disease Registry. 2002. *Jasper County, Missouri Superfund Site Childhood 2000 Lead Exposure Study*. Atlanta, GA: Agency for Toxic Substances and Disease Registry. http://health.mo.gov/living/environment/hazsubstancesites/pdf/FinalReportAndTOC.pdf.

6. Agency for Toxic Substances and Disease Registry. 2007. *Public Health Assessment for: Bunker Hill Mining and Metallurgical Complex Operable Unit 3 (aka Coeur D'Alene River Basin) - Kootenai and Shoshone Counties, Idaho Westward to Spokane and Stevens Counties, Washington*. Atlanta, GA: Agency for Toxic Substances and Disease Registry. http://www.atsdr.cdc.gov/HAC/pha/bunkerhillmining/bunkerhillpha032607.pdf.

7. Agency for Toxic Substances and Disease Registry. *Tar Creek Superfund Site - Ottawa, County, OK*. Agency for Toxic Substances and Disease Registry. Retrieved July 29, 2011 from http://www.atsdr.cdc.gov/sites/tarcreek/tarcreekreport-p1.html.

8. Casteel, S.W., C.P. Weis, and W.J. Brattin. 1998. *Bioavailability of Lead in a Slag Sample from the Midvale Slag NPL Site, Midvale, Utah: Phase II Swine Bioavailability Investigations*. Washington, DC: U.S. Environmental Protection Agency. 908R98003. http://www.epa.gov/region8/r8risk/pdf/rba-pb_midvale.pdf.

9. Centers for Disease Control and Prevention. 1995. Mercury exposure among residents of a building formerly used for industrial purposes - New Jersey, 1995. *Morbidity and Mortality Weekly Report* 45 (20):422-424.

10. New Jersey Department of Health and Senior Services. 2011. *Dover Township Childhood Cancer Investigation*. New Jersey Department of Health and Senior Services. Retrieved July 30, 2011 from http://www.state.nj.us/health/eoh/hhazweb/dovertwp.shtml.

11. Brunner, E., and M.G. Marmot. 2006. Social organization, stress, and health. In *Social Determinants of Health, 2nd Edition*, edited by M. G. Marmot and R. G. Wilkinson. Oxford: Oxford University Press.

12. Centers for Disease Control and Prevention. 2005. *Social Determinants of Health*. CDC. Retrieved December 1, 2008 from http://www.cdc.gov/socialdeterminants/.

13. Evans, R.G., M.L. Barer, and T.R. Marmor. 1994. *Why Are Some People Healthy and Others Not? The Determinants of Health of Populations*. New York: Aldine de Gruyter.

14. Gee, G.C. 2002. A multilevel analysis of the relationship between institutional and individual racial discrimination and health status. *American Journal of Public Health* 92 (4):615-623.

15. Geronimus, A.T., J. Bound, and C.G. Colem. 2011. Excess black mortality in the United States and in selected Black and White high-poverty areas, 1980-2000. *American Journal of Public Health* 101 (4):720-729.

16. Geronimus, A.T., J. Bound, and T.A. Waidmann. 1999. Poverty, time, and place: Variation in excess mortality across selected US populations, 1980-1990. *Journal of Epidemiology and Community Health* 53 (6):325-334.

17. Graham, H. 2004. Social determinants and their unequal distribution: Clarifying policy understandings. *Milbank Quarterly* 82 (1):101-124.

18. Institute of Medicine. 1999. *Toward Environmental Justice: Research, Education, and Health Policy Needs*. Washington, DC: National Academy Press. http://www.nap.edu/catalog.php?record_id=6034.

19. Marmot, M.G., and R.G. Wilkinson, eds. 2006. *Social Determinants of Health, 2nd Edition*. Oxford, UK: Oxford University Press.

20. PBS. 2008. *Unnatural Causes...Is Inequality Making Us Sick?* San Francisco, CA: California Newreel with Vital Pictures, Inc. http://www.unnaturalcauses.org.

Contaminated Lands (continued)

21. Suk, W.A., K. Murray, and M.D. Avakian. 2003. Environmental hazards to children's health in the modern world. *Mutation Research* 544:235-242.

22. Wilkinson, R., and M. Marmot. 2003. *Social Determinants of Health: The Solid Facts: Second Edition.* Copenhagen: World Health Organization. http://www.euro.who.int/__data/assets/pdf_file/0005/98438/e81384.pdf.

23. Adler, N., J. Stewart, S. Cohen, M. Cullen, A.D. Roux, W. Dow, G. Evans, I. Kawachi, M. Marmot, K. Matthews, et al. 2007. *Reaching for a Healthier Life: Facts on Socioeconomic Status and Health in the U.S.* San Francisco, CA: The John D. and Catherine T. MacArthur Foundation Research Network on SES and Health. http://www.macses.ucsf.edu/downloads/Reaching_for_a_Healthier_Life.pdf.

24. Graham, H. 2007. *Unequal Lives: Health and Socioeconomic Inequalities.* New York: Open University Press.

25. Mackenbach, J.P., P. Martikainen, C.W. Looman, J.A. Dalstra, A.E. Kunst, and E. Lahelma. 2005. The shape of the relationship between income and self-assessed health: An international study. *International Journal of Epidemiology* 34 (2):286-93.

26. Stronks, K., H. van de Mheen, J. van den Bos, and J.P. Mackenbach. 1997. The interrelationship between income, health and employment status. *International Journal of Epidemiology* 26 (3):592-600.

Climate Change

1. U.S. Environmental Protection Agency. 2010. *Climate Change Science Facts.* Washington, DC: U.S. EPA, Office of Air and Radiation. EPA 430-F-10-002. http://www.epa.gov/climatechange/downloads/Climate_Change_Science_Facts.pdf.

2. U.S. Environmental Protection Agency. 2009. *Technical Support Document for Endangerment and Cause or Contribute Findings for Greenhouse Gases under Section 202 (a) of the Clean Air Act.* Washington, DC: U.S. EPA, Office of Atmospheric Programs, Climate Change Division. http://epa.gov/climatechange/Downloads/endangerment/Endangerment_TSD.pdf.

3. U.S. Global Change Research Program. 2008. *Analyses of the Effects of Global Change on Human Health and Welfare and Human Systems (SAP 4.6).* Washington, DC: U.S. Environmental Protection Agency. http://downloads.climatescience.gov/sap/sap4-6/sap4-6-final-report-all.pdf.

4. Field, C.B., L.D. Mortsch, M. Brklacich, D.L. Forbes, P. Kovacs, J.A. Patz, S.W. Running, and M.J. Scott. 2007. North America. In *Climate Change 2007: Impacts, Adaptation and Vulnerability. Contribution of Working Group II to the Fourth Assessment Report of the Intergovernmental Panel on Climate Change.* Edited by M. L. Parry, O. F. Canziani, J. P. Palutikof, P. J. van der Linden and C. E. Hanson. Cambridge, UK: Cambridge University Press, 617-652.

5. National Research Council. 2010. Advancing the Science of Climate Change. In *America's Climate Choices.* Washington, DC: The National Academies Press.

6. The Interagency Working Group on Climate Change and Health. 2010. *A Human Health Perspective on Climate Change.* Research Triangle Park, NC: Environmental Health Perspectives/National Institute of Environmental Health Sciences. http://www.niehs.nih.gov/health/assets/docs_a_e/climatereport2010.pdf.

7. U.S. Environmental Protection Agency. 2009. *Climate Change and Children's Health.* Washington, DC: U.S. EPA, Office of Children's Health Protection. EPA-100-K-09-008. http://yosemite.epa.gov/ochp/ochpweb.nsf/content/OCHP_Climate_Brochure.htm/$File/OCHP_Climate_Brochure.pdf.

8. U.S. Global Change Research Program. 2009. *Global Climate Change Impacts in the United States.* Edited by T. R. Karl, J. M. Melillo and T. C. Peterson. Cambridge, UK: Cambridge University Press. http://downloads.globalchange.gov/usimpacts/pdfs/climate-impacts-report.pdf.

9. Shea, K.M. 2007. Global climate change and children's health. *Pediatrics* 120 (5):1149-52.

10. Committee on Sports Medicine and Fitness. 2000. Climatic heat stress and the exercising child and adolescent: American Academy of Pediatrics policy statement. *Pediatrics* 106:158-9.

11. Knowlton, K., M. Rotkin-Ellman, G. King, H.G. Margolis, D. Smith, G. Solomon, R. Trent, and P. English. 2009. The 2006 California heat wave: Impacts on hospitalizations and emergency department visits. *Environmental Health Perspectives* 117 (1):61-7.

12. Basu, R., and B.D. Ostro. 2008. A multicounty analysis identifying the populations vulnerable to mortality associated with high ambient temperature in California. *American Journal of Epidemiology* 168 (6):632-637.

13. U.S. Environmental Protection Agency. 2006. *Excessive Heat Events Guidebook.* Washington, DC: U.S. EPA, Office of Atmospheric Programs. EPA 430-B-06-005. http://www.epa.gov/heatisld/about/pdf/EHEguide_final.pdf.

14. Sheffield, P.E., and P.J. Landrigan. 2011. Global climate change and children's health: threats and strategies for prevention. *Environmental Health Perspectives* 119 (3):291-8.

15. Confalonieri, U., B. Menne, R. Akhtar, K.L. Ebi, M. Hauengue, R.S. Kovats, B. Revich, and A. Woodward. 2007. Human Health. In *Climate Change 2007: Impacts, Adaptation and Vulnerability. Contribution of Working Group II to the Fourth Assessment Report of the Intergovernmental Panel on Climate Change,* edited by M. L. Parry, O. F. Canziani, J. P. Palutikof, P. J. van der Linden and C. E. Hanson. Cambridge, UK: Cambridge University Press, 391-431.

Climate Change (continued)

16. Baccini, M., A. Biggeri, G. Accetta, T. Kosatsky, K. Katsouyanni, A. Analitis, H.R. Anderson, L. Bisanti, D. D'Ippoliti, J. Danova, et al. 2008. Heat effects on mortality in 15 European cities. *Epidemiology* 19 (5):711-9.

17. Curriero, F.C., K.S. Heiner, J.M. Samet, S.L. Zeger, L. Strug, and J.A. Patz. 2002. Temperature and mortality in 11 cities of the eastern United States. *American Journal of Epidemiology* 155 (1):80-87.

18. Ye, X., R. Wolff, W. Yu, P. Vaneckova, X. Pan, and S. Tong. 2011. Ambient temperature and morbidity: A review of epidemiological evidence. *Environmental Health Perspectives* 120 (1):19-28.

19. O'Neill, M.S., and K.L. Ebi. 2009. Temperature extremes and health: Impacts of climate variability and change in the United States. *Journal of Occupational and Environmental Medicine* 51 (1):13-25.

20. Agency for Toxic Substances and Disease Registry. 1998. *Toxicological Profile for Sulfur Dioxide*. Atlanta, GA: U.S. Department of Health and Human Services, Public Health Service. www.atsdr.cdc.gov/toxprofiles/tp116-c2.pdf.

21. Andersen, Z.J., P. Wahlin, O. Raaschou-Nielsen, T. Scheike, and S. Loft. 2007. Ambient particle source apportionment and daily hospital admissions among children and elderly in Copenhagen. *Journal of Exposure Science and Environmental Epidemiology* 17 (7):625-36.

22. Annesi-Maesano, I., D. Moreau, D. Caillaud, F. Lavaud, Y. Le Moullec, A. Taytard, G. Pauli, and D. Charpin. 2007. Residential proximity fine particles related to allergic sensitisation and asthma in primary school children. *Respiratory Medicine* 101 (8):1721-9.

23. Gauderman, W.J., G.F. Gilliland, H. Vora, E. Avol, D. Stram, R. McConnell, D. Thomas, F. Lurmann, H.G. Margolis, E.B. Rappaport, et al. 2002. Association between air pollution and lung function growth in southern California children: Results from a second cohort. *American Journal of Respiratory and Critical Care Medicine* 166 (1):76-84.

24. Gent, J.F., E.W. Triche, T.R. Holford, K. Belanger, M.B. Bracken, W.S. Beckett, and B.P. Leaderer. 2003. Association of low-level ozone and fine particles with respiratory symptoms in children with asthma. *Journal of the American Medical Association* 290 (14):1859-67.

25. Karr, C.J., P.A. Demers, M.W. Koehoorn, C.C. Lencar, L. Tamburic, and M. Brauer. 2009. Influence of ambient air pollutant sources on clinical encounters for infant bronchiolitis. *American Journal of Respiratory and Critical Care Medicine* 180 (10):995-1001.

26. McConnell, R., T. Islam, K. Shankardass, M. Jerrett, F. Lurmann, F. Gilliland, J. Gauderman, E. Avol, N. Kuenzli, L. Yao, et al. 2010. Childhood incident asthma and traffic-related air pollution at home and school. *Environmental Health Perspectives* 118:1021-26.

27. Mortimer, K., R. Neugebauer, F. Lurmann, S. Alcorn, J. Balmes, and I. Tager. 2008. Air pollution and pulmonary function in asthmatic children: Effects of prenatal and lifetime exposures. *Epidemiology* 19 (4):550-7.

28. Norris, G., S.N. YoungPong, J.Q. Koenig, T.V. Larson, L. Sheppard, and J.W. Stout. 1999. An association between fine particles and asthma emergency department visits for children in Seattle. *Environmental Health Perspectives* 107 (6):489-93.

29. Ostro, B., L. Roth, B. Malig, and M. Marty. 2009. The effects of fine particle components on respiratory hospital admissions in children. *Environmental Health Perspectives* 117 (3):475-80.

30. Romieu, I., F. Meneses, S. Ruiz, J.J. Sienra, J. Huerta, M.C. White, and R.A. Etzel. 1996. Effects of air pollution on the respiratory health of asthmatic children living in Mexico City. *American Journal of Respiratory and Critical Care Medicine* 154 (2 Pt 1):300-7.

31. Tang, C.S., L.T. Chang, H.C. Lee, and C.C. Chan. 2007. Effects of personal particulate matter on peak expiratory flow rate of asthmatic children. *The Science of the Total Environment* 382 (1):43-51.

32. U.S. Environmental Protection Agency. 2006. *Air Quality Criteria for Ozone and Related Photochemical Oxidants*. Research Triangle Park, NC: U.S. EPA, National Center for Environmental Assessment. EPA/600/R-05/004aF. http://cfpub.epa.gov/ncea/cfm/recordisplay.cfm?deid=149923.

33. U.S. Environmental Protection Agency. 2008. *Integrated Science Assessment for Oxides of Nitrogen--Health Criteria*. Research Triangle Park, NC: U.S. EPA. http://cfpub.epa.gov/ncea/cfm/recordisplay.cfm?deid=194645.

34. U.S. Environmental Protection Agency. 2009. *Integrated Science Assessment for Particulate Matter (Final Report)*. Washington, DC: U.S. EPA, National Center for Environmental Assessment. EPA/600/R-08/139F. http://cfpub.epa.gov/ncea/CFM/recordisplay.cfm?deid=216546.

35. Villeneuve, P.J., L. Chen, B.H. Rowe, and F. Coates. 2007. Outdoor air pollution and emergency department visits for asthma among children and adults: A case-crossover study in northern Alberta, Canada. *Environmental Health* 6:40.

36. Intergovernmental Panel on Climate Change. 2012. *Managing the Risks of Extreme Events and Disasters to Advance Climate Change Adaptation. A Special Report of Working Groups I and II of the Intergovernmental Panel on Climate Change*. Cambridge, UK: Cambridge University Press. http://www.ipcc-wg2.gov/SREX/images/uploads/SREX-All_FINAL.pdf.

37. Drayna, P., S.L. McLellan, P. Simpson, S.H. Li, and M.H. Gorelick. 2010. Association between rainfall and pediatric emergency department visits for acute gastrointestinal illness. *Environmental Health Perspectives* 118 (10):1439-43.

38. Ziska, L., K. Knowlton, C. Rogers, D. Dalan, N. Tierney, M.A. Elder, W. Filley, J. Shropshire, L.B. Ford, C. Hedberg, et al. 2011. Recent warming by latitude associated with increased length of ragweed pollen season in central North America. *Proceedings of the National Academy of Sciences* 108 (10):4248-51.

Climate Change (continued)

39. Héguy, L., M. Garneau, M.S. Goldberg, M. Raphoz, F. Guay, and M.F. Valois. 2008. Associations between grass and weed pollen and emergency department visits for asthma among children in Montreal. *Environmental Research* 106 (2):203-11.

40. Schmier, J.K., and K.L. Ebi. 2009. The impact of climate change and aeroallergens on children's health. *Allergy and Asthma Proceedings* 30 (3):229-37.

41. Ziska, L.H., P.R. Epstein, and C.A. Rogers. 2008. Climate change, aerobiology, and public health in the northeast United States. *Mitigation and Adaptation Strategies for Global Change* 13:607-613.

42. U.S. Environmental Protection Agency. 2009. *Climate Change: Water Resources.* U.S. EPA, Climate Change Division. Retrieved February 11, 2011 from http://www.epa.gov/climatechange/effects/water/index.html.

43. Bogdal, C., P. Schmid, M. Zennegg, F.S. Anselmetti, M. Scheringer, and K. Hungerbuhler. 2009. Blast from the past: Melting glaciers as a relevant source for persistent organic pollutants. *Environmental Science and Technology* 43 (21):8173-7.

44. Carrie, J., F. Wang, H. Sanei, R.W. Macdonald, P.M. Outridge, and G.A. Stern. 2010. Increasing contaminant burdens in an Arctic fish, Burbot (Lota lota), in a warming climate. *Environmental Science and Technology* 44 (1):316-322.

45. Institute of Medicine. 2011. *Climate Change, the Indoor Environment, and Health.* Washington, DC: The National Academies Press. http://www.nap.edu/catalog.php?record_id=13115.

Biomonitoring

Introduction

1. Centers for Disease Control and Prevention. 2009. *Fourth National Report on Human Exposure to Environmental Chemicals*. Atlanta, GA: CDC. http://www.cdc.gov/exposurereport/.

2. Centers for Disease Control and Prevention. 2011. *National Health and Nutrition Examination Survey*. CDC. Retrieved January 18, 2012 from http://www.cdc.gov/nchs/nhanes.htm.

3. Miller, M.D., K.M. Crofton, D.C. Rice, and R.T. Zoeller. 2009. Thyroid-disrupting chemicals: Interpreting upstream biomarkers of adverse outcomes. *Environmental Health Perspectives* 117 (7):1033-1041.

4. Sexton, K., A.D. Ryan, J.L. Adgate, D.B. Barr, and L.L. Needham. 2011. Biomarker measurements of concurrent exposure to multiple environmental chemicals and chemical classes in children. *Journal of Toxicology and Environmental Health, Part A* 74 (14):927-42.

5. Woodruff, T.J., A.R. Zota, and J.M. Schwartz. 2011. Environmental chemicals in pregnant women in the United States: NHANES 2003-2004. *Environmental Health Perspectives* 119 (6):878-85.

6. National Research Council. 2008. *Phthalates and Cumulative Risk Assessment: The Tasks Ahead*. Washington, DC: The National Academies Press. http://www.nap.edu/catalog/12528.html.

7. U.S. Census Bureau. *Poverty Thresholds by Size of Family and Number of Related Children Under 18 Years: 2010*. U.S. Census Bureau. Retrieved August 20, 2011 from http://www.census.gov/hhes/www/cpstables/032011/pov/new35_000.htm.

8. Axelrad, D.A., and J. Cohen. 2011. Calculating summary statistics for population chemical biomonitoring in women of childbearing age with adjustment for age-specific natality. *Environmental Research* 111 (1):149-155.

Lead

1. Centers for Disease Control and Prevention. 2005. *Preventing Lead Poisoning in Young Children*. Atlanta, GA.

2. Lanphear, B.P., R. Hornung, M. Ho, C.R. Howard, S. Eberly, and K. Knauf. 2002. Environmental lead exposure during early childhood. *The Journal of Pediatrics* 140 (1):40-7.

3. Lanphear, B.P., and K.J. Roghmann. 1997. Pathways of lead exposure in urban children. *Environmental Research* 74 (1):67-73.

4. Rabinowitz, M., A. Leviton, H. Needleman, D. Bellinger, and C. Waternaux. 1985. Environmental correlates of infant blood lead levels in Boston. *Environmental Research* 38 (1):96-107.

5. Levin, R., M.J. Brown, M.E. Kashtock, D.E. Jacobs, E.A. Whelan, J. Rodman, M.R. Schock, A. Padilla, and T. Sinks. 2008. Lead exposures in U.S. Children, 2008: implications for prevention. *Environmental Health Perspectives* 116 (10):1285-93.

6. Gaitens, J.M., S.L. Dixon, D.E. Jacobs, J. Nagaraja, W. Strauss, J.W. Wilson, and P. Ashley. 2008. U.S. Children's Exposure to Residential dust lead, 1999-2004: I. Housing and Demographic Factors. *Environmental Health Perspectives* 117 (3):461-7.

7. Jacobs, D.E., R.P. Clickner, J.Y. Zhou, S.M. Viet, D.A. Marker, J.W. Rogers, D.C. Zeldin, P. Broene, and W. Friedman. 2002. The prevalence of lead-based paint hazards in U.S. housing. *Environmental Health Perspectives* 110 (10):A599-606.

8. Centers for Disease Control and Prevention. 2009. Children with elevated blood lead levels related to home renovation, repair, and painting activities - New York State, 2006-2007. *Morbidity and Mortality Weekly Report* 58 (3):55-58.

9. Centers for Disease Control and Prevention. 1997. Children with elevated blood lead levels attributed to home renovation and remodeling activities - New York, 1993-1994. *Morbidity and Mortality Weekly Report* 45 (51-52):1120-1123.

10. Adgate, J.L., G.G. Rhoads, and P.J. Lioy. 1998. The use of isotope ratios to apportion sources of lead in Jersey City, NJ, house dust wipe samples. *Science of the Total Environment* 221 (2-3):171-80.

11. Clark, S., W. Menrath, M. Chen, P. Succop, R. Bornschein, W. Galke, and J. Wilson. 2004. The influence of exterior dust and soil lead on interior dust lead levels in housing that had undergone lead-based paint hazard control. *The Journal of Occupational and Environmental Hygiene* 1 (5):273-82.

12. von Lindern, I., S. Spalinger, V. Petroysan, and M. von Braun. 2003. Assessing remedial effectiveness through the blood lead:soil/dust lead relationship at the Bunker Hill superfund site in the Silver Valley of Idaho. *Science of the Total Environment* 303 ((1-2)):139-70.

13. Lanphear, B.P., T.D. Matte, J. Rogers, R.P. Clickner, B. Dietz, R.L. Bornschein, P. Succop, K.R. Mahaffey, S. Dixon, W. Galke, et al. 1998. The contribution of lead-contaminated house dust and residential soil to children's blood lead levels. A pooled analysis of 12 epidemiologic studies. *Environmental Research* 79 (1):51-68.

14. Mielke, H.W., and P.L. Reagan. 1998. Soil is an important pathway of human lead exposure. *Environmental Health Perspectives* 106 Suppl 1:217-29.

15. U.S. Environmental Protection Agency. 2006. *Air Quality Criteria for Lead. Volume I of II*. Washington, DC: United States Environmental Protection Agency. EPA/600/R-5/144aF.

Lead (continued)

16. McElvaine, M.D., E.G. DeUngria, T.D. Matte, C.G. Copley, and S. Binder. 1992. Prevalence of radiographic evidence of paint chip ingestion among children with moderate to severe lead poisoning, St Louis, Missouri, 1989 through 1990. *Pediatrics* 89 (4 Pt 2):740-2.

17. Edwards, M., S. Triantafyllidou, and D. Best. 2009. Elevated blood lead in young children due to lead-contaminated drinking water: Washington, DC, 2001-2004. *Environmental Science and Technology* 43 (5):1618-1623.

18. Miranda, M.L., D. Kim, A.P. Hull, C.J. Paul, and M.A. Galeano. 2007. Changes in blood lead levels associated with use of chloramines in water treatment systems. *Environmental Health Perspectives* 115 (2):221-5.

19. VanArsdale, J.L., R.D. Leiker, M. Kohn, T.A. Merritt, and B.Z. Horowitz. 2004. Lead poisoning from a toy necklace. *Pediatrics* 114 (4):1096-9.

20. Weidenhamer, J.D., and M.L. Clement. 2007. Widespread lead contamination of imported low-cost jewelry in the US. *Chemosphere* 67 (5):961-5.

21. Mannino, D.M., R. Albalak, S. Grosse, and J. Repace. 2003. Second-hand smoke exposure and blood lead levels in U.S. children. *Epidemiology* 14 (6):719-27.

22. Gorospe, E.C., and S.L. Gerstenberger. 2008. Atypical sources of childhood lead poisoning in the United States: a systematic review from 1966-2006. *Clinical Toxicology (Philadelphia)* 46 (8):728-37.

23. Saper, R.B., S.N. Kales, J. Paquin, M.J. Burns, D.M. Eisenberg, R.B. Davis, and R.S. Phillips. 2004. Heavy metal content of ayurvedic herbal medicine products. *The Journal of the American Medical Association* 292 (23):2868-73.

24. Woolf, A.D., J. Hussain, L. McCullough, M. Petranovic, and C. Chomchai. 2008. Infantile lead poisoning from an Asian tongue powder: a case report & subsequent public health inquiry. *Clinical Toxicology (Philadelphia, PA)* 46 (9):841-4.

25. Agency for Toxic Substances and Disease Registry. 2007. *Toxicological Profile for Lead*. Atlanta, GA: ATSDR, Division of Toxicology and Environmental Medicine/Applied Toxicology Branch. http://www.atsdr.cdc.gov/ToxProfiles/tp13.pdf.

26. Pirkle, J.L., R.B. Kaufmann, D.J. Brody, T. Hickman, E.W. Gunter, and D.C. Paschal. 1998. Exposure of the U.S. population to lead, 1991-1994. *Environmental Health Perspectives* 106 (11):745-50.

27. Dixon, S.L., J.M. Gaitens, D.E. Jacobs, W. Strauss, J. Nagaraja, T. Pivetz, J.W. Wilson, and P. Ashley. 2009. U.S. Children's exposure to residential dust lead, 1999-2004: II. The contribution of lead-contaminated dust to children's blood lead levels. *Environmental Health Perspectives* 117 (3):468-74.

28. Kim, D.Y., F. Staley, G. Curtis, and S. Buchanan. 2002. Relation between housing age, housing value, and childhood blood lead levels in children in Jefferson County, Ky. *American Journal of Public Health* 92 (5):769-72.

29. Tehranifar, P., J. Leighton, A.H. Auchincloss, A. Faciano, H. Alper, A. Paykin, and S. Wu. 2008. Immigration and risk of childhood lead poisoning: findings from a case control study of New York City children. *American Journal of Public Health* 98 (1):92-7.

30. U.S. Environmental Protection Agency. 2000. *National Air Quality and Emissions Trends Report, 1998*. Research Triangle Park, NC: EPA Office of Air Quality Planning and Standards. http://epa.gov/airtrends/aqtrnd98/.

31. National Toxicology Program. 2012. *NTP Monograph on Health Effects of Low-Level Lead*. Research Triangle Park, NC: National Institute of Environmental Health Sciences, National Toxicology Program. http://ntp.niehs.nih.gov/go/36443.

32. Bellinger, D., J. Sloman, A. Leviton, M. Rabinowitz, H.L. Needleman, and C. Waternaux. 1991. Low-level lead exposure and children's cognitive function in the preschool years. *Pediatrics* 87 (2):219-27.

33. Canfield, R.L., C.R. Henderson, Jr., D.A. Cory-Slechta, C. Cox, T.A. Jusko, and B.P. Lanphear. 2003. Intellectual impairment in children with blood lead concentrations below 10 microg per deciliter. *New England Journal of Medicine* 348 (16):1517-26.

34. Jusko, T.A., C.R. Henderson, B.P. Lanphear, D.A. Cory-Slechta, P.J. Parsons, and R.L. Canfield. 2008. Blood lead concentrations < 10 microg/dL and child intelligence at 6 years of age. *Environmental Health Perspectives* 116 (2):243-8.

35. Lanphear, B.P., K. Dietrich, P. Auinger, and C. Cox. 2000. Cognitive deficits associated with blood lead concentrations <10 microg/dL in US children and adolescents. *Public Health Reports* 115 (6):521-9.

36. Lanphear, B.P., R. Hornung, J. Khoury, K. Yolton, P. Baghurst, D.C. Bellinger, R.L. Canfield, K.N. Dietrich, R. Bornschein, T. Greene, et al. 2005. Low-level environmental lead exposure and children's intellectual function: an international pooled analysis. *Environmental Health Perspectives* 113 (7):894-9.

37. Schnaas, L., S.J. Rothenberg, M.F. Flores, S. Martinez, C. Hernandez, E. Osorio, S.R. Velasco, and E. Perroni. 2006. Reduced intellectual development in children with prenatal lead exposure. *Environmental Health Perspectives* 114 (5):791-7.

38. Surkan, P.J., A. Zhang, F. Trachtenberg, D.B. Daniel, S. McKinlay, and D.C. Bellinger. 2007. Neuropsychological function in children with blood lead levels <10 microg/dL. *Neurotoxicology* 28 (6):1170-7.

Lead (continued)

39. Calderon, J., M.E. Navarro, M.E. Jimenez-Capdeville, M.A. Santos-Diaz, A. Golden, I. Rodriguez-Leyva, V. Borja-Aburto, and F. Diaz-Barriga. 2001. Exposure to arsenic and lead and neuropsychological development in Mexican children. *Environmental Research* 85 (2):69-76.

40. Chiodo, L.M., C. Covington, R.J. Sokol, J.H. Hannigan, J. Jannise, J. Ager, M. Greenwald, and V. Delaney-Black. 2007. Blood lead levels and specific attention effects in young children. *Neurotoxicology and Teratology* 29:538-546.

41. Chiodo, L.M., S.W. Jacobson, and J.L. Jacobson. 2004. Neurodevelopmental effects of postnatal lead exposure at very low levels. *Neurotoxicology and Teratology* 26 (3):359-71.

42. Nicolescu, R., C. Petcu, A. Cordeanu, K. Fabritius, M. Schlumpf, R. Krebs, U. Kramer, and G. Winneke. 2010. Environmental exposure to lead, but not other neurotoxic metals, relates to core elements of ADHD in Romanian children: performance and questionnaire data. *Environmental Research* 110 (5):476-83.

43. Ris, M.D., K.N. Dietrich, P.A. Succop, O.G. Berger, and R.L. Bornschein. 2004. Early exposure to lead and neuropsychological outcome in adolescence. *Journal of the International Neuropsychological Society* 10 (2):261-70.

44. Nigg, J.T., G.M. Knottnerus, M.M. Martel, M. Nikolas, K. Cavanagh, W. Karmaus, and M.D. Rappley. 2008. Low blood lead levels associated with clinically diagnosed attention-deficit/hyperactivity disorder and mediated by weak cognitive control. *Biological Psychiatry* 63 (3):325-31.

45. Braun, J.M., R.S. Kahn, T. Froehlich, P. Auinger, and B.P. Lanphear. 2006. Exposures to environmental toxicants and attention deficit hyperactivity disorder in U.S. children. *Environmental Health Perspectives* 114 (12):1904-9.

46. Eubig, P.A., A. Aguiar, and S.L. Schantz. 2010. Lead and PCBs as risk factors for attention deficit/hyperactivity disorder. *Environmental Health Perspectives* 118 (12):1654-1667.

47. Froehlich, T.E., B.P. Lanphear, P. Auinger, R. Hornung, J.N. Epstein, J. Braun, and R.S. Kahn. 2009. Association of tobacco and lead exposures with attention-deficit/hyperactivity disorder. *Pediatrics* 124 (6):e1054-63.

48. Ha, M., H.J. Kwon, M.H. Lim, Y.K. Jee, Y.C. Hong, J.H. Leem, J. Sakong, J.M. Bae, S.J. Hong, Y.M. Roh, et al. 2009. Low blood levels of lead and mercury and symptoms of attention deficit hyperactivity in children: a report of the children's health and environment research (CHEER). *Neurotoxicology* 30 (1):31-6.

49. Nigg, J.T., M. Nikolas, G. Mark Knottnerus, K. Cavanagh, and K. Friderici. 2010. Confirmation and extension of association of blood lead with attention-deficit/hyperactivity disorder (ADHD) and ADHD symptom domains at population-typical exposure levels. *The Journal of Child Psychology and Psychiatry* 51 (1):58-65.

50. Roy, A., D. Bellinger, H. Hu, J. Schwartz, A.S. Ettinger, R.O. Wright, M. Bouchard, K. Palaniappan, and K. Balakrishnan. 2009. Lead exposure and behavior among young children in Chennai, India. *Environmental Health Perspectives* 117 (10):1607-11.

51. Tuthill, R.W. 1996. Hair lead levels related to children's classroom attention-deficit behavior. *Archives of Environmental Health* 51 (3):214-20.

52. Wang, H., X. Chen, B. Yang, M. Hao, and D. Ruan. 2008. Case-Control study of blood lead levels and Attention-Deficit Hyperactivity Disorder in Chinese children. *Environmental Health Perspectives* 116 (10):1401-06.

53. Braun, J.M., T.E. Froehlich, J.L. Daniels, K.N. Dietrich, R. Hornung, P. Auinger, and B.P. Lanphear. 2008. Association of environmental toxicants and conduct disorder in U.S. children: NHANES 2001-2004. *Environmental Health Perspectives* 116 (7):956-62.

54. Marcus, D.K., J.J. Fulton, and E.J. Clarke. 2010. Lead and conduct problems: a meta-analysis. *Journal of Clinical Child and Adolescent Psychology* 39 (2):234-41.

55. Dietrich, K.N., M.D. Ris, P.A. Succop, O.G. Berger, and R.L. Bornschein. 2001. Early exposure to lead and juvenile delinquency. *Neurotoxicology and Teratology* 23 (6):511-8.

56. Needleman, H.L., C. McFarland, R.B. Ness, S.E. Fienberg, and M.J. Tobin. 2002. Bone lead levels in adjudicated delinquents. A case control study. *Neurotoxicology and Teratology* 24 (6):711-7.

57. Needleman, H.L., J.A. Riess, M.J. Tobin, G.E. Biesecker, and J.B. Greenhouse. 1996. Bone lead levels and delinquent behavior. *The Journal of the American Medical Association* 275 (5):363-9.

58. Nevin, R. 2007. Understanding international crime trends: the legacy of preschool lead exposure. *Environmental Research* 104 (3):315-36.

59. Wright, J.P., K.N. Dietrich, M.D. Ris, R.W. Hornung, S.D. Wessel, B.P. Lanphear, M. Ho, and M.N. Rae. 2008. Association of prenatal and childhood blood lead concentrations with criminal arrests in early adulthood. *PLoS Medicine* 5 (5):e101.

60. Bellinger, D.C. 2008. Lead neurotoxicity and socioeconomic status: conceptual and analytical issues. *Neurotoxicology* 29 (5):828-32.

61. Weiss, B., and D.C. Bellinger. 2006. Social ecology of children's vulnerability to environmental pollutants. *Environmental Health Perspectives* 114 (10):1479-1485.

Lead (continued)

62. Chuang, H.Y., J. Schwartz, T. Gonzales-Cossio, M.C. Lugo, E. Palazuelos, A. Aro, H. Hu, and M. Hernandez-Avila. 2001. Interrelations of lead levels in bone, venous blood, and umbilical cord blood with exogenous lead exposure through maternal plasma lead in peripartum women. *Environmental Health Perspectives* 109 (5):527-32.

63. Ettinger, A.S., M.M. Tellez-Rojo, C. Amarasiriwardena, T. Gonzalez-Cossio, K.E. Peterson, A. Aro, H. Hu, and M. Hernandez-Avila. 2004. Levels of lead in breast milk and their relation to maternal blood and bone lead levels at one month postpartum. *Environmental Health Perspectives* 112 (8):926-31.

64. Advisory Committee on Childhood Lead Poisoning Prevention. 2010. *Guidelines for the Identification and Management of Lead Exposure in Pregnant and Lactating Women.* Atlanta, GA: Centers for Disease Control and Prevention. http://www.cdc.gov/nceh/lead/publications/leadandpregnancy2010.pdf.

65. Chen, A., K.N. Dietrich, J.H. Ware, J. Radcliffe, and W.J. Rogan. 2005. IQ and blood lead from 2 to 7 years of age:are the effects in older children the residual of high blood lead concentration in 2-year-olds? *Environmental Health Perspectives* 113:597-601.

66. Hornung, R.W., B.P. Lanphear, and K.N. Dietrich. 2009. Age of greatest susceptibility to childhood lead exposure: A new statistical approach. *Environmental Health Perspectives* 117 (8):1309-12.

67. Brubaker, C.J., K.N. Dietrich, B.P. Lanphear, and K.M. Cecil. 2010. The influence of age of lead exposure on adult gray matter volume. *Neurotoxicology* 31 (3):259-66.

68. Cecil, K.M., C.J. Brubaker, C.M. Adler, K.N. Dietrich, M. Altaye, J.C. Egelhoff, S. Wessel, I. Elangovan, R. Hornung, K. Jarvis, et al. 2008. Decreased brain volume in adults with childhood lead exposure. *PLoS Medicine* 5 (5):e112.

69. Mazumdar, M., D.C. Bellinger, M. Gregas, K. Abanilla, J. Bacic, and H.L. Needleman. 2011. Low-level environmental lead exposure in childhood and adult intellectual function: a follow-up study. *Environmental Health* 10 (1):24.

70. Gulson, B.L., K.J. Mizon, M.J. Korsch, J.M. Palmer, and J.B. Donnelly. 2003. Mobilization of lead from human bone tissue during pregnancy and lactation--a summary of long-term research. *Science of the Total Environment* 303 (1-2):79-104.

71. Stein, J., T. Schettler, B. Rohrer, and M. Valenti. 2008. *Environmental Threats to Health Aging: With a Closer Look at Alzheimer's and Parkinson's Diseases.* Boston, MA: Greater Boston Physicians for Social Responsibility and Science and Environmental Health Network. http://www.agehealthy.org/pdf/GBPSRSEHN_HealthyAging1017.pdf.

72. Centers for Disease Control and Prevention. 2002. *Managing Elevated Blood Lead Levels Among Young Children: Recommendations from the Advisory Committee on Childhood Lead Poisoning Prevention.* Atlanta, GA.

73. Tellez-Rojo, M.M., D.C. Bellinger, C. Arroyo-Quiroz, H. Lamadrid-Figueroa, A. Mercado-Garcia, L. Schnaas-Arrieta, R.O. Wright, M. Hernandez-Avila, and H. Hu. 2006. Longitudinal associations between blood lead concentrations lower than 10 microg/dL and neurobehavioral development in environmentally exposed children in Mexico City. *Pediatrics* 118 (2):e323-30.

74. Centers for Disease Control and Prevention. 1997. *Screening Young Children for Lead Poisoning: Guidance for State and Local Public Health Officials.* Atlanta, GA.

75. Advisory Committee on Childhood Lead Poisoning Prevention. 2012. *Low Level Lead Exposure Harms Children: A Renewed Call for Primary Prevention.* Atlanta, GA: Centers for Disease Control and Prevention. http://www.cdc.gov/nceh/lead/ACCLPP/Final_Document_030712.pdf.

76. Centers for Disease Control and Prevention. 2012. *CDC Response to Advisory Committee on Childhood Lead Poisoning Prevention Recommendations in Low Level Lead Exposure Harms Children: A Renewed Call for Primary Prevention.* Atlanta, GA: Centers for Disease Control and Prevention. http://www.cdc.gov/nceh/lead/acclpp/cdc_response_lead_exposure_recs.pdf.

77. Centers for Disease Control and Prevention. 2009. *Fourth National Report on Human Exposure to Environmental Chemicals.* Atlanta, GA: CDC. http://www.cdc.gov/exposurereport/.

Mercury

1. U.S. Environmental Protection Agency. 2007. *Organic Mercury: TEACH Chemical Summary.* Retrieved January 26, 2010 from http://www.epa.gov/teach/chem_summ/mercury_org_summary.pdf.

2. U.S. Environmental Protection Agency. 2006. *National Vehicle Mercury Switch Recovery Program.* U.S. EPA. Retrieved October 5, 2011 from http://www.epa.gov/hg/switchfs.htm.

3. U.S. Environmental Protection Agency. 2011. Proposed Rule: National Emission Standards for Hazardous Air Pollutants from Coal- and Oil-Fired Electric Utility Steam Generating Units and Standards of Performance for Fossil-Fuel-Fired Electric Utility, Industrial-Commercial-Institutional, and Small Industrial-Commercial-Institutional Steam Generating Units. *Federal Register* 76 (85):4976-25147. http://federalregister.gov/a/2011-7237.

4. U.S. Environmental Protection Agency. 1997. *Mercury Study Report to Congress Volumes I to VII.* Washington DC: U.S. Environmental Protection Agency Office of Air Quality Planning and Standards and Office of Research and Development. EPA-452/R-97-003. http://www.epa.gov/hg/report.htm.

Mercury (continued)

5. Fitzgerald, W.F., D.R. Engstrom, R.P. Mason, and E.A. Nater. 1998. The case for atmospheric mercury contamination in remote areas. *Environmental Science and Technology* 32 (1):1-7.

6. Carrie, J., F. Wang, H. Sanei, R.W. Macdonald, P.M. Outridge, and G.A. Stern. 2010. Increasing contaminant burdens in an arctic fish, Burbot (Lota lota), in a warming climate. *Environmental Science and Technology* 44 (1):316-22.

7. Lindberg, S.E., S. Brooks, C.J. Lin, K.J. Scott, M.S. Landis, R.K. Stevens, M. Goodsite, and A. Richter. 2002. Dynamic oxidation of gaseous mercury in the Arctic troposphere at polar sunrise. *Environmental Science and Technology* 36 (6):1245-56.

8. Lindberg, S.E., S. Brooks, C.-J. Lin, K. Scott, T. Meyers, L. Chambers, M. Landis, and R. Stevens. 2001. Formation of reactive gaseous mercury in the Arctic: evidence of oxidation of Hg° to gas-phase HG-II compounds after Arctic sunrise. *Water, Air, and Soil Pollution; Focus* 1 (5-6):295-302.

9. Lee, R., D. Middleton, K. Caldwell, S. Dearwent, S. Jones, B. Lewis, C. Monteilh, M.E. Mortensen, R. Nickle, K. Orloff, et al. 2009. A review of events that expose children to elemental mercury in the United States. *Environmental Health Perspectives* 117 (6):871-878.

10. U.S. Environmental Protection Agency. 2002. *Task Force on Ritualistic Uses of Mercury Report*. Washington, DC: U.S. EPA, Office of Emergency and Remedial Response. EPA/540-R-01-005. http://www.epa.gov/superfund/community/pdfs/mercury.pdf.

11. Agency for Toxic Substances and Disease Registry. 2009. *Children's Exposure to Elemental Mercury: A National Review of Exposure Events*. Atlanta, GA: Agency for Toxic Substances and Disease Registry.

12. Agency for Toxic Substances and Disease Registry. 2006. *Health Consultation: Mercury-Containing Polyurethane Floors in Minnesota Schools*. Atlanta, GA: U.S. Department of Health and Human Services. http://www.atsdr.cdc.gov/HAC/pha/MercuryVaporReleaseAthleticPolymerFloors/MercuryVaporRelease-FloorsHC092806.pdf.

13. Bellinger, D.C., F. Trachtenberg, L. Barregard, M. Tavares, E. Cernichiari, D. Daniel, and S. McKinlay. 2006. Neuropsychological and renal effects of dental amalgam in children: a randomized clinical trial. *JAMA* 295 (15):1775-83.

14. DeRouen, T.A., M.D. Martin, B.G. Leroux, B.D. Townes, J.S. Woods, J. Leitao, A. Castro-Caldas, H. Luis, M. Bernardo, G. Rosenbaum, et al. 2006. Neurobehavioral effects of dental amalgam in children: a randomized clinical trial. *JAMA* 295 (15):1784-92.

15. National Research Council. 2000. *Toxicological Effects of Methylmercury*. Washington, DC: National Academy Press.

16. Institute of Medicine. 2004. *Immunization Safety Review: Vaccines and Autism*. Washington, DC: National Academies Press. http://www.nap.edu/catalog.php?record_id=10997.

17. Price, C.S., W.W. Thompson, B. Goodson, E.S. Weintraub, L.A. Croen, V.L. Hinrichsen, M. Marcy, A. Robertson, E. Eriksen, E. Lewis, et al. 2010. Prenatal and infant exposure to thimerosal from vaccines and immunoglobulins and risk of autism. *Pediatrics* 126 (4):656-64.

18. Thompson, W.W., C. Price, B. Goodson, D.K. Shay, P. Benson, V.L. Hinrichsen, E. Lewis, E. Eriksen, P. Ray, S.M. Marcy, et al. 2007. Early thimerosal exposure and neuropsychological outcomes at 7 to 10 years. *New England Journal of Medicine* 357 (13):1281-92.

19. Centers for Disease Control and Prevention. *Mercury and Thimerosal: Vaccine Safety*. CDC. Retrieved October 12, 2010 from http://www.cdc.gov/vaccinesafety/Concerns/thimerosal/index.html.

20. Canadian Council of Ministers of the Environment. 2000. *Methylmercury: Canadian Tissue Residue Guidelines for the Protection of Wildlife Consumers of Aquatic Biota*. Ottawa, Ontario: Environment Canada.

21. Harada, M. 1995. Minamata disease: methylmercury poisoning in Japan caused by environmental pollution. *Critical Reviews in Toxicology* 25 (1):1-24.

22. Amin-Zaki, L., S. Elhassani, M.A. Majeed, T.W. Clarkson, R.A. Doherty, and M. Greenwood. 1974. Intra-uterine methylmercury poisoning in Iraq. *Pediatrics* 54 (5):587-95.

23. Budtz-Jorgensen, E., P. Grandjean, and P. Weihe. 2007. Separation of risks and benefits of seafood intake. *Environmental Health Perspectives* 115 (3):323-7.

24. Crump, K.S., T. Kjellstrom, A.M. Shipp, A. Silvers, and A. Stewart. 1998. Influence of prenatal mercury exposure upon scholastic and psychological test performance: benchmark analysis of a New Zealand cohort. *Risk Analysis* 18 (6):701-13.

25. Debes, F., E. Budtz-Jorgensen, P. Weihe, R.F. White, and P. Grandjean. 2006. Impact of prenatal methylmercury exposure on neurobehavioral function at age 14 years. *Neurotoxicology and Teratology* 28 (3):363-75.

26. Grandjean, P., P. Weihe, R.F. White, F. Debes, S. Araki, K. Yokoyama, K. Murata, N. Sorensen, R. Dahl, and P.J. Jorgensen. 1997. Cognitive deficit in 7-year-old children with prenatal exposure to methylmercury. *Neurotoxicology and Teratology* 19 (6):417-28.

27. Kjellstrom, T., P. Kennedy, S. Wallis, and C. Mantell. 1986. *Physical and mental development of children with prenatal exposure to mercury from fish. Stage 1: Preliminary tests at age 4*. Sweden: Swedish National Environmental Protection Board.

28. Oken, E., and D.C. Bellinger. 2008. Fish consumption, methylmercury and child neurodevelopment. *Current Opinion in Pediatrics* 20 (2):178-83.

Mercury (continued)

29. Myers, G.J., P.W. Davidson, C. Cox, C.F. Shamlaye, D. Palumbo, E. Cernichiari, J. Sloane-Reeves, G.E. Wilding, J. Kost, L.S. Huang, et al. 2003. Prenatal methylmercury exposure from ocean fish consumption in the Seychelles child development study. *Lancet* 361 (9370):1686-92.

30. Davidson, P.W., J.J. Strain, G.J. Myers, S.W. Thurston, M.P. Bonham, C.F. Shamlaye, A. Stokes-Riner, J.M. Wallace, P.J. Robson, E.M. Duffy, et al. 2008. Neurodevelopmental effects of maternal nutritional status and exposure to methylmercury from eating fish during pregnancy. *Neurotoxicology* 29 (5):767-75.

31. Lynch, M.L., L.S. Huang, C. Cox, J.J. Strain, G.J. Myers, M.P. Bonham, C.F. Shamlaye, A. Stokes-Riner, J.M. Wallace, E.M. Duffy, et al. 2011. Varying coefficient function models to explore interactions between maternal nutritional status and prenatal methylmercury toxicity in the Seychelles Child Development Nutrition Study. *Environmental Research* 111 (1):75-80.

32. Strain, J.J., P.W. Davidson, M.P. Bonham, E.M. Duffy, A. Stokes-Riner, S.W. Thurston, J.M. Wallace, P.J. Robson, C.F. Shamlaye, L.A. Georger, et al. 2008. Associations of maternal long-chain polyunsaturated fatty acids, methyl mercury, and infant development in the Seychelles Child Development Nutrition Study. *Neurotoxicology* 5:776-82.

33. Lederman, S.A., R.L. Jones, K.L. Caldwell, V. Rauh, S.E. Sheets, D. Tang, S. Viswanathan, M. Becker, J.L. Stein, R.Y. Wang, et al. 2008. Relation between cord blood mercury levels and early child development in a World Trade Center cohort. *Environmental Health Perspectives* 116 (8):1085-91.

34. Oken, E., J.S. Radesky, R.O. Wright, D.C. Bellinger, C.J. Amarasiriwardena, K.P. Kleinman, H. Hu, and M.W. Gillman. 2008. Maternal fish intake during pregnancy, blood mercury levels, and child cognition at age 3 years in a US cohort. *American Journal of Epidemiology* 167 (10):1171-81.

35. Oken, E., R.O. Wright, K.P. Kleinman, D. Bellinger, C.J. Amarasiriwardena, H. Hu, J.W. Rich-Edwards, and M.W. Gillman. 2005. Maternal fish consumption, hair mercury, and infant cognition in a U.S. Cohort. *Environmental Health Perspectives* 113 (10):1376-80.

36. Cao, Y., A. Chen, R.L. Jones, J. Radcliffe, K.L. Caldwell, K.N. Dietrich, and W.J. Rogan. 2010. Does background postnatal methyl mercury exposure in toddlers affect cognition and behavior? *Neurotoxicology* 31 (1):1-9.

37. Davidson, P.W., G.J. Myers, C. Cox, C. Axtell, C. Shamlaye, J. Sloane-Reeves, E. Cernichiari, L. Needham, A. Choi, Y. Wang, et al. 1998. Effects of prenatal and postnatal methylmercury exposure from fish consumption on neurodevelopment: outcomes at 66 months of age in the Seychelles Child Development Study. *JAMA* 280 (8):701-7.

38. Freire, C., R. Ramos, M.J. Lopez-Espinosa, S. Diez, J. Vioque, F. Ballester, and M.F. Fernandez. 2010. Hair mercury levels, fish consumption, and cognitive development in preschool children from Granada, Spain. *Environmental Research* 110 (1):96-104.

39. Karagas, M.R., A.L. Choi, E. Oken, M. Horvat, R. Schoeny, E. Kamai, W. Cowell, P. Grandjean, and S. Korrick. 2012. Evidence on the human health effects of low-level methylmercury exposure. *Environmental Health Perspectives* 120 (6):799-806.

40. Grandjean, P., K. Murata, E. Budtz-Jorgensen, and P. Weihe. 2004. Cardiac autonomic activity in methylmercury neurotoxicity: 14-year follow-up of a Faroese birth cohort. *The Journal of Pediatrics* 144 (2):169-76.

41. Sorensen, N., K. Murata, E. Budtz-Jorgensen, P. Weihe, and P. Grandjean. 1999. Prenatal methylmercury exposure as a cardiovascular risk factor at seven years of age. *Epidemiology* 10 (4):370-5.

42. Brenden, N., H. Rabbani, and M. Abedi-Valugerdi. 2001. Analysis of mercury-induced immune activation in nonobese diabetic (NOD) mice. *Clinical and Experimental Immunology* 125 (2):202-10.

43. Sweet, L.I., and J.T. Zelikoff. 2001. Toxicology and immunotoxicology of mercury: a comparative review in fish and humans. *Journal of Toxicology and Environmental Health. Part B, Critical Reviews* 4 (2):161-205.

44. Institute of Medicine. 2007. *Seafood Choices. Balancing Benefits and Risks.* Washington, DC: Committee on Nutrient Relationships in Seafood: Selections to Balance Benefits and Risks. Food and Nutrition Board. Institute of Medicine.

45. U.S. Environmental Protection Agency, and U.S. Food and Drug Administration. 2004. *What you need to know about mercury in fish and shellfish. Advice for women who might become pregnant, women who are pregnant, nursing mothers and children.* Washington DC: U.S. Environmental Protection Agency and U.S. Food and Drug Administration. EPA-823-F-04-009. http://www.epa.gov/waterscience/fish/files/MethylmercuryBrochure.pdf.

46. U.S. Department of Agriculture, and U.S. Department of Health and Human Services. 2010. *Dietary Guidelines for Americans, 2010.* Washington, DC: U.S. Government Printing Office. http://www.cnpp.usda.gov/Publications/DietaryGuidelines/2010/PolicyDoc/PolicyDoc.pdf.

47. Mahaffey, K.R., R.P. Clickner, and C.C. Bodurow. 2004. Blood organic mercury and dietary mercury intake: National Health and Nutrition Examination Survey, 1999 and 2000. *Environmental Health Perspectives* 112 (5):562-70.

48. Mahaffey, K.R., R.P. Clickner, and R.A. Jeffries. 2009. Adult women's blood mercury concentrations vary regionally in the United States: association with patterns of fish consumption (NHANES 1999-2004). *Environmental Health Perspectives* 117 (1):47-53.

49. Hightower, J.M., A. O'Hare, and G.T. Hernandez. 2006. Blood mercury reporting in NHANES: identifying Asian, Pacific Islander, Native American, and multiracial groups. *Environmental Health Perspectives* 114 (2):173-5.

Mercury (continued)

50. McKelvey, W., R.C. Gwynn, N. Jeffery, D. Kass, L.E. Thorpe, R.K. Garg, C.D. Palmer, and P.J. Parsons. 2007. A biomonitoring study of lead, cadmium, and mercury in the blood of New York city adults. *Environmental Health Perspectives* 115 (10):1435-41.

51. Schober, S.E., T.H. Sinks, R.L. Jones, P.M. Bolger, M. McDowell, J. Osterloh, E.S. Garrett, R.A. Canady, C.F. Dillon, Y. Sun, et al. 2003. Blood mercury levels in US children and women of childbearing age, 1999-2000. *The Journal of the American Medical Association* 289 (13):1667-74.

52. Knobeloch, L., H.A. Anderson, P. Imm, D. Peters, and A. Smith. 2005. Fish consumption, advisory awareness, and hair mercury levels among women of childbearing age. *Environmental Research* 97 (2):220-7.

53. Centers for Disease Control and Prevention. 2009. *Fourth National Report on Human Exposure to Environmental Chemicals*. Atlanta, GA: CDC. http://www.cdc.gov/exposurereport/.

54. Clarkson, T.W. 2002. The three modern faces of mercury. *Environmental Health Perspectives* 110 Suppl 1:11-23.

55. Tollefson, L., and F. Cordle. 1986. Methylmercury in fish: a review of residue levels, fish consumption and regulatory action in the United States. *Environmental Health Perspectives* 68:203-8.

56. Caldwell, K.L., M.E. Mortensen, R.L. Jones, S.P. Caudill, and J.D. Osterloh. 2009. Total blood mercury concentrations in the U.S. population: 1999-2006. *International Journal of Hygiene and Environmental Health* 212 (6):588-98.

57. National Center for Health Statistics. *Vital Statistics Natality Birth Data*. Retrieved June 15, 2009 from http://www.cdc.gov/nchs/data_access/Vitalstatsonline.htm. .

58. Axelrad, D.A., and J. Cohen. 2011. Calculating summary statistics for population chemical biomonitoring in women of childbearing age with adjustment for age-specific natality. *Environmental Research* 111 (1):149-155.

Cotinine

1. U.S. Department of Health and Human Services. 2006. *The Health Consequences of Involuntary Exposure to Tobacco Smoke: A Report of the Surgeon General*. Atlanta, GA: U.S. Department of Health and Human Services, Centers for Disease Control and Prevention, Coordinating Center for Health Promotion, National Center for Chronic Disease Prevention and Health Promotion, Office on Smoking and Health. http://www.surgeongeneral.gov/library/reports/secondhandsmoke/fullreport.pdf.

2. National Toxicology Program. 2011. *Report on Carcinogens, 12th Edition*. Research Triangle Park, NC: U.S. Department of Health and Human Services, National Toxicology Program. http://ntp.niehs.nih.gov/ntp/roc/twelfth/roc12.pdf.

3. U.S. Environmental Protection Agency. 1992. *Respiratory Health Effects of Passive Smoking: Lung Cancer and Other Disorders*. Washington, DC: EPA Office of Research and Development. http://cfpub.epa.gov/ncea/cfm/ets/etsindex.cfm.

4. Gergen, P.J., J.A. Fowler, K.R. Maurer, W.W. Davis, and M.D. Overpeck. 1998. The burden of environmental tobacco smoke exposure on the respiratory health of children 2 months through 5 years of age in the United States: Third National Health and Nutrition Examination Survey, 1988 to 1994. *Pediatrics* 101 (2):E8.

5. Institute of Medicine. 2000. *Clearing the Air: Asthma and Indoor Air Exposures*. Washington, DC: National Academy Press. http://books.nap.edu/catalog/9610.html.

6. Lovasi, G.S., A.V. Diez Roux, E.A. Hoffman, S.M. Kawut, D.R. Jacobs, Jr., and R.G. Barr. Association of environmental tobacco smoke exposure in childhood with early emphysema in adulthood among nonsmokers: the MESA-lung study. *American Journal of Epidemiology* 171 (1):54-62.

7. Yousey, Y.K. 2006. Household characteristics, smoking bans, and passive smoke exposure in young children. *Journal of Pediatric Health Care* 20 (2):98-105.

8. Wamboldt, F.S., R.C. Balkissoon, A.E. Rankin, S.J. Szefler, S.K. Hammond, R.E. Glasgow, and W.P. Dickinson. 2008. Correlates of household smoking bans in low-income families of children with and without asthma. *Family Process* 47 (1):81-94.

9. Pirkle, J.L., J.T. Bernert, S.P. Caudill, C.S. Sosnoff, and T.F. Pechacek. 2006. Trends in the exposure of nonsmokers in the U.S. population to secondhand smoke: 1988-2002. *Environmental Health Perspectives* 114 (6):853-8.

10. Centers for Disease Control and Prevention. 2007. Cigarette smoking among adults—United States, 2006. *Morbidity and Mortality Weekly Report* 56 (44):1157-1161.

11. Centers for Disease Control and Prevention. 2011. Vital Signs: Current Cigarette Smoking Among Adults Aged ≥18 Years --- United States, 2005--2010. *Morbidity and Mortality Weekly Report* 60 (35):1207-12.

12. Centers for Disease Control and Prevention. 2007. State-specific prevalence of smoke-free home rules - United States, 1992-2003. *Morbidity and Mortality Weekly Report* 56 (20):501-504.

13. King, K., M. Martynenko, M.H. Bergman, Y.-H. Liu, J.P. Winickoff, and M. Weitzman. 2009. Family Composition and Children's Exposure to Adult Smokers in Their Homes. *Pediatrics* 123 (4):559-64.

Cotinine (continued)

14. Centers for Disease Control and Prevention. 2011. *Smoke-Free Policies Reduce Secondhand Smoke Exposure* CDC, National Center for Chronic Disease Prevention and Health Promotion, Office on Smoking and Health. Retrieved May 24, 2011 from http://www.cdc.gov/tobacco/data_statistics/fact_sheets/secondhand_smoke/protection/shs_exposure/index.htm.

15. Dove, M.S., D.W. Dockery, and G.N. Connolly. 2010. Smoke-free air laws and secondhand smoke exposure among nonsmoking youth. *Pediatrics* 126 (1):80-7.

16. Dove, M.S., D.W. Dockery, and G.N. Connolly. 2011. Smoke-free air laws and asthma prevalence, symptoms, and severity among nonsmoking youth. *Pediatrics* 127 (1):102-9.

17. Mackay, D., S. Haw, J.G. Ayres, C. Fischbacher, and J.P. Pell. 2010. Smoke-free legislation and hospitalizations for childhood asthma. *New England Journal of Medicine* 363 (12):1139-45.

18. Rayens, M.K., P.V. Burkhart, M. Zhang, S. Lee, D.K. Moser, D. Mannino, and E.J. Hahn. 2008. Reduction in asthma-related emergency department visits after implementation of a smoke-free law. *Journal of Allergy and Clinical Immunology* 122 (3):537-41 e3.

19. Centers for Disease Control and Prevention. 2009. *Fourth National Report on Human Exposure to Environmental Chemicals* Atlanta (GA): CDC. http://www.cdc.gov/exposurereport/.

20. Jarvis, M.J., H. Tunstall-Pedoe, C. Feyerabend, C. Vesey, and Y. Saloojee. 1987. Comparison of tests used to distinguish smokers from nonsmokers. *American Journal of Public Health* 77 (11):1435-8.

21. Watts, R.R., J.J. Langone, G.J. Knight, and J. Lewtas. 1990. Cotinine analytical workshop report: consideration of analytical methods for determining cotinine in human body fluids as a measure of passive exposure to tobacco smoke. *Environmental Health Perspectives* 84:173-82.

22. Benowitz, N.L. 1999. Biomarkers of environmental tobacco smoke exposure. *Environmental Health Perspectives* 107 Suppl 2:349-55.

23. Benowitz, N.L. 1996. Cotinine as a biomarker of environmental tobacco smoke exposure. *Epidemiologic Reviews* 18 (2):188-204.

24. Wagenknecht, L.E., G.R. Cutter, N.J. Haley, S. Sidney, T.A. Manolio, G.H. Hughes, and D.R. Jacobs. 1990. Racial differences in serum cotinine levels among smokers in the Coronary Artery Risk Development in (Young) Adults study. *American Journal of Public Health* 80 (9):1053-6.

25. Caraballo, R.S., G.A. Giovino, T.F. Pechacek, P.D. Mowery, P.A. Richter, W.J. Strauss, D.J. Sharp, M.P. Eriksen, J.L. Pirkle, and K.R. Maurer. 1998. Racial and ethnic differences in serum cotinine levels of cigarette smokers: Third National Health and Nutrition Examination Survey, 1988-1991. *Journal of the American Medical Association* 280 (2):135-9.

26. Perez-Stable, E.J., B. Herrera, P. Jacob, 3rd, and N.L. Benowitz. 1998. Nicotine metabolism and intake in black and white smokers. *Journal of the American Medical Association* 280 (2):152-6.

27. Benowitz, N.L., E.J. Perez-Stable, I. Fong, G. Modin, B. Herrera, and P. Jacob, 3rd. 1999. Ethnic differences in N-glucuronidation of nicotine and cotinine. *Journal of Pharmacology and Experimental Therapeutics* 291 (3):1196-203.

28. Benowitz, N.L., E.J. Perez-Stable, B. Herrera, and P. Jacob, 3rd. 2002. Slower metabolism and reduced intake of nicotine from cigarette smoking in Chinese-Americans. *Journal of the National Cancer Institute* 94 (2):108-15.

29. Pirkle, J.L., K.M. Flegal, J.T. Bernert, D.J. Brody, R.A. Etzel, and K.R. Maurer. 1996. Exposure of the US population to environmental tobacco smoke: the Third National Health and Nutrition Examination Survey, 1988 to 1991. *Journal of the American Medical Association* 275 (16):1233-40.

30. National Center for Health Statistics. *Vital Statistics Natality Birth Data*. Retrieved June 15, 2009 from http://www.cdc.gov/nchs/data_access/Vitalstatsonline.htm. .

31. Axelrad, D.A., and J. Cohen. 2011. Calculating summary statistics for population chemical biomonitoring in women of childbearing age with adjustment for age-specific natality. *Environmental Research* 111 (1):149-155.

Perfluorochemicals (PFCs)

1. Agency for Toxic Substances and Disease Registry (ATSDR). 2009. *Toxicological Profile for Perfluoroalkyls. (Draft for Public Comment)*. Atlanta, GA: U.S. Department of Health and Human Services, Public Health Service. http://www.atsdr.cdc.gov/toxprofiles/tp.asp?id=1117&tid=237.

2. Centers for Disease Control and Prevention. 2009. *Fourth National Report on Human Exposure to Environmental Chemicals*. Atlanta, GA: CDC. http://www.cdc.gov/exposurereport/.

3. Conder, J.M., R.A. Hoke, W. De Wolf, M.H. Russell, and R.C. Buck. 2008. Are PFCAs bioaccumulative? A critical review and comparison with regulatory criteria and persistent lipophilic compounds. *Environmental Science and Technology* 42 (4):995-1003.

4. Fromme, H., S.A. Tittlemier, W. Volkel, M. Wilhelm, and D. Twardella. 2009. Perfluorinated compounds--exposure assessment for the general population in Western countries. *International Journal of Hygiene and Environmental Health* 212 (3):239-70.

Perfluorochemicals (PFCs) (continued)

5. Kelly, B.C., M.G. Ikonomou, J.D. Blair, A.E. Morin, and F.A. Gobas. 2007. Food web-specific biomagnification of persistent organic pollutants. *Science* 317 (5835):236-9.

6. Kelly, B.C., M.G. Ikonomou, J.D. Blair, B. Surridge, D. Hoover, R. Grace, and F.A. Gobas. 2009. Perfluoroalkyl contaminants in an Arctic marine food web: trophic magnification and wildlife exposure. *Environmental Science and Technology* 43 (11):4037-43.

7. Lau, C., K. Anitole, C. Hodes, D. Lai, A. Pfahles-Hutchens, and J. Seed. 2007. Perfluoroalkyl acids: a review of monitoring and toxicological findings. *Toxicological Sciences* 99 (2):366-94.

8. Martin, J.W., S.A. Mabury, K.R. Solomon, and D.C. Muir. 2003. Bioconcentration and tissue distribution of perfluorinated acids in rainbow trout (Oncorhynchus mykiss). *Environmental Toxicology and Chemistry* 22 (1):196-204.

9. Bartell, S.M., A.M. Calafat, C. Lyu, K. Kato, P.B. Ryan, and K. Steenland. 2010. Rate of decline in serum PFOA concentrations after granular activated carbon filtration at two public water systems in Ohio and West Virginia. *Environmental Health Perspectives* 118 (2):222-8.

10. Brede, E., M. Wilhelm, T. Goen, J. Muller, K. Rauchfuss, M. Kraft, and J. Holzer. 2010. Two-year follow-up biomonitoring pilot study of residents' and controls' PFC plasma levels after PFOA reduction in public water system in Arnsberg, Germany. *International Journal of Hygiene and Environmental Health* 213 (3):217-23.

11. Harada, K., K. Inoue, A. Morikawa, T. Yoshinaga, N. Saito, and A. Koizumi. 2005. Renal clearance of perfluorooctane sulfonate and perfluorooctanoate in humans and their species-specific excretion. *Environmental Research* 99 (2):253-61.

12. Olsen, G.W., J.M. Burris, D.J. Ehresman, J.W. Froehlich, A.M. Seacat, J.L. Butenhoff, and L.R. Zobel. 2007. Half-life of serum elimination of perfluorooctanesulfonate, perfluorohexanesulfonate, and perfluorooctanoate in retired fluorochemical production workers. *Environmental Health Perspectives* 115 (9):1298-305.

13. Seals, R., S.M. Bartell, and K. Steenland. 2011. Accumulation and clearance of perfluorooctanoic acid (PFOA) in current and former residents of an exposed community. *Environmental Health Perspectives* 119 (1):119-24.

14. Kato, K., L.Y. Wong, L.T. Jia, Z. Kuklenyik, and A.M. Calafat. 2011. Trends in exposure to polyfluoroalkyl chemicals in the U.S. population: 1999-2008. *Environmental Science and Technology* 45 (19):8037-45.

15. U.S. Environmental Protection Agency. 2009. *Long-Chain Perfluorinated Chemicals (PFCs) Action Plan.* Washington, DC: U.S. EPA, Office of Pollution Prevention and Toxics. http://www.epa.gov/oppt/existingchemicals/pubs/pfcs_action_plan1230_09.pdf.

16. 3M. 2010. *What is 3M Doing?* Retrieved January 18, 2010 from http://solutions.3m.com/wps/portal/3M/en_US/PFOS/PFOA/Information/Action/.

17. U.S. Environmental Protection Agency. 2010. *News Release: EPA Announces Substantial Decrease of PFOA* Retrieved January 20, 2010 from http://yosemite.epa.gov/opa/admpress.nsf/68b5f2d54f3eefd28525701500517fbf/8f9dbdd044050f71852573e50064439f!OpenDocument.

18. Egeghy, P.P., and M. Lorber. 2011. An assessment of the exposure of Americans to perfluorooctane sulfonate: a comparison of estimated intake with values inferred from NHANES data. *Journal of Exposure Science and Environmental Epidemiology* 21 (2):150-68.

19. Trudel, D., L. Horowitz, M. Wormuth, M. Scheringer, I.T. Cousins, and K. Hungerbuhler. 2008. Estimating consumer exposure to PFOS and PFOA. *Risk Analysis* 28 (2):251-69.

20. Begley, T.H., K. White, P. Honigfort, M.L. Twaroski, R. Neches, and R.A. Walker. 2005. Perfluorochemicals: potential sources of and migration from food packaging. *Food Additives and Contaminants* 22 (10):1023-31.

21. Tittlemier, S.A., K. Pepper, C. Seymour, J. Moisey, R. Bronson, X.L. Cao, and R.W. Dabeka. 2007. Dietary exposure of Canadians to perfluorinated carboxylates and perfluorooctane sulfonate via consumption of meat, fish, fast foods, and food items prepared in their packaging. *Journal of Agricultural and Food Chemistry* 55 (8):3203-10.

22. Ericson, I., R. Marti-Cid, M. Nadal, B. Van Bavel, G. Lindstrom, and J.L. Domingo. 2008. Human exposure to perfluorinated chemicals through the diet: intake of perfluorinated compounds in foods from the Catalan (Spain) market. *Journal of Agricultural and Food Chemistry* 56 (5):1787-94.

23. Schecter, A., J. Colacino, D. Haffner, K. Patel, M. Opel, O. Papke, and L. Birnbaum. 2010. Perfluorinated Compounds, Polychlorinated Biphenyl, and Organochlorine Pesticide Contamination in Composite Food Samples from Dallas, Texas. *Environmental Health Perspectives* 118:796-802.

24. Young, W.M., P. South, T.H. Begley, G.W. Diachenko, and G.O. Noonan. 2012. Determination of perfluorochemicals in cow's milk using liquid chromatography-tandem mass spectrometry. *Journal of Agricultural and Food Chemistry* 60 (7):1652-8.

25. Konwick, B.J., G.T. Tomy, N. Ismail, J.T. Peterson, R.J. Fauver, D. Higginbotham, and A.T. Fisk. 2008. Concentrations and patterns of perfluoroalkyl acids in Georgia, USA surface waters near and distant to a major use source. *Environmental Toxicology and Chemistry* 27 (10):2011-8.

26. Moody, C.A., G.N. Hebert, S.H. Strauss, and J.A. Field. 2003. Occurrence and persistence of perfluorooctanesulfonate and other perfluorinated surfactants in groundwater at a fire-training area at Wurtsmith Air Force Base, Michigan, USA. *Journal of Environmental Monitoring* 5 (2):341-5.

Perfluorochemicals (PFCs) (continued)

27. Post, G.B., J.B. Louis, K.R. Cooper, B.J. Boros-Russo, and R.L. Lippincott. 2009. Occurrence and potential significance of perfluorooctanoic acid (PFOA) detected in New Jersey public drinking water systems. *Environmental Science and Technology* 43 (12):4547-54.

28. Shin, H.M., V.M. Vieira, P.B. Ryan, R. Detwiler, B. Sanders, K. Steenland, and S.M. Bartell. 2011. Environmental Fate and Transport Modeling for Perfluorooctanoic Acid Emitted from the Washington Works Facility in West Virginia. *Environmental Science and Technology* 45 (4):1435-42.

29. Sinclair, E., D.T. Mayack, K. Roblee, N. Yamashita, and K. Kannan. 2006. Occurrence of perfluoroalkyl surfactants in water, fish, and birds from New York State. *Archives of Environmental Contamination and Toxicology* 50 (3):398-410.

30. Skutlarek, D., M. Exner, and H. Farber. 2006. Perfluorinated surfactants in surface and drinking waters. *Environmental Science and Pollution Research International* 13 (5):299-307.

31. Steenland, K., C. Jin, J. MacNeil, C. Lally, A. Ducatman, V. Vieira, and T. Fletcher. 2009. Predictors of PFOA Levels in a Community Surrounding a Chemical Plant *Environmental Health Perspectives* 117 (7):1083-1088.

32. Karrman, A., I. Ericson, B. van Bavel, P.O. Darnerud, M. Aune, A. Glynn, S. Lignell, and G. Lindstrom. 2007. Exposure of perfluorinated chemicals through lactation: levels of matched human milk and serum and a temporal trend, 1996-2004, in Sweden. *Environmental Health Perspectives* 115 (2):226-30.

33. Llorca, M., M. Farre, Y. Pico, M.L. Teijon, J.G. Alvarez, and D. Barcelo. 2010. Infant exposure of perfluorinated compounds: levels in breast milk and commercial baby food. *Environment International* 36 (6):584-92.

34. Tao, L., K. Kannan, C.M. Wong, K.F. Arcaro, and J.L. Butenhoff. 2008. Perfluorinated compounds in human milk from Massachusetts, U.S.A. *Environmental Science and Technology* 42 (8):3096-101.

35. Thomsen, C., L.S. Haug, H. Stigum, M. Froshaug, S.L. Broadwell, and G. Becher. 2010. Changes in concentrations of perfluorinated compounds, polybrominated diphenyl ethers, and polychlorinated biphenyls in Norwegian breast-milk during twelve months of lactation. *Environmental Science and Technology* 44 (24):9550-6.

36. Volkel, W., O. Genzel-Boroviczeny, H. Demmelmair, C. Gebauer, B. Koletzko, D. Twardella, U. Raab, and H. Fromme. 2008. Perfluorooctane sulphonate (PFOS) and perfluorooctanoic acid (PFOA) in human breast milk: results of a pilot study. *International Journal of Hygiene and Environmental Health* 211 (3-4):440-6.

37. Bjorklund, J.A., K. Thuresson, and C.A. De Wit. 2009. Perfluoroalkyl compounds (PFCs) in indoor dust: concentrations, human exposure estimates, and sources. *Environmental Science and Technology* 43 (7):2276-81.

38. Strynar, M.J., and A.B. Lindstrom. 2008. Perfluorinated compounds in house dust from Ohio and North Carolina, USA. *Environmental Science and Technology* 42 (10):3751-6.

39. Kato, K., A.M. Calafat, and L.L. Needham. 2009. Polyfluoroalkyl chemicals in house dust. *Environmental Research* 109 (5):518-23.

40. Kubwabo, C., B. Stewart, J. Zhu, and L. Marro. 2005. Occurrence of perfluorosulfonates and other perfluorochemicals in dust from selected homes in the city of Ottawa, Canada. *Journal of Environmental Monitoring* 7 (11):1074-8.

41. Harrad, S., C.A. de Wit, M.A. Abdallah, C. Bergh, J.A. Bjorklund, A. Covaci, P.O. Darnerud, J. de Boer, M. Diamond, S. Huber, et al. 2010. Indoor contamination with hexabromocyclododecanes, polybrominated diphenyl ethers, and perfluoroalkyl compounds: an important exposure pathway for people? *Environmental Science and Technology* 44 (9):3221-31.

42. U.S. Environmental Protection Agency. 2008. *Child-Specific Exposure Factors Handbook (Final Report)*. Washington, DC. EPA/600/R-06/096F. http://cfpub.epa.gov/ncea/cfm/recordisplay.cfm?deid=199243#Download.

43. Kato, K., A.M. Calafat, L.Y. Wong, A.A. Wanigatunga, S.P. Caudill, and L.L. Needham. 2009. Polyfluoroalkyl compounds in pooled sera from children participating in the National Health and Nutrition Examination Survey 2001-2002. *Environmental Science and Technology* 43 (7):2641-7.

44. Toms, L.M., A.M. Calafat, K. Kato, J. Thompson, F. Harden, P. Hobson, A. Sjodin, and J.F. Mueller. 2009. Polyfluoroalkyl chemicals in pooled blood serum from infants, children, and adults in Australia. *Environmental Science and Technology* 43 (11):4194-9.

45. Woodruff, T.J., A.R. Zota, and J.M. Schwartz. 2011. Environmental Chemicals in Pregnant Women in the US: NHANES 2003-2004. *Environmental Health Perspectives* 119 (6):878-85.

46. Apelberg, B.J., L.R. Goldman, A.M. Calafat, J.B. Herbstman, Z. Kuklenyik, J. Heidler, L.L. Needham, R.U. Halden, and F.R. Witter. 2007. Determinants of fetal exposure to polyfluoroalkyl compounds in Baltimore, Maryland. *Environmental Science and Technology* 41 (11):3891-7.

47. Inoue, K., F. Okada, R. Ito, S. Kato, S. Sasaki, S. Nakajima, A. Uno, Y. Saijo, F. Sata, Y. Yoshimura, et al. 2004. Perfluorooctane sulfonate (PFOS) and related perfluorinated compounds in human maternal and cord blood samples: assessment of PFOS exposure in a susceptible population during pregnancy. *Environmental Health Perspectives* 112 (11):1204-7.

Perfluorochemicals (PFCs) (continued)

48. Calafat, A.M., Z. Kuklenyik, J.A. Reidy, S.P. Caudill, J.S. Tully, and L.L. Needham. 2007. Serum concentrations of 11 polyfluoroalkyl compounds in the U.S. population: data from the national health and nutrition examination survey (NHANES). *Environmental Science and Technology* 41 (7):2237-42.

49. Apelberg, B.J., F.R. Witter, J.B. Herbstman, A.M. Calafat, R.U. Halden, L.L. Needham, and L.R. Goldman. 2007. Cord serum concentrations of perfluorooctane sulfonate (PFOS) and perfluorooctanoate (PFOA) in relation to weight and size at birth. *Environmental Health Perspectives* 115 (11):1670-6.

50. Fei, C., J.K. McLaughlin, R.E. Tarone, and J. Olsen. 2007. Perfluorinated chemicals and fetal growth: a study within the Danish National Birth Cohort. *Environmental Health Perspectives* 115 (11):1677-82.

51. Fei, C., J.K. McLaughlin, R.E. Tarone, and J. Olsen. 2008. Fetal growth indicators and perfluorinated chemicals: a study in the Danish National Birth Cohort. *American Journal of Epidemiology* 168 (1):66-72.

52. Washino, N., Y. Saijo, S. Sasaki, S. Kato, S. Ban, K. Konishi, R. Ito, A. Nakata, Y. Iwasaki, K. Saito, et al. 2009. Correlations between prenatal exposure to perfluorinated chemicals and reduced fetal growth. *Environmental Health Perspectives* 117 (4):660-7.

53. Hamm, M.P., N.M. Cherry, E. Chan, J.W. Martin, and I. Burstyn. 2010. Maternal exposure to perfluorinated acids and fetal growth. *Journal of Exposure Science and Environmental Epidemiology* 20 (7):589-97.

54. Monroy, R., K. Morrison, K. Teo, S. Atkinson, C. Kubwabo, B. Stewart, and W.G. Foster. 2008. Serum levels of perfluoroalkyl compounds in human maternal and umbilical cord blood samples. *Environmental Research* 108 (1):56-62.

55. Butenhoff, J.L., G.L. Kennedy, Jr., S.R. Frame, J.C. O'Connor, and R.G. York. 2004. The reproductive toxicology of ammonium perfluorooctanoate (APFO) in the rat. *Toxicology* 196 (1-2):95-116.

56. Era, S., K.H. Harada, M. Toyoshima, K. Inoue, M. Minata, N. Saito, T. Takigawa, K. Shiota, and A. Koizumi. 2009. Cleft palate caused by perfluorooctane sulfonate is caused mainly by extrinsic factors. *Toxicology* 256 (1-2):42-7.

57. Fuentes, S., M.T. Colomina, J. Rodriguez, P. Vicens, and J.L. Domingo. 2006. Interactions in developmental toxicology: concurrent exposure to perfluorooctane sulfonate (PFOS) and stress in pregnant mice. *Toxicology Letters* 164 (1):81-9.

58. Grasty, R.C., D.C. Wolf, B.E. Grey, C.S. Lau, and J.M. Rogers. 2003. Prenatal window of susceptibility to perfluorooctane sulfonate-induced neonatal mortality in the Sprague-Dawley rat. *Birth Defects Research Part B: Developmental and Reproductive Toxicology* 68 (6):465-71.

59. Hines, E.P., S.S. White, J.P. Stanko, E.A. Gibbs-Flournoy, C. Lau, and S.E. Fenton. 2009. Phenotypic dichotomy following developmental exposure to perfluorooctanoic acid (PFOA) in female CD-1 mice: Low doses induce elevated serum leptin and insulin, and overweight in mid-life. *Molecular and Cellular Endocrinology* 304 (1-2):97-105.

60. Lau, C., J.L. Butenhoff, and J.M. Rogers. 2004. The developmental toxicity of perfluoroalkyl acids and their derivatives. *Toxicology and Applied Pharmacology* 198 (2):231-41.

61. Lau, C., J.R. Thibodeaux, R.G. Hanson, M.G. Narotsky, J.M. Rogers, A.B. Lindstrom, and M.J. Strynar. 2006. Effects of perfluorooctanoic acid exposure during pregnancy in the mouse. *Toxicological Sciences* 90 (2):510-8.

62. Lau, C., J.R. Thibodeaux, R.G. Hanson, J.M. Rogers, B.E. Grey, M.E. Stanton, J.L. Butenhoff, and L.A. Stevenson. 2003. Exposure to perfluorooctane sulfonate during pregnancy in rat and mouse. II: postnatal evaluation. *Toxicological Sciences* 74 (2):382-92.

63. Luebker, D.J., M.T. Case, R.G. York, J.A. Moore, K.J. Hansen, and J.L. Butenhoff. 2005. Two-generation reproduction and cross-foster studies of perfluorooctanesulfonate (PFOS) in rats. *Toxicology* 215 (1-2):126-48.

64. Luebker, D.J., R.G. York, K.J. Hansen, J.A. Moore, and J.L. Butenhoff. 2005. Neonatal mortality from in utero exposure to perfluorooctanesulfonate (PFOS) in Sprague-Dawley rats: dose-response, and biochemical and pharamacokinetic parameters. *Toxicology* 215 (1-2):149-69.

65. Thibodeaux, J.R., R.G. Hanson, J.M. Rogers, B.E. Grey, B.D. Barbee, J.H. Richards, J.L. Butenhoff, L.A. Stevenson, and C. Lau. 2003. Exposure to perfluorooctane sulfonate during pregnancy in rat and mouse. I: maternal and prenatal evaluations. *Toxicological Sciences* 74 (2):369-81.

66. Olsen, G.W., and L.R. Zobel. 2007. Assessment of lipid, hepatic, and thyroid parameters with serum perfluorooctanoate (PFOA) concentrations in fluorochemical production workers. *International Archives of Occupational and Environmental Health* 81 (2):231-46.

67. Dallaire, R., E. Dewailly, D. Pereg, S. Dery, and P. Ayotte. 2009. Thyroid function and plasma concentrations of polyhalogenated compounds in Inuit adults. *Environmental Health Perspectives* 117 (9):1380-6.

68. Melzer, D., N. Rice, M.H. Depledge, W.E. Henley, and T.S. Galloway. 2010. Association Between Serum Perfluoroctanoic Acid (PFOA) and Thyroid Disease in the NHANES Study. *Environmental Health Perspectives* 118 (686-692).

69. Chan, E., I. Burstyn, N. Cherry, F. Bamforth, and J.W. Martin. 2011. Perfluorinated acids and hypothyroxinemia in pregnant women. *Environmental Research* 111 (4):559-64.

Perfluorochemicals (PFCs) (continued)

70. Chang, S.C., J.R. Thibodeaux, M.L. Eastvold, D.J. Ehresman, J.A. Bjork, J.W. Froehlich, C. Lau, R.J. Singh, K.B. Wallace, and J.L. Butenhoff. 2008. Thyroid hormone status and pituitary function in adult rats given oral doses of perfluorooctanesulfonate (PFOS). *Toxicology* 243 (3):330-9.

71. Martin, M.T., R.J. Brennan, W. Hu, E. Ayanoglu, C. Lau, H. Ren, C.R. Wood, J.C. Corton, R.J. Kavlock, and D.J. Dix. 2007. Toxicogenomic study of triazole fungicides and perfluoroalkyl acids in rat livers predicts toxicity and categorizes chemicals based on mechanisms of toxicity. *Toxicological Sciences* 97 (2):595-613.

72. Seacat, A.M., P.J. Thomford, K.J. Hansen, L.A. Clemen, S.R. Eldridge, C.R. Elcombe, and J.L. Butenhoff. 2003. Sub-chronic dietary toxicity of potassium perfluorooctanesulfonate in rats. *Toxicology* 183 (1-3):117-31.

73. Seacat, A.M., P.J. Thomford, K.J. Hansen, G.W. Olsen, M.T. Case, and J.L. Butenhoff. 2002. Subchronic toxicity studies on perfluorooctanesulfonate potassium salt in cynomolgus monkeys. *Toxicological Sciences* 68 (1):249-64.

74. Yu, W.G., W. Liu, and Y.H. Jin. 2009. Effects of perfluorooctane sulfonate on rat thyroid hormone biosynthesis and metabolism. *Environmental Toxicology and Chemistry* 28 (5):990-6.

75. Morreale de Escobar, G., M.J. Obregon, and F. Escobar del Rey. 2000. Is neuropsychological development related to maternal hypothyroidism or to maternal hypothyroxinemia? *The Journal of Clinical Endocrinology and Metabolism* 85 (11):3975-87.

76. Vamecq, J., and N. Latruffe. 1999. Medical significance of peroxisome proliferator-activated receptors. *Lancet* 354 (9173):141-8.

77. Nelson, J.W., E.E. Hatch, and T.F. Webster. 2010. Exposure to polyfluoroalkyl chemicals and cholesterol, body weight, and insulin resistance in the general U.S. population. *Environmental Health Perspectives* 118:197-202.

78. Costa, G., S. Sartori, and D. Consonni. 2009. Thirty years of medical surveillance in perfluooctanoic acid production workers. *Journal of Occupational and Environmental Medicine* 51 (3):364-72.

79. Gilliland, F.D., and J.S. Mandel. 1996. Serum perfluorooctanoic acid and hepatic enzymes, lipoproteins, and cholesterol: a study of occupationally exposed men. *American Journal of Industrial Medicine* 29 (5):560-8.

80. Haughom, B., and O. Spydevold. 1992. The mechanism underlying the hypolipemic effect of perfluorooctanoic acid (PFOA), perfluorooctane sulphonic acid (PFOSA) and clofibric acid. *Biochimica et Biophysica Acta* 1128 (1):65-72.

81. Lin, C.Y., P.C. Chen, Y.C. Lin, and L.Y. Lin. 2009. Association among serum perfluoroalkyl chemicals, glucose homeostasis, and metabolic syndrome in adolescents and adults. *Diabetes Care* 32 (4):702-7.

82. Olsen, G.W., J.M. Burris, M.M. Burlew, and J.H. Mandel. 2003. Epidemiologic assessment of worker serum perfluorooctanesulfonate (PFOS) and perfluorooctanoate (PFOA) concentrations and medical surveillance examinations. *Journal of Occupational and Environmental Medicine* 45 (3):260-70.

83. Olsen, G.W., J.M. Burris, J.H. Mandel, and L.R. Zobel. 1999. Serum perfluorooctane sulfonate and hepatic and lipid clinical chemistry tests in fluorochemical production employees. *Journal of Occupational and Environmental Medicine* 41 (9):799-806.

84. Sakr, C.J., K.H. Kreckmann, J.W. Green, P.J. Gillies, J.L. Reynolds, and R.C. Leonard. 2007. Cross-sectional study of lipids and liver enzymes related to a serum biomarker of exposure (ammonium perfluorooctanoate or APFO) as part of a general health survey in a cohort of occupationally exposed workers. *Journal of Occupational and Environmental Medicine* 49 (10):1086-96.

85. Woollett, L.A. 2001. The origins and roles of cholesterol and fatty acids in the fetus. *Current Opinion in Lipidology* 12 (3):305-12.

86. Keil, D.E., T. Mehlmann, L. Butterworth, and M.M. Peden-Adams. 2008. Gestational exposure to perfluorooctane sulfonate suppresses immune function in B6C3F1 mice. *Toxicological Sciences* 103 (1):77-85.

87. Fang, X., L. Zhang, Y. Feng, Y. Zhao, and J. Dai. 2008. Immunotoxic effects of perfluorononanoic acid on BALB/c mice. *Toxicological Sciences* 105 (2):312-21.

88. Peden-Adams, M.M., J.M. Keller, J.G. Eudaly, J. Berger, G.S. Gilkeson, and D.E. Keil. 2008. Suppression of humoral immunity in mice following exposure to perfluorooctane sulfonate. *Toxicological Sciences* 104 (1):144-54.

89. Ehresman, D.J., J.W. Froehlich, G.W. Olsen, S.C. Chang, and J.L. Butenhoff. 2007. Comparison of human whole blood, plasma, and serum matrices for the determination of perfluorooctanesulfonate (PFOS), perfluorooctanoate (PFOA), and other fluorochemicals. *Environ Res* 103 (2):176-84.

90. National Center for Health Statistics. *Vital Statistics Natality Birth Data, 2003-2004*. Retrieved June 15, 2009 from http://www.cdc.gov/nchs/data_access/Vitalstatsonline.htm. .

91. Axelrad, D.A., and J. Cohen. 2011. Calculating summary statistics for population chemical biomonitoring in women of childbearing age with adjustment for age-specific natality. *Environmental Research* 111 (1):149-155.

Polychlorinated Biphenyls (PCBs)

1. U.S. Environmental Protection Agency. 2010. *Basic Information: Polychlorinated Biphenyl (PCB)* U.S. EPA. Retrieved November 1, 2010 from http://www.epa.gov/epawaste/hazard/tsd/pcbs/pubs/about.htm.

2. Agency for Toxic Substances and Disease Registry (ATSDR). 2000. *Toxicological Profile for Polychlorinated Biphenyls (PCBs)*. Atlanta, GA: U.S. Department of Health and Human Services, Public Health Service. http://www.atsdr.cdc.gov/toxprofiles/tp.asp?id=142&tid=26.

3. U.S. Environmental Protection Agency. 2011. *Polychlorinated Biphenyls (PCBs): Aroclor and Other PCB Mixtures*. Retrieved August 15, 2011 from http://www.epa.gov/epawaste/hazard/tsd/pcbs/pubs/aroclor.htm.

4. Simon, T., J.K. Britt, and R.C. James. 2007. Development of a neurotoxic equivalence scheme of relative potency for assessing the risk of PCB mixtures. *Regulatory Toxicology and Pharmacology* 48 (2):148-70.

5. Van den Berg, M., L.S. Birnbaum, M. Denison, M. De Vito, W. Farland, M. Feeley, H. Fiedler, H. Hakansson, A. Hanberg, L. Haws, et al. 2006. The 2005 World Health Organization reevaluation of human and mammalian toxic equivalency factors for dioxins and dioxin-like compounds. *Toxicological Sciences* 93 (2):223-41.

6. U.S. Environmental Protection Agency. 1979. *EPA Bans PCB Manufacture; Phases Out Uses.* Retrieved March 1, 2010 from http://www.epa.gov/history/topics/pcbs/01.htm.

7. Schecter, A., O. Papke, K.C. Tung, J. Joseph, T.R. Harris, and J. Dahlgren. 2005. Polybrominated diphenyl ether flame retardants in the U.S. population: current levels, temporal trends, and comparison with dioxins, dibenzofurans, and polychlorinated biphenyls. *Journal of Occupational and Environmental Medicine* 47 (3):199-211.

8. Sjodin, A., R.S. Jones, J.F. Focant, C. Lapeza, R.Y. Wang, E.E. McGahee, 3rd, Y. Zhang, W.E. Turner, B. Slazyk, L.L. Needham, et al. 2004. Retrospective time-trend study of polybrominated diphenyl ether and polybrominated and polychlorinated biphenyl levels in human serum from the United States. *Environmental Health Perspectives* 112 (6):654-8.

9. Hickey, J.P., S.A. Batterman, and S.M. Chernyak. 2006. Trends of chlorinated organic contaminants in great lakes trout and walleye from 1970 to 1998. *Archives of Environmental Contamination and Toxicology* 50 (1):97-110.

10. Sun, P., I. Basu, P. Blanchard, K.A. Brice, and R.A. Hites. 2007. Temporal and spatial trends of atmospheric polychlorinated biphenyl concentrations near the Great Lakes. *Environmental Science and Technology* 41 (4):1131-6.

11. Chen, Y.C., Y.L. Guo, C.C. Hsu, and W.J. Rogan. 1992. Cognitive development of Yu-Cheng ("oil disease") children prenatally exposed to heat-degraded PCBs. *Journal of the American Medical Association* 268 (22):3213-8.

12. Chen, Y.J., and C.C. Hsu. 1994. Effects of prenatal exposure to PCBs on the neurological function of children: a neuropsychological and neurophysiological study. *Developmental Medicine & Child Neurology* 36 (4):312-20.

13. Lai, T.J., X. Liu, Y.L. Guo, N.W. Guo, M.L. Yu, C.C. Hsu, and W.J. Rogan. 2002. A cohort study of behavioral problems and intelligence in children with high prenatal polychlorinated biphenyl exposure. *Archives of General Psychiatry* 59 (11):1061-6.

14. Masuda, Y. 2009. Toxic effects of PCB/PCDF to humans observed in Yusho and other poisonings. *Fukuoka Igaku Zasshi* 100 (5):141-55.

15. Rogan, W.J., B.C. Gladen, K.L. Hung, S.L. Koong, L.Y. Shih, J.S. Taylor, Y.C. Wu, D. Yang, N.B. Ragan, and C.C. Hsu. 1988. Congenital poisoning by polychlorinated biphenyls and their contaminants in Taiwan. *Science* 241 (4863):334-6.

16. Wigle, D.T., T.E. Arbuckle, M.C. Turner, A. Berube, Q. Yang, S. Liu, and D. Krewski. 2008. Epidemiologic evidence of relationships between reproductive and child health outcomes and environmental chemical contaminants. *Journal of Toxicology and Environmental Health B Critical Reviews* 11 (5-6):373-517.

17. Jacobson, J.L., and S.W. Jacobson. 2003. Prenatal exposure to polychlorinated biphenyls and attention at school age. *Journal of Pediatrics* 143 (6):780-8.

18. Sagiv, S.K., J.K. Nugent, T.B. Brazelton, A.L. Choi, P.E. Tolbert, L.M. Altshul, and S.A. Korrick. 2008. Prenatal Organochlorine Exposure and Measures of Behavior in Infancy Using the Neonatal Behavioral Assessment Scale (NBAS). *Environmental Health Perspectives* 116 (5):666–673.

19. Stewart, P., S. Fitzgerald, J. Reihman, B. Gump, E. Lonky, T. Darvill, J. Pagano, and P. Hauser. 2003. Prenatal PCB exposure, the corpus callosum, and response inhibition. *Environmental Health Perspectives* 111 (13):1670-7.

20. Stewart, P., E. Lonky, J. Reihman, J. Pagano, B. Gump, and T. Darvill. 2008. The relationship between prenatal PCB exposure and intelligence (IQ) in 9-year-old children. *Environmental Health Perspectives* 116 (10):1416-1422.

21. Stewart, P.W., D.M. Sargent, J. Reihman, B.B. Gump, E. Lonky, T. Darvill, H. Hicks, and J. Pagano. 2006. Response inhibition during Differential Reinforcement of Low Rates (DRL) schedules may be sensitive to low-level polychlorinated biphenyl, methylmercury, and lead exposure in children. *Environmental Health Perspectives* 114 (12):1923-9.

22. Vreugdenhil, H.J., P.G. Mulder, H.H. Emmen, and N. Weisglas-Kuperus. 2004. Effects of perinatal exposure to PCBs on neuropsychological functions in the Rotterdam cohort at 9 years of age. *Neuropsychology* 18 (1):185-93.

Polychlorinated Biphenyls (PCBs) (continued)

23. Sagiv, S.K., S.W. Thurston, D.C. Bellinger, P.E. Tolbert, L.M. Altshul, and S.A. Korrick. 2010. Prenatal Organochlorine Exposure and Behaviors Associated With Attention Deficit Hyperactivity Disorder in School-Aged Children. *American Journal of Epidemiology* 171 (5):593-601.

24. Darvill, T., E. Lonky, J. Reihman, P. Stewart, and J. Pagano. 2000. Prenatal exposure to PCBs and infant performance on the fagan test of infant intelligence. *Neurotoxicology* 21 (6):1029-38.

25. Jacobson, J.L., and S.W. Jacobson. 1996. Intellectual impairment in children exposed to polychlorinated biphenyls in utero. *New England Journal of Medicine* 335 (11):783-9.

26. Jorissen, J. 2007. Literature review. Outcomes associated with postnatal exposure to polychlorinated biphenyls (PCBs) via breast milk. *Advances in Neonatal Care* 7 (5):230-7.

27. Patandin, S., C.I. Lanting, P.G. Mulder, E.R. Boersma, P.J. Sauer, and N. Weisglas-Kuperus. 1999. Effects of environmental exposure to polychlorinated biphenyls and dioxins on cognitive abilities in Dutch children at 42 months of age. *Journal of Pediatrics* 134 (1):33-41.

28. Stewart, P., J. Reihman, E. Lonky, T. Darvill, and J. Pagano. 2000. Prenatal PCB exposure and neonatal behavioral assessment scale (NBAS) performance. *Neurotoxicology and Teratology* 22 (1):21-9.

29. Walkowiak, J., J.A. Wiener, A. Fastabend, B. Heinzow, U. Kramer, E. Schmidt, H.J. Steingruber, S. Wundram, and G. Winneke. 2001. Environmental exposure to polychlorinated biphenyls and quality of the home environment: effects on psychodevelopment in early childhood. *Lancet* 358 (9293):1602-7.

30. Schantz, S.L., J.J. Widholm, and D.C. Rice. 2003. Effects of PCB exposure on neuropsychological function in children. *Environmental Health Perspectives* 111 (3):357-376.

31. Jacobson, J.L., S.W. Jacobson, and H.E. Humphrey. 1990. Effects of exposure to PCBs and related compounds on growth and activity in children. *Neurotoxicology and Teratology* 12 (4):319-26.

32. Ribas-Fito, N., M. Sala, M. Kogevinas, and J. Sunyer. 2001. Polychlorinated biphenyls (PCBs) and neurological development in children: a systematic review. *Journal of Epidemiology and Community Health* 55 (8):537-46.

33. Schantz, S.L., J.C. Gardiner, D.M. Gasior, R.J. McCaffrey, A.M. Sweeney, and H.E.B. Humphrey. 2004. Much Ado About Something: The Weight of Evidence for PCB Effects on Neuropsychological Function. *Psychology in the Schools* 41 (6):669-679.

34. Boucher, O., G. Muckle, and C.H. Bastien. 2009. Prenatal exposure to polychlorinated biphenyls: a neuropsychologic analysis. *Environmental Health Perspectives* 117 (1):7-16.

35. Institute of Medicine. 2003. *Dioxins and Dioxin-like Compounds in the Food Supply*. Washington, DC: National Academy Press. http://books.nap.edu/openbook.php?record_id=10763&page=R1.

36. Rice, D.C. 2000. Parallels between attention deficit hyperactivity disorder and behavioral deficits produced by neurotoxic exposure in monkeys. *Environmental Health Perspectives* 108 Suppl 3:405-8.

37. Heilmann, C., P. Grandjean, P. Weihe, F. Nielsen, and E. Budtz-Jorgensen. 2006. Reduced antibody responses to vaccinations in children exposed to polychlorinated biphenyls. *PLoS Medicine* 3 (8):e311.

38. Dallaire, F., E. Dewailly, G. Muckle, C. Vezina, S.W. Jacobson, J.L. Jacobson, and P. Ayotte. 2004. Acute infections and environmental exposure to organochlorines in Inuit infants from Nunavik. *Environmental Health Perspectives* 112 (14):1359-65.

39. Dallaire, F., E. Dewailly, C. Vezina, G. Muckle, J.P. Weber, S. Bruneau, and P. Ayotte. 2006. Effect of prenatal exposure to polychlorinated biphenyls on incidence of acute respiratory infections in preschool Inuit children. *Environmental Health Perspectives* 114 (8):1301-5.

40. Weisglas-Kuperus, N., S. Patandin, G.A. Berbers, T.C. Sas, P.G. Mulder, P.J. Sauer, and H. Hooijkaas. 2000. Immunologic effects of background exposure to polychlorinated biphenyls and dioxins in Dutch preschool children. *Environmental Health Perspectives* 108 (12):1203-7.

41. Weisglas-Kuperus, N., H.J. Vreugdenhil, and P.G. Mulder. 2004. Immunological effects of environmental exposure to polychlorinated biphenyls and dioxins in Dutch school children. *Toxicology Letters* 149 (1-3):281-5.

42. Park, H., I. Hertz-Picciotto, J. Petrik, L. Palkovicova, A. Kocan, and T. Trnovec. 2008. Prenatal PCB exposure and thymus size at birth in neonates in eastern Slovakia. *Environmental Health Perspectives* 116:104-109.

43. Selgrade, M.K. 2007. Immunotoxicity: the risk is real. *Toxicological Sciences* 100 (2):328-32.

44. Buck Louis, G.M., L.E. Gray, Jr., M. Marcus, S.R. Ojeda, O.H. Pescovitz, S.F. Witchel, W. Sippell, D.H. Abbott, A. Soto, R.W. Tyl, et al. 2008. Environmental factors and puberty timing: expert panel research needs. *Pediatrics* 121 Suppl 3:S192-207.

45. National Toxicology Program. 2011. *Report on Carcinogens, 12th Edition*. Research Triangle Park, NC: U.S. Department of Health and Human Services, National Toxicology Program. http://ntp.niehs.nih.gov/ntp/roc/twelfth/roc12.pdf.

46. Centers for Disease Control and Prevention. 2009. *Fourth National Report on Human Exposure to Environmental Chemicals*. Atlanta, GA: CDC. http://www.cdc.gov/exposurereport/.

Polychlorinated Biphenyls (PCBs) (continued)

47. Windham, G.C., S.M. Pinney, A. Sjodin, R. Lum, R.S. Jones, L.L. Needham, F.M. Biro, R.A. Hiatt, and L.H. Kushi. 2010. Body burdens of brominated flame retardants and other persistent organo-halogenated compounds and their descriptors in US girls. *Environmental Research* 110 (3):251-7.

48. Schecter, A., J. Colacino, D. Haffner, K. Patel, M. Opel, O. Papke, and L. Birnbaum. 2010. Perfluorinated compounds, polychlorinated biphenyls, and organochlorine pesticide contamination in composite food samples from Dallas, Texas, USA. *Environmental Health Perspectives* 118 (6):796-802.

49. Hooper, K., J. She, M. Sharp, J. Chow, N. Jewell, R. Gephart, and A. Holden. 2007. Depuration of polybrominated diphenyl ethers (PBDEs) and polychlorinated biphenyls (PCBs) in breast milk from California first-time mothers (primiparae). *Environmental Health Perspectives* 115 (9):1271-5.

50. Herbstman, J.B., A. Sjodin, B.J. Apelberg, F.R. Witter, D.G. Patterson, R.U. Halden, R.S. Jones, A. Park, Y. Zhang, J. Heidler, et al. 2007. Determinants of prenatal exposure to polychlorinated biphenyls (PCBs) and polybrominated diphenyl ethers (PBDEs) in an urban population. *Environmental Health Perspectives* 115 (12):1794-800.

51. Harrad, S., C. Ibarra, M. Robson, L. Melymuk, X. Zhang, M. Diamond, and J. Douwes. 2009. Polychlorinated biphenyls in domestic dust from Canada, New Zealand, United Kingdom and United States: implications for human exposure. *Chemosphere* 76 (2):232-8.

52. Rudel, R.A., D.E. Camann, J.D. Spengler, L.R. Korn, and J.G. Brody. 2003. Phthalates, alkylphenols, pesticides, polybrominated diphenyl ethers, and other endocrine-disrupting compounds in indoor air and dust. *Environmental Science and Technology* 37 (20):4543-53.

53. Rudel, R.A., L.M. Seryak, and J.G. Brody. 2008. PCB-containing wood floor finish is a likely source of elevated PCBs in residents' blood, household air and dust: a case study of exposure. *Environmental Health* 7:2.

54. Ward, M.H., J.S. Colt, C. Metayer, R.B. Gunier, J. Lubin, V. Crouse, M.G. Nishioka, P. Reynolds, and P.A. Buffler. 2009. Residential exposure to polychlorinated biphenyls and organochlorine pesticides and risk of childhood leukemia. *Environmental Health Perspectives* 117 (6):1007-13.

55. Herrick, R.F., M.D. McClean, J.D. Meeker, L.K. Baxter, and G.A. Weymouth. 2004. An unrecognized source of PCB contamination in schools and other buildings. *Environmental Health Perspectives* 112 (10):1051-3.

56. U.S. Environmental Protection Agency. 2011. *Healthy School Environmental Resources: PCBs*. Retrieved August 15, 2011 from http://cfpub.epa.gov/schools/top_sub.cfm?t_id=41&s_id=32.

57. U.S. Environmental Protection Agency. 2011. *Polychlorinated Biphenyls (PCBs): Proper Maintenance, Removal, and Disposal of PCB-containing Fluorescent Light Ballasts - A Guide for School Administrators and Maintenance Personnel*. Retrieved August 15, 2011 from http://www.epa.gov/osw/hazard/tsd/pcbs/pubs/ballasts.htm.

58. Hu, D., and K.C. Hornbuckle. 2009. Inadvertent Polychlorinated Biphenyls in Commercial Paint Pigments. *Environmental Science and Technology* 44 (8):2822-2827.

59. Rodenburg, L.A., J. Guo, S. Du, and G.J. Cavallo. 2009. Evidence for Unique and Ubiquitous Environmental Sources of 3,3'-Dichlorobiphenyl (PCB 11). *Environmental Science and Technology* 44 (8):2816-2821.

60. Axelrad, D.A., S. Goodman, and T.J. Woodruff. 2009. PCB body burdens in US women of childbearing age 2001-2002: An evaluation of alternate summary metrics of NHANES data. *Environmental Research* 109 (4):368-78.

61. Patterson, D.G., Jr., L.Y. Wong, W.E. Turner, S.P. Caudill, E.S. Dipietro, P.C. McClure, T.P. Cash, J.D. Osterloh, J.L. Pirkle, E.J. Sampson, et al. 2009. Levels in the U.S. population of those persistent organic pollutants (2003-2004) included in the Stockholm Convention or in other long range transboundary air pollution agreements. *Environmental Science and Technology* 43 (4):1211-8.

62. Bogdal, C., P. Schmid, M. Zennegg, F.S. Anselmetti, M. Scheringer, and K. Hungerbuhler. 2009. Blast from the past: melting glaciers as a relevant source for persistent organic pollutants. *Environmental Science and Technology* 43 (21):8173-7.

63. Carrie, J., F. Wang, H. Sanei, R.W. Macdonald, P.M. Outridge, and G.A. Stern. 2010. Increasing Contaminant Burdens in an Arctic Fish, Burbot (Lota lota), in a Warming Climate. *Environmental Science and Technology* 44 (1):316-322.

64. U.S. Environmental Protection Agency. 2010. *EPA Technical Requirements for Phase 2 of Hudson River Dredging Project: Factsheet*. New York, NY. http://www.epa.gov/hudson/Hudson_Phase_2_Fact_Sheet.pdf.

65. U.S. Environmental Protection Agency. 2011. *Hudson River PCBs: Project background*. Retrieved August 15, 2011 from http://www.epa.gov/hudson/.

66. National Center for Health Statistics. *Vital Statistics Natality Birth Data*. Retrieved June 15, 2009 from http://www.cdc.gov/nchs/data_access/Vitalstatsonline.htm. .

67. Axelrad, D.A., and J. Cohen. 2011. Calculating summary statistics for population chemical biomonitoring in women of childbearing age with adjustment for age-specific natality. *Environmental Research* 111 (1):149-155.

Polybrominated Diphenyl Ethers (PBDEs)

1. U.S. Environmental Protection Agency. 2010. *An Exposure Assessment of Polybrominated Diphenyl Ethers*. Washington, DC: U.S. EPA, National Center for Environmental Assessment. EPA/600/R-08/086F. http://cfpub.epa.gov/ncea/cfm/recordisplay.cfm?deid=210404.

2. U.S. Environmental Protection Agency. 2010. *DecaBDE Phase-out Initiative*. U.S. Environmental Protection Agency. Retrieved February 26, 2010 from http://www.epa.gov/oppt/existingchemicals/pubs/actionplans/deccadbe.html.

3. U.S. Environmental Protection Agency. *Environmental Profiles of Chemical Flame-Retardant Alternatives for Low-Density Polyurethane Foam*. U.S. EPA, Design for the Environment. Retrieved November 3, 2010 from http://www.epa.gov/opptintr/dfe/pubs/flameret/ffr-alt.htm.

4. U.S. Environmental Protection Agency. *Polybrominated Diphenyl Ethers (PBDEs) Action Plan Summary*. U.S. EPA, Office of Pollution Prevention and Toxics. Retrieved November 3, 2010 from http://www.epa.gov/oppt/existingchemicals/pubs/actionplans/pbde.html.

5. Fraser, A.J., T.F. Webster, and M.D. McClean. 2009. Diet contributes significantly to the body burden of PBDEs in the general U.S. population. *Environmental Health Perspectives* 117 (10):1520-5.

6. Johnson-Restrepo, B., and K. Kannan. 2009. An assessment of sources and pathways of human exposure to polybrominated diphenyl ethers in the United States. *Chemosphere* 76 (4):542-8.

7. Lorber, M. 2008. Exposure of Americans to polybrominated diphenyl ethers. *Journal of Exposure Science and Environmental Epidemiology* 18 (1):2-19.

8. Schecter, A., O. Papke, T.R. Harris, K.C. Tung, A. Musumba, J. Olson, and L. Birnbaum. 2006. Polybrominated diphenyl ether (PBDE) levels in an expanded market basket survey of U.S. food and estimated PBDE dietary intake by age and sex. *Environmental Health Perspectives* 114 (10):1515-20.

9. Stapleton, H.M., S.M. Kelly, J.G. Allen, M.D. McClean, and T.F. Webster. 2008. Measurement of polybrominated diphenyl ethers on hand wipes: estimating exposure from hand-to-mouth contact. *Environmental Science and Technology* 42 (9):3329-34.

10. Wei, H., M. Turyk, S. Cali, S. Dorevitch, S. Erdal, and A. Li. 2009. Particle size fractionation and human exposure of polybrominated diphenyl ethers in indoor dust from Chicago. *Journal of Environmental Science and Health, Part A: Toxic/Hazardous Substances and Environmental Engineering* 44 (13):1353-61.

11. Wu, N., T. Herrmann, O. Paepke, J. Tickner, R. Hale, L.E. Harvey, M. La Guardia, M.D. McClean, and T.F. Webster. 2007. Human exposure to PBDEs: associations of PBDE body burdens with food consumption and house dust concentrations. *Environmental Science and Technology* 41 (5):1584-9.

12. Frederiksen, M., K. Vorkamp, M. Thomsen, and L.E. Knudsen. 2009. Human internal and external exposure to PBDEs--a review of levels and sources. *International Journal of Hygiene and Environmental Health* 212 (2):109-34.

13. Rose, M., D.H. Bennett, A. Bergman, B. Fangstrom, I.N. Pessah, and I. Hertz-Picciotto. 2010. PBDEs in 2-5 year-old children from California and associations with diet and indoor environment. *Environmental Science and Technology* 44 (7):2648-53.

14. Sjodin, A., O. Papke, E. McGahee, J.F. Focant, R.S. Jones, T. Pless-Mulloli, L.M. Toms, T. Herrmann, J. Muller, L.L. Needham, et al. 2008. Concentration of polybrominated diphenyl ethers (PBDEs) in household dust from various countries. *Chemosphere* 73 (1 Suppl):S131-6.

15. Zota, A.R., R.A. Rudel, R.A. Morello-Frosch, and J.G. Brody. 2008. Elevated house dust and serum concentrations of PBDEs in California: unintended consequences of furniture flammability standards? *Environmental Science and Technology* 42 (21):8158-64.

16. Imm, P., L. Knobeloch, C. Buelow, and H.A. Anderson. 2009. Household exposures to polybrominated diphenyl ethers (PBDEs) in a Wisconsin Cohort. *Environmental Health Perspectives* 117 (12):1890-5.

17. Huwe, J.K., and G.L. Larsen. 2005. Polychlorinated dioxins, furans, and biphenyls, and polybrominated diphenyl ethers in a U.S. meat market basket and estimates of dietary intake. *Environmental Science and Technology* 39 (15):5606-11.

18. Schecter, A., D. Haffner, J. Colacino, K. Patel, O. Papke, M. Opel, and L. Birnbaum. 2010. Polybrominated diphenyl ethers (PBDEs) and hexabromocyclodecane (HBCD) in composite U.S. food samples. *Environmental Health Perspectives* 118 (3):357-62.

19. Yogui, G.T., and J.L. Sericano. 2009. Polybrominated diphenyl ether flame retardants in the U.S. marine environment: a review. *Environment International* 35 (3):655-66.

20. Schecter, A., J. Colacino, K. Patel, K. Kannan, S.H. Yun, D. Haffner, T.R. Harris, and L. Birnbaum. 2010. Polybrominated diphenyl ether levels in foodstuffs collected from three locations from the United States. *Toxicology and Applied Pharmacology* 243 (2):217-24.

21. Harrad, S., S. Hazrati, and C. Ibarra. 2006. Concentrations of polychlorinated biphenyls in indoor air and polybrominated diphenyl ethers in indoor air and dust in Birmingham, United Kingdom: implications for human exposure. *Environmental Science and Technology* 40 (15):4633-8.

22. Jones-Otazo, H.A., J.P. Clarke, M.L. Diamond, J.A. Archbold, G. Ferguson, T. Harner, G.M. Richardson, J.J. Ryan, and B. Wilford. 2005. Is house dust the missing exposure pathway for PBDEs? An analysis of the urban fate and human exposure to PBDEs. *Environmental Science and Technology* 39 (14):5121-30.

Polybrominated Diphenyl Ethers (PBDEs) (continued)

23. Schecter, A., O. Papke, K.C. Tung, J. Joseph, T.R. Harris, and J. Dahlgren. 2005. Polybrominated diphenyl ether flame retardants in the U.S. population: current levels, temporal trends, and comparison with dioxins, dibenzofurans, and polychlorinated biphenyls. *Journal of Occupational and Environmental Medicine* 47 (3):199-211.

24. Sjodin, A., R.S. Jones, J.F. Focant, C. Lapeza, R.Y. Wang, E.E. McGahee, 3rd, Y. Zhang, W.E. Turner, B. Slazyk, L.L. Needham, et al. 2004. Retrospective time-trend study of polybrominated diphenyl ether and polybrominated and polychlorinated biphenyl levels in human serum from the United States. *Environmental Health Perspectives* 112 (6):654-8.

25. Dassanayake, R.M., H. Wei, R.C. Chen, and A. Li. 2009. Optimization of the matrix solid phase dispersion extraction procedure for the analysis of polybrominated diphenyl ethers in human placenta. *Analytical Chemistry* 81 (23):9795-801.

26. Herbstman, J.B., A. Sjodin, B.J. Apelberg, F.R. Witter, R.U. Halden, D.G. Patterson, S.R. Panny, L.L. Needham, and L.R. Goldman. 2008. Birth delivery mode modifies the associations between prenatal polychlorinated biphenyl (PCB) and polybrominated diphenyl ether (PBDE) and neonatal thyroid hormone levels. *Environmental Health Perspectives* 116 (10):1376-82.

27. Mazdai, A., N.G. Dodder, M.P. Abernathy, R.A. Hites, and R.M. Bigsby. 2003. Polybrominated diphenyl ethers in maternal and fetal blood samples. *Environmental Health Perspectives* 111 (9):1249-52.

28. Daniels, J.L., I.J. Pan, R. Jones, S. Anderson, D.G. Patterson, L.L. Needham, and A. Sjodin. 2010. Individual characteristics associated with PBDE levels in U.S. human milk samples. *Environmental Health Perspectives* 118 (1):155-60.

29. Hooper, K., J. She, M. Sharp, J. Chow, N. Jewell, R. Gephart, and A. Holden. 2007. Depuration of polybrominated diphenyl ethers (PBDEs) and polychlorinated biphenyls (PCBs) in breast milk from California first-time mothers (primiparae). *Environmental Health Perspectives* 115 (9):1271-5.

30. Schecter, A., J. Colacino, A. Sjodin, L. Needham, and L. Birnbaum. 2010. Partitioning of polybrominated diphenyl ethers (PBDEs) in serum and milk from the same mothers. *Chemosphere* 78 (10):1279-84.

31. She, J., A. Holden, M. Sharp, M. Tanner, C. Williams-Derry, and K. Hooper. 2007. Polybrominated diphenyl ethers (PBDEs) and polychlorinated biphenyls (PCBs) in breast milk from the Pacific Northwest. *Chemosphere* 67 (9):S307-17.

32. Toms, L.M., A. Sjodin, F. Harden, P. Hobson, R. Jones, E. Edenfield, and J.F. Mueller. 2009. Serum polybrominated diphenyl ether (PBDE) levels are higher in children (2-5 years of age) than in infants and adults. *Environmental Health Perspectives* 117 (9):1461-5.

33. Lunder, S., L. Hovander, I. Athanassiadis, and A. Bergman. 2010. Significantly higher polybrominated diphenyl ether levels in young U.S. children than in their mothers. *Environmental Science and Technology* 44 (13):5256-62.

34. Eskenazi, B., L. Fenster, R. Castorina, A.R. Marks, A. Sjödin, L.G. Rosas, N. Holland, A.G. Guerra, L. López-Carrillo, and A. Bradman. 2011. A comparison of PBDE serum concentrations in Mexican and Mexican-American children living in California. *Environmental Health Perspectives* 119 (10):1442-8.

35. Allen, J.G., M.D. McClean, H.M. Stapleton, and T.F. Webster. 2008. Linking PBDEs in house dust to consumer products using X-ray fluorescence. *Environmental Science and Technology* 42 (11):4222-8.

36. U.S. Environmental Protection Agency. 2008. *Child-Specific Exposure Factors Handbook (Final Report)* Washington, DC: U.S. EPA. National Center for Environmental Assessment. EPA/600/R-06/096. http://cfpub.epa.gov/ncea/cfm/recordisplay.cfm?deid=199243.

37. Costa, L.G., G. Giordano, S. Tagliaferri, A. Caglieri, and A. Mutti. 2008. Polybrominated diphenyl ether (PBDE) flame retardants: environmental contamination, human body burden and potential adverse health effects. *Acta Biomed* 79 (3):172-83.

38. Gee, J.R., and V.C. Moser. 2008. Acute postnatal exposure to brominated diphenylether 47 delays neuromotor ontogeny and alters motor activity in mice. *Neurotoxicology and Teratology* 30 (2):79-87.

39. Rice, D.C., E.A. Reeve, A. Herlihy, R.T. Zoeller, W.D. Thompson, and V.P. Markowski. 2007. Developmental delays and locomotor activity in the C57BL6/J mouse following neonatal exposure to the fully-brominated PBDE, decabromodiphenyl ether. *Neurotoxicology and Teratology* 29 (4):511-20.

40. Herbstman, J.B., A. Sjodin, M. Kurzon, S.A. Lederman, R.S. Jones, V. Rauh, L.L. Needham, D. Tang, M. Niedzwiecki, R.Y. Wang, et al. 2010. Prenatal exposure to PBDEs and neurodevelopment. *Environmental Health Perspectives* 118 (5):712-719.

41. Roze, E., L. Meijer, A. Bakker, K.N. Van Braeckel, P.J. Sauer, and A.F. Bos. 2009. Prenatal exposure to organohalogens, including brominated flame retardants, influences motor, cognitive, and behavioral performance at school age. *Environmental Health Perspectives* 117 (12):1953-8.

42. Diamanti-Kandarakis, E., J.P. Bourguignon, L.C. Giudice, R. Hauser, G.S. Prins, A.M. Soto, R.T. Zoeller, and A.C. Gore. 2009. Endocrine-disrupting chemicals: an Endocrine Society scientific statement. *Endocrine Reviews* 30 (4):293-342.

43. Kavlock, R.J., G.P. Daston, C. DeRosa, P. Fenner-Crisp, L.E. Gray, S. Kaattari, G. Lucier, M. Luster, M.J. Mac, C. Maczka, et al. 1996. Research needs for the risk assessment of health and environmental effects of endocrine disruptors: a report of the U.S. EPA-sponsored workshop. *Environmental Health Perspectives* 104 Suppl 4:715-40.

Polybrominated Diphenyl Ethers (PBDEs) (continued)

44. Chevrier, J., K.G. Harley, A. Bradman, M. Gharbi, A. Sjodin, and B. Eskenazi. 2010. Polybrominated diphenyl ether (PBDE) flame retardants and thyroid hormone during pregnancy. *Environmental Health Perspectives* 118 (10):1444-9.

45. Meeker, J.D., P.I. Johnson, D. Camann, and R. Hauser. 2009. Polybrominated diphenyl ether (PBDE) concentrations in house dust are related to hormone levels in men. *Science of the Total Environment* 407 (10):3425-9.

46. Turyk, M.E., V.W. Persky, P. Imm, L. Knobeloch, R. Chatterton, and H.A. Anderson. 2008. Hormone disruption by PBDEs in adult male sport fish consumers. *Environmental Health Perspectives* 116 (12):1635-41.

47. van der Ven, L.T., T. van de Kuil, A. Verhoef, P.E. Leonards, W. Slob, R.F. Canton, S. Germer, T. Hamers, T.J. Visser, S. Litens, et al. 2008. A 28-day oral dose toxicity study enhanced to detect endocrine effects of a purified technical pentabromodiphenyl ether (pentaBDE) mixture in Wistar rats. *Toxicology* 245 (1-2):109-22.

48. Morreale de Escobar, G., M.J. Obregon, and F. Escobar del Rey. 2000. Is neuropsychological development related to maternal hypothyroidism or to maternal hypothyroxinemia? *Journal of Clinical Endocrinology & Metabolism* 85 (11):3975-87.

49. Kuriyama, S.N., C.E. Talsness, K. Grote, and I. Chahoud. 2005. Developmental Exposure to Low Dose PBDE 99: 1--Effects on Male Fertility and Neurobehavior in Rat Offspring. *Environmental Health Perspectives* 113 (2):149-54.

50. Lilienthal, H., A. Hack, A. Roth-Harer, S.W. Grande, and C.E. Talsness. 2006. Effects of developmental exposure to 2,2,4,4,5-pentabromodiphenyl ether (PBDE-99) on sex steroids, sexual development, and sexually dimorphic behavior in rats. *Environmental Health Perspectives* 114 (2):194-201.

51. Stoker, T.E., R.L. Cooper, C.S. Lambright, V.S. Wilson, J. Furr, and L.E. Gray. 2005. In vivo and in vitro anti-androgenic effects of DE-71, a commercial polybrominated diphenyl ether (PBDE) mixture. *Toxicology and Applied Pharmacology* 207 (1):78-88.

52. Main, K.M., H. Kiviranta, H.E. Virtanen, E. Sundqvist, J.T. Tuomisto, J. Tuomisto, T. Vartiainen, N.E. Skakkebaek, and J. Toppari. 2007. Flame retardants in placenta and breast milk and cryptorchidism in newborn boys. *Environmental Health Perspectives* 115 (10):1519-26.

53. Sharpe, R. 2009. *Male Reproductive Health Disorders and the Potential Role of Environmental Chemical Exposures*. London, UK: CHEM Trust. http://www.chemicalshealthmonitor.org/IMG/pdf/PROFSHARPE-MaleReproductiveHealth-CHEMTrust09.pdf.

54. Skakkebaek, N.E., E. Rajpert-De Meyts, and K.M. Main. 2001. Testicular dysgenesis syndrome: an increasingly common developmental disorder with environmental aspects. *Human Reproduction* 16 (5):972-8.

55. Harley, K.G., A.R. Marks, J. Chevrier, A. Bradman, A. Sjodin, and B. Eskenazi. 2010. PBDE concentrations in women's serum and fecundability. *Environmental Health Perspectives* 118 (5):699-704.

56. Centers for Disease Control and Prevention. 2009. *Fourth National Report on Human Exposure to Environmental Chemicals*. Atlanta, GA: CDC. http://www.cdc.gov/exposurereport/.

57. National Center for Health Statistics. *Vital Statistics Natality Birth Data*. Retrieved June 15, 2009 from http://www.cdc.gov/nchs/data_access/Vitalstatsonline.htm..

58. Axelrad, D.A., and J. Cohen. 2011. Calculating summary statistics for population chemical biomonitoring in women of childbearing age with adjustment for age-specific natality. *Environmental Research* 111 (1):149-155.

Phthalates

1. U.S. Environmental Protection Agency. 2012. *Phthalates Action Plan*. Washington, DC: U.S. EPA. http://www.epa.gov/oppt/existingchemicals/pubs/actionplans/phthalates_actionplan_revised_2012-03-14.pdf.

2. Thornton, J. 2000. *Pandora's Poison: Chlorine, Health, and a New Environmental Strategy*. Cambridge, Massachusetts: MIT Press.

3. Center for Health Environment and Justice, and The Environmental Health Strategy Center. 2004. *PVC: Bad News Comes in 3s*. Falls Church, VA and Portland, ME: Center for Health, Environment, and Justice; The Environmental Health Strategy Center.

4. National Research Council. 2008. *Phthalates and Cumulative Risk Assessment: The Tasks Ahead*. Washington, DC: The National Academies Press. http://www.nap.edu/catalog/12528.html.

5. Agency for Toxic Substances and Disease Registry (ATSDR). 1995. *Toxicological Profile for Diethyl Phthalate*. Atlanta, GA: U.S. Department of Health and Human Services, Public Health Service. http://www.atsdr.cdc.gov/toxprofiles/tp73.pdf.

6. Agency for Toxic Substances and Disease Registry (ATSDR). 1997. *Toxicological Profile for Di-n-octylphthalate (DNOP)*. Atlanta, GA: U.S. Department of Health and Human Services, Public Health Service. http://www.atsdr.cdc.gov/toxprofiles/tp95.pdf.

7. Agency for Toxic Substances and Disease Registry (ATSDR). 2001. *Toxicological Profile for Di-n-butyl Phthalate. Update*. Atlanta, GA: U.S. Department of Health and Human Services, Public Health Service. http://www.atsdr.cdc.gov/toxprofiles/tp135.pdf.

8. Agency for Toxic Substances and Disease Registry (ATSDR). 2002. *Toxicological Profile for Di(2-ethylhexyl)phthalate (DEHP)*. Atlanta, GA: U.S. Department of Health and Human Services, Public Health Service. http://www.atsdr.cdc.gov/toxprofiles/tp9.pdf.

Phthalates (continued)

9. Nassberger, L., A. Arbin, and J. Ostelius. 1987. Exposure of patients to phthalates from polyvinyl chloride tubes and bags during dialysis. *Nephron* 45 (4):286-90.

10. Sathyanarayana, S. 2008. Phthalates and children's health. *Current Problems in Pediatric and Adolescent Health Care* 38 (2):34-49.

11. Duty, S.M., R.M. Ackerman, A.M. Calafat, and R. Hauser. 2005. Personal care product use predicts urinary concentrations of some phthalate monoesters. *Environmental Health Perspectives* 113 (11):1530-5.

12. Kwapniewski, R., S. Kozaczka, R. Hauser, M.J. Silva, A.M. Calafat, and S.M. Duty. 2008. Occupational exposure to dibutyl phthalate among manicurists. *Journal of Occupational and Environmental Medicine* 50 (6):705-11.

13. Biron, M. 2009. *Phthalate ousting: Not so easy but some alternatives are viable*. SpecialChem. Retrieved August 11, 2011 from http://www.specialchem4polymers.com/resources/print.aspx?id=3980.

14. Lowell Center for Sustainable Production. 2011. *Phthalates and Their alternatives: Health and Environmental Concerns*. Lowell, MA: University of Massachusetts Lowell. http://www.sustainableproduction.org/downloads/PhthalateAlternatives-January2011.pdf.

15. Scientific Committee on Emerging and Newly Identified Health Risks. 2007. *Preliminary Report on the Safety of Medical Devices Containing DEHP-plasticized PVC or Other Plasticizers on Neonates and Other Groups Possibly at Risk*. Brussels, Belgium: European Commission. http://ec.europa.eu/health/ph_risk/committees/04_scenihr/docs/scenihr_o_008.pdf.

16. Calafat, A.M., and R.H. McKee. 2006. Integrating biomonitoring exposure data into the risk assessment process: phthalates [diethyl phthalate and di(2-ethylhexyl) phthalate] as a case study. *Environmental Health Perspectives* 114 (11):1783-9.

17. Clark, K., I.T. Cousins, and D. Mackay. 2003. Assessment of critical exposure pathways. In *The Handbook of Environmental Chemistry*. New York, NY: Springer.

18. Colacino, J.A., T.R. Harris, and A. Schecter. 2010. Dietary intake is associated with phthalate body burden in a nationally representative sample. *Environmental Health Perspectives* 118 (7):998-1003.

19. Wine, R.N., L.H. Li, L.H. Barnes, D.K. Gulati, and R.E. Chapin. 1997. Reproductive toxicity of di-n-butylphthalate in a continuous breeding protocol in Sprague-Dawley rats. *Environmental Health Perspectives* 105 (1):102-7.

20. Otake, T., J. Yoshinaga, and Y. Yanagisawa. 2004. Exposure to phthalate esters from indoor environment. *Journal of Exposure Analysis and Environmental Epidemiology* 14 (7):524-8.

21. Mortensen, G.K., K.M. Main, A.M. Andersson, H. Leffers, and N.E. Skakkebaek. 2005. Determination of phthalate monoesters in human milk, consumer milk, and infant formula by tandem mass spectrometry (LC-MS-MS). *Anal Bioanal Chem* 382 (4):1084-92.

22. U.S. Environmental Protection Agency. 2008. *Child-specific Exposure Factors Handbook (Final Report)*. Washington, DC: U.S. Environmental Protection Agency. EPA/600/R-06/096F. http://cfpub.epa.gov/ncea/cfm/recordisplay.cfm?deid=199243.

23. Just, A.C., J.J. Adibi, A.G. Rundle, A.M. Calafat, D.E. Camann, R. Hauser, M.J. Silva, and R.M. Whyatt. 2010. Urinary and air phthalate concentrations and self-reported use of personal care products among minority pregnant women in New York city. *Journal of Exposure Science & Environmental Epidemiology* 20 (7):625-33.

24. Romero-Franco, M., R.U. Hernandez-Ramirez, A.M. Calafat, M.E. Cebrian, L.L. Needham, S. Teitelbaum, M.S. Wolff, and L. Lopez-Carrillo. 2011. Personal care product use and urinary levels of phthalate metabolites in Mexican women. *Environment International* 37 (5):867-71.

25. Calafat, A.M., L.L. Needham, M.J. Silva, and G. Lambert. 2004. Exposure to di-(2-ethylhexyl) phthalate among premature neonates in a neonatal intensive care unit. *Pediatrics* 113 (5):e429-34.

26. Green, R., R. Hauser, A.M. Calafat, J. Weuve, T. Schettler, S. Ringer, K. Huttner, and H. Hu. 2005. Use of di(2-ethylhexyl) phthalate-containing medical products and urinary levels of mono(2-ethylhexyl) phthalate in neonatal intensive care unit infants. *Environmental Health Perspectives* 113 (9):1222-5.

27. Weuve, J., B.N. Sanchez, A.M. Calafat, T. Schettler, R.A. Green, H. Hu, and R. Hauser. 2006. Exposure to phthalates in neonatal intensive care unit infants: urinary concentrations of monoesters and oxidative metabolites. *Environmental Health Perspectives* 114 (9):1424-31.

28. Silva, M.J., D.B. Barr, J.A. Reidy, N.A. Malek, C.C. Hodge, S.P. Caudill, J.W. Brock, L.L. Needham, and A.M. Calafat. 2004. Urinary levels of seven phthalate metabolites in the U.S. population from the National Health and Nutrition Examination Survey (NHANES) 1999-2000. *Environmental Health Perspectives* 112 (3):331-8.

29. Becker, K., M. Seiwert, J. Angerer, W. Heger, H.M. Koch, R. Nagorka, E. Rosskamp, C. Schluter, B. Seifert, and D. Ullrich. 2004. DEHP metabolites in urine of children and DEHP in house dust. *International Journal of Hygiene and Environmental Health* 207 (5):409-17.

30. Koch, H.M., H. Drexler, and J. Angerer. 2004. Internal exposure of nursery-school children and their parents and teachers to di(2-ethylhexyl)phthalate (DEHP). *International Journal of Hygiene and Environmental Health* 207 (1):15-22.

31. BKH Consulting Engineers. 2000. Annex 13: List of 146 substances with endocrine disruption classifications prepared in the Expert meeting. In *Towards the Establishment of a Priority List of Substances for Further Evalutation of their Role in Endocrine Disruption*. Delft: The Netherlands. http://ec.europa.eu/environment/docum/pdf/bkh_annex_13.pdf.

Phthalates (continued)

32. Diamanti-Kandarakis, E., J.P. Bourguignon, L.C. Giudice, R. Hauser, G.S. Prins, A.M. Soto, R.T. Zoeller, and A.C. Gore. 2009. Endocrine-disrupting chemicals: an Endocrine Society scientific statement. *Endocrine Reviews* 30 (4):293-342.

33. Gray, L.E., Jr., V.S. Wilson, T. Stoker, C. Lambright, J. Furr, N. Noriega, K. Howdeshell, G.T. Ankley, and L. Guillette. 2006. Adverse effects of environmental antiandrogens and androgens on reproductive development in mammals. *International Journal of Andrology* 29 (1):96-104; discussion 105-8.

34. Patisaul, H.B., and H.B. Adewale. 2009. Long-term effects of environmental endocrine disruptors on reproductive physiology and behavior. *Frontiers in Behavioral Neuroscience* 3:10.

35. Waring, R.H., and R.M. Harris. 2005. Endocrine disrupters: a human risk? *Molecular and Cellular Endocrinology* 244 (1-2):2-9.

36. Kavlock, R.J., G.P. Daston, C. DeRosa, P. Fenner-Crisp, L.E. Gray, S. Kaattari, G. Lucier, M. Luster, M.J. Mac, C. Maczka, et al. 1996. Research needs for the risk assessment of health and environmental effects of endocrine disruptors: a report of the U.S. EPA-sponsored workshop. *Environmental Health Perspectives* 104 Suppl 4:715-40.

37. Andrade, A.J., S.W. Grande, C.E. Talsness, K. Grote, A. Golombiewski, A. Sterner-Kock, and I. Chahoud. 2006. A dose-response study following in utero and lactational exposure to di-(2-ethylhexyl) phthalate (DEHP): effects on androgenic status, developmental landmarks and testicular histology in male offspring rats. *Toxicology* 225 (1):64-74.

38. Barlow, N.J., B.S. McIntyre, and P.M. Foster. 2004. Male reproductive tract lesions at 6, 12, and 18 months of age following in utero exposure to di(n-butyl) phthalate. *Toxicologic Pathology* 32 (1):79-90.

39. Christiansen, S., J. Boberg, M. Axelstad, M. Dalgaard, A.M. Vinggaard, S.B. Metzdorff, and U. Hass. 2010. Low-dose perinatal exposure to di(2-ethylhexyl) phthalate induces anti-androgenic effects in male rats. *Reproductive Toxicology* 30 (2):313-21.

40. Gray, L.E., Jr., J. Ostby, J. Furr, M. Price, D.N. Veeramachaneni, and L. Parks. 2000. Perinatal exposure to the phthalates DEHP, BBP, and DINP, but not DEP, DMP, or DOTP, alters sexual differentiation of the male rat. *Toxicological Sciences* 58 (2):350-65.

41. Howdeshell, K.L., V.S. Wilson, J. Furr, C.R. Lambright, C.V. Rider, C.R. Blystone, A.K. Hotchkiss, and L.E. Gray, Jr. 2008. A mixture of five phthalate esters inhibits fetal testicular testosterone production in the sprague-dawley rat in a cumulative, dose-additive manner. *Toxicological Sciences* 105 (1):153-65.

42. Lehmann, K.P., S. Phillips, M. Sar, P.M. Foster, and K.W. Gaido. 2004. Dose-dependent alterations in gene expression and testosterone synthesis in the fetal testes of male rats exposed to di (n-butyl) phthalate. *Toxicological Sciences* 81 (1):60-8.

43. Mylchreest, E., D.G. Wallace, R.C. Cattley, and P.M. Foster. 2000. Dose-dependent alterations in androgen-regulated male reproductive development in rats exposed to Di(n-butyl) phthalate during late gestation. *Toxicological Sciences* 55 (1):143-51.

44. Parks, L.G., J.S. Ostby, C.R. Lambright, B.D. Abbott, G.R. Klinefelter, N.J. Barlow, and L.E. Gray, Jr. 2000. The plasticizer diethylhexyl phthalate induces malformations by decreasing fetal testosterone synthesis during sexual differentiation in the male rat. *Toxicological Sciences* 58 (2):339-49.

45. Sharpe, R.M. 2008. "Additional" effects of phthalate mixtures on fetal testosterone production. *Toxicological Sciences* 105 (1):1-4.

46. Swan, S.H. 2008. Environmental phthalate exposure in relation to reproductive outcomes and other health endpoints in humans. *Environmental Research* 108 (2):177-84.

47. Swan, S.H., K.M. Main, F. Liu, S.L. Stewart, R.L. Kruse, A.M. Calafat, C.S. Mao, J.B. Redmon, C.L. Ternand, S. Sullivan, et al. 2005. Decrease in anogenital distance among male infants with prenatal phthalate exposure. *Environmental Health Perspectives* 113 (8):1056-61.

48. Macleod, D.J., R.M. Sharpe, M. Welsh, M. Fisken, H.M. Scott, G.R. Hutchison, A.J. Drake, and S. van den Driesche. 2010. Androgen action in the masculinization programming window and development of male reproductive organs. *International Journal of Andrology* 33 (2):279-87.

49. Scott, H.M., G.R. Hutchison, M.S. Jobling, C. McKinnell, A.J. Drake, and R.M. Sharpe. 2008. Relationship between androgen action in the "male programming window," fetal sertoli cell number, and adult testis size in the rat. *Endocrinology* 149 (10):5280-7.

50. Mendiola, J., R.W. Stahlhut, N. Jorgensen, F. Liu, and S.H. Swan. 2011. Shorter anogenital distance predicts poorer semen quality in young men in Rochester, New York. *Environmental Health Perspectives* 119 (7):958-63.

51. Main, K.M., G.K. Mortensen, M.M. Kaleva, K.A. Boisen, I.N. Damgaard, M. Chellakooty, I.M. Schmidt, A.M. Suomi, H.E. Virtanen, D.V. Petersen, et al. 2006. Human breast milk contamination with phthalates and alterations of endogenous reproductive hormones in infants three months of age. *Environmental Health Perspectives* 114 (2):270-6.

52. Engel, S.M., A. Miodovnik, R.L. Canfield, C. Zhu, M.J. Silva, A.M. Calafat, and M.S. Wolff. 2010. Prenatal phthalate exposure is associated with childhood behavior and executive functioning. *Environmental Health Perspectives* 118 (4):565-71.

53. Miodovnik, A., S.M. Engel, C. Zhu, X. Ye, L.V. Soorya, M.J. Silva, A.M. Calafat, and M.S. Wolff. 2011. Endocrine disruptors and childhood social impairment. *Neurotoxicology* 32 (2):261-7.

Phthalates (continued)

54. Cho, S.C., S.Y. Bhang, Y.C. Hong, M.S. Shin, B.N. Kim, J.W. Kim, H.J. Yoo, I.H. Cho, and H.W. Kim. 2010. Relationship between environmental phthalate exposure and the intelligence of school-age children. *Environmental Health Perspectives* 118 (7):1027-32.

55. Kim, B.N., S.C. Cho, Y. Kim, M.S. Shin, H.J. Yoo, J.W. Kim, Y.H. Yang, H.W. Kim, S.Y. Bhang, and Y.C. Hong. 2009. Phthalates exposure and attention-deficit/hyperactivity disorder in school-age children. *Biological Psychiatry* 66 (10):958-63.

56. Latini, G., C. De Felice, G. Presta, A. Del Vecchio, I. Paris, F. Ruggieri, and P. Mazzeo. 2003. In utero exposure to di-(2-ethylhexyl)phthalate and duration of human pregnancy. *Environmental Health Perspectives* 111 (14):1783-5.

57. Meeker, J.D., H. Hu, D.E. Cantonwine, H. Lamadrid-Figueroa, A.M. Calafat, A.S. Ettinger, M. Hernandez-Avila, R. Loch-Caruso, and M.M. Téllez-Rojo. 2009. Urinary phthalate metabolites in relation to preterm birth in Mexico City. *Environmental Health Perspectives* 117 (10):1587-92.

58. Whyatt, R.M., J.J. Adibi, A.M. Calafat, D.E. Camann, V. Rauh, H.K. Bhat, F.P. Perera, H. Andrews, A.C. Just, L. Hoepner, et al. 2009. Prenatal Di(2-ethylhexyl) phthalate exposure and length of gestation among an inner-city cohort. *Pediatrics* 124 (6):e1213-20.

59. Zhang, Y., L. Lin, Y. Cao, B. Chen, L. Zheng, and R.S. Ge. 2009. Phthalate levels and low birth weight: a nested case-control study of Chinese newborns. *The Journal of Pediatrics* 155 (4):500-4.

60. Adibi, J.J., R. Hauser, P.L. Williams, R.M. Whyatt, A.M. Calafat, H. Nelson, R. Herrick, and S.H. Swan. 2009. Maternal urinary metabolites of Di-(2-Ethylhexyl) phthalate in relation to the timing of labor in a US multicenter pregnancy cohort study. *American Journal of Epidemiology* 169 (8):1015-24.

61. Jaakkola, J.J., and T.L. Knight. 2008. The role of exposure to phthalates from polyvinyl chloride products in the development of asthma and allergies: a systematic review and meta-analysis. *Environmental Health Perspectives* 116 (7):845-53.

62. Bornehag, C.G., J. Sundell, C.J. Weschler, T. Sigsgaard, B. Lundgren, M. Hasselgren, and L. Hagerhed-Engman. 2004. The association between asthma and allergic symptoms in children and phthalates in house dust: a nested case-control study. *Environmental Health Perspectives* 112 (14):1393-7.

63. Jaakkola, J.J., L. Oie, P. Nafstad, G. Botten, S.O. Samuelsen, and P. Magnus. 1999. Interior surface materials in the home and the development of bronchial obstruction in young children in Oslo, Norway. *American Journal of Public Health* 89 (2):188-92.

64. National Toxicology Program. 2006. NTP-CERHR Monograph on the Potential Human Reproductive and Developmental Effects of Di(2-Ethylhexyl) Phthalate (DEHP), edited by U.S. Department of Health and Human Services. Research Triangle Park, NC: NIH.

65. National Toxicology Program. 2003. *NTP-CERHR Monograph on the Potential Human Reproductive and Developmental Effects of Di-n-Butyl Phthalate (DBP)*. Research Triangle Park, NC: National Institute of Environmental Health Sciences, National Toxicology Program. http://ntp.niehs.nih.gov/ntp/ohat/phthalates/dbp/DBP_Monograph_Final.pdf.

66. Koch, H.M., R. Preuss, H. Drexler, and J. Angerer. 2005. Exposure of nursery school children and their parents and teachers to di-n-butylphthalate and butylbenzylphthalate. *International Archives of Occupational and Environmental Health* 78 (3):223-9.

67. National Toxicology Program. 2003. *NTP-CERHR Monograph on the Potential Human Reproductive and Developmental Effects of Butyl Benzyl Phthalate (BBP)*. Research Triangle Park, NC: National Institute of Environmental Health Sciences, National Toxicology Program. http://ntp.niehs.nih.gov/ntp/ohat/phthalates/bb-phthalate/BBP_Monograph_Final.pdf.

68. Huang, P.C., P.L. Kuo, Y.L. Guo, P.C. Liao, and C.C. Lee. 2007. Associations between urinary phthalate monoesters and thyroid hormones in pregnant women. *Human Reproduction* 22 (10):2715-22.

69. Centers for Disease Control and Prevention. 2009. *Fourth National Report on Human Exposure to Environmental Chemicals*. Atlanta, GA: CDC. http://www.cdc.gov/exposurereport/.

70. Albro, P.W., and S.R. Lavenhar. 1989. Metabolism of di(2-ethylhexyl)phthalate. *Drug Metabolism Reviews* 21 (1):13-34.

71. Anderson, W.A., L. Castle, M.J. Scotter, R.C. Massey, and C. Springall. 2001. A biomarker approach to measuring human dietary exposure to certain phthalate diesters. *Food Additives and Contaminants* 18 (12):1068-74.

72. Hauser, R., and A.M. Calafat. 2005. Phthalates and human health. *Occupational and Environmental Medicine* 62 (11):806-18.

73. Herr, C., A. zur Nieden, H.M. Koch, H.C. Schuppe, C. Fieber, J. Angerer, T. Eikmann, and N.I. Stilianakis. 2009. Urinary di(2-ethylhexyl)phthalate (DEHP)--metabolites and male human markers of reproductive function. *International Journal of Hygiene and Environmental Health* 212 (6):648-53.

74. Koch, H.M., R. Preuss, and J. Angerer. 2006. Di(2-ethylhexyl)phthalate (DEHP): human metabolism and internal exposure--an update and latest results. *International Journal of Andrology* 29 (1):155-65; discussion 181-185.

75. Swan, S.H., F. Liu, M. Hines, R.L. Kruse, C. Wang, J.B. Redmon, A. Sparks, and B. Weiss. 2009. Prenatal phthalate exposure and reduced masculine play in boys. *International Journal of Andrology* 33 (2):259-69.

76. Ye, X., L.Y. Wong, A.M. Bishop, and A.M. Calafat. 2011. Variability of urinary concentrations of bisphenol a in spot samples, first morning voids, and 24-hour collections. *Environmental Health Perspectives* 119 (7):983-8.

Phthalates (continued)

77. Mendez, W., E. Dederick, and J. Cohen. 2010. Drinking water contribution to aggregate perchlorate intake of reproductive-age women in the United States estimated by dietary intake simulation and analysis of urinary excretion data. *Journal of Exposure Science and Environmental Epidemiology* 20 (3):288-97.

78. Preau, J.L., Jr., L.Y. Wong, M.J. Silva, L.L. Needham, and A.M. Calafat. 2010. Variability over 1 week in the urinary concentrations of metabolites of diethyl phthalate and di(2-ethylhexyl) phthalate among eight adults: an observational study. *Environmental Health Perspectives* 118 (12):1748-54.

79. Jackson, S. 1966. Creatinine in urine as an index of urinary excretion rate. *Health Physics* 12 (6):843-50.

80. Barr, D.B., L.C. Wilder, S.P. Caudill, A.J. Gonzalez, L.L. Needham, and J.L. Pirkle. 2005. Urinary creatinine concentrations in the U.S. population: implications for urinary biologic monitoring measurements. *Environmental Health Perspectives* 113 (2):192-200.

81. Boeniger, M.F., L.K. Lowry, and J. Rosenberg. 1993. Interpretation of urine results used to assess chemical exposure with emphasis on creatinine adjustments: a review. *American Industrial Hygiene Association Journal* 54 (10):615-27.

82. National Center for Health Statistics. *Vital Statistics Natality Birth Data*. Retrieved June 15, 2009 from http://www.cdc.gov/nchs/data_access/Vitalstatsonline.htm. .

83. Axelrad, D.A., and J. Cohen. 2011. Calculating summary statistics for population chemical biomonitoring in women of childbearing age with adjustment for age-specific natality. *Environmental Research* 111 (1):149-155.

Bisphenol A (BPA)

1. U.S. Food and Drug Administration. 2012. *Bisphenol A (BPA): Use in Food Contact Application*. U.S. Department of Health and Human Services, Food and Drug Administration. Retrieved June 26, 2012 from http://www.fda.gov/NewsEvents/PublicHealthFocus/ucm064437.htm.

2. U.S. Food and Drug Administration. 2009. *Summary of Bisphenol A Biomonitoring Studies*. Washington, DC: U.S. Department of Health and Human Services, Food and Drug Administration. Memorandum from V. Komolprasert, dated November 16, 2009. http://www.regulations.gov/#!documentDetail;D=FDA-2010-N-0100-0010.

3. National Toxicology Program. 2008. *NTP-CERHR Monograph on the Potential Human Reproductive and Developmental Effects of Bisphenol A*. Research Triangle Park, NC: National Institute of Environmental Health Sciences, National Toxicology Program. http://ntp.niehs.nih.gov/ntp/ohat/bisphenol/bisphenol.pdf.

4. Schecter, A., N. Malik, D. Haffner, S. Smith, T.R. Harris, O. Paepke, and L. Birnbaum. 2010. Bisphenol A (BPA) in U.S. food. *Environmental Science and Technology* 44 (24):9425-30.

5. Vandenberg, L.N., R. Hauser, M. Marcus, N. Olea, and W.V. Welshons. 2007. Human exposure to bisphenol A (BPA). *Reproductive Toxicology* 24 (2):139-77.

6. Le, H.H., E.M. Carlson, J.P. Chua, and S.M. Belcher. 2008. Bisphenol A is released from polycarbonate drinking bottles and mimics the neurotoxic actions of estrogen in developing cerebellar neurons. *Toxicology Letters* 176 (2):149-56.

7. Calafat, A.M., X. Ye, L.Y. Wong, J.A. Reidy, and L.L. Needham. 2008. Exposure of the U.S. population to bisphenol A and 4-tertiary-octylphenol: 2003-2004. *Environmental Health Perspectives* 116 (1):39-44.

8. Völkel, W., T. Colnot, G.A. Csanady, J.G. Filser, and W. Dekant. 2002. Metabolism and kinetics of bisphenol a in humans at low doses following oral administration. *Chemical Research in Toxicology* 15 (10):1281-7.

9. Diamanti-Kandarakis, E., J.P. Bourguignon, L.C. Giudice, R. Hauser, G.S. Prins, A.M. Soto, R.T. Zoeller, and A.C. Gore. 2009. Endocrine-disrupting chemicals: an Endocrine Society scientific statement. *Endocrine Reviews* 30 (4):293-342.

10. vom Saal, F.S., B.T. Akingbemi, S.M. Belcher, L.S. Birnbaum, D.A. Crain, M. Eriksen, F. Farabollini, L.J. Guillette, Jr., R. Hauser, J.J. Heindel, et al. 2007. Chapel Hill bisphenol A expert panel consensus statement: integration of mechanisms, effects in animals and potential to impact human health at current levels of exposure. *Reproductive Toxicology* 24 (2):131-8.

11. Kavlock, R.J., G.P. Daston, C. DeRosa, P. Fenner-Crisp, L.E. Gray, S. Kaattari, G. Lucier, M. Luster, M.J. Mac, C. Maczka, et al. 1996. Research needs for the risk assessment of health and environmental effects of endocrine disruptors: a report of the U.S. EPA-sponsored workshop. *Environmental Health Perspectives* 104 Suppl 4:715-40.

12. Kuiper, G.G., J.G. Lemmen, B. Carlsson, J.C. Corton, S.H. Safe, P.T. van der Saag, B. van der Burg, and J.A. Gustafsson. 1998. Interaction of estrogenic chemicals and phytoestrogens with estrogen receptor beta. *Endocrinology* 139 (10):4252-63.

13. Golub, M.S., K.L. Wu, F.L. Kaufman, L.H. Li, F. Moran-Messen, L. Zeise, G.V. Alexeeff, and J.M. Donald. 2010. Bisphenol A: developmental toxicity from early prenatal exposure. *Birth Defects Research. Part B, Developmental and Reproductive Toxicology* 89 (6):441-66.

14. Welshons, W.V., K.A. Thayer, B.M. Judy, J.A. Taylor, E.M. Curran, and F.S. vom Saal. 2003. Large effects from small exposures. I. Mechanisms for endocrine-disrupting chemicals with estrogenic activity. *Environmental Health Perspectives* 111 (8):994-1006.

Bisphenol A (BPA) (continued)

15. Wetherill, Y.B., B.T. Akingbemi, J. Kanno, J.A. McLachlan, A. Nadal, C. Sonnenschein, C.S. Watson, R.T. Zoeller, and S.M. Belcher. 2007. In vitro molecular mechanisms of bisphenol A action. *Reproductive Toxicology* 24 (2):178-98.

16. Tyl, R.W., C.B. Myers, M.C. Marr, B.F. Thomas, A.R. Keimowitz, D.R. Brine, M.M. Veselica, P.A. Fail, T.Y. Chang, J.C. Seely, et al. 2002. Three-generation reproductive toxicity study of dietary bisphenol A in CD Sprague-Dawley rats. *Toxicological Sciences* 68 (1):121-46.

17. Tyl, R.W., C.B. Myers, M.C. Marr, C.S. Sloan, N.P. Castillo, M.M. Veselica, J.C. Seely, S.S. Dimond, J.P. Van Miller, R.N. Shiotsuka, et al. 2008. Two-generation reproductive toxicity study of dietary bisphenol A in CD-1 (Swiss) mice. *Toxicological Sciences* 104 (2):362-84.

18. Morrissey, R.E., J.D. George, C.J. Price, R.W. Tyl, M.C. Marr, and C.A. Kimmel. 1987. The developmental toxicity of bisphenol A in rats and mice. *Fundamental and Applie Toxicology* 8 (4):571-82.

19. Kim, J.C., H.C. Shin, S.W. Cha, W.S. Koh, M.K. Chung, and S.S. Han. 2001. Evaluation of developmental toxicity in rats exposed to the environmental estrogen bisphenol A during pregnancy. *Life Sciences* 69 (22):2611-25.

20. Alonso-Magdalena, P., S. Morimoto, C. Ripoll, E. Fuentes, and A. Nadal. 2006. The estrogenic effect of bisphenol A disrupts pancreatic beta-cell function in vivo and induces insulin resistance. *Environmental Health Perspectives* 114 (1):106-12.

21. Batista, T.M., P. Alonso-Magdalena, E. Vieira, M.E. Amaral, C.R. Cederroth, S. Nef, I. Quesada, E.M. Carneiro, and A. Nadal. 2012. Short-term treatment with bisphenol-A leads to metabolic abnormalities in adult male mice. *PLoS ONE* 7 (3):e33814.

22. Alonso-Magdalena, P., E. Vieira, S. Soriano, L. Menes, D. Burks, I. Quesada, and A. Nadal. 2010. Bisphenol A exposure during pregnancy disrupts glucose homeostasis in mothers and adult male offspring. *Environmental Health Perspectives* 118 (9):1243-50.

23. Ho, S.M., W.Y. Tang, J. Belmonte de Frausto, and G.S. Prins. 2006. Developmental exposure to estradiol and bisphenol A increases susceptibility to prostate carcinogenesis and epigenetically regulates phosphodiesterase type 4 variant 4. *Cancer Research* 66 (11):5624-5632.

24. Soto, A.M., L.N. Vandenberg, M.V. Maffini, and C. Sonnenschein. 2008. Does breast cancer start in the womb? *Basic & Clinical Pharmacology & Toxicology* 102 (2):125-33.

25. Weber Lozada, K., and R.A. Keri. 2011. Bisphenol A increases mammary cancer risk in two distinct mouse models of breast cancer. *Biology of Reproduction* 85 (3):490-7.

26. Beronius, A., C. Rudén, H. Håkansson, and A. Hanberg. 2010. Risk to all or none?: A comparative analysis of controversies in the health risk assessment of Bisphenol A. *Reproductive Toxicology* 29 (2):132-146.

27. Durando, M., L. Kass, J. Piva, C. Sonnenschein, A.M. Soto, E.H. Luque, and M. Munoz-de-Toro. 2007. Prenatal bisphenol A exposure induces preneoplastic lesions in the mammary gland in Wistar rats. *Environmental Health Perspectives* 115 (1):80-6.

28. Gioiosa, L., E. Fissore, G. Ghirardelli, S. Parmigiani, and P. Palanza. 2007. Developmental exposure to low-dose estrogenic endocrine disruptors alters sex differences in exploration and emotional responses in mice. *Hormones and Behavior* 52 (3):307-16.

29. Howdeshell, K.L., J. Furr, C.R. Lambright, V.S. Wilson, B.C. Ryan, and L.E. Gray, Jr. 2008. Gestational and lactational exposure to ethinyl estradiol, but not bisphenol A, decreases androgen-dependent reproductive organ weights and epididymal sperm abundance in the male long evans hooded rat. *Toxicological Sciences* 102 (2):371-82.

30. Howdeshell, K.L., A.K. Hotchkiss, K.A. Thayer, J.G. Vandenbergh, and F.S. vom Saal. 1999. Exposure to bisphenol A advances puberty. *Nature* 401 (6755):763-4.

31. Miyagawa, K., M. Narita, H. Akama, and T. Suzuki. 2007. Memory impairment associated with a dysfunction of the hippocampal cholinergic system induced by prenatal and neonatal exposures to bisphenol-A. *Neuroscience Letters* 418 (3):236-41.

32. Palanza, P., L. Gioiosa, F.S. vom Saal, and S. Parmigiani. 2008. Effects of developmental exposure to bisphenol A on brain and behavior in mice. *Environmental Research* 108 (2):150-7.

33. Palanza, P.L., K.L. Howdeshell, S. Parmigiani, and F.S. vom Saal. 2002. Exposure to a low dose of bisphenol A during fetal life or in adulthood alters maternal behavior in mice. *Environmental Health Perspectives* 110 Suppl 3:415-22.

34. Sharpe, R.M. 2010. Is it time to end concerns over the estrogenic effects of bisphenol A? *Toxicological Sciences* 114 (1):1-4.

35. Vandenberg, L.N., M.V. Maffini, C. Sonnenschein, B.S. Rubin, and A.M. Soto. 2009. Bisphenol-A and the great divide: a review of controversies in the field of endocrine disruption. *Endocrine Reviews* 30 (1):75-95.

36. vom Saal, F.S., and C. Hughes. 2005. An extensive new literature concerning low-dose effects of bisphenol A shows the need for a new risk assessment. *Environmental Health Perspectives* 113 (8):926-33.

37. Clayton, E.M.R., M. Todd, J.B. Dowd, and A.E. Aiello. 2011. The impact of bisphenol A and triclosan on immune parameters in the U.S. population, NHANES 2003–2006. *Environmental Health Perspectives* 119 (3):390-396.

38. Lang, I.A., T.S. Galloway, A. Scarlett, W.E. Henley, M. Depledge, R.B. Wallace, and D. Melzer. 2008. Association of urinary bisphenol A concentration with medical disorders and laboratory abnormalities in adults. *Journal of the American Medical Association* 300 (11):1303-10.

Bisphenol A (BPA) (continued)

39. Melzer, D., N.E. Rice, C. Lewis, W.E. Henley, and T.S. Galloway. 2010. Association of Urinary Bisphenol A Concentration with Heart Disease: Evidence from NHANES 2003/06. *PLoS ONE* 5 (1):e8673.

40. Li, D., Z. Zhou, D. Qing, Y. He, T. Wu, M. Miao, J. Wang, X. Weng, J.R. Ferber, L.J. Herrinton, et al. 2010. Occupational exposure to bisphenol-A (BPA) and the risk of Self-Reported Male Sexual Dysfunction. *Human Reproduction* 25 (2):519-527.

41. Li, D.-K., Z. Zhou, M. Miao, Y. He, D. Qing, T. Wu, J. Wang, X. Weng, J. Ferber, L.J. Herrinton, et al. 2010. Relationship Between Urine Bisphenol-A Level and Declining Male Sexual Function. *Journal of Andrology* 31 (5):500-506.

42. Miao, M., W. Yuan, G. Zhu, X. He, and D.-K. Li. 2011. In utero exposure to bisphenol-A and its effect on birth weight of offspring. *Reproductive Toxicology* 32 (1):64-8.

43. Braun, J.M., K. Yolton, K.N. Dietrich, R. Hornung, X. Ye, A.M. Calafat, and B.P. Lanphear. 2009. Prenatal Bisphenol A Exposure and Early Childhood Behavior. *Environmental Health Perspectives* 117 (12):1945-1952.

44. Ishido, M., Y. Masuo, M. Kunimoto, S. Oka, and M. Morita. 2004. Bisphenol A causes hyperactivity in the rat concomitantly with impairment of tyrosine hydroxylase immunoreactivity. *Journal of Neuroscience Research* 76 (3):423-433.

45. Kawai, K., T. Nozaki, H. Nishikata, S. Aou, M. Takii, and C. Kubo. 2003. Aggressive behavior and serum testosterone concentration during the maturation process of male mice: The effects of fetal exposure to bisphenol A. *Environmental Health Perspectives* 111 (2):175-180.

46. Miodovnik, A., S.M. Engel, C. Zhu, X. Ye, L.V. Soorya, M.J. Silva, A.M. Calafat, and M.S. Wolff. 2011. Endocrine disruptors and childhood social impairment. *Neurotoxicology* 32 (2):261-267.

47. Bushnik, T., D. Haines, P. Levallois, J. Levesque, J. Van Oostdam, and C. Viau. 2010. Lead and bisphenol A concentrations in the Canadian population. *Health Reports, Statistics Canada* 21 (3):7-18.

48. Chapin, R.E., J. Adams, K. Boekelheide, L.E. Gray, Jr., S.W. Hayward, P.S. Lees, B.S. McIntyre, K.M. Portier, T.M. Schnorr, S.G. Selevan, et al. 2008. NTP-CERHR expert panel report on the reproductive and developmental toxicity of bisphenol A. *Birth Defects Research Part B: Developmental and Reproductive Toxicology* 83 (3):157-395.

49. Lakind, J.S., and D.Q. Naiman. 2008. Bisphenol A (BPA) daily intakes in the United States: estimates from the 2003-2004 NHANES urinary BPA data. *Journal of Exposure Science and Environmental Epidemiology* 18 (6):608-15.

50. Lakind, J.S., and D.Q. Naiman. 2011. Daily intake of bisphenol A and potential sources of exposure: 2005-2006 National Health and Nutrition Examination Survey. *Journal of Exposure Science and Environmental Epidemiology* 21 (3):272-9.

51. Morgan, M.K., P.A. Jones, A.M. Calafat, X. Ye, C.W. Croghan, J.C. Chuang, N.K. Wilson, M.S. Clifton, Z. Figueroa, and L.S. Sheldon. 2011. Assessing the quantitative relationships between preschool children's exposures to bisphenol A by route and urinary biomonitoring. *Environmental Science and Technology* 45 (12):5309-16.

52. Völkel, W., M. Kiranoglu, and H. Fromme. 2011. Determination of free and total bisphenol A in urine of infants. *Environmental Research* 111 (1):143-148.

53. Padmanabhan, V., K. Siefert, S. Ransom, T. Johnson, J. Pinkerton, L. Anderson, L. Tao, and K. Kannan. 2008. Maternal bisphenol-A levels at delivery: a looming problem? *Journal of Perinatology* 28 (4):258-63.

54. Balakrishnan, B., K. Henare, E.B. Thorstensen, A.P. Ponnampalam, and M.D. Mitchell. 2010. Transfer of bisphenol A across the human placenta. *American Journal of Obstetrics and Gynecology* 202 (4):393.e1-393.e7.

55. Calafat, A.M., J. Weuve, X. Ye, L.T. Jia, H. Hu, S. Ringer, K. Huttner, and R. Hauser. 2009. Exposure to bisphenol A and other phenols in neonatal intensive care unit premature infants. *Environmental Health Perspectives* 117 (4):639-44.

56. Doerge, D.R., N.C. Twaddle, M. Vanlandingham, and J.W. Fisher. 2010. Pharmacokinetics of bisphenol A in neonatal and adult Sprague-Dawley rats. *Toxicology and Applied Pharmacology* 247 (2):158-165.

57. Doerge, D.R., N.C. Twaddle, M. Vanlandingham, and J.W. Fisher. 2011. Pharmacokinetics of bisphenol A in neonatal and adult CD-1 mice: inter-species comparisons with Sprague-Dawley rats and rhesus monkeys. *Toxicology Letters* 207 (3):298-305.

58. Doerge, D.R., N.C. Twaddle, K.A. Woodling, and J.W. Fisher. 2010. Pharmacokinetics of bisphenol A in neonatal and adult rhesus monkeys. *Toxicology and Applied Pharmacology* 248 (1):1-11.

59. Domoradzki, J.Y., C.M. Thornton, L.H. Pottenger, S.C. Hansen, T.L. Card, D.A. Markham, M.D. Dryzga, R.N. Shiotsuka, and J.M. Waechter, Jr. 2004. Age and dose dependency of the pharmacokinetics and metabolism of bisphenol A in neonatal sprague-dawley rats following oral administration. *Toxicological Sciences* 77 (2):230-42.

60. Fisher, J.W., N.C. Twaddle, M. Vanlandingham, and D.R. Doerge. 2011. Pharmacokinetic modeling: prediction and evaluation of route dependent dosimetry of bisphenol A in monkeys with extrapolation to humans. *Toxicology and Applied Pharmacology* 257 (1):122-36.

61. Taylor, J.A., F.S. vom Saal, W.V. Welshons, B. Drury, G. Rottinghaus, P.A. Hunt, P.-L. Toutain, C.M. Laffont, and C.A. VandeVoort. 2011. Similarity of bisphenol A pharmacokinetics in rhesus monkeys and mice: Relevance for human exposure. *Environmental Health Perspectives* 119 (4):422-430.

Bisphenol A (BPA) (continued)

62. Doerge, D.R., M. Vanlandingham, N.C. Twaddle, and K.B. Delclos. 2010. Lactational transfer of bisphenol A in Sprague-Dawley rats. *Toxicology Letters* 199 (3):372-376.

63. Centers for Disease Control and Prevention. 2009. *Fourth National Report on Human Exposure to Environmental Chemicals*. Atlanta, GA: CDC. http://www.cdc.gov/exposurereport/.

64. Vandenberg, L.N., I. Chahoud, J.J. Heindel, V. Padmanabhan, F.J. Paumgartten, and G. Schoenfelder. 2010. Urinary, circulating, and tissue biomonitoring studies indicate widespread exposure to bisphenol A. *Environmental Health Perspectives* 118 (8):1055-70.

65. Ginsberg, G., and D.C. Rice. 2009. Does rapid metabolism ensure negligible risk from bisphenol A? *Environmental Health Perspectives* 117 (11):1639-43.

66. Nishikawa, M., H. Iwano, R. Yanagisawa, N. Koike, H. Inoue, and H. Yokota. 2010. Placental transfer of conjugated bisphenol A and subsequent reactivation in the rat fetus. *Environmental Health Perspectives* 118 (9):1196-203.

67. Ye, X., L.Y. Wong, A.M. Bishop, and A.M. Calafat. 2011. Variability of urinary concentrations of bisphenol a in spot samples, first morning voids, and 24-hour collections. *Environmental Health Perspectives* 119 (7):983-8.

68. Mendez, W., E. Dederick, and J. Cohen. 2010. Drinking water contribution to aggregate perchlorate intake of reproductive-age women in the United States estimated by dietary intake simulation and analysis of urinary excretion data. *Journal of Exposure Science and Environmental Epidemiology* 20 (3):288-97.

69. Preau, J.L., Jr., L.Y. Wong, M.J. Silva, L.L. Needham, and A.M. Calafat. 2010. Variability over 1 week in the urinary concentrations of metabolites of diethyl phthalate and di(2-ethylhexyl) phthalate among eight adults: an observational study. *Environmental Health Perspectives* 118 (12):1748-54.

70. Jackson, S. 1966. Creatinine in urine as an index of urinary excretion rate. *Health Physics* 12 (6):843-50.

71. Barr, D.B., L.C. Wilder, S.P. Caudill, A.J. Gonzalez, L.L. Needham, and J.L. Pirkle. 2005. Urinary creatinine concentrations in the U.S. population: implications for urinary biologic monitoring measurements. *Environmental Health Perspectives* 113 (2):192-200.

72. Boeniger, M.F., L.K. Lowry, and J. Rosenberg. 1993. Interpretation of urine results used to assess chemical exposure with emphasis on creatinine adjustments: A review. *American Industrial Hygiene Association Journal* 54 (10):615-27.

73. National Center for Health Statistics. *Vital Statistics Natality Birth Data*. Retrieved June 15, 2009 from http://www.cdc.gov/nchs/data_access/Vitalstatsonline.htm.

74. Axelrad, D.A., and J. Cohen. 2011. Calculating summary statistics for population chemical biomonitoring in women of childbearing age with adjustment for age-specific natality. *Environmental Research* 111 (1):149-155.

Perchlorate

1. National Research Council. 2005. *Health Implications of Perchlorate Ingestion*. Washington, D.C.: The National Academies Press. http://www.nap.edu/catalog.php?record_id=11202.

2. U.S. Environmental Protection Agency. 2011. *Perchlorate* Retrieved February 11, 2011 from http://www.epa.gov/safewater/contaminants/unregulated/perchlorate.html.

3. Blount, B.C., L. Valentin-Blasini, J.D. Osterloh, J.P. Mauldin, and J.L. Pirkle. 2007. Perchlorate exposure of the US Population, 2001-2002. *Journal of Exposure Science & Environmental Epidemiology* 17 (4):400-7.

4. English, P., B. Blount, M. Wong, L. Copan, L. Olmedo, S. Patton, R. Haas, R. Atencio, J. Xu, and L. Valentin-Blasini. 2011. Direct measurement of perchlorate exposure biomarkers in a highly exposed population: a pilot study. *PloS One* 6 (3):e17015.

5. Rao, B., T.A. Anderson, G.J. Orris, K.A. Rainwater, S. Rajagopalan, R.M. Sandvig, B.R. Scanlon, D.A. Stonestrom, M.A. Walvoord, and W.A. Jackson. 2007. Widespread natural perchlorate in unsaturated zones of the southwest United States. *Environmental Science & Technology* 41 (13):4522-8.

6. California Environmental Protection Agency. *History of Perchlorate in California Drinking Water*. Retrieved 3/8/2010 from http://www.cdph.ca.gov/certlic/drinkingwater/Pages/Perchloratehistory.aspx.

7. El Aribi, H., Y.J. Le Blanc, S. Antonsen, and T. Sakuma. 2006. Analysis of perchlorate in foods and beverages by ion chromatography coupled with tandem mass spectrometry (IC-ESI-MS/MS). *Analytica Chimica Acta* 567 (1):39-47.

8. Murray, C.W., S.K. Egan, H. Kim, N. Beru, and P.M. Bolger. 2008. US Food and Drug Administration's Total Diet Study: dietary intake of perchlorate and iodine. *Journal of Exposure Science & Environmental Epidemiology* 18 (6):571-80.

9. Sanchez, C.A., L.M. Barraj, B.C. Blount, C.G. Scrafford, L. Valentin-Blasini, K.M. Smith, and R.I. Krieger. 2009. Perchlorate exposure from food crops produced in the lower Colorado River region. *Journal of Exposure Science & Environmental Epidemiology* 19 (4):359-68.

10. Sanchez, C.A., B.C. Blount, L. Valentin-Blasini, S.M. Lesch, and R.I. Krieger. 2008. Perchlorate in the feed-dairy continuum of the southwestern United States. *Journal of Agricultural and Food Chemistry* 56 (13):5443-50.

Perchlorate (continued)

11. Sanchez, C.A., K.S. Crump, R.I. Krieger, N.R. Khandaker, and J.P. Gibbs. 2005. Perchlorate and nitrate in leafy vegetables of North America. *Environmental Science & Technology* 39 (24):9391-7.

12. U.S. Food and Drug Administration. 2007. *2004-2005 Exploratory Survey Data on Perchlorate in Food*. U.S. FDA. Retrieved January 18, 2012 from http://www.fda.gov/Food/FoodSafety/FoodContaminantsAdulteration/ChemicalContaminants/Perchlorate/ucm077685.htm.

13. Dasgupta, P.K., J.V. Dyke, A.B. Kirk, and W.A. Jackson. 2006. Perchlorate in the United States. Analysis of relative source contributions to the food chain. *Environmental Science and Technology* 40 (21):6608-14.

14. Dasgupta, P.K., P.K. Martinelango, W.A. Jackson, T.A. Anderson, K. Tian, R.W. Tock, and S. Rajagopalan. 2005. The origin of naturally occurring perchlorate: the role of atmospheric processes. *Environmental Science and Technology* 39 (6):1569-75.

15. Susarla, S., T.W. Collette, A.W. Garrison, N.L. Wolfe, and S.C. McCutcheon. 1999. Perchlorate identification in fertilizers. *Environmental Science and Technology* 33 (19):3469-3472.

16. Susarla, S., T.W. Collette, A.W. Garrison, N.L. Wolfe, and S.C. McCutcheon. 2000. Additions and corrections: perchlorate identification in fertilizers. *Environmental Science and Technology* 34 (1):1.

17. TRC Environmental Corporation. 1998. *Chemical Fertilizer as a Potential Surce of Perchlorate*. Irvine,CA.

18. Blount, B.C., D.Q. Rich, L. Valentin-Blasini, S. Lashley, C.V. Ananth, E. Murphy, J.C. Smulian, B.J. Spain, D.B. Barr, T. Ledoux, et al. 2009. Perinatal exposure to perchlorate, thiocyanate, and nitrate in New Jersey mothers and newborns. *Environmental Science & Technology* 43 (19):7543-9.

19. Dasgupta, P.K., A.B. Kirk, J.V. Dyke, and S. Ohira. 2008. Intake of iodine and perchlorate and excretion in human milk. *Environmental Science & Technology* 42 (21):8115-21.

20. Kirk, A.B., J.V. Dyke, C.F. Martin, and P.K. Dasgupta. 2007. Temporal patterns in perchlorate, thiocyanate, and iodide excretion in human milk. *Environmental Health Perspectives* 115 (2):182-6.

21. Kirk, A.B., P.K. Martinelango, K. Tian, A. Dutta, E.E. Smith, and P.K. Dasgupta. 2005. Perchlorate and iodide in dairy and breast milk. *Environmental Science & Technology* 39 (7):2011-7.

22. Oldi, J.F., and K. Kannan. 2009. Analysis of perchlorate in human saliva by liquid chromatography-tandem mass spectrometry. *Environmental Science & Technology* 43 (1):142-7.

23. Oldi, J.F., and K. Kannan. 2009. Perchlorate in human blood serum and plasma: Relationship to concentrations in saliva. *Chemosphere* 77 (1):43-7.

24. Centers for Disease Control and Prevention. 2009. *Fourth National Report on Human Exposure to Environmental Chemicals*. Atlanta, GA: CDC. http://www.cdc.gov/exposurereport/.

25. Valentin-Blasini, L., B.C. Blount, S. Otero-Santos, Y. Cao, J.C. Bernbaum, and W.J. Rogan. 2011. Perchlorate exposure and dose estimates in infants. *Environmental Science & Technology* 45 (9):4127-32.

26. Blount, B.C., J.L. Pirkle, J.D. Osterloh, L. Valentin-Blasini, and K.L. Caldwell. 2006. Urinary perchlorate and thyroid hormone levels in adolescent and adult men and women living in the United States. *Environmental Health Perspectives* 114 (12):1865-71.

27. Otero-Santos, S.M., A.D. Delinsky, L. Valentin-Blasini, J. Schiffer, and B.C. Blount. 2009. Analysis of perchlorate in dried blood spots using ion chromatography and tandem mass spectrometry. *Analytical Chemistry* 81 (5):1931-6.

28. Zhang, T., Q. Wu, H.W. Sun, J. Rao, and K. Kannan. 2010. Perchlorate and iodide in whole blood samples from infants, children, and adults in Nanchang, China. *Environmental Science & Technology* 44 (18):6947-53.

29. Blount, B.C., and L. Valentin-Blasini. 2007. Biomonitoring as a method for assessing exposure to perchlorate. *Thyroid* 17 (9):837-41.

30. Centers for Disease Control and Prevention. *Perchlorate in Baby Formula Fact Sheet*. Retrieved 8/13/09 from http://cdc.gov/nceh/features/perchlorate_factsheet.htm.

31. Pearce, E.N., A.M. Leung, B.C. Blount, H.R. Bazrafshan, X. He, S. Pino, L. Valentin-Blasini, and L.E. Braverman. 2007. Breast milk iodine and perchlorate concentrations in lactating Boston-area women. *Journal of Clinical Endocrinology and Metabolism* 92 (5):1673-7.

32. Schier, J.G., A.F. Wolkin, L. Valentin-Blasini, M.G. Belson, S.M. Kieszak, C.S. Rubin, and B.C. Blount. 2009. Perchlorate exposure from infant formula and comparisons with the perchlorate reference dose. *Journal of Exposure Science and Environmental Epidemiology* 20 (3):281-7.

33. Lorber, M. 2009. Use of a simple pharmacokinetic model to characterize exposure to perchlorate. *Journal of Exposure Science and Environmental Epidemiology* 19 (3):260-73.

34. Mendez, W., E. Dederick, and J. Cohen. 2010. Drinking water contribution to aggregate perchlorate intake of reproductive-age women in the United States estimated by dietary intake simulation and analysis of urinary excretion data. *Journal of Exposure Science and Environmental Epidemiology* 20 (3):288-97.

Perchlorate (continued)

35. Blount, B., L. Valentin-Blasini, and D.L. Ashley. 2006. Assessing human exposure to perchlorate using biomonitoring. *Journal of ASTM International* 3 (7):3004-3010.

36. Blount, B.C., K.U. Alwis, R.B. Jain, B.L. Solomon, J.C. Morrow, and W.A. Jackson. 2010. Perchlorate, nitrate, and iodide intake through tap water. *Environmental Science and Technology* 44 (24):9564-70.

37. Huber, D.R., B.C. Blount, D.T. Mage, F.J. Letkiewicz, A. Kumar, and R.H. Allen. 2011. Estimating perchlorate exposure from food and tap water based on US biomonitoring and occurrence data. *Journal of Exposure Science and Environmental Epidemiology* 21 (4):395-407.

38. Greer, M.A., G. Goodman, R.C. Pleus, and S.E. Greer. 2002. Health effects assessment for environmental perchlorate contamination: the dose response for inhibition of thyroidal radioiodine uptake in humans. *Environmental Health Perspectives* 110 (9):927-37.

39. Stanbury, J.B., and J.B. Wyngaarden. 1952. Effect of perchlorate on the human thyroid gland. *Metabolism* 1 (6):533-9.

40. Haddow, J.E., G.E. Palomaki, W.C. Allan, J.R. Williams, G.J. Knight, J. Gagnon, C.E. O'Heir, M.L. Mitchell, R.J. Hermos, S.E. Waisbren, et al. 1999. Maternal thyroid deficiency during pregnancy and subsequent neuropsychological development of the child. *New England Journal of Medicine* 341 (8):549-55.

41. Miller, M.D., K.M. Crofton, D.C. Rice, and R.T. Zoeller. 2009. Thyroid-disrupting chemicals: interpreting upstream biomarkers of adverse outcomes. *Environmental Health Perspectives* 117 (7):1033-41.

42. U.S. Food and Drug Administration. *Perchlorate Questions and Answers*. Retrieved 8/13/09 from http://www.fda.gov/Food/FoodSafety/FoodContaminantsAdulteration/ChemicalContaminants/Perchlorate/ucm077572.htm#effects.

43. U.S. Environmental Protection Agency. 2009. *Perchlorate (ClO4) and Perchlorate Salts Quickview (CASRN 7790-98-9)*. Retrieved 11/19/09 from http://cfpub.epa.gov/ncea/iris/index.cfm?fuseaction=iris.showQuickView&substance_nmbr=1007.

44. Morreale de Escobar, G., M.J. Obregon, and F. Escobar del Rey. 2000. Is neuropsychological development related to maternal hypothyroidism or to maternal hypothyroxinemia? *Journal of Clinical Endocrinology and Metabolism* 85 (11):3975-87.

45. Porterfield, S.P. 1994. Vulnerability of the developing brain to thyroid abnormalities: environmental insults to the thyroid system. *Environmental Health Perspectives* 102 (Suppl 2):125-30.

46. Caldwell, K.L., A. Makhmudov, E. Ely, R.L. Jones, and R.Y. Wang. 2011. Iodine status of the U.S. population, National Health and Nutrition Examination Survey, 2005-2006 and 2007-2008. *Thyroid* 21 (4):419-27.

47. Khan, M.A., S.E. Fenton, A.E. Swank, S.D. Hester, A. Williams, and D.C. Wolf. 2005. A mixture of ammonium perchlorate and sodium chlorate enhances alterations of the pituitary-thyroid axis caused by the individual chemicals in adult male F344 rats. *Toxicologic Pathology* 33 (7):776-83.

48. Steinmaus, C., M.D. Miller, and R. Howd. 2007. Impact of smoking and thiocyanate on perchlorate and thyroid hormone associations in the 2001-2002 National Health and Nutrition Examination Survey. *Environmental Health Perspectives* 115 (9):1333-8.

49. Pearce, E.N., J.H. Lazarus, P.P. Smyth, X. He, D. Dall'amico, A.B. Parkes, R. Burns, D.F. Smith, A. Maina, J.P. Bestwick, et al. 2010. Perchlorate and thiocyanate exposure and thyroid function in first-trimester pregnant women. *Journal of Clinical Endocrinology and Metabolism* 95 (7):3207-15.

50. Téllez Téllez, R., P. Michaud Chacón, C. Reyes Abarca, B.C. Blount, C.B. Van Landingham, K.S. Crump, and J.P. Gibbs. 2005. Long-term environmental exposure to perchlorate through drinking water and thyroid function during pregnancy and the neonatal period. *Thyroid* 15 (9):963-75.

51. Steinmaus, C., M.D. Miller, and A.H. Smith. 2010. Perchlorate in drinking water during pregnancy and neonatal thyroid hormone levels in California. *Journal of Occupational and Environmental Medicine* 52 (12):1217-524.

52. Amitai, Y., G. Winston, J. Sack, J. Wasser, M. Lewis, B.C. Blount, L. Valentin-Blasini, N. Fisher, A. Israeli, and A. Leventhal. 2007. Gestational exposure to high perchlorate concentrations in drinking water and neonatal thyroxine levels. *Thyroid* 17 (9):843-50.

53. Buffler, P.A., M.A. Kelsh, E.C. Lau, C.H. Edinboro, J.C. Barnard, G.W. Rutherford, J.J. Daaboul, L. Palmer, and F.W. Lorey. 2006. Thyroid function and perchlorate in drinking water: an evaluation among California newborns, 1998. *Environmental Health Perspectives* 114 (5):798-804.

54. Crump, C., P. Michaud, R. Téllez, C. Reyes, G. Gonzalez, E.L. Montgomery, K.S. Crump, G. Lobo, C. Becerra, and J.P. Gibbs. 2000. Does perchlorate in drinking water affect thyroid function in newborns or school-age children? *Journal of Occupational and Environmental Medicine* 42 (6):603-12.

55. Lamm, S.H., and M. Doemland. 1999. Has perchlorate in drinking water increased the rate of congenital hypothyroidism? *Journal of Occupational and Environmental Medicine* 41 (5):409-11.

56. Li, F.X., D.M. Byrd, G.M. Deyhle, D.E. Sesser, M.R. Skeels, S.R. Katkowsky, and S.H. Lamm. 2000. Neonatal thyroid-stimulating hormone level and perchlorate in drinking water. *Teratology* 62 (6):429-31.

Perchlorate (continued)

57. U.S. Environmental Protection Agency. 2008. *Interim Drinking Water Health Advisory for Perchlorate*. Washington, DC: U.S. Environmental Protection Agency. EPA 822-R-08-025. http://www.epa.gov/ogwdw/contaminants/unregulated/pdfs/healthadvisory_perchlorate_interim.pdf.

58. U.S. Environmental Protection Agency. 2011. Drinking Water: Regulatory Determination on Perchlorate. *Federal Register* 76:7762-7767.

59. American Water Works Association. 2010. *Perchlorate*. American Water Works Association. Retrieved February 11, 2011 from http://www.awwa.org/Government/Content.cfm?ItemNumber=1065&navItemNumber=3833.

60. Ye, X., L.Y. Wong, A.M. Bishop, and A.M. Calafat. 2011. Variability of urinary concentrations of bisphenol a in spot samples, first morning voids, and 24-hour collections. *Environmental Health Perspectives* 119 (7):983-8.

61. Crump, K.S., and J.P. Gibbs. 2005. Benchmark calculations for perchlorate from three human cohorts. *Environmental Health Perspectives* 113 (8):1001-8.

62. Preau, J.L., Jr., L.Y. Wong, M.J. Silva, L.L. Needham, and A.M. Calafat. 2010. Variability over 1 week in the urinary concentrations of metabolites of diethyl phthalate and di(2-ethylhexyl) phthalate among eight adults: an observational study. *Environmental Health Perspectives* 118 (12):1748-54.

63. Jackson, S. 1966. Creatinine in urine as an index of urinary excretion rate. *Health Physics* 12 (6):843-50.

64. Barr, D.B., L.C. Wilder, S.P. Caudill, A.J. Gonzalez, L.L. Needham, and J.L. Pirkle. 2005. Urinary creatinine concentrations in the U.S. population: implications for urinary biologic monitoring measurements. *Environmental Health Perspectives* 113 (2):192-200.

65. Boeniger, M.F., L.K. Lowry, and J. Rosenberg. 1993. Interpretation of urine results used to assess chemical exposure with emphasis on creatinine adjustments: A review. *American Industrial Hygiene Association Journal* 54 (10):615-27.

66. National Center for Health Statistics. 2003-2004. *Vital Statistics Natality Birth Data*. Retrieved June 15, 2009 from http://www.cdc.gov/nchs/data_access/Vitalstatsonline.htm. .

67. Axelrad, D.A., and J. Cohen. 2011. Calculating summary statistics for population chemical biomonitoring in women of childbearing age with adjustment for age-specific natality. *Environmental Research* 111 (1):149-155.

Health

Introduction

1. U.S. Census Bureau. *Poverty Thresholds by Size of Family and Number of Related Children Under 18 Years: 2010*. U.S. Census Bureau. Retrieved August 20, 2011 from http://www.census.gov/hhes/www/cpstables/032011/pov/new35_000.htm.

Respiratory Diseases

1. Institute of Medicine. 2000. *Clearing the Air: Asthma and Indoor Air Exposures*. Washington DC: National Academy Press. http://books.nap.edu/catalog/9610.html.

2. U.S. Department of Health and Human Services. 2006. *The Health Consequences of Involuntary Exposure to Tobacco Smoke: A Report of the Surgeon General*. Atlanta, GA: Centers for Disease Control and Prevention, Coordinating Center for Health Promotion, National Center for Chronic Disease Prevention and Health Promotion, Office on Smoking and Health. http://www.surgeongeneral.gov/library/secondhandsmoke/report/index.html.

3. U.S. Environmental Protection Agency. 2006. *Air Quality Criteria for Ozone and Related Photochemical Oxidants*. Washington, DC: U.S. EPA. EPA/600/R-05/004aF. http://cfpub.epa.gov/ncea/cfm/recordisplay.cfm?deid=149923.

4. U.S. Environmental Protection Agency. 2008. *Integrated Science Assessment for Oxides of Nitrogen – Health Criteria (Final Report)*. Washington, DC: U.S. EPA, Office of Research and Development. http://oaspub.epa.gov/eims/eimscomm.getfile?p_download_id=475020.

5. U.S. Environmental Protection Agency. 2008. *Integrated Science Assessment for Sulfur Oxides - Health Criteria (Final Report)*. Washington, DC: U.S. EPA. EPA/600/R-08/047F. http://cfpub.epa.gov/ncea/cfm/recordisplay.cfm?deid=198843.

6. U.S. Environmental Protection Agency. 2009. *Integrated Science Assessment for Particulate Matter (Final Report)*. Washington, DC: U.S. EPA. EPA/600/R-08/139F. http://cfpub.epa.gov/ncea/cfm/recordisplay.cfm?deid=216546.

7. U.S. Environmental Protection Agency. 2010. *Integrated Science Assessment for Carbon Monoxide (Final Report)*. Washington, DC: U.S. Environmental Protection Agency. http://cfpub.epa.gov/ncea/cfm/recordisplay.cfm?deid=218686.

8. Fauroux, B., M. Sampil, P. Quénel, and Y. Lemoullec. 2000. Ozone: a trigger for hospital pediatric asthma emergency room visits. *Pediatric Pulmonology* 30 (1):41-6.

9. Parker, J.D., L.J. Akinbami, and T.J. Woodruff. 2009. Air pollution and childhood respiratory allergies in the United States. *Environmental Health Perspectives* 117 (1):140-147.

10. Schildcrout, J.S., L. Sheppard, T. Lumley, J.C. Slaughter, J.Q. Koenig, and G.G. Shapiro. 2006. Ambient air pollution and asthma exacerbations in children: an eight-city analysis. *American Journal of Epidemiology* 164 (6):505-17.

11. Clark, N.A., P.A. Demers, C.J. Karr, M. Koehoorn, C. Lencar, L. Tamburic, and M. Brauer. 2010. Effect of early life exposure to air pollution on development of childhood asthma. *Environmental Health Perspectives* 118 (2):284-90.

12. Gehring, U., A.H. Wijga, M. Brauer, P. Fischer, J.C. de Jongste, M. Kerkhof, M. Oldenwening, H.A. Smit, and B. Brunekreef. 2010. Traffic-related air pollution and the development of asthma and allergies during the first 8 years of life. *American Journal of Respiratory and Critical Care Medicine* 181 (6):596-603.

13. McConnell, R., T. Islam, K. Shankardass, M. Jerrett, F. Lurmann, F. Gilliland, J. Gauderman, E. Avol, N. Kuenzli, L. Yao, et al. 2010. Childhood incident asthma and traffic-related air pollution at home and school. *Environmental Health Perspectives* 118 (7):1021-6.

14. Chauhan, A.J., and S.L. Johnston. 2003. Air pollution and infection in respiratory illness. *British Medical Bulletin* 68:95-112.

15. Ciencewicki, J., and I. Jaspers. 2007. Air pollution and respiratory viral infection. *Inhalation Toxicology* 19 (14):1135-46.

16. Dherani, M., D. Pope, M. Mascarenhas, K.R. Smith, M. Weber, and N. Bruce. 2008. Indoor air pollution from unprocessed solid fuel use and pneumonia risk in children aged under five years: a systematic review and meta-analysis. *Bulletin of the World Health Organization* 86 (5):390-398C.

17. U.S. Environmental Protection Agency. 2010. *National Ambient Air Quality Standards (NAAQS)*. U.S. EPA, Office of Air and Radiation. Retrieved October 20, 2010 from http://www.epa.gov/air/criteria.html.

18. Glinianaia, S.V., J. Rankin, R. Bell, T. Pless-Mulloli, and D. Howel. 2004. Does particulate air pollution contribute to infant death? A systematic review. *Environmental Health Perspectives* 112 (14):1365-71.

19. Woodruff, T.J., J.D. Parker, and K.C. Schoendorf. 2006. Fine particulate matter (PM2.5) air pollution and selected causes of postneonatal infant mortality in California. *Environmental Health Perspectives* 114 (5):786-90.

20. Kajekar, R. 2007. Environmental factors and developmental outcomes in the lung. *Pharmacology & Therapeutics* 114 (2):129-45.

21. Wigle, D.T., T.E. Arbuckle, M. Walker, M.G. Wade, S. Liu, and D. Krewski. 2007. Environmental hazards: evidence for effects on child health. *Toxicology and Environmental Health Part B: Critical Reviews* 10 (1-2):3-39.

22. Islam, T., K. Berhane, R. McConnell, W.J. Gauderman, E. Avol, J.M. Peters, and F.D. Gilliland. 2009. Glutathione-S-transferase (GST) P1, GSTM1, exercise, ozone and asthma incidence in school children. *Thorax* 64 (3):197-202.

Respiratory Diseases (continued)

23. Islam, T., R. McConnell, W.J. Gauderman, E. Avol, J.M. Peters, and F.D. Gilliland. 2008. Ozone, oxidant defense genes, and risk of asthma during adolescence. *American Journal of Respiratory and Critical Care Medicine* 177 (4):388-95.

24. McConnell, R., K. Berhane, F. Gilliland, S.J. London, T. Islam, W.J. Gauderman, E. Avol, H.G. Margolis, and J.M. Peters. 2002. Asthma in exercising children exposed to ozone: a cohort study. *Lancet* 359 (9304):386-91.

25. Villeneuve, P.J., L. Chen, B.H. Rowe, and F. Coates. 2007. Outdoor air pollution and emergency department visits for asthma among children and adults: a case-crossover study in northern Alberta, Canada. *Environmental Health* 6:40.

26. Leikauf, G.D. 2002. Hazardous air pollutants and asthma. *Environmental Health Perspectives* 110 Suppl 4:505-26.

27. Cook, R., M. Strum, J.S. Touma, T. Palma, J. Thurman, D. Ensley, and R. Smith. 2007. Inhalation exposure and risk from mobile source air toxics in future years. *Journal of Exposure Science and Environmental Epidemiology* 17 (1):95-105.

28. U.S. Environmental Protection Agency. 2003. *Toxicological Review of Acrolein (CAS No. 107-02-8)*. Washington, DC: U.S. EPA, National Center for Environmental Assessment. EPA/635/R-03/003. http://www.epa.gov/iris/toxreviews/0364tr.pdf.

29. Woodruff, T.J., D.A. Axelrad, J. Caldwell, R. Morello-Frosch, and A. Rosenbaum. 1998. Public health implications of 1990 air toxics concentrations across the United States. *Environmental Health Perspectives* 106 (5):245-51.

30. Faroon, O., N. Roney, J. Taylor, A. Ashizawa, M.H. Lumpkin, and D.J. Plewak. 2008. Acrolein health effects. *Toxicology and Industrial Health* 24 (7):447-90.

31. Gauderman, W.J., H. Vora, R. McConnell, K. Berhane, F. Gilliland, D. Thomas, F. Lurmann, E. Avol, N. Kunzli, M. Jerrett, et al. 2007. Effect of exposure to traffic on lung development from 10 to 18 years of age: a cohort study. *Lancet* 369 (9561):571-7.

32. Jerrett, M., K. Shankardass, K. Berhane, W.J. Gauderman, N. Künzli, E. Avol, F. Gilliland, F. Lurmann, J.N. Molitor, J.T. Molitor, et al. 2008. Traffic-related air pollution and asthma onset in children: a prospective cohort study with individual exposure measurement. *Environmental Health Perspectives* 116 (10):1433-38.

33. Karr, C.J., P.A. Demers, M.W. Koehoorn, C.C. Lencar, L. Tamburic, and M. Brauer. 2009. Influence of ambient air pollutant sources on clinical encounters for infant bronchiolitis. *American Journal of Respiratory and Critical Care Medicine* 180 (10):995-1001.

34. McConnell, R., K. Berhane, L. Yao, M. Jerrett, F. Lurmann, F. Gilliland, N. Kunzli, J. Gauderman, E. Avol, D. Thomas, et al. 2006. Traffic, susceptibility, and childhood asthma. *Environmental Health Perspectives* 114 (5):766-72.

35. Morgenstern, V., A. Zutavern, J. Cyrys, I. Brockow, U. Gehring, S. Koletzko, C.P. Bauer, D. Reinhardt, H.E. Wichmann, and J. Heinrich. 2007. Respiratory health and individual estimated exposure to traffic-related air pollutants in a cohort of young children. *Occupational and Environmental Medicine* 64 (1):8-16.

36. Salam, M.T., T. Islam, and F.D. Gilliland. 2008. Recent evidence for adverse effects of residential proximity to traffic sources on asthma. *Current Opinion in Pulmonary Medicine* 14 (1):3-8.

37. Health Effects Institute. 2010. *HEI Panel on the Health Effects of Traffic-Related Air Pollution: A Critical Review of the Literature on Emissions, Exposure, and Health Effects*. Boston, MA. HEI Special Report 17. http://pubs.healtheffects.org/view.php?id=334

38. Bartra, J., J. Mullol, A. del Cuvillo, I. Davila, M. Ferrer, I. Jauregui, J. Montoro, J. Sastre, and A. Valero. 2007. Air pollution and allergens. *Journal of Investigational Allergology and Clinical Immunology* 17 Suppl 2:3-8.

39. Bråbäck, L., and B. Forsberg. 2009. Does traffic exhaust contribute to the development of asthma and allergic sensitization in children: findings from recent cohort studies. *Environmental Health* 8:17.

40. Krzyzanowski, M., B. Kuna-Dibbert, and J. Schneider, eds. 2005. *Health Effects of Transport-Related Air Pollution*. Copenhagen, Denmark: World Health Organization, Europe.

41. Dales, R., L. Liu, A.J. Wheeler, and N.L. Gilbert. 2008. Quality of indoor residential air and health. *Canadian Medical Association Journal* 179 (2):147-52.

42. Seltzer, J.M., and M.J. Fedoruk. 2007. Health effects of mold in children. *Pediatric Clinics of North America* 54 (2):309-33, viii-ix.

43. Perez-Padilla, R., A. Schilmann, and H. Riojas-Rodriguez. 2010. Respiratory health effects of indoor air pollution. *The International Journal of Tuberculosis and Lung Disease* 14 (9):1079-86.

44. Dales, R., and M. Raizenne. 2004. Residential exposure to volatile organic compounds and asthma. *Journal of Asthma* 41 (3):259-70.

45. Fuentes-Leonarte, V., J.M. Tenias, and F. Ballester. 2009. Levels of pollutants in indoor air and respiratory health in preschool children: a systematic review. *Pediatric Pulmonology* 44 (3):231-43.

46. McGwin, G., J. Lienert, and J.I. Kennedy. 2010. Formaldehyde exposure and asthma in children: a systematic review. *Environmental Health Perspectives* 118 (3):313-7.

47. Cheraghi, M., and S. Salvi. 2009. Environmental tobacco smoke (ETS) and respiratory health in children. *European Journal of Pediatrics* 168 (8):897-905.

Respiratory Diseases (continued)

48. Li, Y.F., F.D. Gilliland, K. Berhane, R. McConnell, W.J. Gauderman, E.B. Rappaport, and J.M. Peters. 2000. Effects of in utero and environmental tobacco smoke exposure on lung function in boys and girls with and without asthma. *American Journal of Respiratory and Critical Care Medicine* 162 (6):2097-104.

49. Wang, L., and K.E. Pinkerton. 2008. Detrimental effects of tobacco smoke exposure during development on postnatal lung function and asthma. *Birth Defects Research Part C: Embryo Today* 84 (1):54-60.

50. Xepapadaki, P., Y. Manios, T. Liarigkovinos, E. Grammatikaki, N. Douladiris, C. Kortsalioudaki, and N.G. Papadopoulos. 2009. Association of passive exposure of pregnant women to environmental tobacco smoke with asthma symptoms in children. *Pediatric Allergy and Immunology* 20 (5):423-9.

51. Jedrychowski, W., A. Galas, A. Pac, E. Flak, D. Camman, V. Rauh, and F. Perera. 2005. Prenatal ambient air exposure to polycyclic aromatic hydrocarbons and the occurrence of respiratory symptoms over the first year of life. *European Journal of Epidemiology* 20 (9):775-82.

52. Miller, R.L., R. Garfinkel, M. Horton, D. Camann, F.P. Perera, R.M. Whyatt, and P.L. Kinney. 2004. Polycyclic aromatic hydrocarbons, environmental tobacco smoke, and respiratory symptoms in an inner-city birth cohort. *Chest* 126 (4):1071-8.

53. Rosa, M.J., K.H. Jung, M.S. Perzanowski, E.A. Kelvin, K.W. Darling, D.E. Camann, S.N. Chillrud, R.M. Whyatt, P.L. Kinney, F.P. Perera, et al. 2011. Prenatal exposure to polycyclic aromatic hydrocarbons, environmental tobacco smoke and asthma. *Respiratory Medicine* 105 (6):869-76.

54. Mortimer, K., R. Neugebauer, F. Lurmann, S. Alcorn, J. Balmes, and I. Tager. 2008. Air pollution and pulmonary function in asthmatic children: effects of prenatal and lifetime exposures. *Epidemiology* 19 (4):550-7.

55. Mortimer, K., R. Neugebauer, F. Lurmann, S. Alcorn, J. Balmes, and I. Tager. 2008. Early-lifetime exposure to air pollution and allergic sensitization in children with asthma. *Journal of Asthma* 45 (10):874-81.

56. Bloom, B., R.A. Cohen, and G. Freeman. 2010. Summary health statistics for U.S. children: National Health Interview Survey, 2009. *Vital and Health Statistics* 10 (247):1-89.

57. Barnett, S.B., and T.A. Nurmagambetov. 2011. Costs of asthma in the United States: 2002-2007. *The Journal of Allergy and Clinical Immunology* 127 (1):145-52.

58. Rudd, R.A., and J.E. Moorman. 2007. Asthma incidence: data from the National Health Interview Survey, 1980-1996. *Journal of Asthma* 44 (1):65-70.

59. Litonjua, A.A., V.J. Carey, S.T. Weiss, and D.R. Gold. 1999. Race, socioeconomic factors, and area of residence are associated with asthma prevalence. *Pediatric Pulmonology* 28 (6):394-401.

60. Panico, L., M. Bartley, M. Marmot, J.Y. Nazroo, A. Sacker, and Y.J. Kelly. 2007. Ethnic variation in childhood asthma and wheezing illnesses: findings from the Millennium Cohort Study. *International Journal of Epidemiology* 36 (5):1093-102.

61. Pearlman, D.N., S. Zierler, S. Meersman, H.K. Kim, S.I. Viner-Brown, and C. Caron. 2006. Race disparities in childhood asthma: does where you live matter? *Journal of the National Medical Association* 98 (2):239-47.

62. Aligne, C.A., P. Auinger, R.S. Byrd, and M. Weitzman. 2000. Risk factors for pediatric asthma. Contributions of poverty, race, and urban residence. *American Journal of Respiratory and Critical Care Medicine* 162 (3 Pt 1):873-7.

63. Crain, E.F., M. Walter, G.T. O'Connor, H. Mitchell, R.S. Gruchalla, M. Kattan, G.S. Malindzak, P. Enright, R. Evans, 3rd, W. Morgan, et al. 2002. Home and allergic characteristics of children with asthma in seven U.S. urban communities and design of an environmental intervention: the Inner-City Asthma Study. *Environmental Health Perspectives* 110 (9):939-45.

64. Kitch, B.T., G. Chew, H.A. Burge, M.L. Muilenberg, S.T. Weiss, T.A. Platts-Mills, G. O'Connor, and D.R. Gold. 2000. Socioeconomic predictors of high allergen levels in homes in the greater Boston area. *Environmental Health Perspectives* 108 (4):301-7.

65. Leaderer, B.P., K. Belanger, E. Triche, T. Holford, D.R. Gold, Y. Kim, T. Jankun, P. Ren, J.E. McSharry, T.A. Platts-Mills, et al. 2002. Dust mite, cockroach, cat, and dog allergen concentrations in homes of asthmatic children in the northeastern United States: impact of socioeconomic factors and population density. *Environmental Health Perspectives* 110 (4):419-25.

66. Federal Interagency Forum on Child and Family Statistics. 2009. *America's Children: Key National Indicators of Well-Being, 2009: Outdoor and Indoor Air Quality*. Washington, DC: U.S. Government Printing Office. http://childstats.gov/americaschildren09/phenviro1.asp.

67. Arbes, S.J., R.D.Cohn, M. Yin, M.L. Muilenberg, H.A. Burge, W. Friedman, and D.C. Zeldin. 2003. House dust mite allergen in U.S. beds: results from the first national survey of lead and allergens in housing. *Journal of Allergy and Clinical Immunology* 111 (2):408-14.

68. Cohn, R.D., S.J. Arbes, Jr., M. Yin, R. Jaramillo, and D.C. Zeldin. 2004. National prevalence and exposure risk for mouse allergen in US households. *The Journal of Allergy and Clinical Immunology* 113 (6):1167-71.

69. Elliott, L., S.J. Arbes, E.S. Harvey, R.C. Lee, P.M. Salo, R.D. Cohn, S.J. London, and D.C. Zeldin. 2007. Dust weight and asthma prevalence in the National Survey of Lead and Allergens in Housing (NSLAH). *Environmental Health Perspectives* 115 (2):215-20.

Respiratory Diseases (continued)

70. Farber, H.J., C. Johnson, and R.C. Beckerman. 1998. Young inner-city children visiting the emergency room (ER) for asthma: risk factors and chronic care behaviors. *Journal of Asthma* 35 (7):547-52.

71. Halfon, N., and P.W. Newacheck. 1993. Childhood asthma and poverty: differential impacts and utilization of health services. *Pediatrics* 91 (1):56-61.

72. Lozano, P., J.A. Finkelstein, J. Hecht, R. Shulruff, and K.B. Weiss. 2003. Asthma medication use and disease burden in children in a primary care population. *Archives of Pediatrics and Adolescent Medicine* 157 (1):81-8.

73. Price, M.R., J.M. Norris, B. Bucher Bartleson, L.A. Gavin, and M.D. Klinnert. 1999. An investigation of the medical care utilization of children with severe asthma according to their type of insurance. *Journal of Asthma* 36 (3):271-9.

74. Rosenbach, M.L., C. Irvin, and R.F. Coulam. 1999. Access for low-income children: is health insurance enough? *Pediatrics* 103 (6 Pt 1):1167-74.

75. Stanton, M.S., and D. Dougherty. 2005. Chronic Care for Low-Income Children with Asthma: Strategies for Improvement. In *Research in Action Issue 18*. Rockville, MD: Agency for Healthcare Research and Quality.

76. Yoos, H.L., H. Kitzman, and A. McMullen. 2003. Barriers to anti-inflammatory medication use in childhood asthma. *Ambulatory Pediatrics* 3 (4):181-90.

77. Eder, W., M.J. Ege, and E. von Mutius. 2006. The asthma epidemic. *New England Journal of Medicine* 355 (21):2226-35.

78. Zorc, J.J., and C.B. Hall. 2010. Bronchiolitis: recent evidence on diagnosis and management. *Pediatrics* 125 (2):342-9.

79. Coffman, J.M., M.D. Cabana, H.A. Halpin, and E.H. Yelin. 2008. Effects of asthma education on children's use of acute care services: a meta-analysis. *Pediatrics* 121 (3):575-86.

80. Flores, G., M. Abreu, S. Tomany-Korman, and J. Meurer. 2005. Keeping children with asthma out of hospitals: parents' and physicians' perspectives on how pediatric asthma hospitalizations can be prevented. *Pediatrics* 116 (4):957-65.

81. Flores, G., C. Snowden-Bridon, S. Torres, R. Perez, T. Walter, J. Brotanek, H. Lin, and S. Tomany-Korman. 2009. Urban minority children with asthma: substantial morbidity, compromised quality and access to specialists, and the importance of poverty and specialty care. *The Journal of Asthma* 46 (4):392-8.

82. Fox, P., P.G. Porter, S.H. Lob, J.H. Boer, D.A. Rocha, and J.W. Adelson. 2007. Improving asthma-related health outcomes among low-income, multiethnic, school-aged children: results of a demonstration project that combined continuous quality improvement and community health worker strategies. *Pediatrics* 120 (4):e902-11.

83. U.S. Department of Health and Human Services. 2007. *Guidelines for the Diagnosis and Management of Asthma*. Bethesda, MD: National Heart, Lung, and Blood Institute, National Asthma Education and Prevention Program. NIH Publication Number 08-5846. http://www.nhlbi.nih.gov/guidelines/asthma/asthsumm.pdf.

84. Homer, C.J., P. Szilagyi, L. Rodewald, S.R. Bloom, P. Greenspan, S. Yazdgerdi, J.M. Leventhal, D. Finkelstein, and J.M. Perrin. 1996. Does quality of care affect rates of hospitalization for childhood asthma? *Pediatrics* 98 (1):18-23.

85. Russo, M.J., K.M. McConnochie, J.T. McBride, P.G. Szilagyi, A.M. Brooks, and K.J. Roghmann. 1999. Increase in admission threshold explains stable asthma hospitalization rates. *Pediatrics* 104 (3 Pt 1):454-62.

86. Gwynn, R.C., and G.D. Thurston. 2001. The burden of air pollution: impacts among racial minorities. *Environmental Health Perspectives* 109 (Suppl. 4):501-6.

87. Nauenberg, E., and K. Basu. 1999. Effect of insurance coverage on the relationship between asthma hospitalizations and exposure to air pollution. *Public Health Reports* 114 (2):135-48.

88. Akinbami, L.J., J.E. Moorman, P.L. Garbe, and E.J. Sondik. 2009. Status of childhood asthma in the United States, 1980-2007. *Pediatrics* 123 Suppl 3:S131-45.

89. Gupta, R.S., V. Carrion-Carire, and K.B. Weiss. 2006. The widening black/white gap in asthma hospitalizations and mortality. *The Journal of Allergy and Clinical Immunology* 117 (2):351-8.

90. McDaniel, M., C. Paxson, and J. Waldfogel. 2006. Racial disparities in childhood asthma in the United States: evidence from the National Health Interview Survey, 1997 to 2003. *Pediatrics* 117 (5):e868-77.

91. Corburn, J., J. Osleeb, and M. Porter. 2006. Urban asthma and the neighbourhood environment in New York City. *Health & Place* 12 (2):167-79.

92. Maryland Department of Health and Mental Hygiene, and Maryland Department of the Environment. 2008. *Maryland's Children and the Environment*. http://www.dhmh.state.md.us/reports/pdf/MDChildrenEnv08.pdf.

93. Chan, K.S., E. Keeler, M. Schonlau, M. Rosen, and R. Mangione-Smith. 2005. How do ethnicity and primary language spoken at home affect management practices and outcomes in children and adolescents with asthma? *Archives of Pediatrics and Adolescent Medicine* 159 (3):283-9.

Respiratory Diseases (continued)

94. Scott, G., and H. Ni. 2004. Access to health care among Hispanic/Latino children: United States, 1998-2001. *Advance Data from Vital and Health Statistics* (344):1-20.

95. Centers for Disease Control and Prevention. 2008. Youth risk behavior surveillance — United States, 2007. *Morbidity and Mortality Weekly Report* 57 (SS-4).

96. Child and Adolescent Health Measurement Initiative. 2009. *2007 National Survey of Children's Health*. Child and Adolescent Health Measurement Initiative, Data Resource Center for Child and Adolescent Health. Retrieved June 16, 2009 from www.nschdata.org.

Childhood Cancer

1. National Cancer Institute. 2009. *Dictionary of Cancer Terms*. Retrieved January 14, 2009 from http://www.cancer.gov/dictionary.

2. National Cancer Institute. 2010. *A Snapshot of Pediatric Cancers*. Retrieved August 10, 2011 from http://www.cancer.gov/aboutnci/servingpeople/snapshots/pediatric.pdf.

3. Linabery, A.M., and J.A. Ross. 2008. Trends in childhood cancer incidence in the U.S. (1992-2004). *Cancer* 112 (2):416-32.

4. National Cancer Institute. 2012. *Fact Sheet: Childhood Cancers*. National Institutes of Health, National Cancer Institute. Retrieved June 27, 2012 from http://www.cancer.gov/cancertopics/factsheet/Sites-Types/childhood.

5. President's Cancer Panel. 2010. *Reducing Environmental Cancer Risk: What We Can Do Now*. Bethesda, MD: National Cancer Institute, President's Cancer Panel. http://deainfo.nci.nih.gov/advisory/pcp/annualReports/pcp08-09rpt/PCP_Report_08-09_508.pdf.

6. Reis, L.A.G., M.A. Smith, J.G. Gurney, M. Linet, T. Tamra, J.L. Young, and G.R. Bunin. 1999. *Cancer Incidence and Survival among Children and Adolescents: United States SEER Program 1975-1995*. Bethesda, MD: National Cancer Institute, SEER Program. NIH Pub. No. 99-4649. http://www.seer.ims.nci.nih.gov/Publications/PedMono.

7. Jirtle, R.L., and M.K. Skinner. 2007. Environmental epigenomics and disease susceptibility. *Nature Reviews. Genetics* 8 (4):253-62.

8. Bird, A. 2007. Perceptions of epigenetics. *Nature* 447 (7143):396-8.

9. Hanahan, D., and R.A. Weinberg. 2011. Hallmarks of cancer: the next generation. *Cell* 144 (5):646-74.

10. Eyre, R., R.G. Feltbower, E. Mubwandarikwa, T.O. Eden, and R.J. McNally. 2009. Epidemiology of bone tumours in children and young adults. *Pediatric Blood & Cancer* 53 (6):941-52.

11. Holland, N., A. Fucic, D.F. Merlo, R. Sram, and M. Kirsch-Volders. 2011. Micronuclei in neonates and children: effects of environmental, genetic, demographic and disease variables. *Mutagenesis* 26 (1):51-6.

12. Infante-Rivard, C., D. Labuda, M. Krajinovic, and D. Sinnett. 1999. Risk of childhood leukemia associated with exposure to pesticides and with gene polymorphisms. *Epidemiology* 10 (5):481-7.

13. Infante-Rivard, C., G. Mathonnet, and D. Sinnett. 2000. Risk of childhood leukemia associated with diagnostic irradiation and polymorphisms in DNA repair genes. *Environmental Health Perspectives* 108 (6):495-8.

14. Infante-Rivard, C., and S. Weichenthal. 2007. Pesticides and childhood cancer: an update of Zahm and Ward's 1998 review. *Journal of Toxicology and Environmental Health Part B: Critical Reviews* 10 (1-2):81-99.

15. Metayer, C., and P.A. Buffler. 2008. Residential exposures to pesticides and childhood leukaemia. *Radiation Protection Dosimetry* 132 (2):212-9.

16. Institute of Medicine. 2002. *Cancer and the Environment: Gene-Environment Interaction*. Washington, DC: National Academy Press. http://www.nap.edu/catalog.php?record_id=10464.

17. Anderson, L.M., B.A. Diwan, N.T. Fear, and E. Roman. 2000. Critical windows of exposure for children's health: cancer in human epidemiological studies and neoplasms in experimental animal models. *Environmental Health Perspectives* 108 Supplement 3:573-94.

18. Buffler, P.A., M.L. Kwan, P. Reynolds, and K.Y. Urayama. 2005. Environmental and genetic risk factors for childhood leukemia: appraising the evidence. *Cancer Investigation* 23 (1):60-75.

19. Johnson, K.J., N.M. Springer, A.K. Bielinsky, D.A. Largaespada, and J.A. Ross. 2009. Developmental origins of cancer. *Cancer Research* 69 (16):6375-7.

20. Ma, X., P.A. Buffler, R.B. Gunier, G. Dahl, M.T. Smith, K. Reinier, and P. Reynolds. 2002. Critical windows of exposure to household pesticides and risk of childhood leukemia. *Environmental Health Perspectives* 110 (9):955-60.

21. Selevan, S.G., C.A. Kimmel, and P. Mendola. 2000. Identifying critical windows of exposure for children's health. *Environmental Health Perspectives* 108 Supplement 3:451-5.

22. Centers for Disease Control and Prevention. 2009. *Questions and Answers about Leukemia*. Retrieved April 17, 2009 from http://www.cdc.gov/NCEH/RADIATION/phase2/mleukemi.pdf.

Childhood Cancer (continued)

23. Belson, M., B. Kingsley, and A. Holmes. 2007. Risk factors for acute leukemia in children: a review. *Environmental Health Perspectives* 115 (1):138-45.

24. Boice, J., J.D., and R.W. Miller. 1999. Childhood and adult cancer after intrauterine exposure to ionizing radiation. *Teratology* 59 (227-233).

25. Doll, R., and R. Wakeford. 1997. Risk of childhood cancer from fetal irradiation. *British Journal of Radiology* 70:130-139.

26. National Council on Radiation Protection and Measurements. 2009. *Ionizing Radiation Exposure of the Population of the United States (2009)*. Bethesda, MD: NCRP. Report No. 160.

27. Pearce, M.S., J.A. Salotti, M.P. Little, K. McHugh, C. Lee, K.P. Kim, N.L. Howe, C.M. Ronckers, P. Rajaraman, A.W. Craft, et al. 2012. Radiation exposure from CT scans in childhood and subsequent risk of leukaemia and brain tumours: a retrospective cohort study. *The Lancet* 380 (9840):499-505.

28. Linet, M.S., E.E. Hatch, R.A. Kleinerman, L.L. Robison, W.T. Kaune, D.R. Friedman, R.K. Severson, C.M. Haines, C.T. Hartsock, S. Niwa, et al. 1997. Residential exposure to magnetic fields and acute lymphoblastic leukemia in children. *The New England Journal of Medicine* 337 (1):1-7.

29. National Research Council. 1997. *Possible Health Effects of Exposure to Residential Electrical and Magnetic Fields*. Washington, DC: National Academies Press. http://www.nap.edu/openbook.php?isbn=0309054478.

30. Evrard, A.S., D. Hemon, S. Billon, D. Laurier, E. Jougla, M. Tirmarche, and J. Clavel. 2005. Ecological association between indoor radon concentration and childhood leukaemia incidence in France, 1990-1998. *European Journal of Cancer Prevention* 14 (2):147-57.

31. Raaschou-Nielsen, O. 2008. Indoor radon and childhood leukaemia. *Radiation Protection Dosimetry* 132 (2):175-81.

32. Raaschou-Nielsen, O., C.E. Andersen, H.P. Andersen, P. Gravesen, M. Lind, J. Schuz, and K. Ulbak. 2008. Domestic radon and childhood cancer in Denmark. *Epidemiology* 19 (4):536-43.

33. Kendall, G.M., M.P. Little, R. Wakeford, K.J. Bunch, J.C. Miles, T.J. Vincent, J.R. Meara, and M.F. Murphy. 2012. A record-based case-control study of natural background radiation and the incidence of childhood leukaemia and other cancers in Great Britain during 1980-2006. *Leukemia* doi: 10.1038/leu.2012.151.

34. Brown, R.C. 2006. Review: Windows of exposure to pesticides for increased risk of childhood leukemia. *Toxicological & Environmental Chemistry* 88 (3):423-443.

35. Buckley, J.D., L.L. Robison, R. Swotinsky, D.H. Garabrant, M. LeBeau, P. Manchester, M.E. Nesbit, L. Odom, J.M. Peters, and W.G. Woods. 1989. Occupational exposures of parents of children with acute nonlymphocytic leukemia: a report from the Children's Cancer Study Group. *Cancer Research* 49:4030-4037.

36. Carozza, S.E., B. Li, K. Elgethun, and R. Whitworth. 2008. Risk of childhood cancers associated with residence in agriculturally intense areas in the United States. *Environmental Health Perspectives* 116 (4):559-65.

37. Feychting, M., N. Plato, G. Nise, and A. Ahlbom. 2001. Paternal occupational exposures and childhood cancer. *Environmental Health Perspectives* 109 (2):193-6.

38. Rudant, J., F. Menegaux, G. Leverger, A. Baruchel, B. Nelken, Y. Bertrand, C. Patte, H. Pacquement, C. Verite, A. Robert, et al. 2007. Household exposure to pesticides and risk of childhood hematopoietic malignancies: The ESCALE study (SFCE). *Environmental Health Perspectives* 115 (12):1787-93.

39. Turner, M.C., D.T. Wigle, and D. Krewski. 2010. Residential pesticides and childhood leukemia: a systematic review and meta-analysis. *Environmental Health Perspectives* 118 (1):33-41.

40. Van Maele-Fabry, G., A.C. Lantin, P. Hoet, and D. Lison. 2010. Childhood leukaemia and parental occupational exposure to pesticides: a systematic review and meta-analysis. *Cancer Causes & Control* 21 (6):787-809.

41. Van Maele-Fabry, G., A.C. Lantin, P. Hoet, and D. Lison. 2011. Residential exposure to pesticides and childhood leukaemia: a systematic review and meta-analysis. *Environment International* 37 (1):280-91.

42. Vinson, F., M. Merhi, I. Baldi, H. Raynal, and L. Gamet-Payrastre. 2011. Exposure to pesticides and risk of childhood cancer: a meta-analysis of recent epidemiological studies. *Occupational and Environmental Medicine* 68 (9):694-702.

43. Wigle, D.T., T.E. Arbuckle, M.C. Turner, A. Berube, Q. Yang, S. Liu, and D. Krewski. 2008. Epidemiologic evidence of relationships between reproductive and child health outcomes and environmental chemical contaminants. *Journal of Toxicology and Environmental Health Part B: Critical Reviews* 11 (5-6):373-517.

44. Wigle, D.T., M.C. Turner, and D. Krewski. 2009. A systematic review and meta-analysis of childhood leukemia and parental occupational pesticide exposure. *Environmental Health Perspectives* 117:1505-1513.

45. Zahm, S.H., and S.S. Devesa. 1995. Childhood cancer: overview of incidence trends and environmental carcinogens. *Environmental Health Perspectives* 103 (Suppl. 6):177-184.

Childhood Cancer (continued)

46. Zahm, S.H., and M.H. Ward. 1998. Pesticides and childhood cancer. *Environmental Health Perspectives* 106 (Suppl. 3):893-908.

47. Knox, E.G. 2005. Childhood cancers and atmospheric carcinogens. *Journal of Epidemiology and Community Health* 59 (2):101-5.

48. Reynolds, P., J. Von Behren, R.B. Gunier, D.E. Goldberg, A. Hertz, and D.F. Smith. 2003. Childhood cancer incidence rates and hazardous air pollutants in California: an exploratory analysis. *Environmental Health Perspectives* 111 (4):663-8.

49. Whitworth, K.W., E. Symanski, and A.L. Coker. 2008. Childhood lymphohematopoietic cancer incidence and hazardous air pollutants in southeast Texas, 1995-2004. *Environmental Health Perspectives* 116 (11):1576-80.

50. Brosselin, P., J. Rudant, L. Orsi, G. Leverger, A. Baruchel, Y. Bertrand, B. Nelken, A. Robert, G. Michel, G. Margueritte, et al. 2009. Acute childhood leukaemia and residence next to petrol stations and automotive repair garages: the ESCALE study (SFCE). *Occupational and Environmental Medicine* 66 (9):598-606.

51. Pearson, R.L., H. Wachtel, and K.L. Ebi. 2000. Distance-weighted traffic density in proximity to a home is a risk factor for leukemia and other childhood cancers. *Journal of the Air and Waste Management Association* 50 (2):175-80.

52. Weng, H.H., S.S. Tsai, H.F. Chiu, T.N. Wu, and C.Y. Yang. 2008. Association of childhood leukemia with residential exposure to petrochemical air pollution in taiwan. *Inhalation Toxicology* 20 (1):31-6.

53. Weng, H.H., S.S. Tsai, H.F. Chiu, T.N. Wu, and C.Y. Yang. 2009. Childhood leukemia and traffic air pollution in Taiwan: petrol station density as an indicator. *Journal of Toxicology and Environmental Health Part A: Current Issues* 72 (2):83-7.

54. Raaschou-Nielsen, O., O. Hertel, B.L. Thomsen, and J.H. Olsen. 2001. Air pollution from traffic at the residence of children with cancer. *American Journal of Epidemiology* 153 (5):433-43.

55. Reynolds, P., J. Von Behren, R.B. Gunier, D.E. Goldberg, A. Hertz, and D. Smith. 2002. Traffic patterns and childhood cancer incidence rates in California, United States. *Cancer Causes & Control* 13 (7):665-73.

56. Langholz, B., K.L. Ebi, D.C. Thomas, J.M. Peters, and S.J. London. 2002. Traffic density and the risk of childhood leukemia in a Los Angeles case-control study. *Annals of Epidemiology* 12 (7):482-7.

57. Reynolds, P., J. Von Behren, R.B. Gunier, D.E. Goldberg, and A. Hertz. 2004. Residential exposure to traffic in California and childhood cancer. *Epidemiology* 15 (1):6-12.

58. Health Effects Institute. 2010. *HEI Panel on the Health Effects of Traffic-Related Air Pollution: A Critical Review of the Literature on Emissions, Exposure, and Health Effects.* Boston, MA: Health Effects Institute. HEI Special Report 17. http://pubs.healtheffects.org/view.php?id=334.

59. U.S. Department of Health and Human Services. 2006. *The Health Consequences of Involuntary Exposure to Tobacco Smoke: A Report of the Surgeon General.* Atlanta, GA: Centers for Disease Control and Prevention, Coordinating Center for Health Promotion, National Center for Chronic Disease Prevention and Health Promotion, Office on Smoking and Health. http://www.surgeongeneral.gov/library/secondhandsmoke/report/index.html.

60. Baldwin, R.T., and S. Preston-Martin. 2004. Epidemiology of brain tumors in childhood--a review. *Toxicology and Applied Pharmacology* 199 (2):118-31.

61. Streffer, C., R. Shore, G. Konermann, A. Meadows, P. Uma Devi, J. Preston, L.E. Holm, J. Stather, K. Mabuchi, and H.R. Withers. 2003. Biological effects after prenatal irradiation (embryo and fetus). A report of the International Commission on Radiological Protection. *Annals of the International Commission on Radiological Protection* 33 (1-2):5-206.

62. Boice, J.D., Jr., and R.E. Tarone. 2011. Cell phones, cancer, and children. *Journal of the National Cancer Institute* 103 (16):1211-3.

63. Cardis, E., L. Richardson, I. Deltour, B. Armstrong, M. Feychting, C. Johansen, M. Kilkenny, P. McKinney, B. Modan, S. Sadetzki, et al. 2007. The INTERPHONE study: design, epidemiological methods, and description of the study population. *European Journal of Epidemiology* 22 (9):647-64.

64. Hardell, L., M. Carlberg, F. Soderqvist, and K. Hansson Mild. 2008. Meta-analysis of long-term mobile phone use and the association with brain tumours. *International Journal of Oncology* 32 (5):1097-103.

65. Hours, M., M. Bernard, L. Montestrucq, M. Arslan, A. Bergeret, I. Deltour, and E. Cardis. 2007. Cell Phones and risk of brain and acoustic nerve tumours: the French INTERPHONE case-control study. *Revue d'Épidémiologie et de Santé Publique* 55 (5):321-32.

66. Interphone Study Group. 2010. Brain tumour risk in relation to mobile telephone use: results of the INTERPHONE international case-control study. *International Journal of Epidemiology* 39 (3):675-94.

67. Khurana, V.G., C. Teo, M. Kundi, L. Hardell, and M. Carlberg. 2009. Cell phones and brain tumors: a review including the long-term epidemiologic data. *Surgical Neurology* 72 (3):205-14; discussion 214-5.

68. Myung, S.K., W. Ju, D.D. McDonnell, Y.J. Lee, G. Kazinets, C.T. Cheng, and J.M. Moskowitz. 2009. Mobile phone use and risk of tumors: a meta-analysis. *Journal of Clinical Oncology : Official Journal of the American Society of Clinical Oncology* 27 (33):5565-72.

Childhood Cancer (continued)

69. Schoemaker, M.J., and A.J. Swerdlow. 2009. Risk of pituitary tumors in cellular phone users: a case-control study. *Epidemiology* 20 (3):348-54.

70. Minenko, V.F., A.V. Ulanovsky, V.V. Drozdovitch, E.V. Shemiakina, Y.I. Gavrilin, V.T. Khrouch, S.M. Shinkarev, P.G. Voilleque, A. Bouville, L.R. Anspaugh, et al. 2006. Individual thyroid dose estimates for a case-control study of chernobyl-related thyroid cancer among children of Belarus--part II. Contributions from long-lived radionuclides and external radiation. *Health Physics* 90 (4):312-27.

71. Moysich, K.B., R.J. Menezes, and A.M. Michalek. 2002. Chernobyl-related ionising radiation exposure and cancer risk: an epidemiological review. *The Lancet Oncology* 3 (5):269-79.

72. Ron, E. 2007. Thyroid cancer incidence among people living in areas contaminated by radiation from the Chernobyl accident. *Health Physics* 93 (5):502-11.

73. Cooney, M.A., J.L. Daniels, J.A. Ross, N.E. Breslow, B.H. Pollock, and A.F. Olshan. 2007. Household pesticides and the risk of Wilms tumor. *Environmental Health Perspectives* 115 (1):134-7.

74. Armstrong, B.K., and A. Kricker. 2001. The epidemiology of UV induced skin cancer. *Journal of Photochemistry and Photobiology. B, Biology* 63 (1-3):8-18.

75. Balk, S.J. 2011. Ultraviolet radiation: a hazard to children and adolescents. *Pediatrics* 127 (3):e791-817.

76. Strouse, J.J., T.R. Fears, M.A. Tucker, and A.S. Wayne. 2005. Pediatric melanoma: risk factor and survival analysis of the surveillance, epidemiology and end results database. *Journal of Clinical Oncology* 23 (21):4735-41.

77. Narayanan, D.L., R.N. Saladi, and J.L. Fox. 2010. Ultraviolet radiation and skin cancer. *International Journal of Dermatology* 49 (9):978-86.

78. U.S. Environmental Protection Agency. 2010. *Ozone Science: The Facts Behind the Phaseout*. U.S. EPA, Stratospheric Protection Division. Retrieved August 10, 2011 from http://www.epa.gov/ozone/science/sc_fact.html.

79. Demko, C.A., E.A. Borawski, S.M. Debanne, K.D. Cooper, and K.C. Stange. 2003. Use of indoor tanning facilities by white adolescents in the United States. *Archives of Pediatrics & Adolescent Medicine* 157 (9):854-60.

80. Mayer, J.A., S.I. Woodruff, D.J. Slymen, J.F. Sallis, J.L. Forster, E.J. Clapp, K.D. Hoerster, L.C. Pichon, J.R. Weeks, G.E. Belch, et al. 2011. Adolescents' use of indoor tanning: a large-scale evaluation of psychosocial, environmental, and policy-level correlates. *American Journal of Public Health* 101 (5):930-8.

81. Surveillance Epidemiology and End Results Program. 2009. *Population Characteristics*. National Cancer Institute. Retrieved January 28, 2009 from http://seer.cancer.gov/registries/characteristics.html.

82. Surveillance Epidemiology and End Results Program. 2009. *Number of Persons by Race and Hispanic Ethnicity for SEER Participants (2000 Census Data)*. National Cancer Institute. Retrieved January 28, 2009 from http://seer.cancer.gov/registries/data.html.

Neurodevelopmental Disorders

1. Boyle, C.A., S. Boulet, L.A. Schieve, R.A. Cohen, S.J. Blumberg, M. Yeargin-Allsopp, S. Visser, and M.D. Kogan. 2011. Trends in the prevalence of developmental disabilities in US Children, 1997–2008. *Pediatrics* 127 (6):1034-42.

2. Pastor, P.N., and C.A. Reuben. 2008. Diagnosed attention deficit hyperactivity disorder and learning disability: United States, 2004-2006. *Vital and Health Statistics* 10 (237).

3. Grandjean, P., and P.J. Landrigan. 2006. Developmental neurotoxicity of industrial chemicals. *Lancet* 368 (9553):2167-78.

4. Newschaffer, C.J., M.D. Falb, and J.G. Gurney. 2005. National autism prevalence trends from United States special education data. *Pediatrics* 115 (3):e277-82.

5. Prior, M. 2003. Is there an increase in the prevalence of autism spectrum disorders? *Journal of Paediatrics and Child Health* 39 (2):81-2.

6. Rutter, M. 2005. Incidence of autism spectrum disorders: changes over time and their meaning. *Acta Paediatrica* 94 (1):2-15.

7. Centers for Disease Control and Prevention. 2010. Increasing prevalence of parent-reported attention-deficit/hyperactivity disorder among children --- United States, 2003 and 2007. *Morbidity and Mortality Weekly Report* 59 (44):1439-43.

8. Centers for Disease Control and Prevention. 2009. Prevalence of autism spectrum disorders --- autism and developmental disabilities monitoring network, United States, 2006. *Morbidity and Mortality Weekly Report* 58 (SS 10):1-20.

9. Hertz-Picciotto, I., and L. Delwiche. 2009. The rise in autism and the role of age at diagnosis. *Epidemiology* 20 (1):84-90.

10. Newschaffer, C.J. 2006. Investigating diagnostic substitution and autism prevalence trends. *Pediatrics* 117 (4):1436-7.

11. Grupp-Phelan, J., J.S. Harman, and K.J. Kelleher. 2007. Trends in mental health and chronic condition visits by children presenting for care at U.S. emergency departments. *Public Health Reports* 122 (1):55-61.

Neurodevelopmental Disorders (continued)

12. Kelleher, K.J., T.K. McInerny, W.P. Gardner, G.E. Childs, and R.C. Wasserman. 2000. Increasing identification of psychosocial problems: 1979-1996. *Pediatrics* 105 (6):1313-21.

13. U.S. Department of Education. 2007. *27th Annual (2005) Report to Congress on the Implementation of the Individuals with Disabilities Education Act, Vol. 1.* Washington, DC.

14. Aarnoudse-Moens, C.S.H., N. Weisglas-Kuperus, J.B. van Goudoever, and J. Oosterlaan. 2009. Meta-analysis of neurobehavioral outcomes in very preterm and/or very low birth weight children. *Pediatrics* 124 (2):717-728.

15. Banerjee, T.D., F. Middleton, and S.V. Faraone. 2007. Environmental risk factors for attention-deficit hyperactivity disorder. *Acta Pædiatrica* 96 (9):1269-1274.

16. Bhutta, A.T., M.A. Cleves, P.H. Casey, M.M. Cradock, and K.J.S. Anand. 2002. Cognitive and behavioral outcomes of school-aged children who were born preterm. *JAMA: The Journal of the American Medical Association* 288 (6):728-737.

17. Herrmann, M., K. King, and M. Weitzman. 2008. Prenatal tobacco smoke and postnatal secondhand smoke exposure and child neurodevelopment. *Current Opinion in Pediatrics* 20 (2):184-190.

18. Institute of Medicine. 2007. *Preterm Birth: Causes, Consequences, and Prevention.* Edited by R. E. Behrman and A. S. Butler. Washington, DC: The National Academies Press.

19. Linnet, K.M., S. Dalsgaard, C. Obel, K. Wisborg, T.B. Henriksen, A. Rodriguez, A. Kotimaa, I. Moilanen, P.H. Thomsen, J. Olsen, et al. 2003. Maternal lifestyle factors in pregnancy risk of attention deficit hyperactivity disorder and associated behaviors: review of the current evidence. *The American Journal of Psychiatry* 160 (6):1028-40.

20. Nigg, J.T. 2006. *What Causes ADHD? Understanding What Goes Wrong and Why.* New York: The Guilford Press.

21. Weiss, B., and D.C. Bellinger. 2006. Social ecology of children's vulnerability to environmental pollutants. *Environmental Health Perspectives* 114 (10):1479-1485.

22. National Toxicology Program. 2012. *NTP Monograph on Health Effects of Low-Level Lead.* Research Triangle Park, NC: National Institute of Environmental Health Sciences, National Toxicology Program. http://ntp.niehs.nih.gov/go/36443.

23. U.S. Environmental Protection Agency. 1997. *Mercury Study Report to Congress Volumes I to VII.* Washington DC: U.S. Environmental Protection Agency Office of Air Quality Planning and Standards and Office of Research and Development. EPA-452/R-97-003. http://www.epa.gov/hg/report.htm.

24. Amin-Zaki, L., S. Elhassani, M.A. Majeed, T.W. Clarkson, R.A. Doherty, and M. Greenwood. 1974. Intra-uterine methylmercury poisoning in Iraq. *Pediatrics* 54 (5):587-95.

25. Harada, M. 1995. Minamata disease: methylmercury poisoning in Japan caused by environmental pollution. *Critical Reviews in Toxicology* 25 (1):1-24.

26. National Research Council. 2000. *Toxicological Effects of Methylmercury.* Washington, DC: National Academy Press.

27. Budtz-Jorgensen, E., P. Grandjean, and P. Weihe. 2007. Separation of risks and benefits of seafood intake. *Environmental Health Perspectives* 115 (3):323-7.

28. Crump, K.S., T. Kjellstrom, A.M. Shipp, A. Silvers, and A. Stewart. 1998. Influence of prenatal mercury exposure upon scholastic and psychological test performance: benchmark analysis of a New Zealand cohort. *Risk Analysis* 18 (6):701-13.

29. Debes, F., E. Budtz-Jorgensen, P. Weihe, R.F. White, and P. Grandjean. 2006. Impact of prenatal methylmercury exposure on neurobehavioral function at age 14 years. *Neurotoxicology and Teratology* 28 (5):536-47.

30. Grandjean, P., P. Weihe, R.F. White, F. Debes, S. Araki, K. Yokoyama, K. Murata, N. Sorensen, R. Dahl, and P.J. Jorgensen. 1997. Cognitive deficit in 7-year-old children with prenatal exposure to methylmercury. *Neurotoxicology and Teratology* 19 (6):417-28.

31. Kjellstrom, T., P. Kennedy, S. Wallis, and C. Mantell. 1986. *Physical and mental development of children with prenatal exposure to mercury from fish. Stage 1: Preliminary tests at age 4.* Sweden: Swedish National Environmental Protection Board.

32. Oken, E., and D.C. Bellinger. 2008. Fish consumption, methylmercury and child neurodevelopment. *Current Opinion in Pediatrics* 20 (2):178-83.

33. Myers, G.J., P.W. Davidson, C. Cox, C.F. Shamlaye, D. Palumbo, E. Cernichiari, J. Sloane-Reeves, G.E. Wilding, J. Kost, L.S. Huang, et al. 2003. Prenatal methylmercury exposure from ocean fish consumption in the Seychelles child development study. *Lancet* 361 (9370):1686-92.

34. Davidson, P.W., J.J. Strain, G.J. Myers, S.W. Thurston, M.P. Bonham, C.F. Shamlaye, A. Stokes-Riner, J.M. Wallace, P.J. Robson, E.M. Duffy, et al. 2008. Neurodevelopmental effects of maternal nutritional status and exposure to methylmercury from eating fish during pregnancy. *Neurotoxicology* 29 (5):767-75.

Neurodevelopmental Disorders (continued)

35. Lynch, M.L., L.S. Huang, C. Cox, J.J. Strain, G.J. Myers, M.P. Bonham, C.F. Shamlaye, A. Stokes-Riner, J.M. Wallace, E.M. Duffy, et al. 2011. Varying coefficient function models to explore interactions between maternal nutritional status and prenatal methylmercury toxicity in the Seychelles Child Development Nutrition Study. *Environmental Research* 111 (1):75-80.

36. Strain, J.J., P.W. Davidson, M.P. Bonham, E.M. Duffy, A. Stokes-Riner, S.W. Thurston, J.M. Wallace, P.J. Robson, C.F. Shamlaye, L.A. Georger, et al. 2008. Associations of maternal long-chain polyunsaturated fatty acids, methyl mercury, and infant development in the Seychelles Child Development Nutrition Study. *Neurotoxicology* 29 (5):776-782.

37. Lederman, S.A., R.L. Jones, K.L. Caldwell, V. Rauh, S.E. Sheets, D. Tang, S. Viswanathan, M. Becker, J.L. Stein, R.Y. Wang, et al. 2008. Relation between cord blood mercury levels and early child development in a World Trade Center cohort. *Environmental Health Perspectives* 116 (8):1085-91.

30. Oken, E., J.S. Radesky, R.O. Wright, D.C. Bellinger, C.J. Amarasiriwardena, K.P. Kleinman, H. Hu, and M.W. Gillman. 2008. Maternal fish intake during pregnancy, blood mercury levels, and child cognition at age 3 years in a US cohort. *American Journal of Epidemiology* 167 (10):1171-81.

39. Jacobson, J.L., and S.W. Jacobson. 2003. Prenatal exposure to polychlorinated biphenyls and attention at school age. *Journal of Pediatrics* 143 (6):780-8.

40. Sagiv, S.K., S.W. Thurston, D.C. Bellinger, P.E. Tolbert, L.M. Altshul, and S.A. Korrick. 2010. Prenatal organochlorine exposure and behaviors associated with attention deficit hyperactivity disorder in school-aged children. *American Journal of Epidemiology* 171 (5):593-601.

41. Stewart, P., S. Fitzgerald, J. Reihman, B. Gump, E. Lonky, T. Darvill, J. Pagano, and P. Hauser. 2003. Prenatal PCB exposure, the corpus callosum, and response inhibition. *Environmental Health Perspectives* 111 (13):1670-7.

42. Stewart, P., J. Reihman, B. Gump, E. Lonky, T. Darvill, and J. Pagano. 2005. Response inhibition at 8 and 9 1/2 years of age in children prenatally exposed to PCBs. *Neurotoxicology and Teratology* 27 (6):771-80.

43. Stewart, P.W., E. Lonky, J. Reihman, J. Pagano, B.B. Gump, and T. Darvill. 2008. The relationship between prenatal PCB exposure and intelligence (IQ) in 9-year-old children. *Environmental Health Perspectives* 116 (10):1416-22.

44. Stewart, P.W., D.M. Sargent, J. Reihman, B.B. Gump, E. Lonky, T. Darvill, H. Hicks, and J. Pagano. 2006. Response inhibition during Differential Reinforcement of Low Rates (DRL) schedules may be sensitive to low-level polychlorinated biphenyl, methylmercury, and lead exposure in children. *Environmental Health Perspectives* 114 (12):1923-9.

45. Vreugdenhil, H.J., P.G. Mulder, H.H. Emmen, and N. Weisglas-Kuperus. 2004. Effects of perinatal exposure to PCBs on neuropsychological functions in the Rotterdam cohort at 9 years of age. *Neuropsychology* 18 (1):185-93.

46. Darvill, T., E. Lonky, J. Reihman, P. Stewart, and J. Pagano. 2000. Prenatal exposure to PCBs and infant performance on the Fagan test of infant intelligence. *Neurotoxicology* 21 (6):1029-38.

47. Jacobson, J.L., and S.W. Jacobson. 1996. Intellectual impairment in children exposed to polychlorinated biphenyls in utero. *New England Journal of Medicine* 335 (11):783-9.

48. Jacobson, J.L., and S.W. Jacobson. 1997. Teratogen update: polychlorinated biphenyls. *Teratology* 55 (5):338-347.

49. Patandin, S., C.I. Lanting, P.G. Mulder, E.R. Boersma, P.J. Sauer, and N. Weisglas-Kuperus. 1999. Effects of environmental exposure to polychlorinated biphenyls and dioxins on cognitive abilities in Dutch children at 42 months of age. *Journal of Pediatrics* 134 (1):33-41.

50. Stewart, P., J. Reihman, E. Lonky, T. Darvill, and J. Pagano. 2000. Prenatal PCB exposure and neonatal behavioral assessment scale (NBAS) performance. *Neurotoxicology and Teratology* 22 (1):21-9.

51. Walkowiak, J., J.A. Wiener, A. Fastabend, B. Heinzow, U. Kramer, E. Schmidt, H.J. Steingruber, S. Wundram, and G. Winneke. 2001. Environmental exposure to polychlorinated biphenyls and quality of the home environment: effects on psychodevelopment in early childhood. *Lancet* 358 (9293):1602-7.

52. Schantz, S.L., J.J. Widholm, and D.C. Rice. 2003. Effects of PCB exposure on neuropsychological function in children. *Environmental Health Perspectives* 111 (3):357-576.

53. Jacobson, J.L., S.W. Jacobson, and H.E. Humphrey. 1990. Effects of exposure to PCBs and related compounds on growth and activity in children. *Neurotoxicology and Teratology* 12 (4):319-26.

54. Boucher, O., G. Muckle, and C.H. Bastien. 2009. Prenatal exposure to polychlorinated biphenyls: a neuropsychologic analysis. *Environmental Health Perspectives* 117 (1):7-16.

55. Eubig, P.A., A. Aguiar, and S.L. Schantz. 2010. Lead and PCBs as risk factors for attention deficit/hyperactivity disorder. *Environmental Health Perspectives* 118 (12):1654-1667.

56. Ribas-Fito, N., M. Sala, M. Kogevinas, and J. Sunyer. 2001. Polychlorinated biphenyls (PCBs) and neurological development in children: a systematic review. *Journal of Epidemiology and Community Health* 55 (8):537-46.

Neurodevelopmental Disorders (continued)

57. Schantz, S.L., J.C. Gardiner, D.M. Gasior, R.J. McCaffrey, A.M. Sweeney, and H.E.B. Humphrey. 2004. Much Ado About Something: The Weight of Evidence for PCB Effects on Neuropsychological Function. *Psychology in the Schools* 41 (6):669-679.

58. Wigle, D.T., T.E. Arbuckle, M.C. Turner, A. Berube, Q. Yang, S. Liu, and D. Krewski. 2008. Epidemiologic evidence of relationships between reproductive and child health outcomes and environmental chemical contaminants. *Journal of Toxicology and Environmental Health Part B Critical Reviews* 11 (5-6):373-517.

59. Agency for Toxic Substances and Disease Registry (ATSDR). 2000. *Toxicological Profile for Polychlorinated Biphenyls (PCBs)*. Atlanta, GA: U.S. Department of Health and Human Services, Public Health Service. http://www.atsdr.cdc.gov/toxprofiles/tp.asp?id=142&tid=26.

60. Chen, Y.C., Y.L. Guo, C.C. Hsu, and W.J. Rogan. 1992. Cognitive development of Yu-Cheng ("oil disease") children prenatally exposed to heat-degraded PCBs. *Journal of the American Medical Association* 268 (22):3213-8.

61. Chen, Y.C., M.L. Yu, W.J. Rogan, B.C. Gladen, and C.C. Hsu. 1994. A 6-year follow-up of behavior and activity disorders in the Taiwan Yu-cheng children. *American Journal of Public Health* 84 (3):415-21.

62. Lai, T.J., X. Liu, Y.L. Guo, N.W. Guo, M.L. Yu, C.C. Hsu, and W.J. Rogan. 2002. A cohort study of behavioral problems and intelligence in children with high prenatal polychlorinated biphenyl exposure. *Archives of General Psychiatry* 59 (11):1061-6.

63. Rogan, W.J., B.C. Gladen, K.L. Hung, S.L. Koong, L.Y. Shih, J.S. Taylor, Y.C. Wu, D. Yang, N.B. Ragan, and C.C. Hsu. 1988. Congenital poisoning by polychlorinated biphenyls and their contaminants in Taiwan. *Science* 241 (4863):334-6.

64. Centers for Disease Control and Prevention. 2009. *Fourth National Report on Human Exposure to Environmental Chemicals*. Atlanta, GA: CDC. http://www.cdc.gov/exposurereport/.

65. Eskenazi, B., A. Bradman, and R. Castorina. 1999. Exposures of children to organophosphate pesticides and their potential adverse health effects. *Environmental Health Perspectives* 107 (Suppl. 3):409-19.

66. Huen, K., Harley, K., Brooks, J., Hubbard, A., Bradman, A., Eskenazi, B., Holland, N. 2009. Developmental changes in PON1 enzyme activity in young children and effects of PON1 polymorphisms. *Environmental Health Perspectives* 117 (10):1632-8.

67. Marks, A.R., K. Harley, A. Bradman, K. Kogut, D.B. Barr, C. Johnson, N. Calderon, and B. Eskenazi. 2010. Organophosphate pesticide exposure and attention in young Mexican-American children: the CHAMACOS study. *Environmental Health Perspectives* 118 (12):1768-74.

68. Bouchard, M.F., D.C. Bellinger, R.O. Wright, and M.G. Weisskopf. 2010. Attention-Deficit/Hyperactivity Disorder and urinary metabolites of organophosphate pesticides. *Pediatrics* 125 (6):e1270-e1277.

69. Bouchard, M.F., J. Chevrier, K.G. Harley, K. Kogut, M. Vedar, N. Calderon, C. Trujillo, C. Johnson, A. Bradman, D.B. Barr, et al. 2011. Prenatal exposure to organophosphate pesticides and IQ in 7-year old children. *Environmental Health Perspectives* doi: 10.1289/ehp.1003185.

70. Engel, S.M., J. Wetmur, J. Chen, C. Zhu, D.B. Barr, R.L. Canfield, and M.S. Wolff. 2011. Prenatal exposure to organophosphates, paraoxonase 1, and cognitive development in childhood. *Environmental Health Perspectives* doi: 10.1289/ehp.1003183.

71. Rauh, V., S. Arunajadai, M. Horton, F. Perera, L. Hoepner, D.B. Barr, and R. Whyatt. 2011. 7-Year neurodevelopmental scores and prenatal exposure to chlorpyrifos, a common agricultural pesticide. *Environmental Health Perspectives* doi: 10.1289/ehp.1003160.

72. Costa, L.G., G. Giordano, S. Tagliaferri, A. Caglieri, and A. Mutti. 2008. Polybrominated diphenyl ether (PBDE) flame retardants: environmental contamination, human body burden and potential adverse health effects. *Acta Biomed* 79 (3):172-83.

73. Gee, J.R., and V.C. Moser. 2008. Acute postnatal exposure to brominated diphenylether 47 delays neuromotor ontogeny and alters motor activity in mice. *Neurotoxicology and Teratology* 30 (2):79-87.

74. Rice, D.C., E.A. Reeve, A. Herlihy, R.T. Zoeller, W.D. Thompson, and V.P. Markowski. 2007. Developmental delays and locomotor activity in the C57BL6/J mouse following neonatal exposure to the fully-brominated PBDE, decabromodiphenyl ether. *Neurotoxicology and Teratology* 29 (4):511-20.

75. Herbstman, J.B., A. Sjodin, M. Kurzon, S.A. Lederman, R.S. Jones, V. Rauh, L.L. Needham, D. Tang, M. Niedzwiecki, R.Y. Wang, et al. 2010. Prenatal exposure to PBDEs and neurodevelopment. *Environmental Health Perspectives* 118 (5):712-9.

76. Roze, E., L. Meijer, A. Bakker, K.N. Van Braeckel, P.J. Sauer, and A.F. Bos. 2009. Prenatal exposure to organohalogens, including brominated flame retardants, influences motor, cognitive, and behavioral performance at school age. *Environmental Health Perspectives* 117 (12):1953-8.

77. Engel, S.M., A. Miodovnik, R.L. Canfield, C. Zhu, M.J. Silva, A.M. Calafat, and M.S. Wolff. 2010. Prenatal phthalate exposure is associated with childhood behavior and executive functioning. *Environmental Health Perspectives* 118 (4):565-71.

78. Miodovnik, A., S.M. Engel, C. Zhu, X. Ye, L.V. Soorya, M.J. Silva, A.M. Calafat, and M.S. Wolff. 2011. Endocrine disruptors and childhood social impairment. *Neurotoxicology* 32 (2):261-267.

79. Cho, S.-C., S.-Y. Bhang, Y.-C. Hong, M.-S. Shin, B.-N. Kim, J.-W. Kim, H.-J. Yoo, I.H. Cho, and H.-W. Kim. 2010. Relationship between environmental phthalate exposure and the intelligence of school-age children. *Environmental Health Perspectives* 118 (7):1027-1032.

Neurodevelopmental Disorders (continued)

80. Kim, B.N., S.C. Cho, Y. Kim, M.S. Shin, H.J. Yoo, J.W. Kim, Y.H. Yang, H.W. Kim, S.Y. Bhang, and Y.C. Hong. 2009. Phthalates exposure and attention-deficit/hyperactivity disorder in school-age children. *Biological Psychiatry* 66 (10):958-63.

81. National Toxicology Program. 2008. *NTP-CERHR Monograph on the Potential Human Reproductive and Developmental Effects of Bisphenol A*. Research Triangle Park, NC: National Institute of Environmental Health Sciences, National Toxicology Program. http://ntp.niehs.nih.gov/ntp/ohat/bisphenol/bisphenol.pdf.

82. Braun, J.M., K. Yolton, K.N. Dietrich, R. Hornung, X. Ye, A.M. Calafat, and B.P. Lanphear. 2009. Prenatal bisphenol A exposure and early childhood behavior. *Environmental Health Perspectives* 117 (12):1945-1952.

83. Perera, F.P., Z. Li, R. Whyatt, L. Hoepner, S. Wang, D. Camann, and V. Rauh. 2009. Prenatal airborne polycyclic aromatic hydrocarbon exposure and child IQ at age 5 years. *Pediatrics* 124 (2):e195-202.

84. Perera, F.P., S. Wang, J. Vishnevetsky, B. Zhang, K.J. Cole, D. Tang, V. Rauh, and D.H. Phillips. 2011. PAH/Aromatic DNA Adducts in Cord Blood and Behavior Scores in New York City Children. *Environmental Health Perspectives* doi:10.1289/ehp.1002705.

85. Smith, A.H., and C.M. Steinmaus. 2009. Health effects of arsenic and chromium in drinking water: recent human findings. *Annual Review of Public Health* 30:107-22.

86. Sambu, S., and R. Wilson. 2008. Arsenic in food and water--a brief history. *Toxicology and Industrial Health* 24 (4):217-26.

87. U.S. Environmental Protection Agency. 2011. *Perchlorate* Retrieved February 11, 2011 from http://www.epa.gov/safewater/contaminants/unregulated/perchlorate.html.

88. Kirk, A.B., P.K. Martinelango, K. Tian, A. Dutta, E.E. Smith, and P.K. Dasgupta. 2005. Perchlorate and iodide in dairy and breast milk. *Environmental Science & Technology* 39 (7):2011-7.

89. Sanchez, C.A., L.M. Barraj, B.C. Blount, C.G. Scrafford, L. Valentin-Blasini, K.M. Smith, and R.I. Krieger. 2009. Perchlorate exposure from food crops produced in the lower Colorado River region. *Journal of Exposure Science & Environmental Epidemiology* 19 (4):359-68.

90. Greer, M.A., G. Goodman, R.C. Pleus, and S.E. Greer. 2002. Health effects assessment for environmental perchlorate contamination: the dose response for inhibition of thyroidal radioiodine uptake in humans. *Environmental Health Perspectives* 110 (9):927-37.

91. National Research Council. 2005. *Health Implications of Perchlorate Ingestion*. Washington, DC: National Academies Press. http://www.nap.edu/catalog.php?record_id=11202.

92. Haddow, J.E., G.E. Palomaki, W.C. Allan, J.R. Williams, G.J. Knight, J. Gagnon, C.E. O'Heir, M.L. Mitchell, R.J. Hermos, S.E. Waisbren, et al. 1999. Maternal thyroid deficiency during pregnancy and subsequent neuropsychological development of the child. *New England Journal of Medicine* 341 (8):549-55.

93. Miller, M.D., K.M. Crofton, D.C. Rice, and R.T. Zoeller. 2009. Thyroid-disrupting chemicals: interpreting upstream biomarkers of adverse outcomes. *Environmental Health Perspectives* 117 (7):1033-41.

94. Morreale de Escobar, G., M.J. Obregon, and F. Escobar del Rey. 2000. Is neuropsychological development related to maternal hypothyroidism or to maternal hypothyroxinemia? *The Journal of Clinical Endocrinology and Metabolism* 85 (11):3975-87.

95. Bellinger, D.C. 2008. Lead neurotoxicity and socioeconomic status: conceptual and analytical issues. *Neurotoxicology* 29 (5):828-32.

96. Rice, D.C. 2000. Parallels between attention deficit hyperactivity disorder and behavioral deficits produced by neurotoxic exposure in monkeys. *Environmental Health Perspectives* 108 (Suppl. 3):405-408.

97. Rodier, P.M. 1995. Developing brain as a target of toxicity. *Environmental Health Perspectives* 103 Suppl. 6:73-6.

98. American Psychiatric Association. 2000. *Diagnostic and Statistical Manual of Mental Disorders, Fourth Edition Text Revision*. Washington D.C.: American Psychiatric Association.

99. American Psychiatric Association. 1987. *Diagnostic and Statistical Manual of Mental Disorders, Third Edition Text Revision (DSM-III-R)*. Washington, D.C.

100. Larson, K., S.A. Russ, R.S. Kahn, and N. Halfon. 2011. Patterns of comorbidity, functioning, and service use for U.S. children with ADHD, 2007. *Pediatrics* 127 (3):462-470.

101. Aguiar, A., P.A. Eubig, and S.L. Schantz. 2010. Attention deficit/hyperactivity disorder: a focused overview for children's environmental health researchers. *Environmental Health Perspectives* 118 (12):1646-53.

102. Barkley, R.A. 2006. *Attention-Deficit Hyperactivity Disorder: A Handbook for Diagnosis and Treatment, Third Edition*. New York: The Guilford Press.

103. Biederman, J., and S.V. Faraone. 2005. Attention-deficit hyperactivity disorder. *Lancet* 366 (9481):237-48.

104. Faraone, S.V., and E. Mick. 2010. Molecular genetics of attention deficit hyperactivity disorder. *Psychiatric Clinics of North America* 33 (1):159-80.

Neurodevelopmental Disorders (continued)

105. Kieling, C., R.R. Goncalves, R. Tannock, and F.X. Castellanos. 2008. Neurobiology of attention deficit hyperactivity disorder. *Child and Adolescent Psychiatric Clinics of North America* 17 (2):285-307, viii.

106. Thapar, A., K. Langley, P. Asherson, and M. Gill. 2007. Gene-environment interplay in attention-deficit hyperactivity disorder and the importance of a developmental perspective. *The British Journal of Psychiatry* 190:1-3.

107. Langley, K., F. Rice, M.B. van den Bree, and A. Thapar. 2005. Maternal smoking during pregnancy as an environmental risk factor for attention deficit hyperactivity disorder behaviour. A review. *Minerva Pediatrica* 57 (6):359-71.

108. Nigg, J., M. Nikolas, and S.A. Burt. 2010. Measured gene-by-environment interaction in relation to attention-deficit/hyperactivity disorder. *Journal of the American Academy of Child and Adolescent Psychiatry* 49 (9):863-73.

109. Braun, J.M., R.S. Kahn, T. Froehlich, P. Auinger, and B.P. Lanphear. 2006. Exposures to environmental toxicants and attention deficit hyperactivity disorder in U.S. children. *Environmental Health Perspectives* 114 (12):1904-9.

110. Froehlich, T.E., B.P. Lanphear, P. Auinger, R. Hornung, J.N. Epstein, J. Braun, and R.S. Kahn. 2009. Association of tobacco and lead exposures with attention-deficit/hyperactivity disorder. *Pediatrics* 124 (6):e1054-63.

111. Ha, M., H.J. Kwon, M.H. Lim, Y.K. Jee, Y.C. Hong, J.H. Leem, J. Sakong, J.M. Bae, S.J. Hong, Y.M. Roh, et al. 2009. Low blood levels of lead and mercury and symptoms of attention deficit hyperactivity in children: a report of the children's health and environment research (CHEER). *Neurotoxicology* 30 (1):31-6.

112. Nigg, J.T., G.M. Knottnerus, M.M. Martel, M. Nikolas, K. Cavanagh, W. Karmaus, and M.D. Rappley. 2008. Low blood lead levels associated with clinically diagnosed attention-deficit/hyperactivity disorder and mediated by weak cognitive control. *Biological Psychiatry* 63 (3):325-31.

113. Nigg, J.T., M. Nikolas, G. Mark Knottnerus, K. Cavanagh, and K. Friderici. 2010. Confirmation and extension of association of blood lead with attention-deficit/hyperactivity disorder (ADHD) and ADHD symptom domains at population-typical exposure levels. *The Journal of Child Psychology and Psychiatry* 51 (1):58-65.

114. Roy, A., D. Bellinger, H. Hu, J. Schwartz, A.S. Ettinger, R.O. Wright, M. Bouchard, K. Palaniappan, and K. Balakrishnan. 2009. Lead exposure and behavior among young children in Chennai, India. *Environmental Health Perspectives* 117 (10):1607-11.

115. Wang, H.-L., X.-T. Chen, B. Yang, F.-L. Ma, S. Wang, M.-L. Tang, N.-G. Hao, and D.-Y. Ruan. 2008. Case-control study of blood lead levels and attention-deficit hyperactivity disorder in Chinese children *Environmental Health Perspectives* 116 (10):1401-1406.

116. Canfield, R.L., M.H. Gendle, and D.A. Cory-Slechta. 2004. Impaired neuropsychological functioning in lead-exposed children. *Developmental Neuropsychology* 26 (1):513-40.

117. Chiodo, L.M., S.W. Jacobson, and J.L. Jacobson. 2004. Neurodevelopmental effects of postnatal lead exposure at very low levels. *Neurotoxicology and Teratology* 26 (3):359-71.

118. Nicolescu, R., C. Petcu, A. Cordeanu, K. Fabritius, M. Schlumpf, R. Krebs, U. Kramer, and G. Winneke. 2010. Environmental exposure to lead, but not other neurotoxic metals, relates to core elements of ADHD in Romanian children: performance and questionnaire data. *Environmental Research* 110 (5):476-83.

119. Surkan, P.J., A. Zhang, F. Trachtenberg, D.B. Daniel, S. McKinlay, and D.C. Bellinger. 2007. Neuropsychological function in children with blood lead levels <10 microg/dL. *Neurotoxicology* 28 (6):1170-7.

120. Rice, D.C. 1996. Behavioral effects of lead: commonalities between experimental and epidemiologic data. *Environmental Health Perspectives* 104 (Suppl. 2):337-51.

121. Rossi-George, A., M.B. Virgolini, D. Weston, M. Thiruchelvam, and D.A. Cory-Slechta. 2011. Interactions of lifetime lead exposure and stress: behavioral, neurochemical and HPA axis effects. *Neurotoxicology* 32 (1):83-99.

122. Virgolini, M.B., A. Rossi-George, R. Lisek, D.D. Weston, M. Thiruchelvam, and D.A. Cory-Slechta. 2008. CNS effects of developmental Pb exposure are enhanced by combined maternal and offspring stress. *Neurotoxicology* 29 (5):812-27.

123. Gump, B.B., Q. Wu, A.K. Dumas, and K. Kannan. 2011. Perfluorochemical (PFC) exposure in children: associations with impaired response inhibition. *Environmental Science & Technology* 45 (19):8151-9.

124. Hoffman, K., T.F. Webster, M.G. Weisskopf, J. Weinberg, and V.M. Vieira. 2010. Exposure to polyfluoroalkyl chemicals and attention deficit/hyperactivity disorder in U.S. children 12-15 years of age. *Environmental Health Perspectives* 118 (12):1762-7.

125. Stein, C.R., and D.A. Savitz. 2011. Serum perfluorinated compound concentration and attention deficit/hyperactivity disorder in children aged 5 to 18 years. *Environmental Health Perspectives* 119 (10):1466-71.

126. Cheuk, D.K., and V. Wong. 2006. Attention-deficit hyperactivity disorder and blood mercury level: a case-control study in Chinese children. *Neuropediatrics* 37 (4):234-40.

127. Julvez, J., F. Debes, P. Weihe, A. Choi, and P. Grandjean. 2010. Sensitivity of continuous performance test (CPT) at age 14 years to developmental methylmercury exposure. *Neurotoxicology and Teratology* 32 (6):627-632.

Neurodevelopmental Disorders (continued)

128. Plusquellec, P., G. Muckle, E. Dewailly, P. Ayotte, G. Begin, C. Desrosiers, C. Despres, D. Saint-Amour, and K. Poitras. 2010. The relation of environmental contaminants exposure to behavioral indicators in Inuit preschoolers in Arctic Quebec. *Neurotoxicology* 31 (1):17-25.

129. National Dissemination Center for Children with Disabilities. 2010. *Disability Fact Sheet-No. 7: Learning Disabilities*. Retrieved April 6, 2010 from http://www.nichcy.org/InformationResources/Documents/NICHCY%20PUBS/fs7.pdf.

130. National Center for Learning Disabilities. 2010. *LD at a Glance*. Retrieved April 6, 2010 from http://www.ncld.org/ld-basics/ld-explained/basic-facts/learning-disabilities-at-a-glance.

131. Bellinger, D.C. 2008. Very low lead exposures and children's neurodevelopment. *Current Opinion in Pediatrics* 20 (2):172-177.

132. Marlowe, M., A. Cossairt, K. Welch, and J. Errera. 1984. Hair mineral content as a predictor of learning disabilities. *Journal of Learning Disabilities* 17 (7):418-21.

133. Pihl, R.O., and M. Parkes. 1977. Hair element content in learning disabled children. *Science* 198 (4313):204-6.

134. Leviton, A., D. Bellinger, E.N. Allred, M. Rabinowitz, H. Needleman, and S. Schoenbaum. 1993. Pre- and postnatal low-level lead exposure and children's dysfunction in school. *Environmental Research* 60 (1):30-43.

135. Lyngbye, T., O.N. Hansen, A. Trillingsgaard, I. Beese, and P. Grandjean. 1990. Learning disabilities in children: significance of low-level lead-exposure and confounding factors. *Acta Paediatrica Scandinavica* 79 (3):352-60.

136. Needleman, H.L., C. Gunnoe, A. Leviton, R. Reed, H. Peresie, C. Maher, and P. Barrett. 1979. Deficits in psychologic and classroom performance of children with elevated dentine lead levels. *New England Journal of Medicine* 300 (13):689-95.

137. Needleman, H.L., A. Schell, D.C. Bellinger, A. Leviton, and E.N. Allred. 1990. The long term effects of exposure to low doses of lead in childhood, an 11-year follow-up report. *New England Journal of Medicine* 322 (2):83-8.

138. Anderko, L., J. Braun, and P. Auinger. 2010. Contribution of tobacco smoke exposure to learning disabilities. *Journal of Obstetric, Gynecologic, & Neonatal Nursing* 39 (1):111-117.

139. Centers for Disease Control and Prevention. 2010. *Autism Spectrum Disorders: Signs & Symptoms*. Retrieved March 25, 2010 from http://www.cdc.gov/ncbddd/autism/signs.html.

140. Beaudet, A.L. 2007. Autism: highly heritable but not inherited. *Nature Medicine* 13 (5):534-6.

141. Hallmayer, J., S. Cleveland, A. Torres, J. Phillips, B. Cohen, T. Torigoe, J. Miller, A. Fedele, J. Collins, K. Smith, et al. 2011. Genetic heritability and shared environmental factors among twin pairs with autism. *Archives of General Psychiatry* 68 (11):1095-102.

142. Levy, D., M. Ronemus, B. Yamrom, Y.-h. Lee, A. Leotta, J. Kendall, S. Marks, B. Lakshmi, D. Pai, K. Ye, et al. 2011. Rare de novo and transmitted copy-number variation in autistic spectrum disorders. *Neuron* 70 (5):886-897.

143. King, M., and P. Bearman. 2009. Diagnostic change and the increased prevalence of autism. *International Journal of Epidemiology* 38 (5):1224-1234.

144. King, M.D., C. Fountain, D. Dakhlallah, and P.S. Bearman. 2009. Estimated autism risk and older reproductive age. *American Journal of Public Health* 99 (9):1673-1679.

145. Liu, K.Y., M. King, and P.S. Bearman. 2010. Social influence and the autism epidemic. *American Journal of Sociology* 115 (5):1387-434.

146. Shelton, J.F., D.J. Tancredi, and I. Hertz-Picciotto. 2010. Independent and dependent contributions of advanced maternal and paternal ages to autism risk. *Autism Research* 3 (1):30-9.

147. Pessah, I.N., R.F. Seegal, P.J. Lein, J. LaSalle, B.K. Yee, J. Van De Water, and R.F. Berman. 2008. Immunologic and neurodevelopmental susceptibilities of autism. *Neurotoxicology* 29 (3):532-45.

148. Newschaffer, C.J., L.A. Croen, J. Daniels, E. Giarelli, J.K. Grether, S.E. Levy, D.S. Mandell, L.A. Miller, J. Pinto-Martin, J. Reaven, et al. 2007. The epidemiology of autism spectrum disorders. *Annual Review of Public Health* 28:235-58.

149. Sanders, S.J., A.G. Ercan-Sencicek, V. Hus, R. Luo, M.T. Murtha, D. Moreno-De-Luca, S.H. Chu, M.P. Moreau, A.R. Gupta, S.A. Thomson, et al. 2011. Multiple Recurrent De Novo CNVs, Including Duplications of the 7q11.23 Williams Syndrome Region, Are Strongly Associated with Autism. *Neuron* 70 (5):863-85.

150. Sebat, J., B. Lakshmi, D. Malhotra, J. Troge, C. Lese-Martin, T. Walsh, B. Yamrom, S. Yoon, J. Krasnitz, J. Kendall, et al. 2007. Strong association of de novo copy number mutations with autism. *Science* 316 (5823):445-9.

151. Kinney, D.K., D.H. Barch, B. Chayka, S. Napoleon, and K.M. Munir. 2010. Environmental risk factors for autism: do they help cause de novo genetic mutations that contribute to the disorder? *Medical Hypotheses* 74 (1):102-6.

152. James, S.J., P. Cutler, S. Melnyk, S. Jernigan, L. Janak, D.W. Gaylor, and J.A. Neubrander. 2004. Metabolic biomarkers of increased oxidative stress and impaired methylation capacity in children with autism. *American Journal of Clinical Nutrition* 80 (6):1611-7.

Neurodevelopmental Disorders (continued)

153. James, S.J., S. Melnyk, S. Jernigan, M.A. Cleves, C.H. Halsted, D.H. Wong, P. Cutler, K. Bock, M. Boris, J.J. Bradstreet, et al. 2006. Metabolic endophenotype and related genotypes are associated with oxidative stress in children with autism. *American Journal of Medical Genetics Part B: Neuropsychiatric Genetics* 141B (8):947-56.

154. Deth, R., C. Muratore, J. Benzecry, V.A. Power-Charnitsky, and M. Waly. 2008. How environmental and genetic factors combine to cause autism: A redox/methylation hypothesis. *Neurotoxicology* 29 (1):190-201.

155. Croen, L.A., D.V. Najjar, B. Fireman, and J.K. Grether. 2007. Maternal and paternal age and risk of autism spectrum disorders. *Archives of Pediatric & Adolescent Medicine* 161 (4):334-40.

156. Grether, J.K., M.C. Anderson, L.A. Croen, D. Smith, and G.C. Windham. 2009. Risk of autism and increasing maternal and paternal age in a large North American population. *American Journal of Epidemiology* 170 (9):1118-26.

157. Lauritsen, M.B., C.B. Pedersen, and P.B. Mortensen. 2005. Effects of familial risk factors and place of birth on the risk of autism: a nationwide register-based study. *Journal of Child Psychology and Psychiatry* 46 (9):963-71.

158. Chandley, A.C. 1991. On the parental origin of de novo mutation in man. *Journal of Medical Genetics* 28 (4):217-23.

159. Crow, J.F. 2000. The origins, patterns and implications of human spontaneous mutation. *Nature Reviews Genetics* 1 (1):40-7.

160. Adams, J.B., J. Romdalvik, V.M. Ramanujam, and M.S. Legator. 2007. Mercury, lead, and zinc in baby teeth of children with autism versus controls. *Journal of Toxicology and Environmental Health A* 70 (12):1046-51.

161. Bradstreet, J., D.A. Geier, J.J. Kartzinel, J.B. Adams, and M.R. Feier. 2003. A case-control study of mercury burden in children with autistic spectrum disorders. *Journal of American Physicians and Surgeons* 8 (3).

162. Desoto, M.C., and R.T. Hitlan. 2007. Blood levels of mercury are related to diagnosis of autism: a reanalysis of an important data set. *Journal of Child Neurology* 22 (11):1308-11.

163. Hertz-Picciotto, I., P.G. Green, L. Delwiche, R. Hansen, C. Walker, and I.N. Pessah. 2010. Blood mercury concentrations in CHARGE Study children with and without autism. *Environmental Health Perspectives* 118 (1):161-6.

164. Palmer, R.F., S. Blanchard, and R. Wood. 2009. Proximity to point sources of environmental mercury release as a predictor of autism prevalence. *Health Place* 15 (1):18-24.

165. Centers for Disease Control and Prevention. *Mercury and Thimerosal: Vaccine Safety*. CDC. Retrieved October 12, 2010 from http://www.cdc.gov/vaccinesafety/Concerns/thimerosal/index.html.

166. Institute of Medicine. 2004. *Immunization Safety Review: Vaccines and Autism*. Washington, DC: National Academies Press. http://www.nap.edu/catalog.php?record_id=10997.

167. Kalkbrenner, A.E., J.L. Daniels, J.-C. Chen, C. Poole, M. Emch, and J. Morrissey. 2010. Perinatal Exposure to Hazardous Air Pollutants and Autism Spectrum Disorders at Age 8. *Epidemiology* 21 (5):631-41.

168. Windham, G.C., L. Zhang, R. Gunier, L.A. Croen, and J.K. Grether. 2006. Autism spectrum disorders in relation to distribution of hazardous air pollutants in the San Francisco Bay area. *Environmental Health Perspectives* 114 (9):1438-44.

169. Volk, H.E., I. Hertz-Picciotto, L. Delwiche, F. Lurmann, and R. McConnell. 2011. Residential Proximity to Freeways and Autism in the CHARGE Study. *Environmental Health Perspectives* 119 (6):873-7.

170. Larsson, M., B. Weiss, S. Janson, J. Sundell, and C.G. Bornehag. 2009. Associations between indoor environmental factors and parental-reported autistic spectrum disorders in children 6-8 years of age. *Neurotoxicology* 30 (5):822-31.

171. American Association of Intellectual and Developmental Disabilities. 2009. *FAQ on Intellectual Disability*. Retrieved March 23, 2009 from http://www.aamr.org/content_104.cfm?navID=22.

172. Schroeder, S.R. 2000. Mental retardation and developmental disabilities influenced by environmental neurotoxic insults. *Environmental Health Perspectives* 108 (Suppl. 3):395-9.

173. Daily, D.K., H.H. Ardinger, and G.E. Holmes. 2000. Identification and evaluation of mental retardation. *American Family Physician* 61 (4):1059-67, 1070.

174. Flint, J., and A.O. Wilkie. 1996. The genetics of mental retardation. *British Medical Bulletin* 52 (3):453-64.

175. Murphy, C., C. Boyle, D. Schendel, P. Decouflé, and M. Yeargin-Allsopp. 1998. Epidemiology of mental retardation in children. *Mental Retardation and Developmental Disabilities Research Reviews* 4 (1):6-13.

176. Bakir, F., H. Rustam, S. Tikriti, S.F. Al-Damluji, and H. Shihristani. 1980. Clinical and epidemiological aspects of methylmercury poisoning. *Postgraduate Medical Journal* 56 (651):1-10.

177. David, O., S. Hoffman, B. McGann, J. Sverd, and J. Clark. 1976. Low lead levels and mental retardation. *Lancet* 2 (8000):1376-9.

Neurodevelopmental Disorders (continued)

178. McDermott, S., J. Wu, B. Cai, A. Lawson, and C. Marjorie Aelion. 2011. Probability of intellectual disability is associated with soil concentrations of arsenic and lead. *Chemosphere* 84 (1):31-8.

179. U.S. Environmental Protection Agency. 2006. *Air Quality Criteria for Lead (Final Report)*. Washington, DC: U.S. EPA, National Center for Environmental Assessment. EPA/600/R-05/144aF-bF. http://cfpub.epa.gov/ncea/isa/recordisplay.cfm?deid=158823.

180. Edwards, S.C., W. Jedrychowski, M. Butscher, D. Camann, A. Kieltyka, E. Mroz, E. Flak, Z. Li, S. Wang, V. Rauh, et al. 2010. Prenatal exposure to airborne polycyclic aromatic hydrocarbons and children's intelligence at age 5 in a prospective cohort study in Poland. *Environmental Health Perspectives* 118 (9):1326-31.

181. Fewtrell, L.J., A. Pruss-Ustun, P. Landrigan, and J.L. Ayuso-Mateos. 2004. Estimating the global burden of disease of mild mental retardation and cardiovascular diseases from environmental lead exposure. *Environmental Research* 94 (2):120-33.

182. U.S. Environmental Protection Agency. 1997. *The Benefits and Costs of the Clean Air Act, 1970 to 1990*. Washington, DC: U.S. EPA, Office of Air and Radiation. http://www.epa.gov/air/sect812/copy.html.

183. Weiss, B. 2000. Vulnerability of children and the developing brain to neurotoxic hazards. *Environmental Health Perspectives* 108 (Suppl. 3):375-81.

184. De Los Reyes, A., and A.E. Kazdin. 2005. Informant discrepancies in the assessment of childhood psychopathology: a critical review, theoretical framework, and recommendations for further study. *Psychological Bulletin* 131 (4):483-509.

185. Owens, P.L., K. Hoagwood, S.M. Horwitz, P.J. Leaf, J.M. Poduska, S.G. Kellam, and N.S. Ialongo. 2002. Barriers to children's mental health services. *Journal of the American Academy of Child and Adolescent Psychiatry* 41 (6):731-8.

186. U.S. Department of Health and Human Services. 1999. *Mental Health: A Report of the Surgeon General—Executive Summary*. Rockville, MD: U.S. DHS, Substance Abuse and Mental Health Services Administration, Center for Mental Health Services, National Institutes of Health, National Institute of Mental Health. http://www.surgeongeneral.gov/library/mentalhealth/pdfs/ExSummary-Final.pdf.

187. Centers for Disease Control and Prevention. 2007. Prevalence of autism spectrum disorders---autism and developmental disabilities monitoring network, 14 sites, United States, 2002. In: Surveillance Summaries. *Morbidity and Mortality Weekly Report* 56 (No. SS-1):12-28.

188. Kogan, M.D., S.J. Blumberg, L.A. Schieve, C.A. Boyle, J.M. Perrin, R.M. Ghandour, G.K. Singh, B.B. Strickland, E. Trevathan, and P.C. van Dyck. 2009. Prevalence of parent-reported diagnosis of autism spectrum disorder among children in the US, 2007. *Pediatrics* 124 (5):1395-403.

Obesity

1. Centers for Disease Control and Prevention. 2010. *Defining Childhood Overweight and Obesity*. Retrieved April 6, 2010 from http://www.cdc.gov/obesity/childhood/defining.html.

2. Centers for Disease Control and Prevention. 2010. *CDC Growth Charts*. Retrieved May 4, 2010 from http://www.cdc.gov/growthcharts/.

3. Krebs, N.F., J.H. Himes, D. Jacobson, T.A. Nicklas, P. Guilday, and D. Styne. 2007. Assessment of child and adolescent overweight and obesity. *Pediatrics* 120 (Suppl 4):S193-228.

4. Ogden, C.L., R.P. Troiano, R.R. Briefel, R.J. Kuczmarski, K.M. Flegal, and C.L. Johnson. 1997. Prevalence of overweight among preschool children in the United States, 1971 through 1994. *Pediatrics* 99 (4):E1.

5. Ogden, C.L., M.D. Carroll, L.R. Curtin, M.A. McDowell, C.J. Tabak, and K.M. Flegal. 2006. Prevalence of overweight and obesity in the United States, 1999-2004. *Journal of the American Medical Association* 295 (13):1549-55.

6. Ogden, C.L., M.D. Carroll, L.R. Curtin, M.M. Lamb, and K.M. Flegal. 2010. Prevalence of high body mass index in US children and adolescents, 2007-2008. *Journal of the American Medical Association* 303 (3):242-9.

7. Hedley, A.A., C.L. Ogden, C.L. Johnson, M.D. Carroll, L.R. Curtin, and K.M. Flegal. 2004. Prevalence of overweight and obesity among US children, adolescents, and adults, 1999-2002. *Journal of the American Medical Association* 291 (23):2847-50.

8. Flegal, K.M., C.L. Ogden, J.A. Yanovski, D.S. Freedman, J.A. Shepherd, B.I. Graubard, and L.G. Borrud. 2010. High adiposity and high body mass index-for-age in US children and adolescents overall and by race-ethnic group. *The American Journal of Clinical Nutrition* 91 (4):1020-6.

9. Serdula, M.K., D. Ivery, R.J. Coates, D.S. Freedman, D.F. Williamson, and T. Byers. 1993. Do obese children become obese adults? A review of the literature. *Preventive Medicine* 22 (2):167-77.

10. The, N.S., C. Suchindran, K.E. North, B.M. Popkin, and P. Gordon-Larsen. 2010. Association of adolescent obesity with risk of severe obesity in adulthood. *JAMA: The Journal of the American Medical Association* 304 (18):2042-7.

Obesity (continued)

11. Whitaker, R.C., J.A. Wright, M.S. Pepe, K.D. Seidel, and W.H. Dietz. 1997. Predicting obesity in young adulthood from childhood and parental obesity. *New England Journal of Medicine* 337 (13):869-73.

12. Aglony, M., M. Acevedo, and G. Ambrosio. 2009. Hypertension in adolescents. *Expert Review of Cardiovascular Therapy* 7 (12):1595-603.

13. Aguilar, A., V. Ostrow, F. De Luca, and E. Suarez. 2010. Elevated ambulatory blood pressure in a multi-ethnic population of obese children and adolescents. *Journal of Pediatrics* 156 (6):930-5.

14. Bartosh, S.M., and A.J. Aronson. 1999. Childhood hypertension. An update on etiology, diagnosis, and treatment. *Pediatric Clinics of North America* 46 (2):235-52.

15. Biro, F.M., and M. Wien. 2010. Childhood obesity and adult morbidities. *American Journal of Clinical Nutrition* 91 (5):1499S-1505S.

16. de Kroon, M.L., C.M. Renders, J.P. van Wouwe, S. van Buuren, and R.A. Hirasing. 2010. The Terneuzen Birth Cohort: BMI change between 2 and 6 years is most predictive of adult cardiometabolic risk. *PLoS One* 5 (11):e13966.

17. Falkner, B. 2009. Hypertension in children and adolescents: epidemiology and natural history. *Pediatric Nephrology* 25 (7):1219-24.

18. Franks, P.W., R.L. Hanson, W.C. Knowler, M.L. Sievers, P.H. Bennett, and H.C. Looker. 2010. Childhood obesity, other cardiovascular risk factors, and premature death. *New England Journal of Medicine* 362 (6):485-93.

19. Mitsnefes, M.M. 2006. Hypertension in children and adolescents. *Pediatric Clinics of North America* 53 (3):493-512, viii.

20. Raghuveer, G. 2010. Lifetime cardiovascular risk of childhood obesity. *American Journal of Clinical Nutrition* 91 (5):1514S-1519S.

21. Tirosh, A., I. Shai, A. Afek, G. Dubnov-Raz, N. Ayalon, B. Gordon, E. Derazne, D. Tzur, A. Shamis, S. Vinker, et al. 2011. Adolescent BMI trajectory and risk of diabetes versus coronary disease. *The New England Journal of Medicine* 364 (14):1315-25.

22. Williams, C.L., and B.A. Strobino. 2008. Childhood diet, overweight, and CVD risk factors: the Healthy Start project. *Preventive Cardiology* 11 (1):11-20.

23. Brawer, R., N. Brisbon, and J. Plumb. 2009. Obesity and cancer. *Primary Care* 36 (3):509-31.

24. Donohoe, C.L., G.P. Pidgeon, J. Lysaght, and J.V. Reynolds. 2010. Obesity and gastrointestinal cancer. *British Journal of Surgery* 97 (5):628-42.

25. Gale, C.R., G.D. Batty, and I.J. Deary. 2008. Locus of control at age 10 years and health outcomes and behaviors at age 30 years: the 1970 British Cohort Study. *Psychosomatic Medicine* 70 (4):397-403.

26. Gundersen, C., B.J. Lohman, S. Garasky, S. Stewart, and J. Eisenmann. 2008. Food security, maternal stressors, and overweight among low-income US children: results from the National Health and Nutrition Examination Survey (1999-2002). *Pediatrics* 122 (3):e529-40.

27. Koch, F.S., A. Sepa, and J. Ludvigsson. 2008. Psychological stress and obesity. *Journal of Pediatrics* 153 (6):839-44.

28. Stunkard, A.J., M.S. Faith, and K.C. Allison. 2003. Depression and obesity. *Biological Psychiatry* 54 (3):330-7.

29. Ahmad, N., S. Biswas, S. Bae, K.E. Meador, R. Huang, and K.P. Singh. 2009. Association between obesity and asthma in US children and adolescents. *Journal of Asthma* 46 (7):642-6.

30. Bennett, W.D., and K.L. Zeman. 2004. Effect of body size on breathing pattern and fine-particle deposition in children. *Journal of Applied Physiology* 97 (3):821-6.

31. Fiorino, E.K., and L.J. Brooks. 2009. Obesity and respiratory diseases in childhood. *Clinics in Chest Medicine* 30 (3):601-8, x.

32. Chiarelli, F., and M.L. Marcovecchio. 2008. Insulin resistance and obesity in childhood. *European Journal of Endocrinology* 159 (Suppl 1):S67-74.

33. Lamb, M.M., D. Dabelea, X. Yin, L.G. Ogden, G.J. Klingensmith, M. Rewers, and J.M. Norris. 2010. Early-life predictors of higher body mass index in healthy children. *Annals of Nutrition & Metabolism* 56 (1):16-22.

34. Lee, J.M., M.J. Okumura, M.M. Davis, W.H. Herman, and J.G. Gurney. 2006. Prevalence and determinants of insulin resistance among U.S. adolescents: a population-based study. *Diabetes Care* 29 (11):2427-32.

35. Ostro, B., L. Roth, B. Malig, and M. Marty. 2009. The effects of fine particle components on respiratory hospital admissions in children. *Environmental Health Perspectives* 117 (3):475-80.

36. Weigensberg, M.J., and M.I. Goran. 2009. Type 2 diabetes in children and adolescents. *Lancet* 373 (9677):1743-4.

37. Zeitler, P., and O. Pinhas-Hamiel. 2008. Prevention and screening for type 2 diabetes in youth. *Endocrine Research* 33 (1-2):73-91.

38. Aksglaede, L., A. Juul, L.W. Olsen, and T.I. Sorensen. 2009. Age at puberty and the emerging obesity epidemic. *PLoS One* 4 (12):e8450.

39. Kaplowitz, P.B. 2008. Link between body fat and the timing of puberty. *Pediatrics* 121 Suppl 3:S208-17.

Obesity (continued)

40. Slyper, A.H. 1998. Childhood obesity, adipose tissue distribution, and the pediatric practitioner. *Pediatrics* 102 (1):e4.

41. Aksglaede, L., K. Sorensen, J.H. Petersen, N.E. Skakkebaek, and A. Juul. 2009. Recent decline in age at breast development: the Copenhagen Puberty Study. *Pediatrics* 123 (5):e932-9.

42. Gale, E.A. 2005. The myth of the metabolic syndrome. *Diabetologia* 48 (9):1679-83.

43. Pratley, R.E. 2007. Metabolic syndrome: why the controversy? *Current Diabetes Reports* 7 (1):56-9.

44. Reaven, G.M. 2011. The metabolic syndrome: time to get off the merry-go-round? *Journal of Internal Medicine* 269 (2):127-36.

45. Morrison, J.A., L.A. Friedman, P. Wang, and C.J. Glueck. 2008. Metabolic syndrome in childhood predicts adult metabolic syndrome and type 2 diabetes mellitus 25 to 30 years later. *The Journal of Pediatrics* 152 (2):201-6.

46. Morrison, J.A., L.A. Friedman, and C. Gray-McGuire. 2007. Metabolic syndrome in childhood predicts adult cardiovascular disease 25 years later: the Princeton Lipid Research Clinics Follow-up Study. *Pediatrics* 120 (2):340-5.

47. Cruz, M.L., and M.I. Goran. 2004. The metabolic syndrome in children and adolescents. *Current Diabetes Reports* 4 (1):53-62.

48. Ozanne, S.E., and C.N. Hales. 2002. Early programming of glucose-insulin metabolism. *Trends in Endocrinology & Metabolism* 13 (9):368-73.

49. Weiss, R., J. Dziura, T.S. Burgert, W.V. Tamborlane, S.E. Taksali, C.W. Yeckel, K. Allen, M. Lopes, M. Savoye, J. Morrison, et al. 2004. Obesity and the metabolic syndrome in children and adolescents. *New England Journal of Medicine* 350 (23):2362-74.

50. Lassiter, T.L., and S. Brimijoin. 2008. Rats gain excess weight after developmental exposure to the organophosphorothionate pesticide, chlorpyrifos. *Neurotoxicology and Teratology* 30 (2):125-30.

51. Lassiter, T.L., I.T. Ryde, E.A. Mackillop, K.K. Brown, E.D. Levin, F.J. Seidler, and T.A. Slotkin. 2008. Exposure of neonatal rats to parathion elicits sex-selective reprogramming of metabolism and alters the response to a high-fat diet in adulthood. *Environmental Health Perspectives* 116 (11):1456-62.

52. Slotkin, T.A., K.K. Brown, and F.J. Seidler. 2005. Developmental exposure of rats to chlorpyrifos elicits sex-selective hyperlipidemia and hyperinsulinemia in adulthood. *Environmental Health Perspectives* 113 (10):1291-4.

53. La Merrill, M., and L.S. Birnbaum. 2011. Childhood obesity and environmental chemicals. *The Mount Sinai Journal of Medicine* 78 (1):22-48.

54. Newbold, R.R. 2010. Impact of environmental endocrine disrupting chemicals on the development of obesity. *Hormones* 9 (3):206-17.

55. Grun, F., and B. Blumberg. 2006. Environmental obesogens: organotins and endocrine disruption via nuclear receptor signaling. *Endocrinology* 147 (6 Suppl):S50-5.

56. American Diabetes Association. 2000. Type 2 diabetes in children and adolescents. *Pediatrics* 105 (3 Pt 1):671-80.

57. Amed, S., D. Daneman, F.H. Mahmud, and J. Hamilton. 2010. Type 2 diabetes in children and adolescents. *Expert Review of Cardiovascular Therapy* 8 (3):393-406.

58. Karam, J.G., and S.I. McFarlane. 2008. Prevention of type 2 DM: implications for adolescents and young adults. *Pediatric Endocrinology Reviews* 5 (Suppl 4):980-8.

59. Centers for Disease Control and Prevention. 2008. *National Diabetes Fact Sheet: General Information and National Estimates on Diabetes in the United States, 2007.* Atlanta, GA: CDC. http://www.cdc.gov/diabetes/pubs/pdf/ndfs_2007.pdf.

60. Smink, A., N. Ribas-Fito, R. Garcia, M. Torrent, M.A. Mendez, J.O. Grimalt, and J. Sunyer. 2008. Exposure to hexachlorobenzene during pregnancy increases the risk of overweight in children aged 6 years. *Acta Paediatrica* 97 (10):1465-9.

61. Verhulst, S.L., V. Nelen, E.D. Hond, G. Koppen, C. Beunckens, C. Vael, G. Schoeters, and K. Desager. 2009. Intrauterine exposure to environmental pollutants and body mass index during the first 3 years of life. *Environmental Health Perspectives* 117 (1):122-6.

62. Codru, N., M.J. Schymura, S. Negoita, R. Rej, and D.O. Carpenter. 2007. Diabetes in relation to serum levels of polychlorinated biphenyls and chlorinated pesticides in adult Native Americans. *Environmental Health Perspectives* 115 (10):1442-7.

63. Everett, C.J., I.L. Frithsen, V.A. Diaz, R.J. Koopman, W.M. Simpson, Jr., and A.G. Mainous, 3rd. 2007. Association of a polychlorinated dibenzo-p-dioxin, a polychlorinated biphenyl, and DDT with diabetes in the 1999-2002 National Health and Nutrition Examination Survey. *Environmental Research* 103 (3):413-8.

64. Montgomery, M.P., F. Kamel, T.M. Saldana, M.C. Alavanja, and D.P. Sandler. 2008. Incident diabetes and pesticide exposure among licensed pesticide applicators: Agricultural Health Study, 1993-2003. *American Journal of Epidemiology* 167 (10):1235-46.

65. Blanck, H.M., M. Marcus, C. Rubin, P.E. Tolbert, V.S. Hertzberg, A.K. Henderson, and R.H. Zhang. 2002. Growth in girls exposed in utero and postnatally to polybrominated biphenyls and polychlorinated biphenyls. *Epidemiology* 13 (2):205-10.

Obesity (continued)

66. Cupul-Uicab, L.A., M. Hernandez-Avila, E.A. Terrazas-Medina, M.L. Pennell, and M.P. Longnecker. 2010. Prenatal exposure to the major DDT metabolite 1,1-dichloro-2,2-bis(p-chlorophenyl)ethylene (DDE) and growth in boys from Mexico. *Environmental Research* 110 (6):595-603.

67. Gladen, B.C., M.A. Klebanoff, M.L. Hediger, S.H. Katz, D.B. Barr, M.D. Davis, and M.P. Longnecker. 2004. Prenatal DDT exposure in relation to anthropometric and pubertal measures in adolescent males. *Environmental Health Perspectives* 112 (17):1761-7.

68. Jackson, L.W., C.D. Lynch, P.J. Kostyniak, B.M. McGuinness, and G.M. Louis. 2010. Prenatal and postnatal exposure to polychlorinated biphenyls and child size at 24 months of age. *Reproductive Toxicology* 29 (1):25-31.

69. Grun, F., and B. Blumberg. 2009. Endocrine disrupters as obesogens. *Molecular and Cellular Endocrinology* 304 (1-2):19-29.

70. Grun, F., H. Watanabe, Z. Zamanian, L. Maeda, K. Arima, R. Cubacha, D.M. Gardiner, J. Kanno, T. Iguchi, and B. Blumberg. 2006. Endocrine-disrupting organotin compounds are potent inducers of adipogenesis in vertebrates. *Molecular Endocrinology* 20 (9):2141-55.

71. Hugo, E.R., T.D. Brandebourg, J.G. Woo, J. Loftus, J.W. Alexander, and N. Ben-Jonathan. 2008. Bisphenol A at environmentally relevant doses inhibits adiponectin release from human adipose tissue explants and adipocytes. *Environmental Health Perspectives* 116 (12):1642-7.

72. Newbold, R.R., E. Padilla-Banks, W.N. Jefferson, and J.J. Heindel. 2008. Effects of endocrine disruptors on obesity. *International Journal of Andrology* 31 (2):201-8.

73. Newbold, R.R., E. Padilla-Banks, R.J. Snyder, T.M. Phillips, and W.N. Jefferson. 2007. Developmental exposure to endocrine disruptors and the obesity epidemic. *Reproductive Toxicology* 23 (3):290-6.

74. Thayer, K.A., J.J. Heindel, J.R. Bucher, and M.A. Gallo. 2012. Role of environmental chemicals in diabetes and obesity: a National Toxicology Program workshop review. *Environmental Health Perspectives* 120 (6):779-89.

75. U.S. Department of Health and Human Services. 2011. *Strategic Plan for NIH Obesity Research*. Bethesda, MD: National Institutes of Health Obesity Research Task Force. NIH Publication No. 11-5493. http://www.obesityresearch.nih.gov/About/StrategicPlanforNIH_Obesity_Research_Full-Report_2011.pdf.

76. White House Task Force on Childhood Obesity. 2010. *Solving the Problem of Childhood Obesity Within a Generation*. Washington, DC: Executive Office of the President. http://www.letsmove.gov/sites/letsmove.gov/files/TaskForce_on_Childhood_Obesity_May2010_FullReport.pdf.

77. Sun, Q., P. Yue, J.A. Deiuliis, C.N. Lumeng, T. Kampfrath, M.B. Mikolaj, Y. Cai, M.C. Ostrowski, B. Lu, S. Parthasarathy, et al. 2009. Ambient air pollution exaggerates adipose inflammation and insulin resistance in a mouse model of diet-induced obesity. *Circulation* 119 (4):538-46.

78. Chen, J.C., J.M. Cavallari, P.H. Stone, and D.C. Christiani. 2007. Obesity is a modifier of autonomic cardiac responses to fine metal particulates. *Environmental Health Perspectives* 115 (7):1002-6.

79. Shore, S.A., Y.M. Rivera-Sanchez, I.N. Schwartzman, and R.A. Johnston. 2003. Responses to ozone are increased in obese mice. *Journal of Applied Physiology* 95 (3):938-45.

80. Corbo, G.M., F. Forastiere, M. De Sario, L. Brunetti, E. Bonci, M. Bugiani, E. Chellini, S. La Grutta, E. Migliore, R. Pistelli, et al. 2008. Wheeze and asthma in children: associations with body mass index, sports, television viewing, and diet. *Epidemiology* 19 (5):747-55.

81. Arzuaga, X., N. Ren, A. Stromberg, E.P. Black, V. Arsenescu, L.A. Cassis, Z. Majkova, M. Toborek, and B. Hennig. 2009. Induction of gene pattern changes associated with dysfunctional lipid metabolism induced by dietary fat and exposure to a persistent organic pollutant. *Toxicology Letters* 189 (2):96-101.

82. Ghanayem, B.I., R. Bai, G.E. Kissling, G. Travlos, and U. Hoffler. 2010. Diet-induced obesity in male mice is associated with reduced fertility and potentiation of acrylamide-induced reproductive toxicity. *Biology of Reproduction* 82 (1):96-104.

83. La Merrill, M., R. Harper, L.S. Birnbaum, R.D. Cardiff, and D.W. Threadgill. 2010. Maternal dioxin exposure combined with a diet high in fat increases mammary cancer incidence in mice. *Environmental Health Perspectives* 118 (5):596-601.

84. Papas, M.A., A.J. Alberg, R. Ewing, K.J. Helzlsouer, T.L. Gary, and A.C. Klassen. 2007. The built environment and obesity. *Epidemiologic Reviews* 29:129-43.

85. Sallis, J.F., and K. Glanz. 2006. The role of built environments in physical activity, eating, and obesity in childhood. *The Future of Children* 16 (1):89-108.

86. Davison, K.K., and C.T. Lawson. 2006. Do attributes in the physical environment influence children's physical activity? A review of the literature. *The International Journal of Behavioral Nutrition and Physical Activity* 3:19.

87. Dunton, G.F., J. Kaplan, J. Wolch, M. Jerrett, and K.D. Reynolds. 2009. Physical environmental correlates of childhood obesity: a systematic review. *Obesity Reviews* 10 (4):393-402.

Obesity (continued)

88. Maziak, W., K.D. Ward, and M.B. Stockton. 2008. Childhood obesity: are we missing the big picture? *Obesity Reviews* 9 (1):35-42.

89. Rahman, T., R.A. Cushing, and R.J. Jackson. 2011. Contributions of built environment to childhood obesity. *The Mount Sinai Journal of Medicine* 78 (1):49-57.

90. Sallis, J.F., M.F. Floyd, D.A. Rodriguez, and B.E. Saelens. 2012. Role of built environments in physical activity, obesity, and cardiovascular disease. *Circulation* 125 (5):729-37.

91. McDonald, N.C. 2008. Critical factors for active transportation to school among low-income and minority students. Evidence from the 2001 National Household Travel Survey. *American Journal of Preventive Medicine* 34 (4):341-4.

92. Roemmich, J.N., L.H. Epstein, S. Raja, L. Yin, J. Robinson, and D. Winiewicz. 2006. Association of access to parks and recreational facilities with the physical activity of young children. *Preventive Medicine* 43 (6):437-41.

93. American Academy of Pediatrics. 2009. The built environment: designing communities to promote physical activity in children. *Pediatrics* 123 (6):1591-1598.

94. McCurdy, L.E., K.E. Winterbottom, S.S. Mehta, and J.R. Roberts. 2010. Using nature and outdoor activity to improve children's health. *Current Problems in Pediatric and Adolescent Health Care* 40 (5):102-17.

95. Cutts, B.B., K.J. Darby, C.G. Boone, and A. Brewis. 2009. City structure, obesity, and environmental justice: an integrated analysis of physical and social barriers to walkable streets and park access. *Social Science & Medicine* 69 (9):1314-22.

96. Redwood, Y., A.J. Schulz, B.A. Israel, M. Yoshihama, C.C. Wang, and M. Kreuter. 2010. Social, economic, and political processes that create built environment inequities: perspectives from urban African Americans in Atlanta. *Family & Community Health* 33 (1):53-67.

97. Taylor, W.C., J.T. Hepworth, E. Lees, K. Feliz, S. Ahsan, A. Cassells, D.C. Volding, and J.N. Tobin. 2008. Obesity, physical activity, and the environment: is there a legal basis for environmental injustices? *Environmental Justice* 1 (1):45-48.

98. Ewing, R., R.C. Brownson, and D. Berrigan. 2006. Relationship between urban sprawl and weight of United States youth. *American Journal of Preventive Medicine* 31 (6):464-74.

99. Ewing, R., T. Schmid, R. Killingsworth, A. Zlot, and S. Raudenbush. 2003. Relationship between urban sprawl and physical activity, obesity, and morbidity. *American Journal of Health Promotion* 18 (1):47-57.

100. Wang, Y. 2001. Cross-national comparison of childhood obesity: the epidemic and the relationship between obesity and socioeconomic status. *International Journal of Epidemiology* 30 (5):1129-36.

101. Wang, Y., and M.A. Beydoun. 2007. The obesity epidemic in the United States--gender, age, socioeconomic, racial/ethnic, and geographic characteristics: a systematic review and meta-regression analysis. *Epidemiologic Reviews* 29:6-28.

102. Davis, A.M., K.J. Bennett, C. Befort, and N. Nollen. 2011. Obesity and related health behaviors among urban and rural children in the United States: data from the national health and nutrition examination survey 2003-2004 and 2005-2006. *Journal of Pediatric Psychology* 36 (6):669-76.

103. Centers for Disease Control and Prevention. 2010. *About BMI for Children and Teens*. Retrieved April 7, 2010 from http://www.cdc.gov/healthyweight/assessing/bmi/childrens_bmi/about_childrens_bmi.html.

Adverse Birth Outcomes

1. Centers for Disease Control and Prevention. 2009. *Maternal and Infant Health Research: Preterm Birth*. CDC. Retrieved October 10, 2010 from http://www.cdc.gov/reproductivehealth/maternalinfanthealth/PBP.htm.

2. JAMA. 2002. JAMA patient page: Low birth weight. *Journal of the American Medical Association* 287 (2):270.

3. Institute of Medicine. 2007. *Preterm Birth: Causes, Consequences, and Prevention*. Edited by R. E. Behrman and A. S. Butler. Washington, DC: The National Academies Press.

4. Mathews, T.J., and M.F. MacDorman. 2008. Infant mortality statistics from the 2005 period linked birth/infant death data set. *National Vital Statistics Reports* 57 (2).

5. Clark, S.L., D.D. Miller, M.A. Belfort, G.A. Dildy, D.K. Frye, and J.A. Meyers. 2009. Neonatal and maternal outcomes associated with elective term delivery. *American Journal of Obstetrics and Gynecology* 200 (2):156 e1-4.

6. Moster, D., A.J. Wilcox, S.E. Vollset, T. Markestad, and R.T. Lie. 2010. Cerebral Palsy Among Term and Postterm Births. *Journal of the American Medical Association* 304 (9):976-982.

7. Tita, A.T., M.B. Landon, C.Y. Spong, Y. Lai, K.J. Leveno, M.W. Varner, A.H. Moawad, S.N. Caritis, P.J. Meis, R.J. Wapner, et al. 2009. Timing of elective repeat cesarean delivery at term and neonatal outcomes. *New England Journal of Medicine* 360 (2):111-20.

8. Cosmi, E., T. Fanelli, S. Visentin, D. Trevisanuto, and V. Zanardo. 2011. Consequences in infants that were intrauterine growth restricted. *Journal of Pregnancy* 2011:Article ID 364381.

Adverse Birth Outcomes (continued)

9. Rinaudo, P.F., and J. Lamb. 2008. Fetal origins of perinatal morbidity and/or adult disease. *Seminars in Reproductive Medicine* 26 (5):436-45.

10. Martin, J.A., B.E. Hamilton, P.D. Sutton, S.J. Ventura, P.H. Menacker, S. Kirmeyer, and T.J. Mathews. 2009. Births: Final Data for 2006. *National Vital Statistics Reports* 57 (7).

11. Martin, J.A., B.E. Hamilton, P.D. Sutton, S.J. Ventura, T.J. Mathews, S. Kirmeyer, and M.J.K. Osterman. 2010. Births: Final Data for 2007. *National Vital Statistics Reports* 58 (24).

12. American College of Obstetricians and Gynecologists (ACOG). 2000. Intrauterine growth restriction. ACOG practice bulletin, number 12. *Obstetrics and Gynecology* 95 (1).

13. Berghella, V. 2007. Prevention of recurrent fetal growth restriction. *Obstetrics and Gynecology* 110 (4):904-12.

14. Honein, M.A., R.S. Kirby, R.E. Meyer, J. Xing, N.I. Skerrette, N. Yuskiv, L. Marengo, J.R. Petrini, M.J. Davidoff, C.T. Mai, et al. 2009. The association between major birth defects and preterm birth. *Maternal and Child Health Journal* 13 (2):164-75.

15. U.S. Department of Health and Human Services. 2004. *The Health Consequences of Smoking: A Report of the Surgeon General*. Atlanta, GA: Centers for Disease Control and Prevention, Office on Smoking and Health.

16. Goldenberg, R.L., and J.F. Culhane. 2007. Low birth weight in the United States. *American Journal of Clinical Nutrition* 85 (2):584S-590S.

17. Lu, M.C., and N. Halfon. 2003. Racial and ethnic disparities in birth outcomes: a life-course perspective. *Maternal and Child Health Journal* 7 (1):13-30.

18. Collins, J.W., Jr., and A.G. Butler. 1997. Racial differences in the prevalence of small-for-dates infants among college-educated women. *Epidemiology* 8 (3):315-7.

19. McGrady, G.A., J.F. Sung, D.L. Rowley, and C.J. Hogue. 1992. Preterm delivery and low birth weight among first-born infants of black and white college graduates. *American Journal of Epidemiology* 136 (3):266-76.

20. Schoendorf, K.C., C.J. Hogue, J.C. Kleinman, and D. Rowley. 1992. Mortality among infants of black as compared with white college-educated parents. *New England Journal of Medicine* 326 (23):1522-6.

21. Goldenberg, R.L., S.P. Cliver, F.X. Mulvihill, C.A. Hickey, H.J. Hoffman, L.V. Klerman, and M.J. Johnson. 1996. Medical, psychosocial, and behavioral risk factors do not explain the increased risk for low birth weight among black women. *American Journal of Obstetrics & Gynecology* 175 (5):1317-24.

22. Singh, G.K., and S.M. Yu. 1995. Infant mortality in the United States: trends, differentials, and projections, 1950 through 2010. *American Journal of Public Health* 85 (7):957-64.

23. Donahue, S.M., K.P. Kleinman, M.W. Gillman, and E. Oken. 2010. Trends in birth weight and gestational length among singleton term births in the United States: 1990-2005. *Obstetrics and Gynecology* 115 (2 Pt 1):357-64.

24. U.S. Department of Health and Human Services. 2006. *The Health Consequences of Involuntary Exposure to Tobacco Smoke: A Report of the Surgeon General*. Atlanta, GA: U.S. Department of Health and Human Services, Centers for Disease Control and Prevention, Coordinating Center for Health Promotion, National Center for Chronic Disease Prevention and Health Promotion, Office on Smoking and Health.

25. National Toxicology Program. 2012. *NTP Monograph on Health Effects of Low-Level Lead*. Research Triangle Park, NC: National Institute of Environmental Health Sciences, National Toxicology Program. http://ntp.niehs.nih.gov/go/36443.

26. Bobak, M. 2000. Outdoor air pollution, low birth weight, and prematurity. *Environmental Health Perspectives* 108 (2):173-6.

27. Dugandzic, R., L. Dodds, D. Stieb, and M. Smith-Doiron. 2006. The association between low level exposures to ambient air pollution and term low birth weight: a retrospective cohort study. *Environmental Health* 5:3.

28. Ha, E.H., Y.C. Hong, B.E. Lee, B.H. Woo, J. Schwartz, and D.C. Christiani. 2001. Is air pollution a risk factor for low birth weight in Seoul? *Epidemiology* 12 (6):643-8.

29. Huynh, M., T.J. Woodruff, J.D. Parker, and K.C. Schoendorf. 2006. Relationships between air pollution and preterm birth in California. *Paediatric and Perinatal Epidemiology* 20 (6):454-61.

30. Lin, C.M., C.Y. Li, G.Y. Yang, and I.F. Mao. 2004. Association between maternal exposure to elevated ambient sulfur dioxide during pregnancy and term low birth weight. *Environmental Research* 96 (1):41-50.

31. Liu, S., D. Krewski, Y. Shi, Y. Chen, and R.T. Burnett. 2003. Association between gaseous ambient air pollutants and adverse pregnancy outcomes in Vancouver, Canada. *Environmental Health Perspectives* 111 (14):1773-8.

32. Maisonet, M., T.J. Bush, A. Correa, and J.J. Jaakkola. 2001. Relation between ambient air pollution and low birth weight in the Northeastern United States. *Environmental Health Perspectives* 109 (Suppl 3):351-6.

33. Maroziene, L., and R. Grazuleviciene. 2002. Maternal exposure to low-level air pollution and pregnancy outcomes: a population-based study. *Environmental Health* 1 (1):6.

Adverse Birth Outcomes (continued)

34. Parker, J.D., T.J. Woodruff, R. Basu, and K.C. Schoendorf. 2005. Air pollution and birth weight among term infants in California. *Pediatrics* 115 (1):121-8.

35. Sagiv, S.K., P. Mendola, D. Loomis, A.H. Herring, L.M. Neas, D.A. Savitz, and C. Poole. 2005. A time-series analysis of air pollution and preterm birth in Pennsylvania, 1997-2001. *Environmental Health Perspectives* 113 (5):602-6.

36. U.S. Environmental Protection Agency. 2009. *Integrated Science Assessment for Particulate Matter (Final Report).* Washington, DC: U.S. EPA. EPA/600/R-08/139F. http://cfpub.epa.gov/ncea/cfm/recordisplay.cfm?deid=216546.

37. Wang, X., H. Ding, L. Ryan, and X. Xu. 1997. Association between air pollution and low birth weight: a community-based study. *Environmental Health Perspectives* 105 (5):514-20.

38. Wilhelm, M., and B. Ritz. 2005. Local variations in CO and particulate air pollution and adverse birth outcomes in Los Angeles County, California, USA. *Environmental Health Perspectives* 113 (9):1212-21.

39. Xu, X., H. Ding, and X. Wang. 1995. Acute effects of total suspended particles and sulfur dioxides on preterm delivery: a community-based cohort study. *Archives of Environmental Health* 50 (6):407-15.

40. Liu, S., D. Krewski, Y. Shi, Y. Chen, and R.T. Burnett. 2007. Association between maternal exposure to ambient air pollutants during pregnancy and fetal growth restriction. *Journal of Exposure Science and Environmental Epidemiology* 17 (5):426-32.

41. Stillerman, K.P., D.R. Mattison, L.C. Giudice, and T.J. Woodruff. 2008. Environmental exposures and adverse pregnancy outcomes: a review of the science. *Reproductive Sciences* 15 (7):631-50.

42. Glinianaia, S.V., J. Rankin, R. Bell, T. Pless-Mulloli, and D. Howel. 2004. Particulate air pollution and fetal health: a systematic review of the epidemiologic evidence. *Epidemiology* 15 (1):36-45.

43. Parker, J.D., D.Q. Rich, S.V. Glinianaia, J.H. Leem, D. Wartenberg, M.L. Bell, M. Bonzini, M. Brauer, L. Darrow, U. Gehring, et al. 2011. The International Collaboration on Air Pollution and Pregnancy Outcomes: initial results. *Environmental Health Perspectives* 119 (7):1023-8.

44. Ritz, B., and M. Wilhelm. 2008. Ambient air pollution and adverse birth outcomes: methodologic issues in an emerging field. *Basic and Clinical Pharmacology and Toxicology* 102 (2):182-90.

45. Sram, R.J., B. Binkova, J. Dejmek, and M. Bobak. 2005. Ambient air pollution and pregnancy outcomes: a review of the literature. *Environmental Health Perspectives* 113 (4):375-82.

46. Choi, H., W. Jedrychowski, J. Spengler, D.E. Camann, R.M. Whyatt, V. Rauh, W.Y. Tsai, and F.P. Perera. 2006. International studies of prenatal exposure to polycyclic aromatic hydrocarbons and fetal growth. *Environmental Health Perspectives* 114 (11):1744-50.

47. Choi, H., V. Rauh, R. Garfinkel, Y. Tu, and F.P. Perera. 2008. Prenatal exposure to airborne polycyclic aromatic hydrocarbons and risk of intrauterine growth restriction. *Environmental Health Perspectives* 116 (5):658-65.

48. Perera, F.P., V. Rauh, R.M. Whyatt, W.Y. Tsai, J.T. Bernert, Y.H. Tu, H. Andrews, J. Ramirez, L. Qu, and D. Tang. 2004. Molecular evidence of an interaction between prenatal environmental exposures and birth outcomes in a multiethnic population. *Environmental Health Perspectives* 112 (5):626-30.

49. Perera, F.P., D. Tang, V. Rauh, K. Lester, W.Y. Tsai, Y.H. Tu, L. Weiss, L. Hoepner, J. King, G. Del Priore, et al. 2005. Relationships among polycyclic aromatic hydrocarbon-DNA adducts, proximity to the World Trade Center, and effects on fetal growth. *Environmental Health Perspectives* 113 (8):1062-7.

50. Brauer, M., C. Lencar, L. Tamburic, M. Koehoorn, P. Demers, and C. Karr. 2008. A cohort study of traffic-related air pollution impacts on birth outcomes. *Environmental Health Perspectives* 116 (5):680-6.

51. Genereux, M., N. Auger, M. Goneau, and M. Daniel. 2008. Neighbourhood socioeconomic status, maternal education and adverse birth outcomes among mothers living near highways. *Journal of Epidemiology and Community Health* 62 (8):695-700.

52. Health Effects Institute. 2010. *HEI Panel on the Health Effects of Traffic-Related Air Pollution: A Critical Review of the Literature on Emissions, Exposure, and Health Effects.* Boston, MA. HEI Special Report 17. http://pubs.healtheffects.org/view.php?id=334

53. Ponce, N.A., K.J. Hoggatt, M. Wilhelm, and B. Ritz. 2005. Preterm birth: the interaction of traffic-related air pollution with economic hardship in Los Angeles neighborhoods. *American Journal of Epidemiology* 162 (2):140-8.

54. Wilhelm, M., and B. Ritz. 2003. Residential proximity to traffic and adverse birth outcomes in Los Angeles county, California, 1994-1996. *Environmental Health Perspectives* 111 (2):207-16.

55. Adibi, J.J., R. Hauser, P.L. Williams, R.M. Whyatt, A.M. Calafat, H. Nelson, R. Herrick, and S.H. Swan. 2009. Maternal urinary metabolites of Di-(2-Ethylhexyl) phthalate in relation to the timing of labor in a US multicenter pregnancy cohort study. *American Journal of Epidemiology* 169 (8):1015-24.

56. Latini, G., C. De Felice, G. Presta, A. Del Vecchio, I. Paris, F. Ruggieri, and P. Mazzeo. 2003. In utero exposure to di-(2-ethylhexyl)phthalate and duration of human pregnancy. *Environmental Health Perspectives* 111 (14):1783-5.

Adverse Birth Outcomes (continued)

57. Meeker, J.D., H. Hu, D.E. Cantonwine, H. Lamadrid-Figueroa, A.M. Calafat, A.S. Ettinger, M. Hernandez-Avila, R. Loch-Caruso, and M.M. Tellez-Rojo. 2009. Urinary phthalate metabolites in relation to preterm birth in Mexico City. *Environmental Health Perspectives* 117 (10):1587-92.

58. Whyatt, R.M., J.J. Adibi, A.M. Calafat, D.E. Camann, V. Rauh, H.K. Bhat, F.P. Perera, H. Andrews, A.C. Just, L. Hoepner, et al. 2009. Prenatal di(2-ethylhexyl)phthalate exposure and length of gestation among an inner-city cohort. *Pediatrics* 124 (6):e1213-20.

59. Zhang, Y., L. Lin, Y. Cao, B. Chen, L. Zheng, and R.S. Ge. 2009. Phthalate levels and low birth weight: a nested case-control study of Chinese newborns. *Journal of Pediatrics* 155 (4):500-4.

60. Murphy, L.E., A.L. Gollenberg, G.M. Buck Louis, P.J. Kostyniak, and R. Sundaram. 2010. Maternal serum preconception polychlorinated biphenyl concentrations and infant birth weight. *Environmental Health Perspectives* 118 (2):297-302.

61. Wigle, D.T., T.E. Arbuckle, M.C. Turner, A. Berube, Q. Yang, S. Liu, and D. Krewski. 2008. Epidemiologic evidence of relationships between reproductive and child health outcomes and environmental chemical contaminants. *Journal of Toxicology and Environmental Health Part B Crit Reviews* 11 (5-6):373-517.

62. Hertz-Picciotto, I., M.J. Charles, R.A. James, J.A. Keller, E. Willman, and S. Teplin. 2005. In utero polychlorinated biphenyl exposures in relation to fetal and early childhood growth. *Epidemiology* 16 (5):648-56.

63. Baibergenova, A., R. Kudyakov, M. Zdeb, and D.O. Carpenter. 2003. Low birth weight and residential proximity to PCB-contaminated waste sites. *Environmental Health Perspectives* 111 (10):1352-7.

64. Halldorsson, T.I., I. Thorsdottir, H.M. Meltzer, F. Nielsen, and S.F. Olsen. 2008. Linking exposure to polychlorinated biphenyls with fatty fish consumption and reduced fetal growth among Danish pregnant women: a cause for concern? *American Journal of Epidemiology* 168 (8):958-65.

65. Longnecker, M.P., M.A. Klebanoff, J.W. Brock, and X. Guo. 2005. Maternal levels of polychlorinated biphenyls in relation to preterm and small-for-gestational-age birth. *Epidemiology* 16 (5):641-7.

66. Apelberg, B.J., F.R. Witter, J.B. Herbstman, A.M. Calafat, R.U. Halden, L.L. Needham, and L.R. Goldman. 2007. Cord serum concentrations of perfluorooctane sulfonate (PFOS) and perfluorooctanoate (PFOA) in relation to weight and size at birth. *Environmental Health Perspectives* 115 (11):1670-6.

67. Fei, C., J.K. McLaughlin, R.E. Tarone, and J. Olsen. 2007. Perfluorinated chemicals and fetal growth: a study within the Danish National Birth Cohort. *Environmental Health Perspectives* 115 (11):1677-82.

68. Fei, C., J.K. McLaughlin, R.E. Tarone, and J. Olsen. 2008. Fetal growth indicators and perfluorinated chemicals: a study in the Danish National Birth Cohort. *American Journal of Epidemiology* 168 (1):66-72.

69. Stein, C.R., D.A. Savitz, and M. Dougan. 2009. Serum levels of perfluorooctanoic acid and perfluorooctane sulfonate and pregnancy outcome. *American Journal of Epidemiology* 170 (7):837-46.

70. Washino, N., Y. Saijo, S. Sasaki, S. Kato, S. Ban, K. Konishi, R. Ito, A. Nakata, Y. Iwasaki, K. Saito, et al. 2009. Correlations between prenatal exposure to perfluorinated chemicals and reduced fetal growth. *Environmental Health Perspectives* 117 (4):660-7.

71. Hamm, M.P., N.M. Cherry, E. Chan, J.W. Martin, and I. Burstyn. 2010. Maternal exposure to perfluorinated acids and fetal growth. *Journal of Exposure Science and Environmental Epidemiology* 20 (7):589-97.

72. Monroy, R., K. Morrison, K. Teo, S. Atkinson, C. Kubwabo, B. Stewart, and W.G. Foster. 2008. Serum levels of perfluoroalkyl compounds in human maternal and umbilical cord blood samples. *Environmental Research* 108 (1):56-62.

73. Bove, F., Y. Shim, and P. Zeitz. 2002. Drinking water contaminants and adverse pregnancy outcomes: a review. *Environmental Health Perspectives* 110 (Suppl 1):61-74.

74. Hoffman, C.S., P. Mendola, D.A. Savitz, A.H. Herring, D. Loomis, K.E. Hartmann, P.C. Singer, H.S. Weinberg, and A.F. Olshan. 2008. Drinking water disinfection by-product exposure and fetal growth. *Epidemiology* 19 (5):729-37.

75. Hoffman, C.S., P. Mendola, D.A. Savitz, A.H. Herring, D. Loomis, K.E. Hartmann, P.C. Singer, H.S. Weinberg, and A.F. Olshan. 2008. Drinking water disinfection by-product exposure and duration of gestation. *Epidemiology* 19 (5):738-46.

76. Myers, S.L., D.T. Lobdell, Z. Liu, Y. Xia, H. Ren, Y. Li, R.K. Kwok, J.L. Mumford, and P. Mendola. 2010. Maternal drinking water arsenic exposure and perinatal outcomes in inner Mongolia, China. *Journal of Epidemiology and Community Health* 64 (4):325-9.

77. Rahman, A., M. Vahter, A.H. Smith, B. Nermell, M. Yunus, S. El Arifeen, L.A. Persson, and E.C. Ekstrom. 2009. Arsenic exposure during pregnancy and size at birth: a prospective cohort study in Bangladesh. *American Journal of Epidemiology* 169 (3):304-12.

78. Smith, A.H., and C.M. Steinmaus. 2009. Health effects of arsenic and chromium in drinking water: recent human findings. *Annual Review of Public Health* 30:107-22.

79. Slama, R., O. Thiebaugeorges, V. Goua, L. Aussel, P. Sacco, A. Bohet, A. Forhan, B. Ducot, I. Annesi-Maesano, J. Heinrich, et al. 2009. Maternal personal exposure to airborne benzene and intrauterine growth. *Environmental Health Perspectives* 117 (8):1313-21.

Adverse Birth Outcomes (continued)

80. Ochoa-Acuna, H., J. Frankenberger, L. Hahn, and C. Carbajo. 2009. Drinking-water herbicide exposure in Indiana and prevalence of small-for-gestational-age and preterm delivery. *Environmental Health Perspectives* 117 (10):1619-24.

81. Ranjit, N., K. Siefert, and V. Padmanabhan. 2010. Bisphenol-A and disparities in birth outcomes: a review and directions for future research. *Journal of Perinatology* 30 (1):2-9.

82. Konishi, K., S. Sasaki, S. Kato, S. Ban, N. Washino, J. Kajiwara, T. Todaka, H. Hirakawa, T. Hori, D. Yasutake, et al. 2009. Prenatal exposure to PCDDs/PCDFs and dioxin-like PCBs in relation to birth weight. *Environmental Research* 109 (7):906-13.

83. Zota, A.R., A.S. Ettinger, M. Bouchard, C.J. Amarasiriwardena, J. Schwartz, H. Hu, and R.O. Wright. 2009. Maternal blood manganese levels and infant birth weight. *Epidemiology* 20 (3):367-73.

84. Davidoff, M.J., T. Dias, K. Damus, R. Russell, V.R. Bettegowda, S. Dolan, R.H. Schwarz, N.S. Green, and J. Petrini. 2006. Changes in the gestational age distribution among U.S. singleton births: impact on rates of late preterm birth, 1992 to 2002. *Seminars in Perinatology* 30 (1):8-15.

85. Heron, M., P.D. Sutton, J. Xu, S.J. Ventura, D.M. Strobino, and B. Guyer. 2010. Annual summary of vital statistics: 2007. *Pediatrics* 125 (1):4-15.

Supplementary Topics

Birth Defects

1. California Birth Defects Monitoring Program. *Overview: The Problem of Birth Defects*. Retrieved August 19, 2009 from http://www.cdph.ca.gov/programs/CBDMP/Pages/default.aspx.

2. National Institute of Child Health and Human Development. 2007. *Birth Defects*. Retrieved August 19, 2009 from http://www.nichd.nih.gov/health/topics/birth_defects.cfm.

3. National Center for Health Statistics. 2009. *Health, United States 2008, with Chartbook*. Hyattsville, MD: U.S. Department of Health and Human Services. Centers for Disease Control and Prevention. National Center for Health Statistics.

4. Centers for Disease Control and Prevention. 2011. *Guidance for Preventing Birth Defects*. Retrieved August 30, 2011 from http://www.cdc.gov/ncbddd/birthdefects/prevention.html.

5. Garcia-Bournissen, F., L. Tsur, L.H. Goldstein, A. Staroselsky, M. Avner, F. Asrar, M. Berkovitch, G. Straface, G. Koren, and M. De Santis. 2008. Fetal exposure to isotretinoin-an international problem. *Reproductive Toxicology* 25 (1):124-8.

6. American Academy of Pediatrics. Committee on Substance Abuse and Committee on Children With Disabilities. 2000. Fetal alcohol syndrome and alcohol-related neurodevelopmental disorders. *Pediatrics* 106 (2 Pt 1):358-61.

7. Williams, L.J., S.A. Rasmussen, A. Flores, R.S. Kirby, and L.D. Edmonds. 2005. Decline in the prevalence of spina bifida and anencephaly by race/ethnicity: 1995-2002. *Pediatrics* 116 (3):580-6.

8. Brent, R.L. 2004. Environmental causes of human congenital malformations: the pediatrician's role in dealing with these complex clinical problems caused by a multiplicity of environmental and genetic factors. *Pediatrics* 113 (4 Suppl):957-68.

9. Landrigan, P.J., and A. Garg. 2002. Chronic effects of toxic environmental exposures on children's health. *Journal of Toxicology - Clinical Toxicology* 40 (4):449-56.

10. Wigle, D.T., T.E. Arbuckle, M.C. Turner, A. Berube, Q. Yang, S. Liu, and D. Krewski. 2008. Epidemiologic evidence of relationships between reproductive and child health outcomes and environmental chemical contaminants. *Journal of Toxicology and Environmental Health Part B: Critical Reviews* 11 (5-6):373-517.

11. Harada, M., H. Akagi, T. Tsuda, T. Kizaki, and H. Ohno. 1999. Methylmercury level in umbilical cords from patients with congenital Minamata disease. *Science of the Total Environment* 234 (1-3):59-62.

12. Rogan, W.J. 1982. PCBs and cola-colored babies: Japan, 1968, and Taiwan, 1979. *Teratology* 26 (3):259-61.

13. McMartin, K.I., M. Chu, E. Kopecky, T.R. Einarson, and G. Koren. 1998. Pregnancy outcome following maternal organic solvent exposure: a meta-analysis of epidemiologic studies. *American Journal of Industrial Medicine* 34 (3):288-92.

14. Aschengrau, A., J.M. Weinberg, P.A. Janulewicz, L.G. Gallagher, M.R. Winter, V.M. Vieira, T.F. Webster, and D.M. Ozonoff. 2009. Prenatal exposure to tetrachloroethylene-contaminated drinking water and the risk of congenital anomalies: a retrospective cohort study. *Environmental Health* 8:44.

15. Bell, E.M., I. Hertz-Picciotto, and J.J. Beaumont. 2001. A case-control study of pesticides and fetal death due to congenital anomalies. *Epidemiology* 12 (2):148-56.

16. Blatter, B.M., R. Hermens, M. Bakker, N. Roeleveld, A.L. Verbeek, and G.A. Zielhuis. 1997. Paternal occupational exposure around conception and spina bifida in offspring. *American Journal of Industrial Medicine* 32 (3):283-91.

17. Blatter, B.M., and N. Roeleveld. 1996. Spina bifida and parental occupation in a Swedish register-based study. *Scandinavian Journal of Work, Environment, and Health* 22 (6):433-7.

18. Calvert, G.M., W.A. Alarcon, A. Chelminski, M.S. Crowley, R. Barrett, A. Correa, S. Higgins, H.L. Leon, J. Correia, A. Becker, et al. 2007. Case report: three farmworkers who gave birth to infants with birth defects closely grouped in time and place-Florida and North Carolina, 2004-2005. *Environmental Health Perspectives* 115 (5):787-91.

19. Dimich-Ward, H., C. Hertzman, K. Teschke, R. Hershler, S.A. Marion, A. Ostry, and S. Kelly. 1996. Reproductive effects of paternal exposure to chlorophenate wood preservatives in the sawmill industry. *Scandinavian Journal of Work, Environment, and Health* 22 (4):267-73.

20. Dugas, J., M.J. Nieuwenhuijsen, D. Martinez, N. Iszatt, P. Nelson, and P. Elliott. 2010. Use of biocides and insect repellents and risk of hypospadias. *Occupational and Environmental Medicine* 67 (3):196-200.

21. Engel, L.S., E.S. O'Meara, and S.M. Schwartz. 2000. Maternal occupation in agriculture and risk of limb defects in Washington State, 1980-1993. *Scandinavian Journal of Work, Environment, and Health* 26 (3):193-8.

22. Fernandez, M.F., B. Olmos, A. Granada, M.J. Lopez-Espinosa, J.M. Molina-Molina, J.M. Fernandez, M. Cruz, F. Olea-Serrano, and N. Olea. 2007. Human exposure to endocrine-disrupting chemicals and prenatal risk factors for cryptorchidism and hypospadias: a nested case-control study. *Environmental Health Perspectives* 115 Suppl 1:8-14.

Birth Defects (continued)

23. Garcia, A.M., F.G. Benavides, T. Fletcher, and E. Orts. 1998. Paternal exposure to pesticides and congenital malformations. *Scandinavian Journal of Work, Environment, and Health* 24 (6):473-80.

24. Garcia, A.M., T. Fletcher, F.G. Benavides, and E. Orts. 1999. Parental agricultural work and selected congenital malformations. *American Journal of Epidemiology* 149 (1):64-74.

25. Garry, V.F., M.E. Harkins, L.L. Erickson, L.K. Long-Simpson, S.E. Holland, and B.L. Burroughs. 2002. Birth defects, season of conception, and sex of children born to pesticide applicators living in the Red River Valley of Minnesota, USA. *Environmental Health Perspectives* 110 Suppl 3:441-9.

26. Heeren, G.A., J. Tyler, and A. Mandeya. 2003. Agricultural chemical exposures and birth defects in the Eastern Cape Province, South Africa: a case-control study. *Environmental Health* 2 (1):11.

27. Irgens, A., K. Kruger, A.H. Skorve, and L.M. Irgens. 2000. Birth defects and paternal occupational exposure. Hypotheses tested in a record linkage based dataset. *Acta Obstetricia et Gynecologica Scandinavica* 79 (6):465-70.

28. Loffredo, C.A., E.K. Silbergeld, C. Ferencz, and J. Zhang. 2001. Association of transposition of the great arteries in infants with maternal exposures to herbicides and rodenticides. *American Journal of Epidemiology* 153 (6):529-36.

29. Schreinemachers, D.M. 2003. Birth malformations and other adverse perinatal outcomes in four U.S. Wheat-producing states. *Environmental Health Perspectives* 111 (9):1259-64.

30. Shaw, G.M., C.R. Wasserman, C.D. O'Malley, V. Nelson, and R.J. Jackson. 1999. Maternal pesticide exposure from multiple sources and selected congenital anomalies. *Epidemiology* 10 (1):60-6.

31. Weselak, M., T.E. Arbuckle, D.T. Wigle, M.C. Walker, and D. Krewski. 2008. Pre- and post-conception pesticide exposure and the risk of birth defects in an Ontario farm population. *Reproductive Toxicology* 25 (4):472-80.

32. Bove, F., Y. Shim, and P. Zeitz. 2002. Drinking water contaminants and adverse pregnancy outcomes: a review. *Environmental Health Perspectives* 110 Suppl 1:61-74.

33. Hwang, B.F., J.J. Jaakkola, and H.R. Guo. 2008. Water disinfection by-products and the risk of specific birth defects: A population-based cross-sectional study in Taiwan. *Environmental Health* 7 (1):23.

34. Hwang, B.F., P. Magnus, and J.J. Jaakkola. 2002. Risk of specific birth defects in relation to chlorination and the amount of natural organic matter in the water supply. *American Journal of Epidemiology* 156 (4):374-82.

35. Nieuwenhuijsen, M.J., D. Martinez, J. Grellier, J. Bennett, N. Best, N. Iszatt, M. Vrijheid, and M.B. Toledano. 2009. Chlorination disinfection by-products in drinking water and congenital anomalies: review and meta-analyses. *Environmental Health Perspectives* 117 (10):1486-93.

36. Baskin, L.S., K. Himes, and T. Colborn. 2001. Hypospadias and endocrine disruption: is there a connection? *Environmental Health Perspectives* 109 (11):1175-83.

37. Bornman, R., C. de Jager, Z. Worku, P. Farias, and S. Reif. 2010. DDT and urogenital malformations in newborn boys in a malarial area. *British Journal of Urology International* 106 (3):405-11.

38. Dolk, H., M. Vrijheid, B. Armstrong, L. Abramsky, F. Bianchi, E. Garne, V. Nelen, E. Robert, J.E. Scott, D. Stone, et al. 1998. Risk of congenital anomalies near hazardous-waste landfill sites in Europe: the EUROHAZCON study. *Lancet* 352 (9126):423-7.

39. Kristensen, P., L.M. Irgens, A. Andersen, A.S. Bye, and L. Sundheim. 1997. Birth defects among offspring of Norwegian farmers, 1967-1991. *Epidemiology* 8 (5):537-44.

40. Main, K.M., H. Kiviranta, H.E. Virtanen, E. Sundqvist, J.T. Tuomisto, J. Tuomisto, T. Vartiainen, N.E. Skakkebaek, and J. Toppari. 2007. Flame retardants in placenta and breast milk and cryptorchidism in newborn boys. *Environmental Health Perspectives* 115 (10):1519-26.

41. Meyer, K.J., J.S. Reif, D.N. Veeramachaneni, T.J. Luben, B.S. Mosley, and J.R. Nuckols. 2006. Agricultural pesticide use and hypospadias in eastern Arkansas. *Environmental Health Perspectives* 114 (10):1589-95.

42. Nassar, N., P. Abeywardana, A. Barker, and C. Bower. 2010. Parental occupational exposure to potential endocrine disrupting chemicals and risk of hypospadias in infants. *Occupational and Environmental Medicine* 67 (9):585-9.

43. Pierik, F.H., A. Burdorf, J.A. Deddens, R.E. Juttmann, and R.F. Weber. 2004. Maternal and paternal risk factors for cryptorchidism and hypospadias: a case-control study in newborn boys. *Environmental Health Perspectives* 112 (15):1570-6.

44. Swan, S.H., K.M. Main, F. Liu, S.L. Stewart, R.L. Kruse, A.M. Calafat, C.S. Mao, J.B. Redmon, C.L. Ternand, S. Sullivan, et al. 2005. Decrease in anogenital distance among male infants with prenatal phthalate exposure. *Environmental Health Perspectives* 113 (8):1056-61.

45. Nelson, C.P., J.M. Park, J. Wan, D.A. Bloom, R.L. Dunn, and J.T. Wei. 2005. The increasing incidence of congenital penile anomalies in the United States. *The Journal of Urology* 174 (4 Pt 2):1573-6.

46. Paulozzi, L.J., J.D. Erickson, and R.J. Jackson. 1997. Hypospadias trends in two US surveillance systems. *Pediatrics* 100 (5):831-4.

Birth Defects (continued)

47. Sharpe, R. 2009. *Male Reproductive Health Disorders and the Potential Role of Environmental Chemical Exposures.* London, UK: CHEM Trust. http://www.chemicalshealthmonitor.org/IMG/pdf/PROFSHARPE-MaleReproductiveHealth-CHEMTrust09.pdf.

48. Dadvand, P., J. Rankin, S. Rushton, and T. Pless-Mulloli. 2011. Ambient air pollution and congenital heart disease: a register-based study. *Environmental Research* 111 (3):435-41.

49. Dolk, H., B. Armstrong, K. Lachowycz, M. Vrijheid, J. Rankin, L. Abramsky, P.A. Boyd, and D. Wellesley. 2010. Ambient air pollution and risk of congenital anomalies in England, 1991-1999. *Occupational and Environmental Medicine* 67 (4):223-7.

50. Gilboa, S.M., P. Mendola, A.F. Olshan, P.H. Langlois, D.A. Savitz, D. Loomis, A.H. Herring, and D.E. Fixler. 2005. Relation between ambient air quality and selected birth defects, seven county study, Texas, 1997-2000. *American Journal of Epidemiology* 162 (3):238-52.

51. Hansen, C.A., A.G. Barnett, B.B. Jalaludin, and G.G. Morgan. 2009. Ambient air pollution and birth defects in brisbane, australia. *PLoS One* 4 (4):e5408.

52. Hwang, B.F., and J.J. Jaakkola. 2008. Ozone and other air pollutants and the risk of oral clefts. *Environmental Health Perspectives* 116 (10):1411-5.

53. Kim, O.J., E.H. Ha, B.M. Kim, J.H. Seo, H.S. Park, W.J. Jung, B.E. Lee, Y.J. Suh, Y.J. Kim, J.T. Lee, et al. 2007. PM10 and pregnancy outcomes: a hospital-based cohort study of pregnant women in Seoul. *Journal of Occupational and Environmental Medicine* 49 (12):1394-402.

54. Marshall, E.G., G. Harris, and D. Wartenberg. 2010. Oral cleft defects and maternal exposure to ambient air pollutants in New Jersey. *Birth Defects Research Part A: Clinical and Molecular Teratology* 88 (4):205-15.

55. Rankin, J., T. Chadwick, M. Natarajan, D. Howel, M.S. Pearce, and T. Pless-Mulloli. 2009. Maternal exposure to ambient air pollutants and risk of congenital anomalies. *Environmental Research* 109 (2):181-7.

56. Ritz, B., F. Yu, S. Fruin, G. Chapa, G.M. Shaw, and J.A. Harris. 2002. Ambient air pollution and risk of birth defects in Southern California. *American Journal of Epidemiology* 155 (1):17-25.

57. Strickland, M.J., M. Klein, A. Correa, M.D. Reller, W.T. Mahle, T.J. Riehle-Colarusso, L.D. Botto, W.D. Flanders, J.A. Mulholland, C. Siffel, et al. 2009. Ambient air pollution and cardiovascular malformations in Atlanta, Georgia, 1986-2003. *American Journal of Epidemiology* 169 (8):1004-14.

58. Vrijheid, M., D. Martinez, S. Manzanares, P. Dadvand, A. Schembari, J. Rankin, and M. Nieuwenhuijsen. 2011. Ambient air pollution and risk of congenital anomalies: a systematic review and meta-analysis. *Environmental Health Perspectives* 119 (5):598-606.

59. Goldman, L.R., B. Paigen, M.M. Magnant, and J.H. Highland. 1985. Low birth-weight, prematurity and birth-defects in children living near the hazardous waste site, Love Canal. *Hazardous Waste and Hazardous Materials* 2 (2):209-223.

60. New York State Department of Health. 2008. *Love Canal Follow-up Health Study.* New York City, NY: Division of Environmental Health Assessment Center for Environmental Health http://www.health.state.ny.us/environmental/investigations/love_canal/docs/report_public_comment_final.pdf.

61. Croen, L.A., G.M. Shaw, L. Sanbonmatsu, S. Selvin, and P.A. Buffler. 1997. Maternal residential proximity to hazardous waste sites and risk for selected congenital malformations. *Epidemiology* 8 (4):347-54.

62. Geschwind, S.A., J.A. Stolwijk, M. Bracken, E. Fitzgerald, A. Stark, C. Olsen, and J. Melius. 1992. Risk of congenital malformations associated with proximity to hazardous waste sites. *American Journal of Epidemiology* 135 (11):1197-207.

63. Shaw, G.M., J. Schulman, J.D. Frisch, S.K. Cummins, and J.A. Harris. 1992. Congenital malformations and birthweight in areas with potential environmental contamination. *Archives of Environmental Health* 47 (2):147-54.

64. Yauck, J.S., M.E. Malloy, K. Blair, P.M. Simpson, and D.G. McCarver. 2004. Proximity of residence to trichloroethylene-emitting sites and increased risk of offspring congenital heart defects among older women. *Birth Defects Research. Part A, Clinical and Molecular Teratology* 70 (10):808-14.

65. Brender, J.D., F.B. Zhan, P.H. Langlois, L. Suarez, and A. Scheuerle. 2008. Residential proximity to waste sites and industrial facilities and chromosomal anomalies in offspring. *International Journal of Hygiene and Environmental Health* 211 (1-2):50-8.

66. Vrijheid, M., H. Dolk, B. Armstrong, L. Abramsky, F. Bianchi, I. Fazarinc, E. Garne, R. Ide, V. Nelen, E. Robert, et al. 2002. Chromosomal congenital anomalies and residence near hazardous waste landfill sites. *Lancet* 359 (9303):320-2.

67. Kuehn, C.M., B.A. Mueller, H. Checkoway, and M. Williams. 2007. Risk of malformations associated with residential proximity to hazardous waste sites in Washington State. *Environmental Research* 103 (3):405-12.

68. Suarez, L., J.D. Brender, P.H. Langlois, F.B. Zhan, and K. Moody. 2007. Maternal exposures to hazardous waste sites and industrial facilities and risk of neural tube defects in offspring. *Annals of Epidemiology* 17 (10):772-7.

69. Currie, J., M. Greenstone, and E. Moretti. 2011. Superfund cleanups and infant health. *American Economic Review: Papers and Proceedings* 101 (3):435-441.

Birth Defects (continued)

70. Wilson, J.G. 1973. *Environment and Birth Defects, Environmental Science Series.* London: Academic Press.

71. Winchester, P.D., J. Huskins, and J. Ying. 2009. Agrichemicals in surface water and birth defects in the United States. *Acta Pædiatrica* 98 (4):664-9.

72. Boulet, S.L., M. Shin, R.S. Kirby, D. Goodman, and A. Correa. 2011. Sensitivity of birth certificate reports of birth defects in Atlanta, 1995-2005: effects of maternal, infant, and hospital characteristics. *Public Health Reports* 126 (2):186-94.

73. Marengo, L. 2010. Results from an in-house quality control report conducted by the Texas Department of State Health Services. Email from Lisa Marengo, Texas Birth Defects Epidemiology and Surveillance Branch, to Julie Sturza, U.S. EPA, November 4, 2010.

74. National Birth Defects Prevention Network. 2008. State birth defects surveillance program directory. *Birth Defects Research Part A: Clinical and Molecular Teratology* 82:906-961.

75. Parker, S.E., C.T. Mai, M.A. Canfield, R. Rickard, Y. Wang, R.E. Meyer, P. Anderson, C.A. Mason, J.S. Collins, R.S. Kirby, et al. 2010. Updated national birth prevalence estimates for selected birth defects in the United States, 2004-2006. *Birth Defects Research. Part A, Clinical and Molecular Teratology* 88 (12):1008-16.

76. Trust for America's Health. 2002. *Birth Defects Tracking and Prevention: Too Many States are Not Making the Grade.* Washington, DC: Trust for America's Health. http://healthyamericans.org/reports/birthdefects02/bdreport.pdf.

77. Texas Birth Defects Epidemiology & Surveillance Branch. 2010. *Texas Birth Defects Registry.* Texas Birth Defects Epidemiology & Surveillance Branch. Retrieved August 10, 2010 from http://www.dshs.state.tx.us/birthdefects/default.shtm.

78. Centers for Disease Control and Prevention. 2010. *VitalStats: Birth Data Files.* National Center for Health Statistics. Retrieved August 10, 2010 from http://www.cdc.gov/nchs/data_access/vitalstats/VitalStats_Births.htm.

79. Botto, L.D., A. Correa, and J.D. Erickson. 2001. Racial and temporal variations in the prevalence of heart defects. *Pediatrics* 107 (3):E32.

80. Centers for Disease Control and Prevention. 2008. Update on overall prevalence of major birth defects--Atlanta, Georgia, 1978-2005. *Morbidity and Mortality Weekly Report* 57 (1):1-5.

81. Langlois, P.H., L.K. Marengo, and M.A. Canfield. 2011. Time trends in the prevalence of birth defects in Texas 1999-2007: real or artifactual? *Birth Defects Research. Part A, Clinical and Molecular Teratology* 91 (10):902-17.

82. Langlois, P.H., and A. Scheuerle. 2007. Using registry data to suggest which birth defects may be more susceptible to artifactual clusters and trends. *Birth Defects Research. Part A, Clinical and Molecular Teratology* 79 (11):798-805.

83. Texas Department of State Health Services. *Report of Birth Defects Among 1999-2007 Deliveries.* Texas Birth Defects Epidemiology and Surveillance Branch. Retrieved July 27, 2012 from http://www.dshs.state.tx.us/birthdefects/data/annl99-07.shtm.

Contaminants in Schools and Child Care Facilities

1. U.S. Environmental Protection Agency. 2003. *IAQ Tools for Schools.* Washington, DC: U.S. EPA, Indoor Environments Division. EPA 402-F-03-011. http://www.epa.gov/iaq/schools/pdfs/publications/iaqtfs_factsheet.pdf.

2. Breysse, P.N., G.B. Diette, E.C. Matsui, A.M. Butz, N.N. Hansel, and M.C. McCormack. 2010. Indoor air pollution and asthma in children. *Proceedings of the American Thoracic Society* 7 (2):102-6.

3. Rudel, R.A., and L.J. Perovich. 2009. Endocrine disrupting chemicals in indoor and outdoor air. *Atmospheric Environment* 43 (1):170-181.

4. U.S. Department of Health and Human Services. 2006. *The Health Consequences of Involuntary Exposure to Tobacco Smoke: A Report of the Surgeon General.* Atlanta, GA: U.S. Department of Health and Human Services, Centers for Disease Control and Prevention, Coordinating Center for Health Promotion, National Center for Chronic Disease Prevention and Health Promotion, Office on Smoking and Health.

5. Faustman, E.M., S.M. Silbernagel, R.A. Fenske, T.M. Burbacher, and R.A. Ponce. 2000. Mechanisms underlying children's susceptibility to environmental toxicants. *Environmental Health Perspectives* 108 (Suppl 1):13-21.

6. Landrigan, P.J., L. Claudio, S.B. Markowitz, G.S. Berkowitz, B.L. Brenner, H. Romero, J.G. Wetmur, T.D. Matte, A.C. Gore, J.H. Godbold, et al. 1999. Pesticides and inner-city children: exposures, risks, and prevention. *Environmental Health Perspectives* 107 (Suppl 3):431-7.

7. National Research Council. 1993. *Pesticides in the Diets of Infants and Children.* Washington, DC: National Academies Press. http://www.nap.edu/openbook.php?isbn=0309048753.

8. U.S. Environmental Protection Agency. 2008. *Child-Specific Exposure Factors Handbook (Final Report)* Washington, DC: U.S. EPA. National Center for Environmental Assessment. EPA/600/R-06/096. http://cfpub.epa.gov/ncea/cfm/recordisplay.cfm?deid=199243.

9. U.S. Environmental Protection Agency. 2011. *School Advanced Ventilation Engineering Software (SAVES).* U.S. EPA, Indoor Environments Division. Retrieved September 8, 2011 from http://www.epa.gov/iaq/schooldesign/saves.html.

10. Wilson, N.K., J.C. Chuang, and C. Lyu. 2001. Levels of persistent organic pollutants in several child day care centers. *Journal of Exposure Analysis and Environmental Epidemiology* 11 (6):449-458.

Contaminants in Schools and Child Care Facilities (continued)

11. Marker, D., J. Rogers, A. Fraser, and S.M. Viet. 2003. *First National Environmental Health Survey of Child Care Centers: Final Report.* Rockville, MD: Westat, Inc. http://www.nmic.org/nyccelp/documents/HUD_NEHSCCC.pdf.

12. U.S. Consumer Product Safety Commission. 1999. *CPSC Staff Study of Safety Hazards in Child Care Settings.* Washington, DC: Consumer Product Safety Commission. http://www.cpsc.gov/library/ccstudy.html.

13. Branham, D. 2004. The wise man builds his house upon the rock: The effects of inadequate school building infrastructure on student attendance. *Social Science Quarterly* 85 (5):1112-1128.

14. Mendell, M.J., and G.A. Heath. 2005. Do indoor pollutants and thermal conditions in schools influence student performance? A critical review of the literature. *Indoor Air* 15 (1):27-52.

15. National Research Council. 2006. *Green Schools: Attributes for Health and Learning.* Washington, DC: The National Academies Press. http://www.nap.edu/catalog.php?record_id=11756.

16. Somers, T.S., M.L. Harvey, and S.M. Rusnak. 2011. Making child care centers SAFER: a non-regulatory approach to improving child care center siting. *Public Health Reports* 126 Suppl 1:34-40.

17. U.S. Environmental Protection Agency. 2011. *School Siting Guidelines.* Washington, DC: U.S. EPA, Office of Children's Health Protection. EPA-100-K-11-004. http://www.epa.gov/schools/siting/downloads/School_Siting_Guidelines.pdf.

18. U.S. Environmental Protection Agency, and U.S. Consumer Product Safety Commission. 1993. *The Inside Story: A Guide to Indoor Air Quality.* Washington, DC: U.S. EPA, Office of Radiation and Indoor Air. EPA-402-R-93-013. http://www.epa.gov/iaq/pubs/insidestory.html.

19. Daisey, J.M., W.J. Angell, and M.G. Apte. 2003. Indoor air quality, ventilation and health symptoms in schools: an analysis of existing information. *Indoor Air* 13 (1):53-64.

20. U.S. Environmental Protection Agency. 2010. *Basic Information: Polychlorinated Biphenyl (PCB)* U.S. EPA. Retrieved November 1, 2010 from http://www.epa.gov/epawaste/hazard/tsd/pcbs/pubs/about.htm.

21. U.S. Environmental Protection Agency. 2010. *Managing School IAQ.* U.S. EPA, Indoor Environments Division. Retrieved September 8, 2011 from http://www.epa.gov/iaq/schools/symptoms.html.

22. Shendell, D.G., R. Prill, W.J. Fisk, M.G. Apte, D. Blake, and D. Faulkner. 2004. Associations between classroom CO_2 concentrations and student attendance in Washington and Idaho. *Indoor Air* 14 (5):333-41.

23. Silverstein, M.D., J.E. Mair, S.K. Katusic, P.C. Wollan, J. O'Connell E, and J.W. Yunginger. 2001. School attendance and school performance: a population-based study of children with asthma. *Journal of Pediatrics* 139 (2):278-83.

24. Shaughnessy RJ, Haverinen-Shaughnessy U, Nevalainen A, and M. D. 2006. A preliminary study on the association between ventilation rates in classrooms and student performance. *Indoor Air* 16 (6):465-8.

25. Myhrvold, A.N., E. Olsen, and O. Lauridsen. 1996. Indoor environment in schools - Pupils' health and performance in regard to CO_2 concentrations. *Proceedings: Indoor Air '96, The 7th International Conference on Indoor Air Quality and Climate, Nagoya, Japan* 1:369-74.

26. Smedje, G., D. Norback, and C. Edling. 1996. Mental performance by secondary school pupils in relation to the quality of indoor air. *Proceedings: Indoor Air '96, The 7th International Conference on Indoor Air Quality and Climate, Nagoya, Japan* 1:413-18.

27. U.S. General Accounting Office. 1995. *School Facilities: Condition of America's Schools.* Washington, DC: GAO. GAO/HEHS-95-61. http://www.gao.gov/archive/1995/he95061.pdf.

28. Sexton, K., I.A.N.A. Greaves, T.R. Church, J.L. Adgate, G. Ramachandran, R.L. Tweedie, A. Fredrickson, M. Geisser, M. Sikorski, and G. Fischer. 2000. A school-based strategy to assess children's environmental exposures and related health effects in economically disadvantaged urban neighborhoods. *Journal of Exposure Science and Environmental Epidemiology* 10:682-694.

29. Agency for Toxic Substances and Disease Registry (ATSDR). 2010. *Case Studies in Environmental Medicine (CSEM) Environmental Triggers of Asthma, Environmental Factors.* Retrieved September 28, 2010 from http://www.atsdr.cdc.gov/csem/asthma/envfactors.html.

30. National Toxicology Program. 2012. *NTP Monograph on Health Effects of Low-Level Lead.* Research Triangle Park, NC: National Institute of Environmental Health Sciences, National Toxicology Program. http://ntp.niehs.nih.gov/go/36443.

31. Advisory Committee on Childhood Lead Poisoning Prevention. 2012. *Low Level Lead Exposure Harms Children: A Renewed Call for Primary Prevention.* Atlanta, GA: Centers for Disease Control and Prevention, Advisory Committee on Childhood Lead Poisoning Prevention. http://www.cdc.gov/nceh/lead/acclpp/final_document_010412.pdf.

32. U.S. Environmental Protection Agency. *Lead in Paint, Dust, and Soil.* U.S. EPA, Office of Pollution Prevention and Toxics. Retrieved October 7, 2010 from http://www.epa.gov/lead/.

33. Levin, R., M.J. Brown, M.E. Kashtock, D.E. Jacobs, E.A. Whelan, J. Rodman, M.R. Schock, A. Padilla, and T. Sinks. 2008. Lead exposures in U.S. children, 2008: implications for prevention. *Environmental Health Perspectives* 116 (10):1285-93.

Contaminants in Schools and Child Care Facilities (continued)

34. Lambrinidou, Y., S. Triantafyllidou, and M. Edwards. 2010. Failing our children: lead in US school drinking water. *New Solutions: A Journal of Environmental and Occupational Health Policy* 20 (1):25-47.

35. U.S. Environmental Protection Agency. *Basic Information: Lead in Paint, Dust, and Soil*. U.S. EPA, Office of Pollution Prevention and Toxics. Retrieved September 29, 2010 from http://www.epa.gov/lead/pubs/leadinfo.htm#where.

36. Jacobs, D.E., R.P. Clickner, J.Y. Zhou, S.M. Viet, D.A. Marker, J.W. Rogers, D.C. Zeldin, P. Broene, and W. Friedman. 2002. The prevalence of lead-based paint hazards in U.S. housing. *Environmental Health Perspectives* 110 (10):A599-606.

37. Braun, J.M., R.S. Kahn, T. Froehlich, P. Auinger, and B.P. Lanphear. 2006. Exposures to environmental toxicants and attention deficit hyperactivity disorder in U.S. children. *Environmental Health Perspectives* 114 (12):1904-9.

38. Eubig, P.A., A. Aguiar, and S.L. Schantz. 2010. Lead and PCBs as risk factors for attention deficit/hyperactivity disorder. *Environmental Health Perspectives* 118 (12):1654-67.

39. Froehlich, T.E., B.P. Lanphear, P. Auinger, R. Hornung, J.N. Epstein, J. Braun, and R.S. Kahn. 2009. Association of tobacco and lead exposures with attention-deficit/hyperactivity disorder. *Pediatrics* 124 (6):e1054-63.

40. Ha, M., H.J. Kwon, M.H. Lim, Y.K. Jee, Y.C. Hong, J.H. Leem, J. Sakong, J.M. Bae, S.J. Hong, Y.M. Roh, et al. 2009. Low blood levels of lead and mercury and symptoms of attention deficit hyperactivity in children: a report of the children's health and environment research (CHEER). *Neurotoxicology* 30 (1):31-6.

41. Nigg, J.T., G.M. Knottnerus, M.M. Martel, M. Nikolas, K. Cavanagh, W. Karmaus, and M.D. Rappley. 2008. Low blood lead levels associated with clinically diagnosed attention-deficit/hyperactivity disorder and mediated by weak cognitive control. *Biological Psychiatry* 63 (3):325-31.

42. Nigg, J.T., M. Nikolas, G. Mark Knottnerus, K. Cavanagh, and K. Friderici. 2010. Confirmation and extension of association of blood lead with attention-deficit/hyperactivity disorder (ADHD) and ADHD symptom domains at population-typical exposure levels. *The Journal of Child Psychology and Psychiatry* 51 (1):58-65.

43. Roy, A., D. Bellinger, H. Hu, J. Schwartz, A.S. Ettinger, R.O. Wright, M. Bouchard, K. Palaniappan, and K. Balakrishnan. 2009. Lead exposure and behavior among young children in Chennai, India. *Environmental Health Perspectives* 117 (10):1607-11.

44. Wang, H.L., X.T. Chen, B. Yang, F.L. Ma, S. Wang, M.L. Tang, M.G. Hao, and D.Y. Ruan. 2008. Case-control study of blood lead levels and attention deficit hyperactivity disorder in Chinese children. *Environmental Health Perspectives* 116 (10):1401-6.

45. Needleman, H.L., A. Schell, D. Bellinger, A. Leviton, and E.N. Allred. 1990. The long-term effects of exposure to low doses of lead in childhood. An 11-year follow-up report. *New England Journal of Medicine* 322 (2):83-8.

46. Dietrich, K.N., M.D. Ris, P.A. Succop, O.G. Berger, and R.L. Bornschein. 2001. Early exposure to lead and juvenile delinquency. *Neurotoxicology and Teratology* 23 (6):511-8.

47. Marcus, D.K., J.J. Fulton, and E.J. Clarke. 2010. Lead and conduct problems: a meta-analysis. *Journal of Clinical Child and Adolescent Psychology* 39 (2):234-41.

48. Needleman, H.L., C. McFarland, R.B. Ness, S.E. Fienberg, and M.J. Tobin. 2002. Bone lead levels in adjudicated delinquents. A case control study. *Neurotoxicology and Teratology* 24 (6):711-7.

49. Needleman, H.L., J.A. Riess, M.J. Tobin, G.E. Biesecker, and J.B. Greenhouse. 1996. Bone lead levels and delinquent behavior. *The Journal of the American Medical Association* 275 (5):363-9.

50. Nevin, R. 2007. Understanding international crime trends: the legacy of preschool lead exposure. *Environmental Research* 104 (3):315-36.

51. Wright, J.P., K.N. Dietrich, M.D. Ris, R.W. Hornung, S.D. Wessel, B.P. Lanphear, M. Ho, and M.N. Rae. 2008. Association of prenatal and childhood blood lead concentrations with criminal arrests in early adulthood. *PLoS Medicine* 5 (5):e101.

52. U.S. Environmental Protection Agency. 2010. *Healthy School Environment Resources: PCBs*. U.S. EPA. Retrieved September 23, 2010 from http://cfpub.epa.gov/schools/top_sub.cfm?t_id=41&s_id=32.

53. U.S. Environmental Protection Agency. 2010. *EPA Issues National Guidance to Address Proper Maintenance, Removal, and Disposal of PCB-Containing Fluorescent Lights (Press Release)*. U.S. EPA, Office of Public Affairs. Retrieved December 29, 2010 from http://yosemite.epa.gov/opa/admpress.nsf/d0cf6618525a9efb85257359003fb69d/6c03fdec1e63274c8525780800693d7d!OpenDocument.

54. Newman, D.M. 2010. PCBs in schools: What about school maintenance workers? *New Solutions: A Journal of Environmental and Occupational Health Policy* 20 (2):189-191.

55. U.S. Environmental Protection Agency. 2010. *Healthy School Environment Resources: Siting*. U.S. EPA. Retrieved September 28, 2010 from http://cfpub.epa.gov/schools/top_sub.cfm?t_id=45&s_id=64.

56. Herrick, R.F., D.J. Lefkowitz, and G.A. Weymouth. 2007. Soil contamination from PCB-containing buildings. *Environmental Health Perspectives* 115 (2):173-175.

Contaminants in Schools and Child Care Facilities (continued)

57. Herrick, R.F., M.D. McClean, J.D. Meeker, L.K. Baxter, and G.A. Weymouth. 2004. An unrecognized source of PCB contamination in schools and other buildings. *Environmental Health Perspectives* 112 (10):1051-3.

58. Boucher, O., G. Muckle, and C.H. Bastien. 2009. Prenatal exposure to polychlorinated biphenyls: a neuropsychologic analysis. *Environmental Health Perspectives* 117 (1):7-16.

59. Ribas-Fito, N., M. Sala, M. Kogevinas, and J. Sunyer. 2001. Polychlorinated biphenyls (PCBs) and neurological development in children: a systematic review. *Journal of Epidemiology and Community Health* 55 (8):537-46.

60. Schantz, S.L., J.C. Gardiner, D.M. Gasior, R.J. McCaffrey, A.M. Sweeney, and H.E.B. Humphrey. 2004. Much ado about something: The weight of evidence for PCB effects on neuropsychological function. *Psychology in the Schools* 41 (6):669-679.

61. Schantz, S.L., J.J. Widholm, and D.C. Rice. 2003. Effects of PCB exposure on neuropsychological function in children. *Environmental Health Perspectives* 111 (3):357-576.

62. Wigle, D.T., T.E. Arbuckle, M.C. Turner, A. Berube, Q. Yang, S. Liu, and D. Krewski. 2008. Epidemiologic evidence of relationships between reproductive and child health outcomes and environmental chemical contaminants. *Journal of Toxicology and Environmental Health Part B Critical Reviews* 11 (5-6):373-517.

63. U.S. Environmental Protection Agency. 2010. *Asbestos: Basic Information*. U.S. EPA, Office of Pollution Prevention and Toxics. Retrieved September 23, 2010 from http://www.epa.gov/asbestos/pubs/help.html.

64. U.S. Environmental Protection Agency. 2010. *Asbestos: Laws and Regulations*. U.S. EPA, Office of Pollution Prevention and Toxics. Retrieved September 23, 2010 from http://www.epa.gov/asbestos/pubs/asbreg.html.

65. United States Code. 1986. *Asbestos Hazard Emergency Response*. Title 15, Chapter 53, Subchapter II. http://www.gpo.gov/fdsys/pkg/USCODE-2009-title15/html/USCODE-2009-title15-chap53-subchapII.htm.

66. Agency for Toxic Substances and Disease Registry (ATSDR). 2001. *Toxicological Profile for Asbestos. Update*. Atlanta, GA: U.S. Department of Health and Human Services, Public Health Service.

67. U.S. Environmental Protection Agency. 2010. *20 Frequently Asked Questions About Asbestos in Schools*. U.S. EPA, Office of Pollution Prevention and Toxics. Retrieved September 23, 2010 from http://www.epa.gov/asbestos/pubs/ais20quests.pdf.

68. Sutton, R. 2009. *Greener School Cleaning Supplies = Fresh Air + Healthier Kids: New Research Links School Air Quality to School Cleaning Supplies*. Washington, DC: Environmental Working Group. http://www.ewg.org/files/2009/10/school-cleaners/EWGschoolcleaningsupplies.pdf.

69. Nazaroff, W.W., and C.J. Weschler. 2004. Cleaning products and air fresheners: exposure to primary and secondary air pollutants. *Atmospheric Environment* 38 (18):2841-2865.

70. U.S. Environmental Protection Agency. 2010. *An Introduction to Indoor Air Quality: Volatile Organic Compounds (VOCs)*. U.S. EPA. Retrieved September 23, 2010 from http://www.epa.gov/iaq/voc.html.

71. Redlich, C.A., J. Sparer, and M.R. Cullen. 1997. Sick-building syndrome. *Lancet* 349 (9057):1013-6.

72. Tortolero, S.R., L.K. Bartholomew, S. Tyrrell, S.L. Abramson, M.M. Sockrider, C.M. Markham, L.W. Whitehead, and G.S. Parcel. 2002. Environmental allergens and irritants in schools: a focus on asthma. *Journal of School Health* 72 (1):33-8.

73. Institute of Medicine. 2000. *Clearing the Air: Asthma and Indoor Air Exposures*. Washington DC: National Academy Press. http://books.nap.edu/catalog/9610.html.

74. Samet, J.M., M.C. Marbury, and J.D. Spengler. 1987. Health effects and sources of indoor air pollution. Part I. *The American Review of Respiratory Disease* 136 (6):1486-508.

75. California Air Resources Board and California Department of Health Services. 2004. *Environmental Health Conditions in California's Portable Classrooms*. Sacramento, CA: California Air Resources Board, California Department of Health Services. http://www.arb.ca.gov/research/indoor/pcs/leg_rpt/pcs_r2l.pdf.

76. U.S. Environmental Protection Agency. 2010. *Radon in Schools*. U.S. EPA. Retrieved September 28, 2010 from http://www.epa.gov/radon/pubs/schoolrn.html.

77. Egeghy, P.P., L. Sheldon, R.C. Fortmann, D.M. Stout II, N.S. Tulve, E. Cohen-Hubal, L.J. Melnyk, M.K. Morgan, P.A. Jones, D.A. Whitaker, et al. 2007. *Important Exposure Factors to Children: An Analysis of Laboratory and Observational Field Data Characterizing Cumulative Exposures to Pesticides*. Research Triangle Park, NC: U.S. EPA, Office of Research and Development. EPA 600/R-07/013. EPA 600/R-07/013. http://www.epa.gov/nerl/research/data/exposure-factors.pdf.

78. Gurunathan, S., M. Robson, N. Freeman, B. Buckley, A. Roy, R. Meyer, J. Bukowski, and P.J. Lioy. 1998. Accumulation of chlorpyrifos on residential surfaces and toys accessible to children. *Environmental Health Perspectives* 106 (1):9-16.

Contaminants in Schools and Child Care Facilities (continued)

79. Hore, P., M. Robson, N. Freeman, J. Zhang, D. Wartenberg, H. Ozkaynak, N. Tulve, L. Sheldon, L. Needham, D. Barr, et al. 2005. Chlorpyrifos accumulation patterns for child-accessible surfaces and objects and urinary metabolite excretion by children for 2 weeks after crack-and-crevice application. *Environmental Health Perspectives* 113 (2):211-9.

80. Nishioka, M.G., R.G. Lewis, M.C. Brinkman, H.M. Burkholder, C.E. Hines, and J.R. Menkedick. 2001. Distribution of 2,4-D in air and on surfaces inside residences after lawn applications: comparing exposure estimates from various media for young children. *Environmental Health Perspectives* 109 (11):1185-91.

81. Roberts, J.W., L.A. Wallace, D.E. Camann, P. Dickey, S.G. Gilbert, R.G. Lewis, and T.K. Takaro. 2009. Monitoring and reducing exposure of infants to pollutants in house dust. *Reviews of Environmental Contamination & Toxicology* 201:1-39.

82. Stout, D.M., 2nd, and R.B. Leidy. 2000. A preliminary examination of the translocation of microencapsulated cyfluthrin following applications to the perimeter of residential dwellings. *Journal of Environmental Science and Health Part B* 35 (4):477-89.

83. Wright, C.G., R.B. Leidy, and H.E. Dupree, Jr. 1993. Cypermethrin in the ambient air and on surfaces of rooms treated for cockroaches. *Bulletin of Environmental Contamination and Toxicology* 51 (3):356-60.

84. Tulve, N.S., P.P. Egeghy, R.C. Fortmann, D.A. Whitaker, M.G. Nishioka, L.P. Naeher, and A. Hilliard. 2008. Multimedia measurements and activity patterns in an observational pilot study of nine young children. *Journal of Exposure Science and Environmental Epidemiology* 18 (1):31-44.

85. Whitmore, R.W., F.W. Immerman, D.E. Camann, A.E. Bond, R.G. Lewis, and J.L. Schaum. 1994. Non-occupational exposures to pesticides for residents of two U.S. cities. *Archives of Environmental Contamination and Toxicology* 26 (1):47-59.

86. Whyatt, R.M., D.B. Barr, D.E. Camann, P.L. Kinney, J.R. Barr, H.F. Andrews, L.A. Hoepner, R. Garfinkel, Y. Hazi, A. Reyes, et al. 2003. Contemporary-use pesticides in personal air samples during pregnancy and blood samples at delivery among urban minority mothers and newborns. *Environmental Health Perspectives* 111 (5):749-56.

87. Williams, M.K., D.B. Barr, D.E. Camann, L.A. Cruz, E.J. Carlton, M. Borjas, A. Reyes, D. Evans, P.L. Kinney, R.D. Whitehead, Jr., et al. 2006. An intervention to reduce residential insecticide exposure during pregnancy among an inner-city cohort. *Environmental Health Perspectives* 114 (11):1684-9.

88. Matoba, Y., Y. Takimoto, and T. Kato. 1998. Indoor behavior and risk assessment following residual spraying of d-phenothrin and d-tetramethrin. *American Industrial Hygiene Association Journal* 59 (3):191-9.

89. Leng, G., E. Berger-Preiss, K. Levsen, U. Ranft, D. Sugiri, W. Hadnagy, and H. Idel. 2005. Pyrethroids used indoor-ambient monitoring of pyrethroids following a pest control operation. *International Journal of Hygiene and Environmental Health* 208 (3):193-9.

90. Julien, R., G. Adamkiewicz, J.I. Levy, D. Bennett, M. Nishioka, and J.D. Spengler. 2008. Pesticide loadings of select organophosphate and pyrethroid pesticides in urban public housing. *Journal of Exposure Science and Environmental Epidemiology* 18 (2):167-74.

91. Colt, J.S., J. Lubin, D. Camann, S. Davis, J. Cerhan, R.K. Severson, W. Cozen, and P. Hartge. 2004. Comparison of pesticide levels in carpet dust and self-reported pest treatment practices in four US sites. *Journal of Exposure Analysis and Environmental Epidemiology* 14 (1):74-83.

92. Stout, D.M., 2nd, K.D. Bradham, P.P. Egeghy, P.A. Jones, C.W. Croghan, P.A. Ashley, E. Pinzer, W. Friedman, M.C. Brinkman, M.G. Nishioka, et al. 2009. American Healthy Homes Survey: a national study of residential pesticides measured from floor wipes. *Environmental Science & Technology* 43 (12):4294-300.

93. Berger-Preiß, E., A. Preiß, K. Sielaff, M. Raabe, B. Ilgen, and K. Levsen. 1997. The behavior of pyrethroids indoors: A model study. *Indoor Air* 7:248-261.

94. Weschler, C.J. 2009. Changes in indoor pollutants since the 1950s. *Atmospheric Environment* 43 (1):153-169.

95. Harnly, M.E., A. Bradman, M. Nishioka, T.E. McKone, D. Smith, R. McLaughlin, G. Kavanagh-Baird, R. Castorina, and B. Eskenazi. 2009. Pesticides in dust from homes in an agricultural area. *Environmental Science & Technology* 43 (23):8767-74.

96. Morgan, M.K., et al. 2004. *A Pilot Study of Children's Total Exposure to Persistent Pesticides and Other Persistent Organic Pollutrants (CTEPP) Volume I and II.* Research Triangle Park, NC: U.S. EPA, Office of Research and Development. http://www.epa.gov/heasd/ctepp/ctepp_report.pdf.

97. Tulve, N.S., P.A. Jones, M.G. Nishioka, R.C. Fortmann, C.W. Croghan, J.Y. Zhou, A. Fraser, C. Cavel, and W. Friedman. 2006. Pesticide measurements from the first national environmental health survey of child care centers using a multi-residue GC/MS analysis method. *Environmental Science and Technology* 40 (20):6269-74.

98. Morgan, M.K., D.M. Stout, P.A. Jones, and D.B. Barr. 2008. An observational study of the potential for human exposures to pet-borne diazinon residues following lawn applications. *Environmental Research* 107 (3):336-42.

99. California Department of Health Services and California Air Resources Board. 2004. *Report to the California Legislature: Environmental Health Conditions in California's Portable Classrooms.* Sacramento, CA: California Department of Health Services, California Air Resources Board. http://www.arb.ca.gov/research/apr/reports/l3006.pdf.

Contaminants in Schools and Child Care Facilities (continued)

100. Carozza, S.E., B. Li, K. Elgethun, and R. Whitworth. 2008. Risk of childhood cancers associated with residence in agriculturally intense areas in the United States. *Environmental Health Perspectives* 116 (4):559-65.

101. Daniels, J.L., A.F. Olshan, and D.A. Savitz. 1997. Pesticides and childhood cancers. *Environmental Health Perspectives* 105 (10):1068-77.

102. Ma, X., P.A. Buffler, R.B. Gunier, G. Dahl, M.T. Smith, K. Reinier, and P. Reynolds. 2002. Critical windows of exposure to household pesticides and risk of childhood leukemia. *Environmental Health Perspectives* 110 (9):955-60.

103. Turner, M.C., D.T. Wigle, and D. Krewski. 2010. Residential pesticides and childhood leukemia: a systematic review and meta-analysis. *Environmental Health Perspectives* 118 (1):33-41.

104. U.S. Environmental Protection Agency. 2008. *Chemicals Evaluated for Carcinogenic Potential by the Office of Pesticide Programs.* Washington, DC: U.S. EPA, Office of Pesticide Programs.

105. Bouchard, M.F., D.C. Bellinger, R.O. Wright, and M.G. Weisskopf. 2010. Attention-deficit/hyperactivity disorder and urinary metabolites of organophosphate pesticides. *Pediatrics* 125 (6):e1270-7.

106. Eskenazi, B., A.R. Marks, A. Bradman, K. Harley, D.B. Barr, C. Johnson, N. Morga, and N.P. Jewell. 2007. Organophosphate pesticide exposure and neurodevelopment in young Mexican-American children. *Environmental Health Perspectives* 115 (5):792-8.

107. Lovasi, G.S., J.W. Quinn, V.A. Rauh, F.P. Perera, H.F. Andrews, R. Garfinkel, L. Hoepner, R. Whyatt, and A. Rundle. 2011. Chlorpyrifos Exposure and Urban Residential Environment Characteristics as Determinants of Early Childhood Neurodevelopment. *American Journal of Public Health* 101 (1):63-70.

108. Salam, M.T., Y.F. Li, B. Langholz, and F.D. Gilliland. 2004. Early-life environmental risk factors for asthma: findings from the Children's Health Study. *Environmental Health Perspectives* 112 (6):760-5.

109. U.S. General Accounting Office. 1999. *Pesticides: Use, Effects, and Alternatives to Pesticides in Schools.* Washington, DC: U.S. General Accounting Office. http://www.gao.gov/archive/2000/rc00017.pdf.

110. Owens, K. 2010. Schooling of state pesticide laws: 2010 update. *Pesticides and You* 29 (3):9-20.

111. Mir, D.F., Y. Finkelstein, and G.D. Tulipano. 2010. Impact of integrated pest management (IPM) training on reducing pesticide exposure in Illinois childcare centers. *Neurotoxicology* 31 (5):621-626.

112. Williams, M.K., A. Rundle, D. Holmes, M. Reyes, L.A. Hoepner, D.B. Barr, D.E. Camann, F.P. Perera, and R.M. Whyatt. 2008. Changes in pest infestation levels, self-reported pesticide use, and permethrin exposure during pregnancy after the 2000-2001 U.S. Environmental Protection Agency restriction of organophosphates. *Environmental Health Perspectives* 116 (12):1681-8.

113. Beyond Pesticides. 2010. *State and Local School Pesticide Policies.* Beyond Pesticides. Retrieved July 8, 2010 from http://www.beyondpesticides.org/schools/schoolpolicies/index.htm.

114. Wilson, N.K., J.C. Chuang, R. Iachan, C. Lyu, S.M. Gordon, M.K. Morgan, H. Ozkaynak, and L.S. Sheldon. 2004. Design and sampling methodology for a large study of preschool children's aggregate exposures to persistent organic pollutants in their everyday environments. *Journal of Exposure Analysis and Environmental Epidemiology* 14 (3):260-274.

115. California Department of Pesticide Regulation. 2010. *The Healthy Schools Act of 2000 (AB 2260) Frequently Asked Questions.* Retrieved September 22 from http://apps.cdpr.ca.gov/schoolipm/overview/faq2000.cfm.

Appendices

Appendix A: Data Tables

Appendix A: Data Tables

Environments and Contaminants

Criteria Air Pollutants

Table E1: Percentage of children ages 0 to 17 years living in counties with pollutant concentrations above the levels of the current air quality standards, 1999-2009*

1999-2004						
Pollutant	1999	2000	2001	2002	2003	2004
Any standard	74.9	76.1	76.3	75.9	77.4	73.8
Ozone (8-hour)	65.2	64.9	66.3	66.1	67.8	61.6
$PM_{2.5}$ (24-hour)	55.0	62.5	60.8	60.9	56.8	56.0
Sulfur dioxide (1-hour)	31.1	28.8	26.6	25.6	21.6	20.5
$PM_{2.5}$ (annual)	24.2	29.6	24.7	20.9	19.1	16.4
Nitrogen dioxide (1-hour)	23.2	19.4	17.4	18.9	17.5	16.3
PM_{10} (24-hour)	7.9	6.3	6.0	4.8	7.8	5.2
Carbon monoxide (8-hour)	5.7	4.4	0.7	4.1	<0.1	0.1
Lead (3-month)	2.3	1.6	2.2	1.2	1.6	1.2
2005-2009						
Pollutant	2005	2006	2007	2008	2009	
Any standard	75.9	72.9	74.4	69.2	58.6	
Ozone (8-hour)	66.2	65.3	64.1	59.2	48.9	
$PM_{2.5}$ (24-hour)	60.2	45.7	53.6	37.1	32.2	
Sulfur dioxide (1-hour)	20.9	16.6	15.5	16.7	11.4	
$PM_{2.5}$ (annual)	24.3	12.5	16.1	7.3	2.1	
Nitrogen dioxide (1-hour)	13.9	12.5	10.9	12.6	8.7	
PM_{10} (24-hour)	5.0	5.1	12.5	4.0	2.8	
Carbon monoxide (8-hour)	0.2	0.3	0.1	0.2	0.0	
Lead (3-month)	1.6	1.2	5.0	5.0	4.2	

DATA: U.S. Environmental Protection Agency, Office of Air and Radiation, Air Quality System

* EPA periodically reviews air quality standards and may change them based on updated scientific findings. Measuring concentrations above the level of a standard is not equivalent to violating the standard. The level of a standard may be exceeded on multiple days before the exceedance is considered a violation of the standard. See the indicator text for additional discussion. The indicator is calculated with reference to the current levels of the air quality standards for all years shown. Note that EPA promulgated a revised annual $PM_{2.5}$ standard in December 2012, which has not been incorporated into this analysis.

Table E1a: Percentage of children ages 0 to 17 years living in counties with pollutant concentrations above the levels of the current air quality standards, by race/ethnicity, 2009*

Pollutant	All Races/ Ethnicities	White non-Hispanic	Black non-Hispanic	American Indian/Alaska Native non-Hispanic	Asian or Pacific Islander non-Hispanic	Hispanic
Any standard	58.6	51.9	62.7	38.1	70.0	71.4
Ozone (8-hour)	48.9	40.9	52.2	29.5	60.7	65.2
PM$_{2.5}$ (24-hour)	32.2	24.0	34.0	19.8	47.7	48.8
Sulfur dioxide (1-hour)	11.4	9.9	17.4	5.2	10.2	11.5
PM$_{2.5}$ (annual)	2.1	1.1	0.8	1.7	2.7	5.3
Nitrogen dioxide (1-hour)	8.7	4.0	9.8	2.8	12.8	19.4
PM$_{10}$ (24-hour)	2.8	2.5	0.9	5.8	1.8	4.8
Carbon monoxide (8-hour)	0.0	0.0	0.0	0.0	0.0	0.0
Lead (3-month)	4.2	2.0	2.6	1.2	8.2	10.0

DATA: U.S. Environmental Protection Agency, Office of Air and Radiation, Air Quality System

* EPA periodically reviews air quality standards and may change them based on updated scientific findings. Measuring concentrations above the level of a standard is not equivalent to violating the standard. The level of a standard may be exceeded on multiple days before the exceedance is considered a violation of the standard. See the indicator text for additional discussion. The indicator is calculated with reference to the current levels of the air quality standards for all years shown. Note that EPA promulgated a revised annual PM$_{2.5}$ standard in December 2012, which has not been incorporated into this analysis.

Table E1b: Percentage of children ages 0 to 17 years living in counties with pollutant concentrations above the levels of the current air quality standards, by family income, 2009*

Pollutant	All Incomes	< Poverty Level	≥ Poverty Level	≥ Poverty (Detail) 100-200% of Poverty Level	≥ 200% of Poverty Level
Any standard	58.6	59.0	58.6	56.3	59.3
Ozone (8-hour)	48.9	49.9	48.7	47.2	49.3
PM$_{2.5}$ (24-hour)	32.2	36.5	31.3	32.9	30.8
Sulfur dioxide (1-hour)	11.4	12.5	11.1	11.2	11.1
PM$_{2.5}$ (annual)	2.1	3.1	1.9	2.6	1.6
Nitrogen dioxide (1-hour)	8.7	12.2	8.1	10.1	7.4
PM$_{10}$ (24-hour)	2.8	2.6	2.8	3.1	2.7
Carbon monoxide (8-hour)	0.0	0.0	0.0	0.0	0.0
Lead (3-month)	4.2	5.7	3.9	4.9	3.6

DATA: U.S. Environmental Protection Agency, Office of Air and Radiation, Air Quality System

* EPA periodically reviews air quality standards and may change them based on updated scientific findings. Measuring concentrations above the level of a standard is not equivalent to violating the standard. The level of a standard may be exceeded on multiple days before the exceedance is considered a violation of the standard. See the indicator text for additional discussion. The indicator is calculated with reference to the current levels of the air quality standards for all years shown. Note that EPA promulgated a revised annual PM$_{2.5}$ standard in December 2012, which has not been incorporated into this analysis.

Table E2: Percentage of children ages 0 to 17 years living in counties with 8-hour ozone and 24-hour PM$_{2.5}$ concentrations above the levels of air quality standards, by frequency of occurrence, 2009*

Ozone (8-hour)							
1999-2005	**1999**	**2000**	**2001**	**2002**	**2003**	**2004**	**2005**
No days with concentrations above the standard	2.9	4.4	4.2	4.6	3.7	9.4	5.1
1-3 days	4.6	9.6	6.9	6.7	8.7	22.8	9.3
4-10 days	10.8	22.9	16.2	9.6	28.5	21.0	17.7
11-25 days	26.7	16.2	29.5	21.5	18.1	10.0	28.1
26 or more days	23.2	16.2	13.7	28.4	12.5	7.8	11.2
No monitoring data	31.8	30.7	29.6	29.2	28.4	29.0	28.7
2006-2009	**2006**	**2007**	**2008**	**2009**			
No days with concentrations above the standard	6.4	8.0	13.6	24.3			
1-3 days	10.6	11.2	18.6	27.7			
4-10 days	24.8	19.8	23.9	11.8			
11-25 days	19.6	25.9	8.5	3.0			
26 or more days	10.4	7.2	8.2	6.4			
No monitoring data	28.3	28.0	27.2	26.8			
PM$_{2.5}$ (24-hour)							
1999-2005	**1999**	**2000**	**2001**	**2002**	**2003**	**2004**	**2005**
No days with concentrations above the standard	13.4	10.6	12.5	12.9	16.4	14.6	11.1
1-7 days	36.3	41.5	39.1	37.5	37.4	40.0	41.9
8-10 days	3.3	2.5	1.7	4.3	3.8	5.3	4.7
11-25 days	9.2	11.2	12.6	11.1	9.8	8.3	10.7
26 or more days	6.2	7.2	7.4	7.8	5.4	2.2	2.4
No monitoring data	31.6	27.0	26.8	26.5	27.2	29.6	29.1
2006-2009	**2006**	**2007**	**2008**	**2009**			
No days with concentrations above the standard	25.4	17.5	33.9	38.4			
1-7 days	34.9	38.4	29.3	28.4			
8-10 days	6.5	1.8	4.8	0.8			
11-25 days	1.0	10.2	1.3	1.9			
26 or more days	1.8	1.9	1.0	0.9			
No monitoring data	30.4	30.2	29.7	29.6			

DATA: U.S. Environmental Protection Agency, Office of Air and Radiation, Air Quality System

* EPA periodically reviews air quality standards and may change them based on updated scientific findings. Measuring concentrations above the level of a standard is not equivalent to violating the standard. The level of a standard may be exceeded on multiple days before the exceedance is considered a violation of the standard. See the indicator text for additional discussion. The indicator is calculated with reference to the current levels of the air quality standards for all years shown.

Table E3: Percentage of days with good, moderate, or unhealthy air quality for children ages 0 to 17 years, 1999-2009

Pollution Level							
1999-2005	**1999**	**2000**	**2001**	**2002**	**2003**	**2004**	**2005**
Good	41.2	43.2	44.0	45.5	47.1	49.4	47.7
Moderate	22.1	23.3	23.3	21.6	21.5	20.4	21.2
Unhealthy	8.8	7.2	7.0	7.6	6.0	4.8	5.7
No monitoring data	27.9	26.3	25.7	25.3	25.4	25.4	25.4
2006-2009	**2006**	**2007**	**2008**	**2009**			
Good	48.9	48.6	51.9	56.6			
Moderate	20.5	20.5	18.3	15.5			
Unhealthy	5.0	4.9	3.7	2.8			
No monitoring data	25.7	26.0	26.0	25.1			

DATA: U.S. Environmental Protection Agency, Office of Air and Radiation, Air Quality System

NOTE: Good, moderate, and unhealthy air quality are defined using EPA's Air Quality Index (AQI). The health information that supports EPA's periodic reviews of the air quality standards informs decisions on the AQI breakpoints and may change based on updated scientific findings. See text for additional discussion.

Table E3a: Percentage of days with good, moderate, or unhealthy air quality for children ages 0 to 17 years, by race/ethnicity, 2009

Pollution Level	All Races/ Ethnicities	White non-Hispanic	Black non-Hispanic	American Indian/ Alaska Native	Asian or Pacific Islander	Hispanic
Good	56.6	54.5	60.8	50.3	65.4	57.3
Moderate	15.5	12.1	16.0	11.6	20.1	23.2
Unhealthy	2.8	1.6	1.9	1.7	4.5	6.0
No monitoring data	25.1	31.8	21.3	36.5	10.0	13.5

DATA: U.S. Environmental Protection Agency, Office of Air and Radiation, Air Quality System

NOTE: Good, moderate, and unhealthy air quality are defined using EPA's Air Quality Index (AQI). The health information that supports EPA's periodic reviews of the air quality standards informs decisions on the AQI breakpoints and may change based on updated scientific findings. See text for additional discussion.

Table E3b: Percentage of days with good, moderate, or unhealthy air quality for children ages 0 to 17 years, by family income, 2009

Pollution Level	All Incomes	< Poverty Level	≥ Poverty Level	≥ Poverty (Detail) 100-200% of Poverty Level	≥ 200% of Poverty Level
Good	56.6	53.6	57.2	52.8	58.7
Moderate	15.5	16.9	15.3	15.9	15.1
Unhealthy	2.8	3.6	2.6	3.2	2.4
No monitoring data	25.1	26.0	24.9	28.1	23.8

DATA: U.S. Environmental Protection Agency, Office of Air and Radiation, Air Quality System

NOTE: Good, moderate, and unhealthy air quality are defined using EPA's Air Quality Index (AQI). The health information that supports EPA's periodic reviews of the air quality standards informs decisions on the AQI breakpoints and may change based on updated scientific findings. See text for additional discussion.

Hazardous Air Pollutants

Table E4: Percentage of children ages 0 to 17 years living in census tracts where estimated hazardous air pollutant concentrations were greater than health benchmarks in 2005

Health Benchmark	
Cancer, one in 100,000	99.9
Cancer, one in 10,000	6.6
Other health effects	56.4

DATA: U.S. Environmental Protection Agency, National Air Toxics Assessment

Table E4a: Percentage of schoolchildren attending schools in census tracts where estimated hazardous air pollutant concentrations were greater than health benchmarks in 2005

Health Benchmark	
Cancer, one in 100,000	100.0
Cancer, one in 10,000	6.2
Other health effects	56.6

DATA: U.S. Environmental Protection Agency, National Air Toxics Assessment

Table E4b: Percentage of children ages 0 to 17 years living in census tracts where the cancer risk from estimated hazardous air pollutant concentrations was at least one in 10,000 in 2005, by race/ethnicity and family income

Race / Ethnicity	All Incomes	< Poverty Level	≥ Poverty Level
All Races/Ethnicities	6.6	9.3	5.9
White	4.1	6.5	3.7
Black	7.2	7.4	7.0
Asian	14.4	21.1	13.5
American Indian/Alaska Native	4.1	3.5	4.4
Native Hawaiian or Other Pacific Islander	8.2	9.1	7.8
Hispanic	16.2	16.9	15.9
All Other Races†	17.0	19.9	16.0

DATA: U.S. Environmental Protection Agency, National Air Toxics Assessment

NOTE: Race categories include children of Hispanic ethnicity. Hispanic children may be of any race.

† The "All Other Races" category includes all other races not specified, together with those individuals who report more than one race.

Table E4c: Percentage of children ages 0 to 17 years living in census tracts where the non-cancer risk from estimated hazardous air pollutant concentrations exceeded health benchmarks in 2005, by race/ethnicity and family income

Race / Ethnicity	All Incomes	< Poverty Level	≥ Poverty Level
All Races/Ethnicities	56.4	57.3	56.2
White	49.3	46.0	50.0
Black	72.9	71.8	73.5
Asian	81.4	84.4	81.0
American Indian/Alaska Native	32.1	28.0	34.2
Native Hawaiian or Other Pacific Islander	57.2	57.1	57.3
Hispanic	68.7	63.8	70.8
All Other Races†	70.3	68.5	71.0

DATA: U.S. Environmental Protection Agency, National Air Toxics Assessment

NOTE: Race categories include children of Hispanic ethnicity. Hispanic children may be of any race.

† The "All Other Races" category includes all other races not specified, together with those individuals who report more than one race.

Indoor Environments

Table E5: Percentage of children ages 0 to 6 years regularly exposed to environmental tobacco smoke in the home, by family income, 1994, 2005, and 2010

Year	All Incomes	< Poverty Level	100-200% of Poverty Level	≥ 200% of Poverty Level
1994	27.3	37.1	32.7	18.5
2005	8.4	14.6	11.7	4.7
2010	6.1	10.2	8.1	3.0

DATA: Centers for Disease Control and Prevention, National Center for Health Statistics, National Health Interview Survey

Table E5a: Percentage of children ages 0 to 6 years regularly exposed to environmental tobacco smoke in the home, by race/ethnicity and family income, 2010

Race / Ethnicity	All Incomes (n=6,890)	< Poverty Level (n=2,072)	≥ Poverty Level (n=4,818)	> Poverty (Detail) 100-200% of Poverty Level (n=1,787)	> 200% of Poverty Level (n=3,030)
All Races/Ethnicities (n=6,890)		10.2	4.7	8.1	3.0
White non-Hispanic (n=2,662)	7.5	19.9	5.2	11.5	3.1
Black or African-American non-Hispanic (n=1,049)	8.5	10.4	7.0	7.8	6.3
Asian non-Hispanic (n=381)	NA**	NA**	NA**	NA**	NA**
Hispanic (n=2,492)	2.2	2.5*	2.1	2.5*	1.6*
Mexican (n=1,687)	2.2	2.6*	1.9*	NA**	NA**
Puerto Rican (n=209)	4.8*	NA**	NA**	NA**	NA**
All Other Races† (n=306)	9.5	13.7*	8.3*	2.5*	NA**
American Indian/Alaska Native non-Hispanic (n=22)	NA**	NA**	NA**	NA**	NA**

DATA: Centers for Disease Control and Prevention, National Center for Health Statistics, National Health Interview Survey

† The "All Other Races" category includes all other races not specified, together with those individuals who report more than one race.

* The estimate should be interpreted with caution because the standard error of the estimate is relatively large: the relative standard error, RSE, is at least 30% but is less than 40% (RSE = standard error divided by the estimate).

**Not available. The estimate is not reported because it has large uncertainty: the relative standard error, RSE, is 40% or greater (RSE = standard error divided by the estimate).

Table E6: Percentage of children ages 0 to 5 years living in homes with interior lead hazards, 1998-1999 and 2005-2006

Year	Interior Lead Dust	Interior Deteriorated Lead-Based Paint	Either Interior Lead Dust or Interior Deteriorated Lead-Based Paint
1998-1999	16.2	11.9	21.6
2005-2006	12.5	10.6	14.6

DATA: U.S. Department of Housing and Urban Development, National Survey of Lead and Allergens in Housing, American Healthy Homes Survey

NOTE: Lead hazards are defined here by current federal standards indicating that floor and window lead dust should not exceed 40 micrograms of lead per square foot ($\mu g/ft^2$) and 250 $\mu g/ft^2$, respectively, in order to protect children from developing "elevated" blood lead levels as defined by the CDC at the time the standards were issued. EPA is currently reviewing the lead dust standards to determine whether they should be lowered.

Drinking Water Contaminants

Table E7: Estimated percentage of children ages 0 to 17 years served by community water systems that did not meet all applicable health-based drinking water standards, 1993-2009

1993-1997					
Type of standard violated	1993	1994	1995	1996	1997
Any health-based standard	19.2	15.7	10.7	9.7	9.4
Total coliforms	10.1	8.6	4.0	4.2	3.5
Surface water treatment[#]	6.3	5.5	4.1	3.7	3.4
Lead and copper[†]	2.8	1.7	1.9	1.8	1.8
Chemical and radionuclide	1.1	0.9	1.3	0.8	1.0
Nitrate/nitrite	0.3	0.1	0.2	0.2	0.4

1998-2003					
Type of standard violated	1998	1999	2000	2001	2002	2003
Any health-based standard	8.2	7.6	8.1	5.1	11.8	8.4
Total coliforms	2.9	3.1	2.9	2.1	2.5	3.0
Surface water treatment[#]	2.8	2.4	3.2	1.3	6.3	1.5
Disinfectants and disinfection byproducts	NA‡	NA‡	NA‡	NA‡	1.5	3.0
Lead and copper[†]	1.5	1.4	1.2	1.1	0.8	0.6
Chemical and radionuclide	0.9	0.7	0.8	0.7	0.8	0.8
Nitrate/nitrite	0.6	0.3	0.5	0.2	0.2	0.3

2004-2009					
Type of standard violated	2004	2005	2006	2007	2008	2009
Any health-based standard	10.4	12.6	10.6	7.8	6.7	7.4
Total coliforms	3.5	3.3	2.3	2.4	2.3	2.5
Surface water treatment[#]	3.3	5.2	5.0	2.6	1.5	2.0
Disinfectants and disinfection byproducts	2.6	2.6	1.6	1.4	1.4	1.3
Lead and copper[†]	0.9	0.8	0.4	0.4	0.5	0.8
Chemical and radionuclide	1.1	0.9	1.2	1.2	1.1	1.1
Nitrate/nitrite	0.1	0.1	0.5	0.2	0.1	0.1

DATA: U.S. Environmental Protection Agency, Office of Water, Safe Drinking Water Information System Federal Version

[#] "Surface water treatment" includes violations of the Surface Water Treatment Rule and of the Interim Enhanced Surface Water Treatment Rule.

† Lead and copper represents the lead and copper rule, which is a set of standards and implementation measures.

‡ The standard for disinfectants and disinfection byproducts was first implemented in 2002.

NOTE: A new standard for disinfection byproducts was implemented beginning in 2002 for larger drinking water systems and 2004 for smaller systems.[1] Revisions to the standard for surface water treatment took effect in 2002.[2] A revised standard for radionuclides went into effect in 2003.[3] A revised standard for arsenic went into effect in 2006.[4] No other revisions to the standards have taken effect during the period of trend data.

Table E8: Estimated percentage of children ages 0 to 17 years served by community water systems with violations of drinking water monitoring and reporting requirements, 1993-2009

1993-1997

Type of standard violated	1993	1994	1995	1996	1997
Any violation	19.4	14.1	12.6	10.8	10.5
Chemical and radionuclide	7.9	5.9	6.3	4.8	4.3
Lead and copper	6.4	3.8	3.0	3.4	3.9
Total coliforms	5.5	5.8	4.4	3.7	3.6
Surface water treatment#	1.9	1.1	0.4	0.3	0.3

1998-2003

Type of standard violated	1998	1999	2000	2001	2002	2003
Any violation	16.6	15.6	14.8	16.8	20.6	19.1
Chemical and radionuclide	6.3	5.7	4.8	6.4	8.2	8.2
Lead and copper	7.3	7.3	7.9	6.9	6.5	6.7
Total coliforms	4.0	3.5	3.1	4.4	2.7	4.3
Disinfectants and disinfection byproducts	NA‡	NA‡	NA‡	NA‡	1.8	1.7
Surface water treatment#	0.3	0.8	0.4	0.3	3.3	2.6

2004-2009

Type of standard violated	2004	2005	2006	2007	2008	2009
Any violation	19.6	22.5	20.6	22.1	19.2	13.4
Chemical and radionuclide	7.6	6.8	7.9	8.8	5.6	4.1
Lead and copper	7.4	8.0	8.0	7.7	7.4	3.1
Total coliforms	4.7	4.4	4.7	4.4	4.1	3.1
Disinfectants and disinfection byproducts	6.5	6.5	4.7	5.6	3.5	3.5
Surface water treatment#	2.9	3.8	2.4	2.1	2.4	1.2

DATA: U.S. Environmental Protection Agency, Office of Water, Safe Drinking Water Information System Federal Version

"Surface water treatment" includes violations of the Surface Water Treatment Rule and of the Interim Enhanced Surface Water Treatment Rule.

‡ The standard for disinfectants and disinfection byproducts was first implemented in 2002.

NOTE: A new standard for disinfection byproducts was implemented beginning in 2002 for larger drinking water systems and 2004 for smaller systems.[1] Revisions to the standard for surface water treatment took effect in 2002.[2] A revised standard for radionuclides went into effect in 2003.[3] A revised standard for arsenic went into effect in 2006.[4] No other revisions to the standards have taken effect during the period of trend data.

Chemicals in Food

Table E9: Percentage of sampled apples, carrots, grapes, and tomatoes with detectable residues of organophosphate pesticides, 1998-2009

	1998	1999	2000	2001	2002	2003	2004	2005	2006	2007	2008	2009
Apples	NA	80.7	54.9	49.3	45.5	NA	50.5	45.0	NA	NA	NA	34.7
Carrots	NA	NA	10.3	6.2	8.3	NA	NA	NA	3.5	5.4	NA	NA
Grapes	NA	NA	20.6	14.8	NA	NA	16.5	16.2	NA	NA	NA	7.7
Tomatoes	37.4	29.9	NA	NA	NA	14.6	11.8	NA	NA	9.7	9.5	NA

DATA: U.S. Department of Agriculture, Pesticide Data Program

NOTE: For purposes of indicator calculation, only the 43 organophosphate pesticides measured by the pesticide data program in each year 1998-2009 were considered, so that indicator data are comparable over time. "NA" indicate that the food was not sampled by the Pesticide Data Program in a particular year. Improvements in measurement technology increase the capability to detect pesticide residues in more recent samples. In this analysis, limits of detection are held constant so that indicator data are comparable over time. A separate analysis found that actual detections of pesticide residues were similar or only slightly greater than the values shown in this table.

Contaminated Lands

Table E10: Percentage of children ages 0 to 17 years living within one mile of Superfund and Corrective Action sites that may not have all human health protective measures in place, 2009

Race / Ethnicity	All Incomes	< Poverty Level	≥ Poverty Level
All Races/Ethnicities	5.8	7.7	5.4
White	4.7	5.9	4.6
Black	8.1	9.6	7.4
American Indian/Alaska Native	5.1	3.7	5.5
Asian	8.6	10.5	8.3
Native Hawaiian or Other Pacific Islander	10.4	11.8	10.2
Hispanic	8.0	8.5	7.8
All Other Races†	8.5	9.3	8.3

DATA: U.S. Environmental Protection Agency, Office of Solid Waste and Emergency Response, CERCLIS, and RCRAInfo

NOTE: Race categories include children of Hispanic ethnicity. Hispanic children may be of any race.

† The "All Other Races" category includes all other races not specified, together with those individuals who report more than one race.

Table E11: Distribution by race/ethnicity and family income of children living near selected contaminated lands* in 2009, compared with the distribution by race/ethnicity and income of children in the general U.S. population

Race / Ethnicity	Population	All Incomes	< Poverty Level
White	Children living near selected sites	55.6	36.0
	All children	68.6	47.3
Black	Children living near selected sites	21.1	37.6
	All children	15.1	30.1
American Indian/	Children living near selected sites	1.0	0.8
Alaska Native	All children	1.2	1.7
Asian	Children living near selected sites	5.1	3.5
	All children	3.4	2.6
Native Hawaiian or	Children living near selected sites	0.3	0.2
Other Pacific Islander	All children	0.2	0.1
All Other Races†	Children living near selected sites	17.0	21.8
	All children	11.6	18.1
Hispanic	Children living near selected sites	23.5	31.7
	All children	17.1	28.7

DATA: U.S. Environmental Protection Agency, Office of Solid Waste and Emergency Response, CERCLIS, and RCRAInfo

*Within one mile of Superfund and Corrective Action sites that may not have all human health protective measures in place.

NOTE: Race categories include children of Hispanic ethnicity. Hispanic children may be of any race.

† The "All Other Races" category includes all other races not specified, together with those individuals who report more than one race.

Biomonitoring

Lead

Table B1: Lead in children ages 1 to 5 years: Median and 95th percentile concentrations in blood, 1976-2010

	Blood lead concentration (µg/dL)								
	1976-1980	1988-1991	1991-1994	1999-2000	2001-2002	2003-2004	2005-2006	2007-2008	2009-2010
Median	15.0	3.5	2.6	2.2	1.6	1.6	1.4	1.4	1.2
95th percentile	29.0	12.1	9.7	7.0	5.8	5.1	3.8	4.1	3.4

DATA: Centers for Disease Control and Prevention, National Center for Health Statistics and National Center for Environmental Health, National Health and Nutrition Examination Survey

Table B1a: Lead in children ages 1 to 17 years: Blood lead concentrations by age group, 2009-2010

	Blood lead concentration (µg/dL)						
	Ages 1 to 17 years	Age 1 year	Age 2 years	Ages 3 to 5 years	Ages 6 to 10 years	Ages 11 to 15 years	Ages 16 to 17 years
Median	0.8	1.2	1.2	1.1	0.8	0.7	0.7
95th percentile	2.2	4.2	3.5	2.8	2.1	1.7	1.4

DATA: Centers for Disease Control and Prevention, National Center for Health Statistics and National Center for Environmental Health, National Health and Nutrition Examination Survey

Table B2. Lead in children ages 1 to 5 years: Median concentrations in blood, by race/ethnicity and family income, 2007-2010

	Median blood lead concentration (µg/dL)		
Race / Ethnicity	All Incomes‡ (n=1,653)	< Poverty Level (n=642)	≥ Poverty Level (n=898)
All Races/Ethnicities (n=1,653)	1.3	1.5	1.2
White Non-Hispanic (n=536)	1.2	1.5*	1.2
Black Non-Hispanic (n=338)	1.6	1.7*	1.4
Mexican-American (n=490)	1.2	1.3	1.1
All Other Races/Ethnicities† (n=289)	1.2	1.6	1.1

DATA: Centers for Disease Control and Prevention, National Center for Health Statistics and National Center for Environmental Health, National Health and Nutrition Examination Survey

† The "All Other Races/Ethnicities" category includes all other races or ethnicities not specified, together with those individuals who report more than one race.

‡ Includes sampled individuals for whom income information is missing.

*The estimate should be interpreted with caution because the standard error of the estimate is relatively large: the relative standard error, RSE, is at least 30% but is less than 40% (RSE = standard error divided by the estimate), or the RSE may be underestimated.

Table B2a. Lead in children ages 1 to 5 years: 95th percentile concentrations in blood, by race/ethnicity and family income, 2007-2010

Race / Ethnicity	95th percentile blood lead concentration (µg/dL)		
	All Incomes‡ (n=1,653)	< Poverty Level (n=642)	≥ Poverty Level (n=898)
All Races/Ethnicities (n=1,653)	3.9	4.7	3.3
White non-Hispanic (n=536)	3.5	4.5*	3.4
Black non-Hispanic (n=338)	5.8	6.8*	4.2
Mexican-American (n=490)	3.3	4.1	3.2
All Other Races/Ethnicities† (n=289)	3.5	4.2	2.7

DATA: Centers for Disease Control and Prevention, National Center for Health Statistics and National Center for Environmental Health, National Health and Nutrition Examination Survey

† The "All Other Races/Ethnicities" category includes all other races or ethnicities not specified, together with those individuals who report more than one race.

‡ Includes sampled individuals for whom income information is missing.

*The estimate should be interpreted with caution because the standard error of the estimate is relatively large: the relative standard error, RSE, is at least 30% but is less than 40% (RSE = standard error divided by the estimate), or the RSE may be underestimated.

Table B2b. Lead in children ages 1 to 5 years: Median concentrations in blood, by race/ethnicity and family income, 1991-1994

Race / Ethnicity	Median blood lead concentration (µg/dL)		
	All Incomes‡ (n=2,367)	< Poverty Level (n=974)	≥ Poverty Level (n=1,253)
All Races/Ethnicities (n=2,367)	2.6	4.0	2.2
White Non-Hispanic (n=623)	2.3	3.2*	2.1
Black Non-Hispanic (n=773)	4.3	5.1	3.5
Mexican-American (n=822)	3.1	3.7	2.6
All Other Races/Ethnicities† (n=149)	2.5	NA**	2.0*

DATA: Centers for Disease Control and Prevention, National Center for Health Statistics and National Center for Environmental Health, National Health and Nutrition Examination Survey

† The "All Other Races/Ethnicities" category includes all other races or ethnicities not specified, together with those individuals who report more than one race.

‡ Includes sampled individuals for whom income information is missing.

*The estimate should be interpreted with caution because the standard error of the estimate is relatively large: the relative standard error, RSE, is at least 30% but is less than 40% (RSE = standard error divided by the estimate), or the RSE may be underestimated.

** Not available. The estimate is not reported because it has large uncertainty: the relative standard error, RSE, is 40% or greater (RSE = standard error divided by the estimate), or the RSE cannot be reliably estimated.

Mercury

Table B3: Mercury in women ages 16 to 49 years: Median and 95th percentile concentrations in blood, 1999-2010

	Concentration of mercury in blood (µg/L)					
	1999-2000	2001-2002	2003-2004	2005-2006	2007-2008	2009-2010
Median	0.9	0.7	0.8	0.8	0.7	0.8
95th percentile	7.4	3.7	4.5	4.0	3.7	4.2

DATA: Centers for Disease Control and Prevention, National Center for Health Statistics and National Center for Environmental Health, National Health and Nutrition Examination Survey

NOTE: To reflect exposures to women who are pregnant or may become pregnant, the estimates are adjusted for the probability (by age and race/ethnicity) that a woman gives birth. The intent of this adjustment is to approximate the distribution of exposure to pregnant women. Results will therefore differ from a characterization of exposure to adult women without consideration of birth rates.

Table B3a. Mercury in women ages 16 to 49 years: Median concentrations in blood, by race/ethnicity and family income, 2007-2010

	Median concentration of mercury in blood (µg/L)		
Race / Ethnicity	All Incomes‡ (n=3,456)	< Poverty Level (n=915)	≥ Poverty Level (n=2,261)
All Races/Ethnicities (n=3,456)	0.7	0.6	0.8
White non-Hispanic (n=1,430)	0.7	0.5	0.7
Black non-Hispanic (n=665)	0.8	0.8	0.9
Mexican-American (n=722)	0.6	0.6	0.7
All Other Races/Ethnicities† (n=639)	1.3	0.8	1.5

DATA: Centers for Disease Control and Prevention, National Center for Health Statistics and National Center for Environmental Health, National Health and Nutrition Examination Survey

NOTE: To reflect exposures to women who are pregnant or may become pregnant, the estimates are adjusted for the probability (by age and race/ethnicity) that a woman gives birth. The intent of this adjustment is to approximate the distribution of exposure to pregnant women. Results will therefore differ from a characterization of exposure to adult women without consideration of birth rates.

† The "All Other Races/Ethnicities" category includes all other races or ethnicities not specified, together with those individuals who report more than one race.

‡ Includes sampled individuals for whom income information is missing.

Table B3b. Mercury in women ages 16 to 49 years: 95th percentile concentrations in blood, by race/ethnicity and family income, 2007-2010

Race / Ethnicity	95th percentile concentration of mercury in blood (µg/L)		
	All Incomes‡ (n=3,456)	< Poverty Level (n=915)	≥ Poverty Level (n=2,261)
All Races/Ethnicities (n=3,456)	3.9	2.9	4.0
White non-Hispanic (n=1,430)	3.7	2.9	3.7
Black non-Hispanic (n=665)	2.9	2.3	3.3
Mexican-American (n=722)	2.3	1.9	2.4
All Other Races/Ethnicities† (n=639)	6.7	NA**	6.5

DATA: Centers for Disease Control and Prevention, National Center for Health Statistics and National Center for Environmental Health, National Health and Nutrition Examination Survey

NOTE: To reflect exposures to women who are pregnant or may become pregnant, the estimates are adjusted for the probability (by age and race/ethnicity) that a woman gives birth. The intent of this adjustment is to approximate the distribution of exposure to pregnant women. Results will therefore differ from a characterization of exposure to adult women without consideration of birth rates.

† The "All Other Races/Ethnicities" category includes all other races or ethnicities not specified, together with those individuals who report more than one race.

‡ Includes sampled individuals for whom income information is missing.

** Not available. The estimate is not reported because it has large uncertainty: the relative standard error, RSE, is 40% or greater (RSE = standard error divided by the estimate), or the RSE cannot be reliably estimated.

Table B3c: Mercury in children ages 1 to 5 years: Median and 95th percentile concentrations in blood, 1999-2010

	Concentration of mercury in blood (µg/L)					
	1999-2000	2001-2002	2003-2004	2005-2006	2007-2008	2009-2010
Median	0.3	0.3	0.3	0.2	0.2	0.2
95th percentile	2.3	1.9	1.8	1.4	1.3	1.3

DATA: Centers for Disease Control and Prevention, National Center for Health Statistics and National Center for Environmental Health, National Health and Nutrition Examination Survey

Table B3d: Mercury in children ages 1 to 17 years: Median and 95th percentile concentrations in blood, by age group, 2007-2010

	Concentration of mercury in blood (µg/L)						
	Ages 1 to 17 years	Age 1 year	Age 2 years	Ages 3 to 5 years	Ages 6 to 10 years	Ages 11 to 15 years	Ages 16 to 17 years
Median	0.4	0.2	0.2	0.2	0.4	0.4	0.5
95th percentile	1.9	1.2	1.3	1.4	1.7	2.2	2.8

DATA: Centers for Disease Control and Prevention, National Center for Health Statistics and National Center for Environmental Health, National Health and Nutrition Examination Survey

Cotinine

Table B4: Cotinine in nonsmoking children ages 3 to 17 years: Median and 95th percentile concentrations in blood serum, 1988-2010

	Concentration of cotinine in serum (ng/mL)							
	1988-1991	1991-1994	1999-2000	2001-2002	2003-2004	2005-2006	2007-2008	2009-2010
Median	0.25	0.21	0.11	0.06*	0.10	0.05	0.05	0.03
95th percentile	3.2	3.2	3.1	3.2	3.2	2.3	2.6	2.1

DATA: Centers for Disease Control and Prevention, National Center for Health Statistics and National Center for Environmental Health, National Health and Nutrition Examination Survey

NOTE: Based on children ages 3 to 17 years with serum cotinine ≤ 10 ng/mL (ages 4 to 17 years for 1988-1991 and 1991-1994).

*The estimate should be interpreted with caution because the standard error of the estimate is relatively large: the relative standard error, RSE, is at least 30% but is less than 40% (RSE = standard error divided by the estimate), or the RSE may be underestimated.

Table B4a. Cotinine in nonsmoking children ages 3 to 17 years: Median concentrations in blood serum, by race/ethnicity and family income, 2007-2010

	Median concentration of cotinine in serum (ng/mL)		
Race / Ethnicity	All Incomes‡ (n=4,284)	< Poverty Level (n=1,323)	≥ Poverty Level (n=2,648)
All Races/Ethnicities (n=4,284)	0.04	0.14	0.03
White non-Hispanic (n=1,310)	0.04	NA**	0.04
Black non-Hispanic (n=955)	0.11	0.38	0.06
Mexican-American (n=1,229)	0.02	0.03	0.02
All Other Races/Ethnicities† (n=790)	0.03	0.09	0.02

DATA: Centers for Disease Control and Prevention, National Center for Health Statistics and National Center for Environmental Health, National Health and Nutrition Examination Survey

NOTE: Based on children ages 3 to 17 years with serum cotinine ≤ 10 ng/mL.

† The "All Other Races/Ethnicities" category includes all other races or ethnicities not specified, together with those individuals who report more than one race.

‡ Includes sampled individuals for whom income information is missing.

** Not available. The estimate is not reported because it has large uncertainty: the relative standard error, RSE, is 40% or greater (RSE = standard error divided by the estimate), or the RSE cannot be reliably estimated.

Table B4b. Cotinine in nonsmoking children ages 3 to 17 years: 95th percentile concentrations in blood serum, by race/ethnicity and family income, 2007-2010

Race / Ethnicity	95th percentile concentration of cotinine in serum (ng/mL)		
	All Incomes‡ (n=4,284)	< Poverty Level (n=1,323)	≥ Poverty Level (n=2,648)
All Races/Ethnicities (n=4,284)	2.5	4.1	2.0
White non-Hispanic (n=1,310)	2.9	5.8	2.3
Black non-Hispanic (n=955)	2.6	3.0	2.3
Mexican-American (n=1,229)	0.8	0.9*	0.7*
All Other Races/Ethnicities† (n=790)	1.5	NA**	0.7*

DATA: Centers for Disease Control and Prevention, National Center for Health Statistics and National Center for Environmental Health, National Health and Nutrition Examination Survey

NOTE: Based on children ages 3 to 17 years with serum cotinine ≤ 10 ng/mL.

† The "All Other Races/Ethnicities" category includes all other races or ethnicities not specified, together with those individuals who report more than one race.

‡ Includes sampled individuals for whom income information is missing.

*The estimate should be interpreted with caution because the standard error of the estimate is relatively large: the relative standard error, RSE, is at least 30% but is less than 40% (RSE = standard error divided by the estimate), or the RSE may be underestimated.

** Not available. The estimate is not reported because it has large uncertainty: the relative standard error, RSE, is 40% or greater (RSE = standard error divided by the estimate), or the RSE cannot be reliably estimated.

Table B4c: Cotinine in nonsmoking children ages 3 to 17 years: Median and 95th percentile concentrations in blood serum, by age group, 2007-2010

	Concentration of cotinine in serum (ng/mL)				
	Ages 3 to 17 years	Ages 3 to 5 years	Ages 6 to 10 years	Ages 11 to 15 years	Ages 16 to 17 years
Median	0.04	0.06	0.05	0.03	0.04
95th percentile	2.5	2.9	2.5	2.3	2.9

DATA: Centers for Disease Control and Prevention, National Center for Health Statistics and National Center for Environmental Health, National Health and Nutrition Examination Survey

Table B5: Cotinine in nonsmoking women ages 16 to 49 years: Median and 95th percentile concentrations in blood serum, 1988-2010

	Concentration of cotinine in serum (ng/mL)							
	1988-1991	1991-1994	1999-2000	2001-2002	2003-2004	2005-2006	2007-2008	2009-2010
Median	0.21	0.15	0.06	0.04	0.04	0.04	0.04	0.03
95th percentile	2.3	2.1	1.7	1.6	2.2	1.5	1.9	1.5

DATA: Centers for Disease Control and Prevention, National Center for Health Statistics and National Center for Environmental Health, National Health and Nutrition Examination Survey

NOTES:

- Based on women ages 16 to 49 years with serum cotinine ≤ 10 ng/mL.

- To reflect exposures to women who are pregnant or may become pregnant, the estimates are adjusted for the probability (by age and race/ethnicity) that a woman gives birth. The intent of this adjustment is to approximate the distribution of exposure to pregnant women. Results will therefore differ from a characterization of exposure to adult women without consideration of birth rates.

Table B5a. Cotinine in nonsmoking women ages 16 to 49 years: Median concentrations in blood serum, by race/ethnicity and family income, 2007-2010

Race / Ethnicity	Median concentration of cotinine in serum (ng/mL)		
	All Incomes‡ (n=2,601)	< Poverty Level (n=583)	≥ Poverty Level (n=1,781)
All Races/Ethnicities (n=2,601)	0.03	0.05	0.03
White non-Hispanic (n=949)	0.03	NA**	0.03
Black non-Hispanic (n=475)	0.10	0.33	0.06
Mexican-American (n=654)	0.03	0.04	0.02
All Other Races/Ethnicities† (n=523)	0.03	0.03*	0.03

DATA: Centers for Disease Control and Prevention, National Center for Health Statistics and National Center for Environmental Health, National Health and Nutrition Examination Survey

NOTES:

- Based on women ages 16 to 49 years with serum cotinine ≤ 10 ng/mL.

- To reflect exposures to women who are pregnant or may become pregnant, the estimates are adjusted for the probability (by age and race/ethnicity) that a woman gives birth. The intent of this adjustment is to approximate the distribution of exposure to pregnant women. Results will therefore differ from a characterization of exposure to adult women without consideration of birth rates.

† The "All Other Races/Ethnicities" category includes all other races or ethnicities not specified, together with those individuals who report more than one race.

‡ Includes sampled individuals for whom income information is missing.

*The estimate should be interpreted with caution because the standard error of the estimate is relatively large: the relative standard error, RSE, is at least 30% but is less than 40% (RSE = standard error divided by the estimate), or the RSE may be underestimated.

Table B5b. Cotinine in nonsmoking women ages 16 to 49 years: 95th percentile concentrations in blood serum, by race/ethnicity and family income, 2007-2010

Race / Ethnicity	95th percentile concentration of cotinine in serum (ng/mL)		
	All Incomes‡ (n=2,601)	< Poverty Level (n=583)	≥ Poverty Level (n=1,781)
All Races/Ethnicities (n=2,601)	1.6	3.5	1.4
White non-Hispanic (n=949)	1.4	3.7*	1.0
Black non-Hispanic (n=475)	3.4	8.3	3.0
Mexican-American (n=654)	1.6*	2.5	NA**
All Other Races/Ethnicities† (n=523)	NA**	NA**	NA**

DATA: Centers for Disease Control and Prevention, National Center for Health Statistics and National Center for Environmental Health, National Health and Nutrition Examination Survey

NOTES:

- Based on women ages 16 to 49 years with serum cotinine ≤ 10 ng/mL.

- To reflect exposures to women who are pregnant or may become pregnant, the estimates are adjusted for the probability (by age and race/ethnicity) that a woman gives birth. The intent of this adjustment is to approximate the distribution of exposure to pregnant women. Results will therefore differ from a characterization of exposure to adult women without consideration of birth rates.

† The "All Other Races/Ethnicities" category includes all other races or ethnicities not specified, together with those individuals who report more than one race.

‡ Includes sampled individuals for whom income information is missing.

*The estimate should be interpreted with caution because the standard error of the estimate is relatively large: the relative standard error, RSE, is at least 30% but is less than 40% (RSE = standard error divided by the estimate), or the RSE may be underestimated.

** Not available. The estimate is not reported because it has large uncertainty: the relative standard error, RSE, is 40% or greater (RSE = standard error divided by the estimate), or the RSE cannot be reliably estimated.

Perfluorochemicals (PFCs)

Table B6. Perfluorochemicals in women ages 16 to 49 years: Median concentrations in blood serum, 1999-2008

Year	Median concentration of PFCs in serum (ng/mL)			
	PFOS	PFOA	PFHxS	PFNA
1999-2000	23.8	4.6	1.3	0.5
2003-2004	14.6	3.0	1.4	0.8
2005-2006	11.6	2.9	1.2	0.8
2007-2008	8.7	3.2	1.1	1.2

DATA: Centers for Disease Control and Prevention, National Center for Health Statistics and National Center for Environmental Health, National Health and Nutrition Examination Survey

NOTES:

- PFOS = perfluorooctane sulfonic acid, PFOA = perfluorooctanoic acid, PFHxS = perfluorohexane sulfonic acid, and PFNA = perfluorononanoic acid.

- To reflect exposures to women who are pregnant or may become pregnant, the estimates are adjusted for the probability (by age and race/ethnicity) that a woman gives birth. The intent of this adjustment is to approximate the distribution of exposure to pregnant women. Results will therefore differ from a characterization of exposure to adult women without consideration of birth rates.

Table B6a. Perfluorochemicals in women ages 16 to 49 years: 95th percentile concentrations in blood serum, 1999-2008

Year	95th percentile concentration of PFCs in serum (ng/mL)			
	PFOS	PFOA	PFHxS	PFNA
1999-2000	50.1	8.4	4.9	1.3
2003-2004	42.2	8.4	7.1*	NA**
2005-2006	27.8	6.4	5.4	2.2
2007-2008	22.8	7.9	4.9	3.2

DATA: Centers for Disease Control and Prevention, National Center for Health Statistics and National Center for Environmental Health, National Health and Nutrition Examination Survey

NOTES:

- PFOS = perfluorooctane sulfonic acid, PFOA = perfluorooctanoic acid, PFHxS = perfluorohexane sulfonic acid, and PFNA = perfluorononanoic acid.

- To reflect exposures to women who are pregnant or may become pregnant, the estimates are adjusted for the probability (by age and race/ethnicity) that a woman gives birth. The intent of this adjustment is to approximate the distribution of exposure to pregnant women. Results will therefore differ from a characterization of exposure to adult women without consideration of birth rates.

*The estimate should be interpreted with caution because the standard error of the estimate is relatively large: the relative standard error, RSE, is at least 30% but is less than 40% (RSE = standard error divided by the estimate), or the RSE may be underestimated.

** Not available. The estimate is not reported because it has large uncertainty: the relative standard error, RSE, is 40% or greater (RSE = standard error divided by the estimate), or the RSE cannot be reliably estimated.

Table B6b. Perfluorochemicals in women ages 16 to 49 years: Median concentrations in blood serum, by race/ethnicity and family income, 2005-2008

	Race / Ethnicity	Median concentration of PFCs in serum (ng/mL)		
		All Incomes‡ (n=1,121)	< Poverty Level (n=278)	≥ Poverty Level (n=780)
PFOS	All Races/Ethnicities (n=1,121)	10.1	8.1	11.0
	White non-Hispanic (n=453)	11.4	8.1*	11.6
	Black non-Hispanic (n=255)	11.2	NA**	11.2
	Mexican-American (n=272)	7.4	8.1*	7.3
	All Other Races/Ethnicities† (n=141)	8.3	NA**	10.5
PFOA	All Races/Ethnicities (n=1,121)	3.1	2.7	3.2
	White non-Hispanic (n=453)	3.5	3.3*	3.5
	Black non-Hispanic (n=255)	2.7	NA**	2.6
	Mexican-American (n=272)	2.3	2.1*	2.4
	All Other Races/Ethnicities† (n=141)	2.4	NA**	2.6
PFHxS	All Races/Ethnicities (n=1,121)	1.2	1.0	1.2
	White non-Hispanic (n=453)	1.3	1.1*	1.3
	Black non-Hispanic (n=255)	1.1	NA**	1.1
	Mexican-American (n=272)	0.9	0.9*	1.0
	All Other Races/Ethnicities† (n=141)	0.8	NA**	1.1
PFNA	All Races/Ethnicities (n=1,121)	1.0	1.0	1.0
	White non-Hispanic (n=453)	1.1	1.0*	1.1
	Black non-Hispanic (n=255)	1.1	NA**	1.2
	Mexican-American (n=272)	0.8	0.9*	0.8
	All Other Races/Ethnicities† (n=141)	1.1	NA**	1.2

DATA: Centers for Disease Control and Prevention, National Center for Health Statistics and National Center for Environmental Health, National Health and Nutrition Examination Survey

NOTES:

- PFOS = perfluorooctane sulfonic acid, PFOA = perfluorooctanoic acid, PFHxS = perfluorohexane sulfonic acid, and PFNA = perfluorononanoic acid.

- To reflect exposures to women who are pregnant or may become pregnant, the estimates are adjusted for the probability (by age and race/ethnicity) that a woman gives birth. The intent of this adjustment is to approximate the distribution of exposure to pregnant women. Results will therefore differ from a characterization of exposure to adult women without consideration of birth rates.

† The "All Other Races/Ethnicities" category includes all other races or ethnicities not specified, together with those individuals who report more than one race.

‡ Includes sampled individuals for whom income information is missing.

*The estimate should be interpreted with caution because the standard error of the estimate is relatively large: the relative standard error, RSE, is at least 30% but is less than 40% (RSE = standard error divided by the estimate), or the RSE may be underestimated.

** Not available. The estimate is not reported because it has large uncertainty: the relative standard error, RSE, is 40% or greater (RSE = standard error divided by the estimate), or the RSE cannot be reliably estimated.

Table B6c. Perfluorochemicals in women ages 16 to 49 years: 95th percentile concentrations in blood serum, by race/ethnicity and family income, 2005-2008

	Race / Ethnicity	95th percentile concentration of PFCs in serum (ng/mL)		
		All Incomes‡ (n=1,121)	< Poverty Level (n=278)	≥ Poverty Level (n=780)
PFOS	All Races/Ethnicities (n=1,121)	25.7	22.6	27.8
	White non-Hispanic (n=453)	28.4	23.9*	28.4
	Black non-Hispanic (n=255)	25.7	NA**	25.7
	Mexican-American (n=272)	17.3	17.3*	16.4
	All Other Races/Ethnicities† (n=141)	24.9	NA**	31.0
PFOA	All Races/Ethnicities (n=1,121)	7.5	5.6	7.8
	White non-Hispanic (n=453)	8.1	5.8*	8.1
	Black non-Hispanic (n=255)	6.5	NA**	6.1
	Mexican-American (n=272)	5.5	5.4*	5.6
	All Other Races/Ethnicities† (n=141)	5.8	NA**	5.8
PFHxS	All Races/Ethnicities (n=1,121)	5.1	5.4	5.1
	White non-Hispanic (n=453)	5.6	6.0*	5.6
	Black non-Hispanic (n=255)	4.2	NA**	3.7
	Mexican-American (n=272)	4.6	4.0*	5.2
	All Other Races/Ethnicities† (n=141)	2.1	NA**	2.1
PFNA	All Races/Ethnicities (n=1,121)	2.8	2.7	2.6
	White non-Hispanic (n=453)	2.9	2.0*	2.7
	Black non-Hispanic (n=255)	NA**	NA**	2.3
	Mexican-American (n=272)	2.3	2.3*	2.0
	All Other Races/Ethnicities† (n=141)	2.8	NA**	2.8

DATA: Centers for Disease Control and Prevention, National Center for Health Statistics and National Center for Environmental Health, National Health and Nutrition Examination Survey

NOTES:

- PFOS = perfluorooctane sulfonic acid, PFOA = perfluorooctanoic acid, PFHxS = perfluorohexane sulfonic acid, and PFNA = perfluorononanoic acid.

- To reflect exposures to women who are pregnant or may become pregnant, the estimates are adjusted for the probability (by age and race/ethnicity) that a woman gives birth. The intent of this adjustment is to approximate the distribution of exposure to pregnant women. Results will therefore differ from a characterization of exposure to adult women without consideration of birth rates.

† The "All Other Races/Ethnicities" category includes all other races or ethnicities not specified, together with those individuals who report more than one race.

‡ Includes sampled individuals for whom income information is missing.

*The estimate should be interpreted with caution because the standard error of the estimate is relatively large: the relative standard error, RSE, is at least 30% but is less than 40% (RSE = standard error divided by the estimate), or the RSE may be underestimated.

** Not available. The estimate is not reported because it has large uncertainty: the relative standard error, RSE, is 40% or greater (RSE = standard error divided by the estimate), or the RSE cannot be reliably estimated.

Polychlorinated Biphenyls (PCBs)

Table B7. PCBs in women ages 16 to 49 years: Median concentrations in blood serum, by race/ethnicity and family income, 2001-2004

Race / Ethnicity	Median concentration of PCBs in serum (ng/g lipid)		
	All Incomes‡ (n=1,164)	< Poverty Level (n=299)	≥ Poverty Level (n=810)
All Races/Ethnicities (n=1,164)	30.1	25.8	31.8
White non-Hispanic (n=477)	33.6	29.0*	34.8
Black non-Hispanic (n=281)	32.2	30.3*	37.4
Mexican-American (n=305)	18.0	16.1*	18.9
All Other Races/Ethnicities† (n=101)	31.6	NA**	38.0*

DATA: Centers for Disease Control and Prevention, National Center for Health Statistics and National Center for Environmental Health, National Health and Nutrition Examination Survey

NOTES:

- Values below the limit of detection are assumed equal to the limit of detection divided by the square root of 2.

- To reflect exposures to women who are pregnant or may become pregnant, the estimates are adjusted for the probability (by age and race/ethnicity) that a woman gives birth. The intent of this adjustment is to approximate the distribution of exposure to pregnant women. Results will therefore differ from a characterization of exposure to adult women without consideration of birth rates.

† The "All Other Races/Ethnicities" category includes all other races or ethnicities not specified, together with those individuals who report more than one race.

‡ Includes sampled individuals for whom income information is missing.

*The estimate should be interpreted with caution because the standard error of the estimate is relatively large: the relative standard error, RSE, is at least 30% but is less than 40% (RSE = standard error divided by the estimate), or the RSE may be underestimated.

** Not available. The estimate is not reported because it has large uncertainty: the relative standard error, RSE, is 40% or greater (RSE = standard error divided by the estimate), or the RSE cannot be reliably estimated.

Table B7a. PCBs in women ages 16 to 49 years: 95th percentile concentrations in blood serum, by race/ethnicity and family income, 2001-2004

Race / Ethnicity	95th percentile concentration of PCBs in serum (ng/g lipid)		
	All Incomes‡ (n=1,164)	< Poverty Level (n=299)	≥ Poverty Level (n=810)
All Races/Ethnicities (n=1,164)	106.2	87.6	111.3
White non-Hispanic (n=477)	108.7	87.6*	114.6
Black non-Hispanic (n=281)	101.8	74.3*	118.0
Mexican-American (n=305)	49.1	NA**	58.1
All Other Races/Ethnicities† (n=101)	245.2	NA**	191.3*

DATA: Centers for Disease Control and Prevention, National Center for Health Statistics and National Center for Environmental Health, National Health and Nutrition Examination Survey

NOTES:

- Values below the limit of detection are assumed equal to the limit of detection divided by the square root of 2.

- To reflect exposures to women who are pregnant or may become pregnant, the estimates are adjusted for the probability (by age and race/ethnicity) that a woman gives birth. The intent of this adjustment is to approximate the distribution of exposure to pregnant women. Results will therefore differ from a characterization of exposure to adult women without consideration of birth rates.

† The "All Other Races/Ethnicities" category includes all other races or ethnicities not specified, together with those individuals who report more than one race.

‡ Includes sampled individuals for whom income information is missing.

*The estimate should be interpreted with caution because the standard error of the estimate is relatively large: the relative standard error, RSE, is at least 30% but is less than 40% (RSE = standard error divided by the estimate), or the RSE may be underestimated.

** Not available. The estimate is not reported because it has large uncertainty: the relative standard error, RSE, is 40% or greater (RSE = standard error divided by the estimate), or the RSE cannot be reliably estimated.

Polybrominated Diphenyl Ethers (PBDEs)

Table B8. PBDEs in women ages 16 to 49 years: Median concentrations in blood serum, by race/ethnicity and family income, 2003-2004

Race / Ethnicity	Median concentration of PBDEs in serum (ng/g lipid)
All Races/Ethnicities *(n=540)*	44.2
White non-Hispanic *(n=233)*	48.9
Black non-Hispanic *(n=132)*	47.6*
Mexican-American *(n=131)*	41.0*
All Other Races/Ethnicities† *(n=44)*	NA**
Income	
All Incomes‡ *(n=540)*	44.2
< Poverty Level *(n=156)*	41.8
≥ Poverty Level *(n=352)*	43.9

DATA: Centers for Disease Control and Prevention, National Center for Health Statistics and National Center for Environmental Health, National Health and Nutrition Examination Survey

NOTES:

- Values below the limit of detection are assumed equal to the limit of detection divided by the square root of 2.

- To reflect exposures to women who are pregnant or may become pregnant, the estimates are adjusted for the probability (by age and race/ethnicity) that a woman gives birth. The intent of this adjustment is to approximate the distribution of exposure to pregnant women. Results will therefore differ from a characterization of exposure to adult women without consideration of birth rates.

† The "All Other Races/Ethnicities" category includes all other races or ethnicities not specified, together with those individuals who report more than one race.

‡ Includes sampled individuals for whom income information is missing.

*The estimate should be interpreted with caution because the standard error of the estimate is relatively large: the relative standard error, RSE, is at least 30% but is less than 40% (RSE = standard error divided by the estimate), or the RSE may be underestimated.

** Not available. The estimate is not reported because it has large uncertainty: the relative standard error, RSE, is 40% or greater (RSE = standard error divided by the estimate), or the RSE cannot be reliably estimated.

Table B8a. PBDEs in children ages 12 to 17 years: Median concentrations in blood serum, by race/ethnicity and family income, 2003-2004

Race / Ethnicity	Median concentration of PBDEs in serum (ng/g lipid)
All Races/Ethnicities *(n=464)*	52.9
White non-Hispanic *(n=114)*	47.5*
Black non-Hispanic *(n=176)*	50.4*
Mexican-American *(n=145)*	62.9*
All Other Races/Ethnicities† *(n=29)*	NA**
Income	
All Incomes‡ *(n=464)*	52.9
< Poverty Level *(n=147)*	62.6
≥ Poverty Level *(n=304)*	49.8

DATA: Centers for Disease Control and Prevention, National Center for Health Statistics and National Center for Environmental Health, National Health and Nutrition Examination Survey

NOTE: Values below the limit of detection are assumed equal to the limit of detection divided by the square root of 2.

† The "All Other Races/Ethnicities" category includes all other races or ethnicities not specified, together with those individuals who report more than one race.

‡ Includes sampled individuals for whom income information is missing.

*The estimate should be interpreted with caution because the standard error of the estimate is relatively large: the relative standard error, RSE, is at least 30% but is less than 40% (RSE = standard error divided by the estimate), or the RSE may be underestimated.

** Not available. The estimate is not reported because it has large uncertainty: the relative standard error, RSE, is 40% or greater (RSE = standard error divided by the estimate), or the RSE cannot be reliably estimated.

Phthalates

Table B9: Phthalate metabolites in women ages 16 to 49 years: Median concentrations in urine, 1999-2008

	Median concentration of phthalate metabolites in urine (µg/L)				
	1999-2000	**2001-2002**	**2003-2004**	**2005-2006**	**2007-2008**
DEHP metabolites	Ø	41.9	44.5	40.6	50.6
DBP metabolites[1]	32.6	26.7	32.1	32.2	36.3
BBzP metabolite	13.8	13.6	12.3	9.9	12.4

DATA: Centers for Disease Control and Prevention, National Center for Health Statistics and National Center for Environmental Health, National Health and Nutrition Examination Survey

NOTES:

- DEHP = di-2-ethylhexyl phthalate, DBP = dibutyl phthalate(di-n-butyl phthalate and di-isobutyl phthalate), and BBzP = butyl benzyl phthalate.

- Values below the limit of detection are assumed equal to the limit of detection divided by the square root of 2.

- To reflect exposures to women who are pregnant or may become pregnant, the estimates are adjusted for the probability (by age and race/ethnicity) that a woman gives birth. The intent of this adjustment is to approximate the distribution of exposure to pregnant women. Results will therefore differ from a characterization of exposure to adult women without consideration of birth rates.

Ø The estimate is not reported because the DEHP metabolites MEOHP and MEHHP were not measured in 1999-2000.

---

[1] The primary urinary metabolites of DBP (di-n-butyl phthalate and di-isobutyl phthalate) are mono-n-butyl phthalate (MnBP) and mono-isobutyl phthalate (MiBP). The urinary levels of MnBP and MiBP were measured together for the NHANES 1999–2000 survey cycle, but for the following years were measured separately. Indicators B9 and B10 present the combined urinary levels of MnBP and MiBP for each survey cycle.

Table B9a: Phthalate metabolites in women ages 16 to 49 years: 95th percentile concentrations in urine, 1999-2008

	95th percentile concentration of phthalate metabolites in urine (µg/L)				
	1999-2000	2001-2002	2003-2004	2005-2006	2007-2008
DEHP metabolites	Ø	577.9*	462.2	521.3*	567.2
DBP metabolites	NA**	128.2	139.6	144.9	160.2
BBzP metabolite	73.9	99.7	67.8	67.5	70.5

DATA: Centers for Disease Control and Prevention, National Center for Health Statistics and National Center for Environmental Health, National Health and Nutrition Examination Survey

NOTES:

- DEHP = di-2-ethylhexyl phthalate, DBP = dibutyl phthalate (di-n-butyl phthalate and di-isobutyl phthalate), and BBzP = butyl benzyl phthalate.

- Values below the limit of detection are assumed equal to the limit of detection divided by the square root of 2.

- To reflect exposures to women who are pregnant or may become pregnant, the estimates are adjusted for the probability (by age and race/ethnicity) that a woman gives birth. The intent of this adjustment is to approximate the distribution of exposure to pregnant women. Results will therefore differ from a characterization of exposure to adult women without consideration of birth rates.

- Phthalates do not accumulate in bodily tissues; thus, the distribution of NHANES urinary phthalate metabolite levels may overestimate high-end exposures as a result of collecting one-time urine samples rather than collecting urine for a longer time period.[5-7]

*The estimate should be interpreted with caution because the standard error of the estimate is relatively large: the relative standard error, RSE, is at least 30% but is less than 40% (RSE = standard error divided by the estimate), or the RSE may be underestimated.

** Not available. The estimate is not reported because it has large uncertainty: the relative standard error, RSE, is 40% or greater (RSE = standard error divided by the estimate), or the RSE cannot be reliably estimated.

Ø The estimate is not reported because the DEHP metabolites MEOHP and MEHHP were not measured in 1999-2000.

Table B9b. Phthalate metabolites in women ages 16 to 49 years: Median concentrations in urine by race/ethnicity and family income, 2005-2008

	Race / Ethnicity	Median concentration of phthalate metabolites in urine (μg/L)		
		All Incomes‡ (n=1,187)	< Poverty Level (n=289)	≥ Poverty Level (n=824)
DEHP metabolites	All Races/Ethnicities (n=1,187)	43.9	48.0	41.7
	White non-Hispanic (n=456)	46.5	NA**	41.7
	Black non-Hispanic (n=291)	58.0	49.8*	65.2
	Mexican-American (n=283)	35.5	44.8*	32.3
	All Other Races/Ethnicities† (n=157)	43.3	40.5*	44.6
DBP metabolites	All Races/Ethnicities (n=1,187)	33.2	37.3	31.9
	White non-Hispanic (n=456)	29.9	38.6*	27.5
	Black non-Hispanic (n=291)	48.3	41.2*	51.6
	Mexican-American (n=283)	39.9	32.0*	46.5
	All Other Races/Ethnicities† (n=157)	31.4	29.5*	31.4
BBzP metabolite	All Races/Ethnicities (n=1,187)	10.9	13.3	10.4
	White non-Hispanic (n=456)	10.7	13.4*	10.4
	Black non-Hispanic (n=291)	14.3	14.5*	14.2
	Mexican-American (n=283)	11.5	10.7*	12.0
	All Other Races/Ethnicities† (n=157)	5.8*	11.9*	5.3*

DATA: Centers for Disease Control and Prevention, National Center for Health Statistics and National Center for Environmental Health, National Health and Nutrition Examination Survey

NOTES:

- DEHP = di-2-ethylhexyl phthalate, DBP = dibutyl phthalate (di-n-butyl phthalate and di-isobutyl phthalate), and BBzP = butyl benzyl phthalate.

- Values below the limit of detection are assumed equal to the limit of detection divided by the square root of 2.

- To reflect exposures to women who are pregnant or may become pregnant, the estimates are adjusted for the probability (by age and race/ethnicity) that a woman gives birth. The intent of this adjustment is to approximate the distribution of exposure to pregnant women. Results will therefore differ from a characterization of exposure to adult women without consideration of birth rates.

† The "All Other Races/Ethnicities" category includes all other races or ethnicities not specified, together with those individuals who report more than one race.

‡ Includes sampled individuals for whom income information is missing.

*The estimate should be interpreted with caution because the standard error of the estimate is relatively large: the relative standard error, RSE, is at least 30% but is less than 40% (RSE = standard error divided by the estimate), or the RSE may be underestimated.

** Not available. The estimate is not reported because it has large uncertainty: the relative standard error, RSE, is 40% or greater (RSE = standard error divided by the estimate), or the RSE cannot be reliably estimated.

Table B10: Phthalate metabolites in children ages 6 to 17 years: Median concentrations in urine, 1999-2008

	Median concentration of phthalate metabolites in urine (µg/L)				
	1999-2000	2001-2002	2003-2004	2005-2006	2007-2008
DEHP metabolites	Ø	56.9	59.5	62.4	45.2
DBP metabolites	37.9	36.3	39.7	41.8	41.3
BBzP metabolite	24.8	22.4	22.1	18.5	16.3

DATA: Centers for Disease Control and Prevention, National Center for Health Statistics and National Center for Environmental Health, National Health and Nutrition Examination Survey

NOTES:

- DEHP = di-2-ethylhexyl phthalate, DBP = dibutyl phthalate (di-n-butyl phthalate and di-isobutyl phthalate), and BBzP = butyl benzyl phthalate.

- Values below the limit of detection are assumed equal to the limit of detection divided by the square root of 2.

Ø The estimate is not reported because the DEHP metabolites MEOHP and MEHHP were not measured in 1999-2000.

Table B10a: Phthalate metabolites in children ages 6 to 17 years: 95th percentile concentrations in urine, 1999-2008

	95th percentile concentration of phthalate metabolites in urine (µg/L)				
	1999-2000	2001-2002	2003-2004	2005-2006	2007-2008
DEHP metabolites	Ø	387.4	455.6	524.5	563.9
DBP metabolites	165.7	175.1	191.4	166.2	190.9
BBzP metabolite	122.3	143.1	151.1	104.0	107.1

DATA: Centers for Disease Control and Prevention, National Center for Health Statistics and National Center for Environmental Health, National Health and Nutrition Examination Survey

NOTES:

- DEHP = di-2-ethylhexyl phthalate, DBP = dibutyl phthalate (di-n-butyl phthalate and di-isobutyl phthalate), and BBzP = butyl benzyl phthalate.

- Values below the limit of detection are assumed equal to the limit of detection divided by the square root of 2.

- Phthalates do not accumulate in bodily tissues; thus, the distribution of NHANES urinary phthalate metabolite levels may overestimate high-end exposures as a result of collecting one-time urine samples rather than collecting urine for a longer time period.[5-7]

Ø The estimate is not reported because the DEHP metabolites MEOHP and MEHHP were not measured in 1999-2000.

Table B10b. Phthalate metabolites in children ages 6 to 17 years: Median concentrations in urine, by race/ethnicity and family income, 2005-2008

	Race / Ethnicity	Median concentration of phthalate metabolites in urine (µg/L)		
		All Incomes‡ (n=1,586)	< Poverty Level (n=453)	≥ Poverty Level (n=1,056)
DEHP metabolites	All Races/Ethnicities (n=1,586)	54.0	57.3	53.0
	White non-Hispanic (n=435)	57.9	58.9*	58.1
	Black non-Hispanic (n=465)	55.1	52.2*	56.9
	Mexican-American (n=487)	44.4	53.4	38.4
	All Other Races/Ethnicities† (n=199)	48.1	NA**	47.7
DBP metabolites	All Races/Ethnicities (n=1,586)	41.8	49.9	40.2
	White non-Hispanic (n=435)	40.9	54.1*	40.3
	Black non-Hispanic (n=465)	47.2	46.6*	48.2
	Mexican-American (n=487)	38.5	43.5	33.7
	All Other Races/Ethnicities† (n=199)	41.0	NA**	39.8
BBzP metabolite	All Races/Ethnicities (n=1,586)	17.4	18.7	17.1
	White non-Hispanic (n=435)	18.2	19.4*	18.0
	Black non-Hispanic (n=465)	19.0	23.1*	18.4
	Mexican-American (n=487)	13.7	14.1	13.3
	All Other Races/Ethnicities† (n=199)	15.2	NA**	12.7

DATA: Centers for Disease Control and Prevention, National Center for Health Statistics and National Center for Environmental Health, National Health and Nutrition Examination Survey

NOTES:

- DEHP = di-2-ethylhexyl phthalate, DBP = dibutyl phthalate (di-n-butyl phthalate and di-isobutyl phthalate), and BBzP = butyl benzyl phthalate.

- Values below the limit of detection are assumed equal to the limit of detection divided by the square root of 2.

† The "All Other Races/Ethnicities" category includes all other races or ethnicities not specified, together with those individuals who report more than one race.

‡ Includes sampled individuals for whom income information is missing.

*The estimate should be interpreted with caution because the standard error of the estimate is relatively large: the relative standard error, RSE, is at least 30% but is less than 40% (RSE = standard error divided by the estimate), or the RSE may be underestimated.

** Not available. The estimate is not reported because it has large uncertainty: the relative standard error, RSE, is 40% or greater (RSE = standard error divided by the estimate), or the RSE cannot be reliably estimated.

Table B10c: Phthalate metabolites in children ages 6 to 17 years: Median concentrations in urine by age group, 2005-2008

	Median concentration of phthalate metabolites in urine (µg/L)			
	Ages 6 to 17 years	Ages 6 to 10 years	Ages 11 to 15 years	Ages 16 to 17 years
DEHP metabolites	54.0	57.1	53.6	51.1
DBP metabolites	41.8	41.4	43.8	38.2
BBzP metabolite	17.4	20.1	16.5	13.5

DATA: Centers for Disease Control and Prevention, National Center for Health Statistics and National Center for Environmental Health, National Health and Nutrition Examination Survey

NOTES:

- DEHP = di-2-ethylhexyl phthalate, DBP = dibutyl phthalate (di-n-butyl phthalate and di-isobutyl phthalate), and BBzP = butyl benzyl phthalate.
- Values below the limit of detection are assumed equal to the limit of detection divided by the square root of 2.

Bisphenol A (BPA)

Table B11: Bisphenol A in women ages 16 to 49 years: Median and 95th percentile concentrations in urine, 2003-2010

	Concentration of BPA in urine (µg/L)			
	2003-2004	2005-2006	2007-2008	2009-2010
Median	3.1	2.0	2.5	2.1
95th percentile	15.9	9.8	15.1	9.7

DATA: Centers for Disease Control and Prevention, National Center for Health Statistics and National Center for Environmental Health, National Health and Nutrition Examination Survey

NOTES:

- To reflect exposures to women who are pregnant or may become pregnant, the estimates are adjusted for the probability (by age and race/ethnicity) that a woman gives birth. The intent of this adjustment is to approximate the distribution of exposure to pregnant women. Results will therefore differ from a characterization of exposure to adult women without consideration of birth rates.
- BPA does not appear to accumulate in bodily tissues; thus the distribution of NHANES urinary BPA levels may overestimate high-end exposures as a result of collecting one-time urine samples rather than collecting urine for a longer time period.[6-8]

Table B11a: Bisphenol A in women ages 16 to 49 years: Median concentrations in urine, by race/ethnicity and family income, 2007-2010

	Median concentration of BPA in urine (µg/L)		
Race / Ethnicity	All Incomes‡ (n=1,179)	< Poverty Level (n=329)	≥ Poverty Level (n=755)
All Races/Ethnicities (n=1,179)	2.3	3.0	2.1
White non-Hispanic (n=499)	2.1	3.3	2.0
Black non-Hispanic (n=242)	3.7	3.3*	4.2
Mexican-American (n=227)	2.3	2.2*	2.3
All Other Races/Ethnicities† (n=211)	2.1	3.1*	1.8

DATA: Centers for Disease Control and Prevention, National Center for Health Statistics and National Center for Environmental Health, National Health and Nutrition Examination Survey

NOTES:

- To reflect exposures to women who are pregnant or may become pregnant, the estimates are adjusted for the probability (by age and race/ethnicity) that a woman gives birth. The intent of this adjustment is to approximate the distribution of exposure to pregnant women. Results will therefore differ from a characterization of exposure to adult women without consideration of birth rates.
- The reported measurements of BPA in urine include both BPA itself and biologically inactive metabolites of BPA.

† The "All Other Races/Ethnicities" category includes all other races or ethnicities not specified, together with those individuals who report more than one race.

‡ Includes sampled individuals for whom income information is missing.

*The estimate should be interpreted with caution because the standard error of the estimate is relatively large: the relative standard error, RSE, is at least 30% but is less than 40% (RSE = standard error divided by the estimate), or the RSE may be underestimated.

Table B11b: Bisphenol A in women ages 16 to 49 years: 95th percentile concentrations in urine, by race/ethnicity and family income, 2007-2010

Race / Ethnicity	95th percentile concentration of BPA in urine (μg/L)		
	All Incomes‡ (n=1,179)	< Poverty Level (n=329)	≥ Poverty Level (n=755)
All Races/Ethnicities (n=1,179)	12.2	14.5	10.6
White non-Hispanic (n=499)	9.7	NA**	8.1
Black non-Hispanic (n=242)	15.1	14.8*	15.1
Mexican-American (n=227)	14.7	NA**	17.8
All Other Races/Ethnicities† (n=211)	NA**	23.0*	NA**

DATA: Centers for Disease Control and Prevention, National Center for Health Statistics and National Center for Environmental Health, National Health and Nutrition Examination Survey

NOTES:

- To reflect exposures to women who are pregnant or may become pregnant, the estimates are adjusted for the probability (by age and race/ethnicity) that a woman gives birth. The intent of this adjustment is to approximate the distribution of exposure to pregnant women. Results will therefore differ from a characterization of exposure to adult women without consideration of birth rates.

- The reported measurements of BPA in urine include both BPA itself and biologically inactive metabolites of BPA.

- BPA does not appear to accumulate in bodily tissues; thus the distribution of NHANES urinary BPA levels may overestimate high-end exposures as a result of collecting one-time urine samples rather than collecting urine for a longer time period.[6-8]

† The "All Other Races/Ethnicities" category includes all other races or ethnicities not specified, together with those individuals who report more than one race.

‡ Includes sampled individuals for whom income information is missing.

*The estimate should be interpreted with caution because the standard error of the estimate is relatively large: the relative standard error, RSE, is at least 30% but is less than 40% (RSE = standard error divided by the estimate), or the RSE may be underestimated.

** Not available. The estimate is not reported because it has large uncertainty: the relative standard error, RSE, is 40% or greater (RSE = standard error divided by the estimate), or the RSE cannot be reliably estimated.

Table B12: Bisphenol A in children ages 6 to 17 years: Median and 95th percentile concentrations in urine, 2003-2010

	Concentration of BPA in urine (µg/L)			
	2003-2004	2005-2006	2007-2008	2009-2010
Median	4.0	2.4	2.4	2.0
95th percentile	16.0	16.5	12.2	9.7

DATA: Centers for Disease Control and Prevention, National Center for Health Statistics and National Center for Environmental Health, National Health and Nutrition Examination Survey

NOTE: BPA does not appear to accumulate in bodily tissues; thus the distribution of NHANES urinary BPA levels may overestimate high-end exposures as a result of collecting one-time urine samples rather than collecting urine for a longer time period.[6-8]

Table B12a: Bisphenol A in children ages 6 to 17 years: Median concentrations in urine, by race/ethnicity and family income, 2007-2010

	Median concentration of BPA in urine (µg/L)		
	All Incomes‡	< Poverty Level	≥ Poverty Level
Race / Ethnicity	(n=1,417)	(n=426)	(n=873)
All Races/Ethnicities (n=1,417)	2.2	2.4	2.1
White non-Hispanic (n=425)	2.1	2.7*	2.0
Black non-Hispanic (n=343)	2.8	3.1*	2.7
Mexican-American (n=379)	2.1	2.0	2.2
All Other Races/Ethnicities† (n=270)	1.8	1.9*	2.0

DATA: Centers for Disease Control and Prevention, National Center for Health Statistics and National Center for Environmental Health, National Health and Nutrition Examination Survey

NOTE: The reported measurements of BPA in urine include both BPA itself and biologically inactive metabolites of BPA.

† The "All Other Races/Ethnicities" category includes all other races or ethnicities not specified, together with those individuals who report more than one race.

‡ Includes sampled individuals for whom income information is missing.

*The estimate should be interpreted with caution because the standard error of the estimate is relatively large: the relative standard error, RSE, is at least 30% but is less than 40% (RSE = standard error divided by the estimate), or the RSE may be underestimated.

Table B12b: Bisphenol A in children ages 6 to 17 years: 95th percentile concentrations in urine, by race/ethnicity and family income, 2007-2010

Race / Ethnicity	95th percentile concentration of BPA in urine (µg/L)		
	All Incomes‡ (n=1,417)	< Poverty Level (n=426)	≥ Poverty Level (n=873)
All Races/Ethnicities (n=1,417)	11.9	10.4	12.2
White non-Hispanic (n=425)	12.2	10.4*	12.2
Black non-Hispanic (n=343)	12.6	NA**	12.4
Mexican-American (n=379)	12.3	6.9	15.6*
All Other Races/Ethnicities† (n=270)	9.1	4.7*	9.1

DATA: Centers for Disease Control and Prevention, National Center for Health Statistics and National Center for Environmental Health, National Health and Nutrition Examination Survey

NOTES:

- The reported measurements of BPA in urine include both BPA itself and biologically inactive metabolites of BPA.

- BPA does not appear to accumulate in bodily tissues; thus the distribution of NHANES urinary BPA levels may overestimate high-end exposures as a result of collecting one-time urine samples rather than collecting urine for a longer time period.[6-8]

† The "All Other Races/Ethnicities" category includes all other races or ethnicities not specified, together with those individuals who report more than one race.

‡ Includes sampled individuals for whom income information is missing.

*The estimate should be interpreted with caution because the standard error of the estimate is relatively large: the relative standard error, RSE, is at least 30% but is less than 40% (RSE = standard error divided by the estimate), or the RSE may be underestimated.

** Not available. The estimate is not reported because it has large uncertainty: the relative standard error, RSE, is 40% or greater (RSE = standard error divided by the estimate), or the RSE cannot be reliably estimated.

Table B12c: Bisphenol A in children ages 6 to 17 years: Median and 95th percentile concentrations by age group, 2007-2010

	Concentration of BPA in urine (µg/L)			
	Ages 6 to 17 years	Ages 6 to 10 years	Ages 11 to 15 years	Ages 16 to 17 years
Median	2.2	2.1	2.2	2.2
95th percentile	11.9	10.4	12.2	12.2

DATA: Centers for Disease Control and Prevention, National Center for Health Statistics and National Center for Environmental Health, National Health and Nutrition Examination Survey

NOTES:

- The reported measurements of BPA in urine include both BPA itself and biologically inactive metabolites of BPA.

- BPA does not appear to accumulate in bodily tissues; thus the distribution of NHANES urinary BPA levels may overestimate high-end exposures as a result of collecting one-time urine samples rather than collecting urine for a longer time period.[6-8]

Perchlorate

Table B13. Perchlorate in women ages 16 to 49 years: Median and 95th percentile concentrations in urine, 2001-2008

	Concentration of perchlorate in urine (µg/L)			
	2001-2002	2003-2004	2005-2006	2007-2008
Median	3.3	2.9	3.2	3.4
95th percentile	15.0	NA**	13.0	16.5

DATA: Centers for Disease Control and Prevention, National Center for Health Statistics and National Center for Environmental Health, National Health and Nutrition Examination Survey

NOTES:

- To reflect exposures to women who are pregnant or may become pregnant, the estimates are adjusted for the probability (by age and race/ethnicity) that a woman gives birth. The intent of this adjustment is to approximate the distribution of exposure to pregnant women. Results will therefore differ from a characterization of exposure to adult women without consideration of birth rates.

- Perchlorate does not appear to accumulate in bodily tissues; thus, the distribution of NHANES urinary perchlorate levels may overestimate high-end exposures as a result of collecting one-time urine samples rather than collecting urine for a longer time period.[6,7,9]

** Not available. The estimate is not reported because it has large uncertainty: the relative standard error, RSE, is 40% or greater (RSE = standard error divided by the estimate), or the RSE cannot be reliably estimated.

Table B13a. Perchlorate in women ages 16 to 49 years: Median concentrations in urine, by race/ethnicity and family income, 2005-2008

	Median concentration of perchlorate in urine (µg/L)		
Race / Ethnicity	All Incomes‡ (n=3,529)	< Poverty Level (n=861)	≥ Poverty Level (n=2,453)
All Races/Ethnicities (n=3,529)	3.3	3.4	3.2
White Non-Hispanic (n=1,365)	3.2	3.2	3.2
Black Non-Hispanic (n=858)	3.5	3.5	3.5
Mexican-American (n=843)	3.6	3.5	3.5
All Other Races/Ethnicities† (n=463)	3.3	3.6	3.0

DATA: Centers for Disease Control and Prevention, National Center for Health Statistics and National Center for Environmental Health, National Health and Nutrition Examination Survey

NOTE: To reflect exposures to women who are pregnant or may become pregnant, the estimates are adjusted for the probability (by age and race/ethnicity) that a woman gives birth. The intent of this adjustment is to approximate the distribution of exposure to pregnant women. Results will therefore differ from a characterization of exposure to adult women without consideration of birth rates.

† The "All Other Races/Ethnicities" category includes all other races or ethnicities not specified, together with those individuals who report more than one race.

‡ Includes sampled individuals for whom income information is missing.

Table B13b. Perchlorate in women ages 16 to 49 years: 95th percentile concentrations in urine, by race/ethnicity and family income, 2005-2008

Race / Ethnicity	95th percentile concentration of perchlorate in urine (μg/L)		
	All Incomes‡ (n=3,529)	< Poverty Level (n=861)	≥ Poverty Level (n=2,453)
All Races/Ethnicities (n=3,529)	14.5	14.4	14.4
White Non-Hispanic (n=1,365)	13.2	14.2	13.2
Black Non-Hispanic (n=858)	16.5	14.5*	19.4
Mexican-American (n=843)	16.0	16.5	15.3
All Other Races/Ethnicities† (n=463)	14.7	13.1	14.7

DATA: Centers for Disease Control and Prevention, National Center for Health Statistics and National Center for Environmental Health, National Health and Nutrition Examination Survey

NOTES:

- To reflect exposures to women who are pregnant or may become pregnant, the estimates are adjusted for the probability (by age and race/ethnicity) that a woman gives birth. The intent of this adjustment is to approximate the distribution of exposure to pregnant women. Results will therefore differ from a characterization of exposure to adult women without consideration of birth rates.

- Perchlorate does not appear to accumulate in bodily tissues; thus, the distribution of NHANES urinary perchlorate levels may overestimate high-end exposures as a result of collecting one-time urine samples rather than collecting urine for a longer time period.[6,7,9]

† The "All Other Races/Ethnicities" category includes all other races or ethnicities not specified, together with those individuals who report more than one race.

‡ Includes sampled individuals for whom income information is missing.

*The estimate should be interpreted with caution because the standard error of the estimate is relatively large: the relative standard error, RSE, is at least 30% but is less than 40% (RSE = standard error divided by the estimate), or the RSE may be underestimated.

Table B13c. Perchlorate in children ages 6 to 17 years: Median and 95th percentile concentrations in urine, 2001-2008

	Concentration of perchlorate in urine (μg/L)			
	2001-2002	2003-2004	2005-2006	2007-2008
Median	4.9	4.5	4.6	4.8
95th percentile	15.0	16.0	14.9	18.6

DATA: Centers for Disease Control and Prevention, National Center for Health Statistics and National Center for Environmental Health, National Health and Nutrition Examination Survey

NOTE: Perchlorate does not appear to accumulate in bodily tissues; thus, the distribution of NHANES urinary perchlorate levels may overestimate high-end exposures as a result of collecting one-time urine samples rather than collecting urine for a longer time period.[6,7,9]

Table B13d. Perchlorate in children ages 6 to 17 years: Median concentrations in urine, by race/ethnicity and family income, 2005-2008

Race / Ethnicity	Median concentration of perchlorate in urine (μg/L)		
	All Incomes‡ (n=4,638)	< Poverty Level (n=1,294)	≥ Poverty Level (n=3,096)
All Races/Ethnicities (n=4,638)	4.7	4.5	4.8
White Non-Hispanic (n=1,282)	4.9	4.7	4.9
Black Non-Hispanic (n=1,383)	4.4	4.2	4.5
Mexican-American (n=1,397)	4.9	4.7	5.0
All Other Races/Ethnicities† (n=576)	4.1	3.9*	4.4

DATA: Centers for Disease Control and Prevention, National Center for Health Statistics and National Center for Environmental Health, National Health and Nutrition Examination Survey

† The "All Other Races/Ethnicities" category includes all other races or ethnicities not specified, together with those individuals who report more than one race.

‡ Includes sampled individuals for whom income information is missing.

*The estimate should be interpreted with caution because the standard error of the estimate is relatively large: the relative standard error, RSE, is at least 30% but is less than 40% (RSE = standard error divided by the estimate), or the RSE may be underestimated.

Table B13e. Perchlorate in children ages 6 to 17 years: 95th percentile concentrations in urine, by race/ethnicity and family income, 2005-2008

Race / Ethnicity	95th percentile concentration of perchlorate in urine (μg/L)		
	All Incomes‡ (n=4,638)	< Poverty Level (n=1,294)	≥ Poverty Level (n=3,096)
All Races/Ethnicities (n=4,638)	17.2	16.0	17.5
White Non-Hispanic (n=1,282)	17.6	16.0	17.7
Black Non-Hispanic (n=1,383)	17.5	15.4	17.7
Mexican-American (n=1,397)	15.6	15.9	15.4
All Other Races/Ethnicities† (n=576)	16.9	16.9*	16.9

DATA: Centers for Disease Control and Prevention, National Center for Health Statistics and National Center for Environmental Health, National Health and Nutrition Examination Survey

NOTE: Perchlorate does not appear to accumulate in bodily tissues; thus, the distribution of NHANES urinary perchlorate levels may overestimate high-end exposures as a result of collecting one-time urine samples rather than collecting urine for a longer time period.[6,7,9]

† The "All Other Races/Ethnicities" category includes all other races or ethnicities not specified, together with those individuals who report more than one race.

‡ Includes sampled individuals for whom income information is missing.

*The estimate should be interpreted with caution because the standard error of the estimate is relatively large: the relative standard error, RSE, is at least 30% but is less than 40% (RSE = standard error divided by the estimate), or the RSE may be underestimated.

Table B13f: Perchlorate in children ages 6 to 17 years: Median and 95th percentile concentrations by age group, 2005-2008

	Concentration of perchlorate in urine (µg/L)			
	Ages 6 to 17 years	Ages 6 to 10 years	Ages 11 to 15 years	Ages 16 to 17 years
Median	4.7	4.9	4.7	4.4
95th percentile	17.2	17.1	17.5	16.7

DATA: Centers for Disease Control and Prevention, National Center for Health Statistics and National Center for Environmental Health, National Health and Nutrition Examination Survey

NOTE: Perchlorate does not appear to accumulate in bodily tissues; thus, the distribution of NHANES urinary perchlorate levels may overestimate high-end exposures as a result of collecting one-time urine samples rather than collecting urine for a longer time period.[6,7,9]

Health

Respiratory Diseases

Table H1: Percentage of children ages 0 to 17 years with asthma, 1997-2010

1997-2003							
	1997	1998	1999	2000	2001	2002	2003
Asthma attack prevalence	5.5	5.3	5.3	5.5	5.7	5.8	5.5
Current asthma prevalence‡					8.7	8.3	8.5
2004-2009							
	2004	2005	2006	2007	2008	2009	2010
Asthma attack prevalence	5.5	5.2	5.6	5.2	5.6	5.5	5.7
Current asthma prevalence‡	8.5	8.9	9.3	9.1	9.4	9.6	9.4

DATA: Centers for Disease Control and Prevention, National Center for Health Statistics, National Health Interview Survey

NOTE: Data represent responses to the following survey questions: "Has a doctor or other health professional ever told you that <child's name> had asthma?" and if yes, "During the past 12 months, has <child's name> had an episode of asthma or an asthma attack?" For 2001-2010 the survey included the question: "Does <child's name> still have asthma?" Responses are provided by a parent or other knowledgeable household adult.

‡ This survey question was first asked in 2001.

Table H1a: Percentage of children ages 0 to 17 years with current asthma, 2001-2010, by sex

	2001	2002	2003	2004	2005
All children	8.7	8.3	8.5	8.5	8.9
Boys	9.9	9.5	9.5	10.2	10.0
Girls	7.5	7.2	7.5	6.7	7.8
	2006	2007	2008	2009	2010
All children	9.3	9.1	9.4	9.6	9.4
Boys	11.0	9.7	11.4	11.3	10.5
Girls	7.5	8.5	7.4	7.9	8.2

DATA: Centers for Disease Control and Prevention, National Center for Health Statistics, National Health Interview Survey

NOTE: Data represent responses to the following survey questions: "Has a doctor or other health professional ever told you that <child's name> had asthma?" and if yes, "During the past 12 months, has <child's name> had an episode of asthma or an asthma attack?" For 2001-2010 the survey included the question: "Does <child's name> still have asthma?" Responses are provided by a parent or other knowledgeable household adult.

Table H1b: Percentage of children ages 0 to 17 years with asthma, 1980-1996†

1980-1987							
	1980	1981	1982	1983	1984	1985	1986	1987
Asthma in the past 12 months	3.6	3.7	4.1	4.5	4.3	4.8	5.1	5.3
1988-1996							
	1988	1989	1990	1991	1992	1993	1994	1995	1996
Asthma in the past 12 months	5.0	6.1	5.8	6.4	6.3	7.2	6.9	7.5	6.2

DATA: Centers for Disease Control and Prevention, National Center for Health Statistics, National Health Interview Survey

† The survey questions for asthma changed in 1997; data before 1997 cannot be directly compared to data in 1997 and later, and are thus shown in this separate table. For 1980 to 1996, the asthma survey question was "Did <child's name> have asthma in the past 12 months?"

Table H1c: Percentage of children ages 0 to 17 years with current asthma who reported an asthma attack in the past 12 months, 2001-2010

2001-2008								
	2001	**2002**	**2003**	**2004**	**2005**	**2006**	**2007**	**2008**
Asthma attack prevalence among those with current asthma	61.7	64.9	62.7	61.2	56.7	56.1	54.8	57.2

2009-2010		
	2009	**2010**
Asthma attack prevalence among those with current asthma	53.9	58.3

DATA: Centers for Disease Control and Prevention, National Center for Health Statistics, National Health Interview Survey

NOTE: Data represent responses to the following survey questions: "Has a doctor or other health professional ever told you that <child's name> had asthma?" and if yes, "During the past 12 months, has <child's name> had an episode of asthma or an asthma attack?" For 2001-2010 the survey included the question: "Does <child's name> still have asthma?" Responses are provided by a parent or other knowledgeable household adult.

Table H2: Percentage of children ages 0 to 17 years reported to have current asthma by race/ethnicity and family income, 2007-2010

				≥ Poverty Level (Detail)	
Race / Ethnicity	All (n = 40,569)	< Poverty Level (n = 8,160)	≥Poverty Level (n = 32,409)	100-200% of Poverty Level (n = 9,603)	≥ 200% of Poverty Level (n = 22,806)
All Races/Ethnicities (n = 40,569)	9.4	12.2	8.7	9.9	8.2
White non-Hispanic (n = 17,692)	8.2	10.6	7.9	9.6	7.5
Black or African-American non-Hispanic (n = 6,628)	16.0	18.8	14.4	15.0	13.9
Asian non-Hispanic (n = 2,255)	6.8	NA**	7.2	5.7*	7.7
Hispanic (n = 12,343)	7.9	8.7	7.5	7.3	7.8
Mexican (n = 8,114)	7.0	6.6	7.2	6.5	8.0
Puerto Rican (n=1,116)	16.5	23.3	12.6	15.3	11.0
All Other Races† (n = 1,651)	12.4	15.5	11.4	14.8	9.7
American Indian or Alaska Native non-Hispanic (n = 219)	10.7	13.0*	NA**	NA**	NA**

DATA: Centers for Disease Control and Prevention, National Center for Health Statistics, National Health Interview Survey

† The "All Other Races" category includes all other races not specified, together with those individuals who report more than one race.

* The estimate should be interpreted with caution because the standard error of the estimate is relatively large: the relative standard error, RSE, is at least 30% but is less than 40% (RSE = standard error divided by the estimate).

** Not available. The estimate is not reported because it has large uncertainty: the relative standard error, RSE, is 40% or greater (RSE = standard error divided by the estimate).

Table H2a: Percentage of children ages 0 to 17 years reported to have current asthma by age and sex, 2007-2010

	Ages 0 to 17 years	Ages 0 to 5 years	Ages 6 to 10 years	Ages 11 to 17 years
All children	9.4	7.1	10.0	11.0
Boys	10.7	8.8	11.9	11.7
Girls	8.0	5.2	8.1	10.3

DATA: Centers for Disease Control and Prevention, National Center for Health Statistics, National Health Interview Survey

Table H3: Children's emergency room visits and hospitalizations for asthma and other respiratory causes, ages 0 to 17 years, 1996-2008

	Rate per 10,000 children			
1996-1999	1996	1997	1998	1999
Emergency Room Visits				
Asthma and all other respiratory causes	636.4	631.5	654.7	619.9
All respiratory causes other than asthma	521.9	519.4	530.3	515.4
Upper respiratory	408.4	409.3	426.0	403.0
Pneumonia or influenza	56.3	52.0	58.0	58.8
Other lower respiratory	57.2	58.0	46.3	53.6
Asthma	114.4	112.1	124.4	104.5
Hospitalizations				
Asthma and all other respiratory causes	90.3	102.2	86.3	101.4
All respiratory causes other than asthma	59.9	69.1	61.4	72.5
Upper respiratory	28.9	37.2	27.6	39.5
Pneumonia or influenza	29.6	30.6	33.1	32.0
Other lower respiratory	1.4	1.3	0.7	1.0
Asthma	30.4	33.1	25.0	28.9

	Rate per 10,000 children			
2000-2003	2000	2001	2002	2003
Emergency Room Visits				
Asthma and all other respiratory causes	622.7	624.0	721.1	740.2
All respiratory causes other than asthma	521.8	532.3	621.3	644.8
Upper respiratory	428.1	426.8	494.4	499.1
Pneumonia or influenza	54.1	63.3	79.8	94.3
Other lower respiratory	39.7	42.2	47.1	51.5
Asthma	100.9	91.7	99.9	95.4
Hospitalizations				
Asthma and all other respiratory causes	84.6	85.0	86.7	89.6
All respiratory causes other than asthma	57.3	61.0	62.1	61.1
Upper respiratory	32.5	33.7	33.6	29.8
Pneumonia or influenza	23.9	26.6	27.8	30.2
Other lower respiratory	1.0	NA**	0.6	1.2
Asthma	27.2	24.0	24.6	28.4

2004-2007	Rate per 10,000 children			
	2004	2005	2006	2007
Emergency Room Visits				
Asthma and all other respiratory causes	528.8	639.8	584.3	625.1
All respiratory causes other than asthma	426.0	537.8	504.1	538.5
Upper respiratory	331.6	441.3	396.9	416.2
Pneumonia or influenza	56.9	62.6	61.1	87.6
Other lower respiratory	37.4	33.9	46.1	34.6
Asthma	102.8	102.1	80.2	86.6
Hospitalizations				
Asthma and all other respiratory causes	80.4	72.8	66.3	61.4
All respiratory causes other than asthma	55.8	52.5	47.3	42.3
Upper respiratory	30.5	25.8	23.5	23.1
Pneumonia or influenza	24.2	26.4	22.9	18.9
Other lower respiratory	1.1	0.4*	0.9	NA**
Asthma	24.6	20.3	18.9	19.1

2008	Rate per 10,000 children
	2008
Emergency Room Visits	
Asthma and all other respiratory causes	619.1
All respiratory causes other than asthma	516.6
Upper respiratory	388.2
Pneumonia or influenza	91.3
Other lower respiratory	37.1
Asthma	102.6
Hospitalizations	
Asthma and all other respiratory causes	56.0
All respiratory causes other than asthma	39.9
Upper respiratory	19.1
Pneumonia or influenza	20.3
Other lower respiratory	NA**
Asthma	16.2

DATA: Centers for Disease Control and Prevention, National Center for Health Statistics, National Hospital Ambulatory Medical Care Survey and National Hospital Discharge Survey

* The estimate should be interpreted with caution because the standard error of the estimate is relatively large: the relative standard error, RSE, is at least 30% but is less than 40% (RSE = standard error divided by the estimate).

** Not available. The estimate is not reported because it has large uncertainty: the relative standard error, RSE, exceeds 40% (RSE = standard error divided by the estimate) or there are fewer than 30 sampled hospitalizations.

Table H3a: Children's emergency room visits for asthma and other respiratory causes, by race/ethnicity, ages 0 to 17 years, 2005-2008

	Rate per 10,000 children				
	2005	2006	2007	2008	2005-2008
All Races/Ethnicities (n=5,366)	639.8	584.3	625.1	619.1	617.1
White non-Hispanic (n=2,248)	484.8	442.3	518.8	500.9	486.6
Black non-Hispanic (n=1,557)	1,242.7	1,276.0	1,183.5	1,258.0	1,240.1
American Indian/Alaska Native non-Hispanic (n=33)	NA**	NA**	NA**	NA**	536.2
Asian and Pacific Islander non-Hispanic (n=179)	409.4*	404.7	341.8*	333.1*	371.4
Hispanic (n=1,331)	788.9	600.4	656.4	646.7	671.5

DATA: Centers for Disease Control and Prevention, National Center for Health Statistics, National Hospital Ambulatory Medical Care Survey

* The estimate should be interpreted with caution because the standard error of the estimate is relatively large: the relative standard error, RSE, is at least 30% but is less than 40% (RSE = standard error divided by the estimate).

** Not available. The estimate is not reported because it has large uncertainty: the relative standard error, RSE, exceeds 40% (RSE = standard error divided by the estimate) or there are fewer than 30 sampled emergency room visits.

Table H3b: Children's emergency room visits for asthma and other respiratory causes, by age, 2005-2008

	Rate per 10,000 children				
	2005	2006	2007	2008	2005-2008
Ages 0 to 17 years	639.8	584.3	625.1	619.1	617.1
Age < 1 year	2,344.8	2,040.5	2,098.3	2,090.4	2,142.1
Age 1 year	1,884.3	1,696.4	1,823.1	1,727.5	1,782.3
Age 2 years	1,081.9	957.2	1,015.0	972.7	1,006.3
Ages 3 to 5 years	778.4	668.1	719.8	751.9	729.5
Ages 6 to 10 years	391.6	384.1	389.5	382.7	387.0
Ages 11 to 15 years	252.6	251.0	276.7	268.3	262.0
Ages 16 to 17 years	333.2	310.2	362.9	346.1	338.2

DATA: Centers for Disease Control and Prevention, National Center for Health Statistics, National Hospital Ambulatory Medical Care Survey

Table H3c: Children's hospitalizations for asthma and other respiratory causes, by race,† ages 0 to 17 years, 2005-2008

	Rate per 10,000 children				
	2005	2006	2007	2008	2005-2008
All Races‡ (n=18,088)	72.8	66.3	61.4	56.0	64.1
White (n=9,213)	61.7	56.5	47.7	42.7	52.1
Black (n=4,154)	94.1	91.6	78.0	72.3	84.0

DATA: Centers for Disease Control and Prevention, National Center for Health Statistics, National Hospital Discharge Survey

† Estimates for ethnicity not available. Race categories include children of Hispanic ethnicity.

‡ Includes races other than White and Black.

Table H3d: Children's hospitalizations for asthma and other respiratory causes, by age, 2005-2008

	Rate per 10,000 children				
	2005	**2006**	**2007**	**2008**	**2005-2008**
Ages 0 to 17 years	72.8	66.3	61.4	56.0	64.1
Age < 1 year	477.2	399.6	364.8	344.3	395.5
Age 1 year	232.7	211.9	173.5	152.2	191.9
Age 2 years	115.9	112.2	117.9	89.7	108.8
Ages 3 to 5 years	70.1	68.2	53.9	53.3	61.3
Ages 6 to 10 years	33.0	28.8	29.0	27.6	29.6
Ages 11 to 15 years	15.3	13.8	17.2	13.1	14.9
Ages 16 to 17 years	8.7	15.0	13.9	14.1	12.9

DATA: Centers for Disease Control and Prevention, National Center for Health Statistics, National Hospital Discharge Survey

Childhood Cancer

Table H4: Cancer incidence and mortality for children ages 0 to 19 years, 1992-2009

1992-1997						
	Age-adjusted rate per million children					
	1992	**1993**	**1994**	**1995**	**1996**	**1997**
Incidence	158.4	161.4	153.3	154.8	160.9	154.4
Mortality	33.1	32.6	31.2	29.8	28.7	28.8
1998-2003						
	Age-adjusted rate per million children					
	1998	**1999**	**2000**	**2001**	**2002**	**2003**
Incidence	164.1	158.0	162.3	166.6	171.9	156.6
Mortality	27.5	28.0	28.2	27.5	28.0	27.4
2004-2009						
	Age-adjusted rate per million children					
	2004	**2005**	**2006**	**2007**	**2008**	**2009**
Incidence	167.2	174.6	157.5	172.3	172.5	175.4
Mortality	27.2	26.6	24.6	24.9	24.4	23.5

DATA: National Cancer Institute, Surveillance, Epidemiology, and End Results (SEER) Program

Table H4a: Cancer incidence for children ages 0 to 19 years by race/ethnicity and sex, 2007-2009

	Age-adjusted rate per million children		
	Male	**Female**	**All**
All Races/Ethnicities *(n=5,974)*	183.3	163.0	173.4
White non-Hispanic *(n=2,963)*	199.7	175.2	187.8
Black non-Hispanic *(n=574)*	137.2	129.5	133.4
American Indian/Alaska Native non-Hispanic *(n=69)*	120.8	152.6	136.8
Asian or Pacific Islander non-Hispanic *(n=560)*	155.7	147.5	151.7
Hispanic *(n=1,717)*	181.9	156.1	169.3

DATA: National Cancer Institute, Surveillance, Epidemiology, and End Results (SEER) Program

Table H4b: Cancer mortality for children ages 0 to 19 years by race/ethnicity and sex, 2007-2009

	Age-adjusted rate per million children		
	Male	Female	All
All Races/Ethnicities *(n=6,071)*	26.1	22.4	24.3
White non-Hispanic *(n=3,384)*	25.8	21.7	23.8
Black non-Hispanic *(n=900)*	25.6	23.4	24.5
American Indian/Alaska Native non-Hispanic *(n=55)*	23.6	20.5	22.1
Asian or Pacific Islander non-Hispanic *(n=248)*	26.0	16.5	21.3
Hispanic *(n=1,386)*	27.9	25.4	26.7

DATA: National Cancer Institute, Surveillance, Epidemiology, and End Results (SEER) Program

Following the recommendations of the National Cancer Institute, the mortality rates for all the groups except for "All races/ethnicities" excluded data from the following states, which had large numbers with unknown ethnicity: North Dakota and South Carolina. See http://seer.cancer.gov/seerstat/variables/mort/origin_recode_1990+/index.html.

Table H4c: Cancer incidence for children 0 to 19 years by age, 2007-2009

	Age-adjusted rate per million children
Ages 0 to 4 years	207.6
Ages 5 to 9 years	116.9
Ages 10 to 14 years	139.0
Ages 15 to 19 years	232.3
Ages 0 to 19 years	173.4

DATA: National Cancer Institute, Surveillance, Epidemiology, and End Results (SEER) Program

Table H5: Cancer incidence for children ages 0 to 19 years, by type, 1992-2006

	Age-adjusted rate per million children				
	1992-1994	1995-1997	1998-2000	2001-2003	2004-2006
Acute lymphoblastic leukemia	29.5	32.3	33.4	32.4	34.5
Central nervous system tumors	28.7	26.8	26.9	29.6	27.0
Germ cell tumors	11.3	11.5	10.8	12.0	12.6
Soft tissue sarcomas	10.2	11.5	12.0	11.5	12.3
Hodgkin's lymphoma	12.3	11.6	12.2	11.2	10.8
Acute myeloid leukemia	7.3	7.7	8.3	8.0	8.5
Non-Hodgkin's lymphoma	7.4	7.2	7.7	9.0	8.8
Neuroblastoma	7.4	7.7	6.9	7.3	8.0
Malignant melanoma	4.4	4.7	4.7	5.8	5.7
Thyroid carcinoma	5.2	5.2	6.2	6.1	5.5
Osteosarcoma	4.9	4.8	4.8	5.3	4.5
Wilms' tumor	5.7	5.8	5.5	4.7	4.4
Other and unspecified carcinomas†	3.8	3.9	3.9	3.6	3.3
Ewing's sarcoma	3.2	2.3	2.2	2.5	2.8
Burkitt's lymphoma	2.0	1.9	2.3	2.4	2.2
Hepatoblastoma	1.1	1.2	1.8	1.5	1.7

DATA: National Cancer Institute, Surveillance, Epidemiology, and End Results (SEER) Program

† "Other and unspecified carcinomas" represents all carcinomas and other malignant epithelial neoplasms other than thyroid carcinoma and malignant melanoma.

Table H5a: Cancer incidence rates per million children for malignant cancers by age and type, 2004-2006

	Age-adjusted rate per million children				
	Ages 0 to 4 years	Ages 5 to 9 years	Ages 10 to 14 years	Ages 15 to 19 years	Ages 0 to 19 years
Acute lymphoblastic leukemia	66.3	33.6	22.8	17.0	34.5
Central nervous system tumors	35.1	30.4	23.2	19.7	27.0
Germ cell tumors	7.5	2.9	9.2	30.8	12.6
Soft-tissue sarcomas	11.1	7.1	12.8	18.3	12.3
Hodgkin's lymphoma	NA**	4.1	12.0	26.0	10.8
Acute myeloid leukemia	13.2	4.6	7.3	9.0	8.5
Non-Hodgkin's lymphoma	3.2	5.2	10.5	15.9	8.8
Neuroblastoma	28.5	2.8	1.4	NA**	8.0
Malignant melanoma	0.9*	1.7	4.3	15.5	5.7
Thyroid carcinoma	NA**	1.6	4.6	15.5	5.5
Osteosarcoma	NA**	2.6	7.3	7.9	4.5
Wilms' tumor	13.4	3.8	NA**	NA**	4.4
Other and unspecified carcinomas†	NA**	NA**	3.5	9.0	3.3
Ewing's sarcoma	NA**	1.6	3.5	5.1	2.8
Burkitt's lymphoma	1.5	2.6	2.4	2.2	2.2
Hepatoblastoma	6.9	NA**	NA**	NA**	1.7

DATA: National Cancer Institute, Surveillance, Epidemiology, and End Results (SEER) Program

† "Other and unspecified carcinomas" represents all carcinomas and other malignant epithelial neoplasms other than thyroid carcinoma and malignant melanoma.

* The estimate should be interpreted with caution because the standard error of the estimate is relatively large: the relative standard error, RSE, is at least 30% but is less than 40% (RSE = standard error divided by the estimate).

** Not available. The estimate is not reported because it has large uncertainty: the relative standard error, RSE, is 40% or greater (RSE = standard error divided by the estimate).

Neurodevelopmental Disorders

Table H6: Percentage of children ages 5 to 17 years reported to have attention-deficit/hyperactivity disorder, by sex, 1997-2010

	1997	1998	1999	2000
All children	6.3	6.7	6.4	7.5
Boys	9.5	9.6	9.6	10.6
Girls	3.0	3.7	3.0	4.2
	2001	2002	2003	2004
All children	7.2	8.1	7.2	8.3
Boys	10.3	11.6	10.3	11.5
Girls	3.9	4.4	4.0	4.8
	2005	2006	2007	2008
All children	7.4	8.5	8.1	9.1
Boys	10.4	12.3	11.2	12.5
Girls	4.4	4.5	4.8	5.5
	2009	2010		
All children	9.8	9.5		
Boys	13.2	12.5		
Girls	6.1	6.2		

DATA: Centers for Disease Control and Prevention, National Center for Health Statistics, National Health Interview Survey

NOTE: Data represent responses to the survey question: "Has a doctor or health professional ever told you that <child's name> had Attention Deficit/Hyperactivity Disorder (ADHD) or Attention Deficit Disorder (ADD)?" Responses are provided by a parent or other knowledgeable household adult.

Table H6a: Percentage of children reported to have attention-deficit/hyperactivity disorder, by age and sex, 2007-2010

	Ages 5 to 17 years	Ages 5 to 10 years	Ages 11 to 17 years
All children	9.1	6.7	11.1
Boys	12.4	8.9	15.3
Girls	5.7	4.4	6.7

DATA: Centers for Disease Control and Prevention, National Center for Health Statistics, National Health Interview Survey

NOTE: Data represent responses to the survey question: "Has a doctor or health professional ever told you that <child's name> had Attention Deficit/Hyperactivity Disorder (ADHD) or Attention Deficit Disorder (ADD)?" Responses are provided by a parent or other knowledgeable household adult.

Table H6b: Percentage of children ages 5 to 17 years reported to have attention-deficit/hyperactivity disorder, by race/ethnicity and family income, 2007-2010

	All Incomes (n =28,880)	< Poverty Level (n=5,418)	≥ Poverty Level (n=23,462)	≥ Poverty Level (Detail) 100-200% of Poverty Level (n=6,703)	≥ Poverty Level (Detail) > 200% of Poverty Level (n=16,759)
All Races/Ethnicities (n=28,880)	9.1	11.3	8.6	10.2	7.9
White non-Hispanic (n=12,917)	10.7	16.5	10.1	14.3	9.0
Black or African-American non-Hispanic (n=4,830)	10.2	13.3	8.5	9.7	7.8
Asian non-Hispanic (n=1,589)	1.6	NA**	1.9	NA**	2.0
Hispanic (n=8,450)	4.8	5.6	4.4	4.8	4.1
Mexican (n=5,545)	4.2	4.9	3.8	4.3	3.2
Puerto Rican (n=794)	10.1	12.7	8.7	10.3	7.7
All Other Races† (n=1,094)	11.6	16.2	10.2	14.9	7.8
American Indian or Alaska Native non-Hispanic (n=165)	9.9*	NA**	7.1*	NA**	NA**

DATA: Centers for Disease Control and Prevention, National Center for Health Statistics, National Health Interview Survey

NOTE: Data represent responses to the survey question: "Has a doctor or health professional ever told you that <child's name> had Attention Deficit/Hyperactivity Disorder (ADHD) or Attention Deficit Disorder (ADD)?" Responses are provided by a parent or other knowledgeable household adult.

† The "All Other Races" category includes all other races not specified, together with those individuals who report more than one race.

* The estimate should be interpreted with caution because the standard error of the estimate is relatively large: the relative standard error, RSE, is at least 30% but is less than 40% (RSE = standard error divided by the estimate).

** Not available. The estimate is not reported because it has large uncertainty: the relative standard error, RSE, is 40% or greater (RSE = standard error divided by the estimate).

Table H7: Percentage of children ages 5 to 17 years reported to have a learning disability, by sex, 1997-2010

	1997	1998	1999	2000
All children	8.7	8.2	8.1	8.7
Boys	11.4	10.4	11.0	10.9
Girls	6.0	5.9	5.0	6.4
	2001	**2002**	**2003**	**2004**
All children	8.6	9.2	8.3	8.8
Boys	11.0	11.5	10.2	10.6
Girls	6.1	6.7	6.3	6.9
	2005	**2006**	**2007**	**2008**
All children	7.8	8.6	8.4	9.1
Boys	9.7	10.8	10.8	11.2
Girls	5.8	6.4	5.8	6.9
	2009	**2010**		
All children	9.1	8.6		
Boys	11.7	10.1		
Girls	6.5	7.1		

DATA: Centers for Disease Control and Prevention, National Center for Health Statistics, National Health Interview Survey

NOTE: Data represent responses to the survey question: "Has a representative from a school or a health professional ever told you that <child's name> had a learning disability?" Responses are provided by a parent or other knowledgeable household adult.

Table H7a: Percentage of children reported to have a learning disability, by age and sex, 2007-2010

	Ages 5 to 17 years	Ages 5 to 10 years	Ages 11 to 17 years
All children	8.8	7.5	9.9
Boys	10.9	9.3	12.4
Girls	6.6	5.6	7.4

DATA: Centers for Disease Control and Prevention, National Center for Health Statistics, National Health Interview Survey

NOTE: Data represent responses to the survey question: "Has a representative from a school or a health professional ever told you that <child's name> had a learning disability?" Responses are provided by a parent or other knowledgeable household adult.

Table H7b: Percentage of children ages 5 to 17 years reported to have a learning disability, by race/ethnicity and family income, 2007-2010

	All Incomes (n=28,889)	< Poverty Level (n=5,414)	≥ Poverty Level (n=23,475)	≥ Poverty Level (Detail)	
				100-200% of Poverty Level (n=6,700)	≥ 200% of Poverty Level (n=16,775)
All Races/Ethnicities (n=28,889)	8.8	12.6	7.9	10.3	7.0
White non-Hispanic (n=12,929)	9.3	17.1	8.4	11.7	7.6
Black or African-American non-Hispanic (n=4,830)	10.2	13.6	8.3	10.9	6.5
Asian non-Hispanic (n=1,594)	2.7	NA**	2.9	NA**	3.2
Hispanic (n=8,445)	7.2	8.7	6.6	7.7	5.4
Mexican (n=5,542)	7.1	7.9	6.7	7.6	5.7
Puerto Rican (n=792)	10.8	14.5	8.6	11.1	7.2
All Other Races† (n=1,091)	11.2	15.8	9.8	16.2	6.6
American Indian or Alaska Native non-Hispanic (n=163)	13.9	NA**	12.0*	NA**	NA**

DATA: Centers for Disease Control and Prevention, National Center for Health Statistics, National Health Interview Survey

NOTE: Data represent responses to the survey question: "Has a representative from a school or a health professional ever told you that <child's name> had a learning disability?" Responses are provided by a parent or other knowledgeable household adult.

† The "All Other Races" category includes all other races not specified, together with those individuals who report more than one race.

* The estimate should be interpreted with caution because the standard error of the estimate is relatively large: the relative standard error, RSE, is at least 30% but is less than 40% (RSE = standard error divided by the estimate).

** Not available. The estimate is not reported because it has large uncertainty: the relative standard error, RSE, is 40% or greater (RSE = standard error divided by the estimate).

Table H8: Percentage of children ages 5 to 17 years reported to have autism, 1997-2010

	1997	1998	1999	2000
All children	0.1*	0.2	0.2	0.3
	2001	2002	2003	2004
All children	0.3	0.4	0.4	0.7
	2005	2006	2007	2008
All children	0.6	0.7	0.9	0.7
	2009	2010		
All children	1.2	1.0		

DATA: Centers for Disease Control and Prevention, National Center for Health Statistics, National Health Interview Survey

NOTE: Data represent responses to the survey question: "Has a doctor or health professional ever told you that <child's name> had Autism?" Responses are provided by a parent or other knowledgeable household adult.

* The estimate should be interpreted with caution because the standard error of the estimate is relatively large: the relative standard error, RSE, is at least 30% but is less than 40% (RSE = standard error divided by the estimate).

Table H8a: Percentage of children reported to have autism, by age and sex, 2007-2010

	Ages 5 to 17 years	Ages 5 to 10 years	Ages 11 to 17 years
All children	1.0	1.1	0.8
Boys	1.5	1.7	1.3
Girls	0.4	0.5	0.3

DATA: Centers for Disease Control and Prevention, National Center for Health Statistics, National Health Interview Survey

NOTE: Data represent responses to the survey question: "Has a doctor or health professional ever told you that <child's name> had Autism?" Responses are provided by a parent or other knowledgeable household adult.

Table H8b: Percentage of children ages 5 to 17 years reported to have autism, by race/ethnicity and family income, 2007-2010

	All Incomes (n = 28,919)	< Poverty Level (n = 5,425)	≥ Poverty Level (n = 23,494)	≥ Poverty Level (Detail) 100-200% of Poverty Level (n = 6,703)	≥ 200% of Poverty Level (n = (16,792)
All Races/Ethnicities (n = 28,919)	1.0	1.0	0.9	0.8	1.0
White non-Hispanic (n = 12,938)	1.1	1.8	1.0	1.1	1.0
Black or African-American non-Hispanic (n = 4,840)	0.7	NA**	0.9	1.0*	0.8
Asian non-Hispanic (n = 1,594)	0.8*	NA**	0.7*	NA**	0.9*
Hispanic (n = 8,452)	0.6	0.8	0.5	NA**	0.5
Mexican (n = 5,547)	0.5	0.8	0.4	0.3*	0.4*
Puerto Rican (n = 793)	0.9*	NA**	NA**	NA**	NA**
All Other Races† (n = 1,095)	1.7*	NA**	1.7*	NA**	NA**
American Indian or Alaska Native non-Hispanic (n = 165)	NA**	NA**	NA**	NA**	NA**

DATA: Centers for Disease Control and Prevention, National Center for Health Statistics, National Health Interview Survey

NOTE: Data represent responses to the survey question: "Has a doctor or health professional ever told you that <child's name> had Autism?" Responses are provided by a parent or other knowledgeable household adult.

† The "All Other Races" category includes all other races not specified, together with those individuals who report more than one race.

* The estimate should be interpreted with caution because the standard error of the estimate is relatively large: the relative standard error, RSE, is at least 30% but is less than 40% (RSE = standard error divided by the estimate).

** Not available. The estimate is not reported because it has large uncertainty: the relative standard error, RSE, is 40% or greater (RSE = standard error divided by the estimate).

Table H9: Percentage of children ages 5 to 17 years reported to have intellectual disability (mental retardation), 1997-2010

	1997	1998	1999	2000
All children	0.6	0.7	0.8	0.9
	2001	**2002**	**2003**	**2004**
All children	0.9	0.6	0.7	0.9
	2005	**2006**	**2007**	**2008**
All children	0.7	0.8	0.8	0.7
	2009	**2010**		
All children	0.8	0.7		

DATA: Centers for Disease Control and Prevention, National Center for Health Statistics, National Health Interview Survey

NOTE: Data represent responses to the survey question: "Has a doctor or health professional ever told you that <child's name> had Mental Retardation?" Responses are provided by a parent or other knowledgeable household adult.

Table H9a: Percentage of children reported to have intellectual disability (mental retardation), by age and sex, 2007-2010

	Ages 5 to 17 years	Ages 5 to 10 years	Ages 11 to 17 years
All children	0.8	0.6	0.9
Boys	0.9	0.8	1.0
Girls	0.6	0.5	0.7

DATA: Centers for Disease Control and Prevention, National Center for Health Statistics, National Health Interview Survey

NOTE: Data represent responses to the survey question: "Has a doctor or health professional ever told you that <child's name> had Mental Retardation?" Responses are provided by a parent or other knowledgeable household adult.

Table H9b: Percentage of children ages 5 to 17 years reported to have intellectual disability (mental retardation), by race/ethnicity and family income, 2007-2010

	All Incomes (n = 28,920)	< Poverty Level (n = 5,423)	≥ Poverty Level (n = 23,497)	≥ Poverty Level (Detail) 100-200% of Poverty Level (n = 6,705)	> 200% of Poverty Level (n = 16,791)
All Races/Ethnicities (n = 28,920)	0.8	1.2	0.7	1.0	0.6
White non-Hispanic (n = 12,939)	0.7	1.4*	0.6	1.0	0.5
Black or African-American non-Hispanic (n = 4,836)	1.0	1.2*	0.9	1.1*	0.7*
Asian non-Hispanic (n = 1,594)	0.7*	NA**	0.7*	NA**	0.7*
Hispanic (n =8,456)	0.8	1.0*	0.7	0.9	0.4
Mexican (n = 5,549)	0.8	1.1*	0.6	1.0	0.2*
Puerto Rican (n = 795)	0.6*	NA**	NA**	NA**	NA**
All Other Races† (n = 1,095)	0.8*	NA**	NA**	NA**	NA**
American Indian or Alaska Native non-Hispanic (n = 165)	NA**	NA**	NA**	NA**	NA**

DATA: Centers for Disease Control and Prevention, National Center for Health Statistics, National Health Interview Survey

NOTE: Data represent responses to the survey question: "Has a doctor or health professional ever told you that <child's name> had Mental Retardation?" Responses are provided by a parent or other knowledgeable household adult.

† The "All Other Races" category includes all other races not specified, together with those individuals who report more than one race.

* The estimate should be interpreted with caution because the standard error of the estimate is relatively large: the relative standard error, RSE, is at least 30% but is less than 40% (RSE = standard error divided by the estimate).

** Not available. The estimate is not reported because it has large uncertainty: the relative standard error, RSE, is 40% or greater (RSE = standard error divided by the estimate).

Obesity

Table H10. Percentage of children ages 2 to 17 years who were obese, 1976-2008

	1976-1980	1988-1991	1991-1994	1999-2000	2001-2002	2003-2004	2005-2006	2007-2008
All Races/Ethnicities	5.4	9.4	11.0	13.8	15.2	16.8	15.3	16.9
White non-Hispanic	4.7	8.8	9.7	10.5*	13.4	15.7	13.0	15.4
Black non-Hispanic	7.3	11.2	13.4	18.2	17.9	19.7	20.1	19.9
Mexican-American	10.7*	13.3	15.6	20.7	19.6	19.4	22.7	21.0
All Other Races/Ethnicities†	6.5	6.9*	11.3*	17.5	16.4	16.0	12.3	16.3

DATA: Centers for Disease Control and Prevention, National Center for Health Statistics and National Center for Environmental Health, National Health and Nutrition Examination Survey

† The "All Other Races/Ethnicities" category includes all other races or ethnicities not specified, together with those individuals who report more than one race.

*The estimate should be interpreted with caution because the standard error of the estimate is relatively large: the relative standard error, RSE, is at least 30% but is less than 40% (RSE = standard error divided by the estimate), or the RSE may be underestimated.

Table H10a. Percentage of children who were obese, by age group, 1976-2008

	1976-1980	1988-1991	1991-1994	1999-2000	2001-2002	2003-2004	2005-2006	2007-2008
Ages 2-5 years	4.7	7.3	7.1	10.4	10.5	13.6	10.9	10.1
Ages 6-10 years	6.2	10.1	12.7	14.3	16.0	17.3	14.5	19.3
Ages 11-15 years	5.5	9.1	13.2	15.9	17.0	18.0	18.1	19.5
Ages 16-17 years	4.8	12.3	8.8	13.4	16.3	18.1	17.9	18.2
Ages 2-17 years	5.4	9.4	11.0	13.8	15.2	16.8	15.3	16.9

DATA: Centers for Disease Control and Prevention, National Center for Health Statistics, National Health and Nutrition Examination Survey

Table H11. Percentage of children ages 2-17 who were obese, by race/ethnicity and family income, 2005-2008

Race / Ethnicity	All Incomes‡ (n=6654)	< Poverty Level (n=1,955)	≥ Poverty Level (n=4,314)	≥ Poverty (Detail) 100-200% of Poverty Level (n=1,691)	≥ 200% of Poverty Level (n=2,623)
All Races/Ethnicities (n=6,654)	16.1	19.9	15.1	18.4	13.8
White non-Hispanic (n=1,915)	14.2	17.4	13.7	17.9	12.5
Black non-Hispanic (n=1,874)	20.0	19.7	19.9	21.6	18.8
Mexican-American (n=2,012)	21.9	22.3	21.6	21.0	22.3
All Other Races/Ethnicities† (n=853)	14.5	22.7	11.9	11.9	11.9

DATA: Centers for Disease Control and Prevention, National Center for Health Statistics, National Health and Nutrition Examination Survey

† The "All Other Races/Ethnicities" category includes all other races or ethnicities not specified, together with those individuals who report more than one race.

‡ Includes sampled individuals for whom income information is missing.

Adverse Birth Outcomes

Table H12: Percentage of babies born preterm, by race/ethnicity, 1993-2008

1993-2000	1993	1994	1995	1996	1997	1998	1999	2000
All Races/ Ethnicities	11.0	11.0	11.0	11.0	11.4	11.6	11.8	11.6
White non-Hispanic	9.1	9.3	9.4	9.5	9.9	10.2	10.5	10.4
Black or African-American non-Hispanic	18.6	18.2	17.8	17.5	17.6	17.6	17.6	17.4
Asian or Pacific Islander non-Hispanic	10.0	10.1	9.9	10.0	10.2	10.3	10.4	9.9
American Indian or Alaska Native non-Hispanic	12.3	12.0	12.4	11.9	12.2	12.2	12.7	12.6
Hispanic	11.0	10.9	10.9	10.9	11.2	11.4	11.4	11.2
Mexican	10.6	10.6	10.6	10.5	10.8	11.0	11.1	11.0
Puerto Rican	13.3	13.4	13.4	13.2	13.7	13.9	13.7	13.5
Unknown Ethnicity	10.1	11.0	10.5	9.8	10.7	10.5	10.5	10.8

2001-2008	2001	2002	2003	2004	2005	2006	2007	2008
All Races/Ethnicities	11.9	12.1	12.3	12.5	12.7	12.8	12.7	12.3
White non-Hispanic	10.8	11.0	11.3	11.5	11.7	11.7	11.5	11.1
Black or African-American non-Hispanic	17.6	17.7	17.8	17.9	18.4	18.5	18.3	17.5
Asian or Pacific Islander non-Hispanic	10.3	10.4	10.4	10.5	10.7	10.9	10.8	10.6
American Indian or Alaska Native non-Hispanic	13.2	13.0	13.5	13.7	14.2	14.3	14.1	13.8
Hispanic	11.4	11.6	11.8	12.0	12.1	12.2	12.3	12.1
Mexican	11.2	11.4	11.7	11.8	11.8	11.8	11.9	11.6
Puerto Rican	13.7	14.0	13.8	14.0	14.3	14.4	14.5	14.1
Unknown Ethnicity	11.3	11.2	12.8	12.8	13.2	13.1	13.6	13.9

DATA: Centers for Disease Control and Prevention, National Center for Health Statistics, National Vital Statistics System

Table H12a. Percentage of babies born preterm, by mother's age, 1993-2008

1993-2000	1993	1994	1995	1996	1997	1998	1999	2000
Ages < 20 years	14.3	14.2	13.8	13.6	13.8	14.0	14.1	13.9
Ages 20 to 39 years	10.4	10.5	10.5	10.5	10.9	11.2	11.3	11.2
Ages 40+ years	13.2	13.7	13.7	13.8	14.4	14.9	15.2	15.1
2001-2008	**2001**	**2002**	**2003**	**2004**	**2005**	**2006**	**2007**	**2008**
Ages < 20 years	14.1	14.0	14.3	14.5	14.7	14.8	14.6	14.1
Ages 20 to 39 years	11.6	11.7	12.0	12.1	12.4	12.4	12.3	12.0
Ages 40+ years	15.6	16.0	16.3	16.6	16.8	17.0	17.2	17.1

DATA: Centers for Disease Control and Prevention, National Center for Health Statistics, National Vital Statistics System

Table H12b. Percentage of babies born preterm, by all births, singletons, and multiples, 1993-2008

1993-2000	1993	1994	1995	1996	1997	1998	1999	2000
All births	11.0	11.0	11.0	11.0	11.4	11.6	11.8	11.6
Singletons	9.9	9.9	9.8	9.7	10.0	10.1	10.3	10.1
Multiples	53.1	54.0	54.6	55.6	57.3	58.4	59.4	58.7
2001-2008	**2001**	**2002**	**2003**	**2004**	**2005**	**2006**	**2007**	**2008**
All births	11.9	12.1	12.3	12.5	12.7	12.8	12.7	12.3
Singletons	10.4	10.4	10.6	10.8	11.0	11.1	11.0	10.6
Multiples	59.4	60.1	61.2	61.4	62.1	61.9	61.6	60.4

DATA: Centers for Disease Control and Prevention, National Center for Health Statistics, National Vital Statistics System

Table H13: Percentage of babies born at term with low birth weight, by race/ethnicity, 1993-2008

1993-2000								
	1993	1994	1995	1996	1997	1998	1999	2000
All Races/Ethnicities	2.6	2.6	2.6	2.6	2.6	2.5	2.5	2.5
White non-Hispanic	2.1	2.2	2.2	2.2	2.2	2.1	2.1	2.1
Black or African-American non-Hispanic	4.6	4.5	4.5	4.4	4.3	4.3	4.3	4.3
Asian or Pacific Islander non-Hispanic	2.9	3.0	3.1	3.1	3.0	3.1	3.0	3.1
American Indian or Alaska Native non-Hispanic	2.4	2.3	2.3	2.3	2.3	2.4	2.4	2.2
Hispanic	2.4	2.4	2.3	2.3	2.4	2.3	2.3	2.3
Mexican	2.3	2.2	2.2	2.2	2.2	2.2	2.1	2.2
Puerto Rican	3.4	3.2	3.3	3.3	3.3	3.4	3.1	3.2
Unknown Ethnicity	2.5	2.6	2.4	2.4	2.4	2.2	2.5	2.2
2001-2008								
	2001	2002	2003	2004	2005	2006	2007	2008
All Races/Ethnicities	2.5	2.6	2.6	2.7	2.7	2.7	2.7	2.8
White non-Hispanic	2.2	2.2	2.2	2.3	2.3	2.3	2.3	2.4
Black or African-American non-Hispanic	4.2	4.3	4.4	4.5	4.5	4.5	4.5	4.6
Asian or Pacific Islander non-Hispanic	3.1	3.2	3.2	3.2	3.2	3.3	3.2	3.3
American Indian or Alaska Native non-Hispanic	2.4	2.4	2.5	2.6	2.5	2.5	2.4	2.4
Hispanic	2.3	2.3	2.4	2.4	2.4	2.5	2.4	2.4
Mexican	2.2	2.2	2.2	2.3	2.3	2.4	2.2	2.3
Puerto Rican	3.2	3.2	3.4	3.2	3.3	3.4	3.4	3.3
Unknown Ethnicity	2.2	2.3	2.5	2.6	2.3	2.8	2.9	3.1

DATA: Centers for Disease Control and Prevention, National Center for Health Statistics, National Vital Statistics System

Table H13a. Percentage of babies born at term with low birth weight, by mother's age, 1993-2008

1993-2000								
	1993	1994	1995	1996	1997	1998	1999	2000
Ages < 20 years	3.4	3.5	3.5	3.6	3.6	3.5	3.5	3.5
Ages 20-39 years	2.5	2.5	2.4	2.5	2.4	2.4	2.4	2.4
Ages 40+ years	2.9	3.1	3.1	3.1	3.1	3.0	3.1	3.2
2001-2008								
	2001	2002	2003	2004	2005	2006	2007	2008
Ages < 20 years	3.5	3.5	3.6	3.7	3.7	3.7	3.6	3.7
Ages 20-39 years	2.4	2.4	2.5	2.5	2.5	2.6	2.6	2.6
Ages 40+ years	3.2	3.2	3.2	3.4	3.4	3.4	3.3	3.4

DATA: Centers for Disease Control and Prevention, National Center for Health Statistics, National Vital Statistics System

Table H13b. Percentage of babies born at term with low birth weight, by all births, singletons, and multiples, 1993-2008

1993-2000								
	1993	**1994**	**1995**	**1996**	**1997**	**1998**	**1999**	**2000**
All births	2.6	2.6	2.6	2.6	2.6	2.5	2.5	2.5
Singletons	2.3	2.3	2.3	2.3	2.3	2.2	2.2	2.2
Multiples	13.4	13.0	13.1	13.0	12.3	12.1	11.9	12.0
2001-2008								
	2001	**2002**	**2003**	**2004**	**2005**	**2006**	**2007**	**2008**
All births	2.5	2.6	2.6	2.7	2.7	2.7	2.7	2.8
Singletons	2.2	2.3	2.3	2.3	2.3	2.4	2.4	2.4
Multiples	12.2	12.1	12.0	12.0	12.0	12.3	12.3	12.6

DATA: Centers for Disease Control and Prevention, National Center for Health Statistics, National Vital Statistics System

Supplementary Topics

Birth Defects

Table S1: Birth defects in Texas, 1999-2007

	Cases per 10,000 live births		
	1999-2001	**2002-2004**	**2005-2007**
Musculoskeletal	131.1	148.1	164.8
Cardiac and Circulatory	118.4	137.4	157.9
Genitourinary	91.7	105.1	118.4
Eye and Ear	45.2	57.5	62.1
Gastrointestinal	51.5	51.0	57.8
Central Nervous System	30.5	33.6	40.7
Respiratory	23.5	24.1	25.3
Chromosomal	23.0	22.8	23.9
Oral Cleft	17.0	16.2	16.9

DATA: Texas Birth Defects Registry

Table S1a: Birth defects in Texas, 2005-2007, by race/ethnicity

	Cases per 10,000 live births			
	White non-Hispanic (n=414,420)	**Black non-Hispanic** (n=134,427)	**Hispanic** (n=594,073)	**Other non-Hispanic** (n=48,327)
Musculoskeletal	171.6	163.2	162.1	142.6
Cardiac and Circulatory	154.6	151.1	164.5	125.8
Genitourinary	132.2	115.1	109.6	120.2
Eye and Ear	60.1	48.0	67.3	52.4
Gastrointestinal	60.2	46.1	60.2	39.5
Central Nervous System	41.8	43.7	39.5	35.8
Respiratory	23.1	23.4	27.6	20.5
Chromosomal	23.5	19.9	25.3	18.2
Oral Cleft	18.1	11.1	17.5	15.7

DATA: Texas Birth Defects Registry

Contaminants in Schools and Child Care Facilities

Table S2: Percentage of environmental and personal media samples with detectable pesticides in child care facilities, 2001

	Pentachlorophenol	Chlorpyrifos	*cis*-Permethrin	Diazinon
Indoor Air (Regional Data)	83.2	100.0	40.3	100.0
Hand Wipes (Regional Data)	20.0	65.0	86.5	48.3
Dust (Regional Data)	95.2	100.0	100.0	100.0
Floor Wipes (National Data)	NA	89.0	72.0	67.0

DATA: Children's Total Exposure to Pesticides and Other Persistent Organic Pollutants Study (Regional Data); First National Environmental Health Survey of Child Care Centers (National Data)

NOTE: Data are from both national and regional sources, and are identified accordingly. Regional data are from samples collected in North Carolina and Ohio only.

Table S3: Percentage of environmental and personal media samples with detectable industrial chemicals in child care facilities, 2001

	Dibutyl Phthalate	PCB-52	Polycyclic Aromatic Hydrocarbons	Bisphenol A
Indoor Air	100.0	97.6	45.3	59.7
Hand Wipes	75.0	8.3	65.0	95.2
Dust	100.0	65.1	45.3	62.3

DATA: Children's Total Exposure to Pesticides and Other Persistent Organic Pollutants Study

NOTE: Regional data, from samples collected in North Carolina and Ohio only.

Table S4: Pesticides used inside California schools by commercial applicators, 2002-2007

	Pounds of Pesticide Applied					
	2002	2003	2004	2005	2006	2007
Pyrethrin and Pyrethroid Insecticides	9,452	2,515	2,430	2,274	2,556	1,794
Organophosphate Insecticides	919	244	39	119	36	70
Other Insecticides	2,125	2,037	4,883	2,205	641	142
Herbicides	295	4,031	613	1,099	1,174	701
Fumigants	651	556	3,890	392	149	249
Rodenticides	1	589	219	0.4	0.7	120
Miscellaneous Pesticides	434	52	121	88	76	124

DATA: California Department of Pesticide Regulation, Schools Pesticide Use Reporting Database

References

1. U.S. Environmental Protection Agency. 2010. *Fact Sheet on the Federal Register Notice for Stage 1 Disinfectants and Disinfection Byproducts Rule*. U.S. EPA, Office of Water. Retrieved January 10, 2011 from http://water.epa.gov/lawsregs/rulesregs/sdwa/stage1/factsheet.cfm.

2. U.S. Environmental Protection Agency. 2010. *Fact Sheet on the Interim Enhanced Surface Water Treatment Rule*. U.S. EPA, Office of Water. Retrieved January 10, 2011 from http://water.epa.gov/lawsregs/rulesregs/sdwa/ieswtr/factsheet.cfm.

3. U.S. Environmental Protection Agency. 2001. *Radionuclides Rule: A Quick Reference Guide*. Washington, DC: U.S. EPA, Office of Water. EPA 816-F-01-003. http://www.epa.gov/ogwdw/radionuclides/pdfs/qrg_radionuclides.pdf.

4. U.S. Environmental Protection Agency. 2009. *Technical Fact Sheet: Final Rule for Arsenic in Drinking Water*. U.S. EPA, Office of Water. Retrieved January 10, 2011 from http://water.epa.gov/lawsregs/rulesregs/sdwa/arsenic/regulations_techfactsheet.cfm.

5. Anderson, W.A., L. Castle, M.J. Scotter, R.C. Massey, and C. Springall. 2001. A biomarker approach to measuring human dietary exposure to certain phthalate diesters. *Food Additives and Contaminants* 18 (12):1068-74.

6. Mendez, W., E. Dederick, and J. Cohen. 2010. Drinking water contribution to aggregate perchlorate intake of reproductive-age women in the United States estimated by dietary intake simulation and analysis of urinary excretion data. *Journal of Exposure Science and Environmental Epidemiology* 20 (3):288-97.

7. Preau, J.L., Jr., L.Y. Wong, M.J. Silva, L.L. Needham, and A.M. Calafat. 2010. Variability over 1 week in the urinary concentrations of metabolites of diethyl phthalate and di(2-ethylhexyl) phthalate among eight adults: an observational study. *Environmental Health Perspectives* 118 (12):1748-54.

8. Völkel, W., T. Colnot, G.A. Csanady, J.G. Filser, and W. Dekant. 2002. Metabolism and kinetics of bisphenol a in humans at low doses following oral administration. *Chemical Research in Toxicology* 15 (10):1281-7.

9. Crump, K.S., and J.P. Gibbs. 2005. Benchmark calculations for perchlorate from three human cohorts. *Environmental Health Perspectives* 113 (8):1001-8.

Appendix B: Metadata

Appendix B: Metadata

Air Quality System (AQS)
(Used for Indicators E1, E2, and E3)

Brief description of the data set	The U.S. Environmental Protection Agency compiles air quality monitoring data in the Air Quality System (AQS). Ambient air concentrations are measured at a national network of more than 4,000 monitoring stations and are reported by state, local, and tribal agencies to EPA AQS.
Who provides the data set?	U.S. Environmental Protection Agency, Office of Air Quality Planning and Standards.
How are the data gathered?	Concentrations are measured at a national network of more than 4,000 monitoring stations and are reported by state, local, and tribal agencies to EPA AQS.
What documentation is available describing data collection procedures?	The Ambient Monitoring Technology Information Center (AMTIC) at http:/www.epa.gov/ttn/amtic/ contains information and files on ambient air quality monitoring programs, details on monitoring methods, relevant documents and articles, information on air quality trends and federal regulations related to ambient air quality monitoring. The Air Trends site at http:/www.epa.gov/airtrends contains information on air quality trends. The Green Book site at http://www.epa.gov/air/oaqps/greenbk contains information on nonattainment areas.
What types of data relevant for children's environmental health indicators are available from this database?	Relevant data include measured ambient air pollutant concentrations (lead, carbon monoxide, ozone, PM_{10}, $PM_{2.5}$, sulfur dioxide, and nitrogen dioxide), Air Quality Index, and monitor information (location, monitoring objective).
What is the spatial representation of the database (national or other)?	National. However, not all counties are represented since not all counties have air pollution monitors.
Are raw data (individual measurements or survey responses) available?	Individual hourly or daily measurements by monitor and pollutant are available.
How are database files obtained?	Raw data: http:/www.epa.gov/ttn/airs/aqsdatamart/basic_info.htm. http:/www.epa.gov/ttn/airs/airsaqs/detaildata/downloadaqsdata.htm. Annual summary data (includes annual means and maxima): http://www.epa.gov/ttn/airs/aqsdatamart/. For some indicators additional annual summary data were compiled by EPA staff. This includes annual maximum rolling three-month average lead concentrations, county maximum $PM_{2.5}$ annual means using OAQPS data completeness and weighted average calculations, $PM_{2.5}$ exceedance count data, and air quality index data.
Are there any known data quality or data analysis concerns?	Individual measurements of questionable validity or attributed to exceptional events (e.g., forest fires) are flagged. Monitoring data are not collected in some counties for some pollutants.

Air Quality System (AQS)
(Used for Indicators E1, E2, and E3)

What documentation is available describing quality assurance procedures?	http://www.epa.gov/ttn/amtic/quality.html. http://www.epa.gov/airprogm/oar/oaqps/qa/index.html.
For what years are data available?	1970–present. AQS contains some monitoring data from the late 1950s and early 1960s, but there is not an appreciable amount of data for lead until 1970, sulfur dioxide until 1971, nitrogen dioxide until 1974, carbon monoxide and ozone until 1975, and PM_{10} until 1987. AQS also contains monitoring data for $PM_{2.5}$ beginning with 1999; $PM_{2.5}$ was measured only infrequently prior to 1999.
What is the frequency of data collection?	Hourly or daily. Less frequent monitoring occurs at some monitors (e.g., every three or six days for PM or only in the ozone season for ozone).
What is the frequency of data release?	AIRNow releases ozone and PM2.5 data hourly. Raw data are updated by states approximately monthly. Annual summary data are updated quarterly.
Are the data comparable across time and space?	Counties without air quality monitors cannot be compared with counties with air quality monitors, and some counties are monitored more extensively than others. Although monitor locations and monitoring frequencies change, the network is reasonably stable. An exception occurred for $PM_{2.5}$ in 1999 as the new monitoring network was built up.
Can the data be stratified by race/ethnicity, income, and location (region, state, county or other geographic unit)?	The data can be stratified by region, state, county, and metropolitan area.

American Healthy Homes Survey (AHHS)
(Used for Indicator E6)

Brief description of the data set	A nationally representative sample of homes was selected for this survey. AHHS measured levels of lead, lead hazards, and allergens in homes nationwide. AHHS also surveyed additional potential health hazards such as arsenic, pesticides, and molds. The lead and arsenic data included the levels of lead in paint, dust and soil, and arsenic in dust and soil, and levels of paint deterioration.
Who provides the data set?	U.S. Department of Housing and Urban Development (HUD).
How are the data gathered?	Data were collected from participants in private and public residences. A 3-stage cluster sample was used to select a nationally representative sample of 1,131 homes. Samples were collected via surface wipes from four common living areas, homeowner vacuum bags, and soil samples from outside the home. Lead testing in paint was conducted using a portable X-Ray Fluorescence (XRF) instrument. Demographic and other information was collected using a questionnaire. All samples and survey information were collected during a single day.
What documentation is available describing data collection procedures?	http://www.hud.gov/offices/lead/NHHC/presentations/R-15_Findings_from_AHH_survey.pdf. Slide four and five of the presentation. American Healthy Homes Survey, Draft Final Report. June, 2009.

American Healthy Homes Survey (AHHS)
(Used for Indicator E6)

What types of data relevant for children's environmental health indicators are available from this database?	Relevant environmental contaminant data include measurements of lead paint, lead dust, lead in soils, mold, allergens/endotoxins in dust, arsenic in soil, indoor moisture measurements, and indoor pesticide residues. Other relevant information found in this database includes housing type and age, demographic information on residents (age, race, income group, ethnicity), electrical safety, structural stability, moisture, pest control, ventilation, injury prevention, fire safety, deterioration of carpet, and plumbing facilities.
What is the spatial representation of the database (national or other)?	National.
Are raw data (individual measurements or survey responses) available?	Not currently.
How are database files obtained?	HUD provided data files directly to EPA for purposes of developing an indicator for America's Children and the Environment. Summary tables are available in "American Healthy Homes Survey, Final Report, Lead and Arsenic Findings," June 2009. http://portal.hud.gov/hudportal/documents/huddoc?id=AHHS_REPORT.pdf.
Are there any known data quality or data analysis concerns?	None reported.
What documentation is available describing quality assurance procedures?	"American Healthy Homes Survey, Final Report, Lead and Arsenic Findings," June 2009. http://portal.hud.gov/hudportal/documents/huddoc?id=AHHS_REPORT.pdf
For what years are data available?	2005/2006.
What is the frequency of data collection?	Data were collected once, from June 2005 to March 2006.
What is the frequency of data release?	The final report was released in April 2011 and can be found at http://portal.hud.gov/hudportal/documents/huddoc?id=AHHS_REPORT.pdf.
Are the data comparable across time and space?	As a one-time survey, time comparisons within the AHHS are not possible, but AHHS results can be compared with the earlier NSLAH survey (1999-2000). Geographic comparisons should be possible using the raw data, since the same data were collected at all homes. The Final Report gives some comparisons between the four Census regions.
Can the data be stratified by race/ethnicity, income, and location (region, state, county or other geographic unit)?	The data can be stratified by residents' age, race, and ethnicity, Data can also be stratified by household income, census region, year of home construction, and by the housing type (rented or owned). However, estimates of lead hazards in the home for children ages 0 to 5 years broken out by race/ethnicity and income are not statistically reliable, due to the relatively small number of homes in each group.

California School Pesticide Use Reporting Database
(Used for Measure S4)

Brief description of the data set	A California state-wide database containing the records of pesticide use in California schools and child day care facilities. The records include only pesticides applied by licensed commercial pest management services. Each record contains the name of the school, name of the pesticide product, registration number of the pesticide product, sites of application inside or outside the school, amount of product applied, unit of the measure, and the application date and time. A supplementary dataset giving the percentages of active ingredients in each pesticide product was also obtained from the California Department of Pesticide Regulation (DPR).
Who provides the data set?	California Department of Pesticide Regulation.
How are the data gathered?	As per California pesticide regulations, all businesses engaged in pest control are required to report pesticide use at school sites using a prescribed form to the DPR. More information is available at: http://www.cdpr.ca.gov/docs/legbills/6624fin.pdf.
What documentation is available describing data collection procedures?	The form that pest control companies use to report the pesticide use at school sites is available at: http://www.cdpr.ca.gov/docs/enforce/prenffrm/prenf117.pdf.

The data reported by pest control companies are aggregated by the DPR and provided for the general public. |
| What types of data relevant for children's environmental health indicators are available from this database? | Relevant information includes the amount and type of pesticides used at school sites in California by commercial applicators. This information is relevant to determine exposure of school children to pesticides during their time spent inside the school. |
| What is the spatial representation of the database (national or other)? | State (California). |
| Are raw data (individual measurements or survey responses) available? | Yes. The database contains all instances of pesticide use at school sites that are reported to the DPR. The raw data can be obtained directly from DPR.

The supplementary data files with data on the contents of each pesticide product are available for download at: http://www.cdpr.ca.gov/docs/label/prodtables.htm. |
How are database files obtained?	The database files are obtained from DPR through email correspondence.
Are there any known data quality or data analysis concerns?	The specific gravity for some pesticides is not reported. The amounts used in different school locations are not reported or reported as zero. The database excludes non-commercial pesticide applications such as by school staff.
What documentation is available describing QA procedures?	Not available.
For what years are data available?	2002 – present.
What is the frequency of data collection?	All instances of pesticide use at school and child day care sites by pest management companies need to be reported. The DPR aggregates these data on yearly basis.

California School Pesticide Use Reporting Database
(Used for Measure S4)

What is the frequency of data release?	Yearly.
Are the data comparable across time and space?	Pesticide use can be compared between years or between schools.
Can the data be stratified by race/ethnicity, income, and location (region, state, county or other geographic unit)?	Data can be stratified only by county or at the individual school or child day care facility level. No demographic data are included in this database, although school ID codes are available so that these data can be matched with California or federal school population data.

Census: American Community Survey Data
(Used for Indicator E4)

Brief description of the data set	The U.S. Census Bureau collects detailed population data for a sample of the United States population and provides information for 1-, 3-, and 5-year averages.
Who provides the data set?	U.S. Census Bureau.
How are the data gathered?	The American Community Survey collects detailed population information for a sample of the United States population using a mail survey and/or personal interviews.
What documentation is available describing data collection procedures?	http://www.census.gov/acs/www/data_documentation/summary_file/ http://www.census.gov/acs/www/data_documentation/documentation_main/
What types of data relevant for children's environmental health indicators are available from this database?	Relevant information includes populations by year or group of years, county, census tract, census block group, race, ethnicity, age, and sex.
What is the spatial representation of the database (national or other)?	National.
Are raw data (individual measurements or survey responses) available?	Not publicly released.
How are database files obtained?	http://www.census.gov/acs/www/data_documentation/data_via_ftp/ (all available data tables) http://factfinder2.census.gov/faces/nav/jsf/pages/index.xhtml (selected data tables)
Are there any known data quality or data analysis concerns?	All data are based on a sample and not the entire census. 1-year estimates are only available for areas with populations above 65,000, are less reliable but more current than 3-year or 5-year estimates, and provide the least detailed information. 3-year estimates are only available for areas with populations above 20,000. 5-year estimates are available for all areas, are more reliable but less current than 1-year or 3-year estimates, and provide the most detailed information.

Census: American Community Survey Data
(Used for Indicator E4)

What documentation is available describing quality assurance procedures?	http://factfinder2.census.gov/faces/nav/jsf/pages/index.xhtml
For what years are data available?	1-year ACS files are released annually, beginning with 2002 data. 3-year ACS files are released annually, beginning with 2005-2007 data. 5-year ACS files are released annually, beginning with 2005-2009 data.
What is the frequency of data collection?	Every year.
What is the frequency of data release?	Every year.
Are the data comparable across time and space?	Populations for counties, census tracts, or census block groups may not be comparable between years or periods due to boundary changes.
Can the data be stratified by race/ethnicity, income, and location (region, state, county or other geographic unit)?	The data can be stratified by race, ethnicity, region (state, county, census tract, census block group, MSA, urban area), and income. Stratifications by age, race/ethnicity, and income combined are only available for census tracts in the 5-year data and for higher geographies in the 1- and 3-year data.

Census: Decennial Data
(Used for Indicators E1, E2, E3, E10, E11)

Brief description of the data set	The U.S. Census Bureau collects detailed population data for the entire United States every 10 years.
Who provides the data set?	U.S. Census Bureau.
How are the data gathered?	The decennial census collects detailed population information for the entire United States every 10 years using a mail survey and/or personal interviews. In 1990, 2000, and 2010 the entire population was asked a small set of questions (including age, sex, race, and ethnicity). In 1990 and 2000 about one in six households were also asked more detailed questions (including income).
What documentation is available describing data collection procedures?	http://factfinder2.census.gov/faces/nav/jsf/pages/index.xhtml
What types of data relevant for children's environmental health indicators are available from this database?	Relevant data include populations (by year, county, census tract, census block group, census block,) race, ethnicity, age, sex, and income (not for 2010 and not for census blocks).
What is the spatial representation of the database (national or other)?	National.
Are raw data (individual measurements or survey responses) available?	Not publicly released.

Census: Decennial Data
(Used for Indicators E1, E2, E3, E10, E11)

How are database files obtained?	http://factfinder2.census.gov/faces/nav/jsf/pages/index.xhtml (county and national populations)
	http://geolytics.com/USCensus,Census-2000-Products,Categories.asp (2000 census blocks)
Are there any known data quality or data analysis concerns?	Populations by county, race, and income level are not released for combinations with populations below 100 or where the estimate is based on a sample of 50 or less. Income data are based on a sample and not the entire census. Census block locations are given by the census block centroid (geographical center) which does not account for the shape and size of the census block.
What documentation is available describing quality assurance procedures?	http://factfinder2.census.gov/faces/nav/jsf/pages/index.xhtml
For what years are data available?	1990, 2000, 2010.
What is the frequency of data collection?	Every 10 years.
What is the frequency of data release?	Every 10 years.
Are the data comparable across time and space?	Detailed race data for different decades are not comparable due to changing race group definitions, such as the treatment of respondents with multiple races. Comparisons between populations below reporting thresholds are not possible. Populations for some smaller regions (census blocks, block groups, tracts, and occasionally counties) are not comparable for different decades due to boundary changes.
Can the data be stratified by race/ethnicity, income, and location (region, state, county or other geographic unit)?	Data can be stratified by race, ethnicity, and location (region, state, county, census tract, census block group, census block, MSA, urban area). Income data are available from the American Community Survey (2005 and later) and from samples from the 1990 and 2000 censuses.

Census: Intercensal and Postcensal Data
(Used for Indicators E1, E2, E3, E7, E8, E12, H3)

Brief description of the data set	The U.S. Census Bureau collects detailed population data for the entire United States every 10 years. These data are combined with birth, death, migration, and net international immigration data to estimate populations for the years between (intercensal) or after (postcensal) censuses.
Who provides the data set?	U.S. Census Bureau.

Census: Intercensal and Postcensal Data
(Used for Indicators E1, E2, E3, E7, E8, E12, H3)

How are the data gathered?	The decennial census collects detailed population information for the entire United States every 10 years using a mail survey and/or personal interviews. Intercensal data estimate populations between censuses by combining the decennial census data from the two censuses with birth, death, migration, and net international immigration data. Postcensal data estimate populations after a census by combining the decennial census data from the previous census with birth, death, migration, and net international immigration data. For the 2000s, bridged race estimates of populations in four single race categories were calculated using a statistical regression model with person-level and county-level covariates to estimate the proportion of people in a given detailed multiple race category that would select each single race category.
What documentation is available describing data collection procedures?	http://www.cdc.gov/nchs/nvss/bridged_race.htm (US census populations with bridged race categories) http://www.census.gov/popest/data/historical/index.html
What types of data relevant for children's environmental health indicators are available from this database?	Relevant data include populations by year, county, race, ethnicity, age, and sex.
What is the spatial representation of the database (national or other)?	National.
Are raw data (individual measurements or survey responses) available?	Not publicly released.
How are database files obtained?	http://www.cdc.gov/nchs/nvss/bridged_race.htm (US census populations with bridged race categories) http://www.census.gov/popest/data/historical/index.html (population estimates)
Are there any known data quality or data analysis concerns?	Due to the uncertainties in the statistical methods used to estimate intercensal and postcensal populations, the population counts at the more detailed geographical or demographic stratification levels are less precise.
What documentation is available describing quality assurance procedures?	http://www.census.gov/popest/methodology/2009-stco-char-meth.pdf (methods for bridged race categories including consistency with other population estimates).
For what years are data available?	1977–present.
What is the frequency of data collection?	Varies.
What is the frequency of data release?	Monthly, quarterly, or annually.

Census: Intercensal and Postcensal Data
(Used for Indicators E1, E2, E3, E7, E8, E12, H3)

Are the data comparable across time and space?	Postcensal data for each calendar year between the census and the current year are updated annually using information on the components of population change. Since the components of population change data are revised (e.g., a preliminary natality file is replaced with a final natality file), and since estimation methodologies are improved, population estimates from different annual updates are not comparable. For example, the year 2007 population estimates from the 2008 and 2009 series are not the same because the population change estimates for the years 2001 to 2007 used in the 2008 series were updated for the 2009 series, and the estimation methodologies were also revised (e.g., for international migration and for the effects of hurricanes Katrina and Rita). Race data for different decades may not be comparable due to changing race group definitions.
Can the data be stratified by race/ethnicity, income, and location (region, state, county or other geographic unit)?	Data can be stratified by race, ethnicity, and location (region, state, county). Income data are available from the American Community Survey (2005 and later) and from the 1990 and 2000 censuses.

EPA Superfund Program and the RCRA Corrective Action Program Site Information
(Used for Indicators E10 and E11)

Brief description of the data set	A list of all Superfund sites and RCRA Corrective Action sites that may not have all human health protective measures in place. The list includes the site name, state in which the site is located, whether the site is a federal facility, latitude, longitude, and the acreage.
Who provides the data set?	The U.S. Environmental Protection Agency, Office of Solid Waste and Emergency Response, Superfund Program and the RCRA Corrective Action Program provide data from two independent databases.
	Superfund site information is reported in the Comprehensive Environmental Response, Compensation, and Liability Information System (CERCLIS) Database. CERCLIS includes lists of involved parties and site status (e.g., Human Exposure Under Control, Ground Water Migration Under Control, and Site Wide Ready for Anticipated Use) and the measures Construction Completion and Final Assessment Decisions.
	Information on RCRA Corrective Action sites is maintained in the Resource Conservation and Recovery Act Information (RCRAInfo) Database. RCRAInfo includes site status (e.g., Human Exposure Under Control) among other types of data. For both programs, status designation of Human Exposure Under Control was used as the milestone to determine that a site has all human health protective measures in place.

EPA Superfund Program and the RCRA Corrective Action Program Site Information
(Used for Indicators E10 and E11)

How are the data gathered?	Acreage and latitude/longitude information in RCRAInfo is collected from a variety of sources, such as RCRA permit applications, owners or operators, or public documents. Acreage and latitude/longitude information in CERCLIS is obtained from Preliminary Assessment reports, Site Inspection reports, Records of Decision, Five Year Reviews, or other official site documents.
	Acreage in RCRAInfo refers to the entire site. In CERCLIS, there are a number of types of acreage data. The CERCLIS field labeled "property boundary acreage" was used for calculation of Indicators E10 and E11. Although not meant to serve as estimates of the contaminated acres for Superfund sites, this information is similar to the acreage in RCRAInfo for Corrective Action facilities.
	For Corrective Action facilities, updates and progress are recorded by Regional or authorized State program staff as milestones are achieved. As Superfund site information changes, the CERCLIS database is updated by EPA regional offices.
	EPA undertook a one-time effort to collect site acreage starting in 2007. These data are updated whenever more accurate information is obtained.
What documentation is available describing data collection procedures?	Not applicable.
What types of data relevant for children's environmental health indicators are available from this database?	Relevant data include latitude, longitude, and estimated acres for contaminated sites.
What is the spatial representation of the database (national or other)?	National; each relevant site in the United States is individually identified.
Are raw data (individual measurements or survey responses) available?	Latitude/longitude and acreage are available for each site.
How are database files obtained?	Requests for datasets from CERCLIS or RCRAInfo must be made to EPA offices.
	Summary information on individual Corrective Action or Superfund sites can be found at:
	http://www.epa.gov/osw/hazard/correctiveaction/facility/index.htm
	and
	http://cfpub.epa.gov/supercpad/cursites/srchsites.cfm
	respectively.
	Some of the information in CERCLIS and RCRAInfo (name, address, cleanup progress) is also available on the EPA webpages "Envirofacts" and "Cleanups in My Community," http://www.epa.gov/enviro/ and http://iaspub.epa.gov/Cleanups/.

EPA Superfund Program and the RCRA Corrective Action Program Site Information
(Used for Indicators E10 and E11)

Are there any known data quality or data analysis concerns?	The site latitude and longitude specify a point at the site, which could represent the location of the site entry point or of some other area within the site.
	Actual geographic boundaries of each site (or contaminated areas on each site) are not available in digital form. In absence of geographic boundaries, CERCLIS boundary acres and RCRAInfo site acreage were used to estimate entire site area, fenceline to fenceline. No effort was made to approximate site shape. It is not specified if all site acres are areas of suspected contamination or areas of known contamination. Thus, the area used to represent each site is larger than the area of actual, known contamination.
	The "all human health protective measures in place" designation indicates that there is no complete pathway for human exposures to unacceptable levels of contamination, based on current site conditions. Sites lacking this designation are of three types: sites where a possible exposure route has been identified, sites that have not been fully assessed, or sites that have not been reviewed for the designation. Thus, sites that may not have all human health protective measures in place include both sites where there is a possible route of human exposure and sites where there may be no existing exposure routes.
What documentation is available describing quality assurance procedures?	Not applicable.
For what years are data available?	Data represent site status, including designation of all human health protective measures in place, as of October 2009. Designations are not available for earlier years.
What is the frequency of data collection?	Data collection frequency varies. Information is updated as site information changes.
What is the frequency of data release?	Data are released on a yearly basis.
Are the data comparable across time and space?	Acres used to describe site area are collected differently for sites in each program (see above). Procedures applied within each program will be consistent over time. Contamination level and exposure potential will vary across sites.
Can the data be stratified by race/ethnicity, income, and location (region, state, county or other geographic unit)?	This site list does not contain information on race/ethnicity or income. The data can be stratified by location, specifically by state. Additionally, the latitude and longitude are provided for each site, which allows for more exact location stratifications and for linkage to Census data on local population demographics.

National Air Toxics Assessment (NATA)
(Used for Indicator E4)

Brief description of the data set	The National Air Toxics Assessment is EPA's ongoing comprehensive evaluation of air toxics in the United States. NATA provides estimates of the risk of cancer and other serious health effects from inhaling air toxics in order to inform both national and more localized efforts to identify and prioritize air toxics, emission source types, and locations that are of greatest potential concern in terms of contributing to population risk.
Who provides the data set?	U.S. Environmental Protection Agency, Office of Air Quality Planning and Standards.
How are the data gathered?	Emissions inventory data for individual HAPs are collected from data reported by large individual facilities (point sources) and estimated for area and mobile sources using various emissions inventory models. The compiled inventory is called the National Emissions Inventory. Ambient concentrations are estimated using an air dispersion model. Population exposures are estimated based on a screening-level inhalation exposure model.
What documentation is available describing data collection procedures?	See http:/www.epa.gov/nata2005 for detailed description of NATA organization and data collection practices.
What types of data relevant for children's environmental health indicators are available from this database?	Relevant data include modeled ambient concentrations, exposure concentrations, cancer risks, and non-cancer hazard indices for each HAP in each county and each census tract.
What is the spatial representation of the database (national or other)?	National.
Are raw data (individual measurements or survey responses) available?	Modeled ambient and exposure concentrations for each HAP in each county and census tract are available.
How are database files obtained?	http://www.epa.gov/ttn/atw/nata2005/tables.html.
Are there any known data quality or data analysis concerns?	NATA results provide answers to questions about emissions, ambient air concentrations, exposures and risks across broad geographic areas (such as counties, states, and the nation) at a moment in time. These assessments are based on assumptions and methods that limit the range of questions that can be answered reliably. The results cannot be used to identify exposures and risks for specific individuals, or even to identify exposures and risks in small geographic regions. These estimates reflect chronic exposures resulting from the inhalation of the air toxics emitted and do not consider exposures that may occur indoors or as a results of exposures other than inhalation (i.e., dermal or ingestion). Methods used in NATA were peer reviewed by EPA's Science Advisory Board; the SAB report is available at http://www.epa.gov/ttn/atw/sab/sabrept1201.pdf.
What documentation is available describing quality assurance procedures?	See http:/www.epa.gov/nata2005 and. http://www.epa.gov/ttn/atw/nata2005/05pdf/nata_tmd.pdf

National Air Toxics Assessment (NATA)
(Used for Indicator E4)

For what years are data available?	1996, 1999, 2002, 2005.
What is the frequency of data collection?	Approximately every three years.
What is the frequency of data release?	Approximately every three years.
Are the data comparable across time and space?	Data for different NATA assessments are not comparable across time due to improvements in the estimated national emissions inventory, increases in the numbers of modeled HAPs, and improvements in the health data information. Data may not be comparable over space due to quality differences in emissions inventory reporting.
Can the data be stratified by race/ethnicity, income, and location (region, state, county or other geographic unit)?	Data can be stratified by state, county, and census tract; when combined with census data, NATA estimates can be stratified by race/ethnicity and income.

National Health and Nutrition Examination Survey (NHANES)
(Used in Indicators B1, B2, B3, B4, B5, B6, B7, B8, B9, B10, B11, B12, B13, H10, and H11)

Brief description of the data set	The National Health and Nutrition Examination Survey (NHANES) is a program of studies designed to assess the health and nutritional status of the civilian noninstitutionalized population of the United States, using a combination of interviews, physical examinations, and laboratory analysis of biological specimens.
Who provides the data set?	Centers for Disease Control and Prevention, National Center for Health Statistics.
How are the data gathered?	Laboratory data are obtained by analysis of blood and urine samples collected from survey participants at NHANES Mobile Examination Centers. Health status is assessed by physical examination. Demographic and other survey data regarding health status, nutrition, and health-related behaviors are collected by personal interview, either by self-reporting or, for children under 16 and some others, as reported by an informant.
What documentation is available describing data collection procedures?	See http://www.cdc.gov/nchs/nhanes.htm for detailed survey and laboratory documentation by survey period.
What types of data relevant for children's environmental health indicators are available from this database?	Relevant data include concentrations of environmental chemicals (in urine, blood, and serum), body measurements, health status (as assessed by physical examination, laboratory measurements, and interview responses), and demographic information.
What is the spatial representation of the database (national or other)?	NHANES sampling procedures provide nationally representative data. Analysis of data for any other geographic area (region, state, etc.) is possible only by special arrangement with the NCHS Research Data Center, and such analyses may not be representative of the specified area.
Are raw data (individual measurements or survey responses) available?	Individual laboratory measurements and survey responses are generally available. Individual survey responses for some questions and some measurements are not publicly released.

National Health and Nutrition Examination Survey (NHANES)
(Used in Indicators B1, B2, B3, B4, B5, B6, B7, B8, B9, B10, B11, B12, B13, H10, and H11)

How are database files obtained?	http://www.cdc.gov/nchs/nhanes.htm.
Are there any known data quality or data analysis concerns?	Some environmental chemicals have large percentages of values below the detection limit. Data gathered by interview, including demographic information, and responses regarding health status, nutrition, and health-related behaviors are self-reported, or (for individuals age 16 years and younger) reported by an adult informant. In some cases, the size of a particular sample is too small in an individual 2-year survey cycle to produce statistically reliable estimates; this can be addressed by combining two or more consecutive 2-year survey cycles.
What documentation is available describing quality assurance procedures?	http://www.cdc.gov/nchs/nhanes.htm includes detailed documentation on laboratory and other quality assurance procedures. Data quality information is available at http://www.cdc.gov/nchs/about/policy/quality.htm.
For what years are data available?	Some data elements were collected in predecessors to NHANES beginning in 1959; collection of data on environmental chemicals began with measurement of blood lead in NHANES II, 1976-1980. The range of years for measurement of environmental chemicals varies; apart from lead and cotinine (initiated in NHANES III), measurement of environmental chemicals began with 1999-2000 or later NHANES.
What is the frequency of data collection?	Data are collected on continuous basis, but are grouped into NHANES cycles: NHANES II (1976-1980); NHANES III phase 1 (1988-1991); NHANES III phase 2 (1991-1994); and continuous two-year cycles beginning with 1999-2000 and continuing to the present.
What is the frequency of data release?	Data are released in two-year cycles (e.g. 1999-2000); particular data sets from a two-year NHANES cycle are released as available.
Are the data comparable across time and space?	Detection limits can vary across time, affecting some comparisons. Some contaminants are not measured in every NHANES cycle. Within any NHANES two-year cycle, data are generally collected and analyzed in the same manner for all sampling locations.
Can the data be stratified by race/ethnicity, income, and location (region, state, county or other geographic unit)?	Data are collected to be representative of the U.S. population based on age, sex, and race/ethnicity. The public release files allow stratification by these and other demographic variables, including family income range and poverty income ratio. Data cannot be stratified geographically except by special arrangement with the NCHS Research Data Center.

National Health Interview Survey (NHIS)
(Used for Indicators E5, H1, H2, H6, H7, H8, and H9)

Brief description of the data set	The National Health Interview Survey (NHIS) collects data on a broad range of health topics through personal household interviews. The results of NHIS provide data to track health status, health care access, and progress toward achieving national health objectives.
Who provides the data set?	Centers for Disease Control and Prevention, National Center for Health Statistics.

National Health Interview Survey (NHIS)
(Used for Indicators E5, H1, H2, H6, H7, H8, and H9)

How are the data gathered?	Data are obtained using a health questionnaire through a personal household interview. Interviewers obtain information on health history and demographic characteristics, including age, household income, and race and ethnicity from respondents, or from a knowledgeable household adult for children age 17 years and younger.
What documentation is available describing data collection procedures?	See http://www.cdc.gov/nchs/nhis.htm for detailed survey documentation by survey year.
What types of data relevant for children's environmental health indicators are available from this database?	Relevant data include health history (e.g., asthma, mental health, childhood illnesses), smoking in residences (for selected years), demographic information, and health care use and access information.
What is the spatial representation of the database (national or other)?	NHIS sampling procedures provide nationally representative data, and may also be analyzed by four broad geographic regions: North, Midwest, South and West. Analysis of data for any other smaller geographic areas (state, etc.) is possible only by special arrangement with the NCHS Research Data Center.
Are raw data (individual measurements or survey responses) available?	Data for each year of the NHIS are available for download and analysis (http://www.cdc.gov/nchs/nhis/nhis_questionnaires.htm). Annual reports from the NHIS are also available (http://www.cdc.gov/nchs/nhis/nhis_products.htm) as are interactive data tables (http://www.cdc.gov/nchs/hdi.htm). The files available for download generally contain individual responses to the survey questions; however, for some questions the responses are categorized Some survey responses are not publicly released.
How are database files obtained?	Raw data: http://www.cdc.gov/nchs/nhis.htm.
Are there any known data quality or data analysis concerns?	Data are self-reported, or (for individuals age 17 years and younger) reported by a knowledgeable household adult, usually a parent. Responses to some demographic questions (race/ethnicity, income) are statistically imputed for survey participants lacking a reported response.
What documentation is available describing quality assurance procedures?	http://www.cdc.gov/nchs/data/series/sr_02/sr02_130.pdf provides a summary of quality assurance procedures.
For what years are data available?	Data from the NHIS are available from 1957–present. Availability of data addressing particular issues varies based on when questions were added to the NHIS. The survey is redesigned on a regular basis; many questions of interest for children's environmental health indicators were modified or first asked with the redesign that was implemented in 1997. For environmental tobacco smoke (regular smoking in the home), comparable data are available for 1994, 2005, and 2010.
What is the frequency of data collection?	Continuous throughout the year.
What is the frequency of data release?	Annually.
Are the data comparable across time and space?	Survey design and administration are consistent across locations and from year to year. Many questions were revised or added in 1997, so data for prior years may not be comparable to data from 1997 to present.

National Health Interview Survey (NHIS)
(Used for Indicators E5, H1, H2, H6, H7, H8, and H9)

Can the data be stratified by race/ethnicity, income, and location (region, state, county or other geographic unit)?	Data can be stratified by race, ethnicity, income, and region (four regions only).

National Hospital Ambulatory Medical Care Survey (NHAMCS)
(Used for Indicator H3)

Brief description of the data set	The National Hospital Ambulatory Medical Care Survey (NHAMCS) is designed to collect information on the services provided in hospital emergency and outpatient departments and in ambulatory surgery centers.
Who provides the data set?	Centers for Disease Control and Prevention, National Center for Health Statistics.
How are the data gathered?	Sampled hospitals are noninstitutional general and short-stay hospitals located in all states and Washington DC, but exclude federal, military, and Veteran's Administration hospitals. Data from sampled visits are obtained on the demographic characteristics, expected source(s) of payments, patients' complaints, physician's diagnoses, diagnostic and screening services, procedures, types of health care professionals seen, and causes of injury.
What documentation is available describing data collection procedures?	See http://www.cdc.gov/nchs/ahcd/ahcd_data_collection.htm#nhamcs_collection for data collection documentation.
What types of data relevant for children's environmental health indicators are available from this database?	Relevant data include physicians' diagnoses for ambulatory visits to hospital emergency rooms and outpatient departments as well as demographic information.
What is the spatial representation of the database (national or other)?	NHAMCS sampling procedures provide nationally representative data, and may also be analyzed by four broad geographic regions: North, Midwest, South and West. In addition the database identifies whether or not the hospital is in an MSA. Analysis of data for any other geographic area (state, patient or facility zip code) is possible only by special arrangement with the NCHS Research Data Center.
Are raw data (individual measurements or survey responses) available?	Data for each year of the NHAMCS are available for download and analysis (http://www.cdc.gov/nchs/ahcd/ahcd_questionnaires.htm). Annual reports from the NHAMCS are also available (http://www.cdc.gov/nchs/ahcd/ahcd_products.htm) as are interactive data tables (http://www.cdc.gov/nchs/hdi.htm).
How are database files obtained?	http://www.cdc.gov/nchs/ahcd/ahcd_questionnaires.htm.
Are there any known data quality or data analysis concerns?	Responses to some demographic and other questions (birth year, sex, race, ethnicity, immediacy of being seen) are statistically imputed for survey participants lacking a reported response.
What documentation is available describing quality assurance procedures?	http://www.cdc.gov/nchs/ahcd/ahcd_questionnaires.htm summarizes the quality assurance procedures.

National Hospital Ambulatory Medical Care Survey (NHAMCS)
(Used for Indicator H3)

For what years are data available?	1992–present.
What is the frequency of data collection?	Continuously throughout the year.
What is the frequency of data release?	Annually.
Are the data comparable across time and space?	Changes to some survey questions or to the set of possible responses make their responses non-comparable for different time periods (e.g., reason for visit). Some diagnosis codes are not comparable from year to year due to annual revisions to the International Classification of Diseases (ICD-9).
Can the data be stratified by race/ethnicity, income, and location (region, state, county or other geographic unit)?	Data can be stratified by race, ethnicity, and region (four regions only). For 2006 and later, data can also be stratified by median income, % below poverty, % with college degree or higher level of education, and urban/rural classification for patient's zip code (the zip code itself is not included in the public release version).

National Hospital Discharge Survey (NHDS)
(Used for Indicator H3)

Brief description of the data set	The National Hospital Discharge Survey (NHDS) is an annual probability survey that collects information on the characteristics of inpatients discharged from non-federal short-stay hospitals in the United States.
Who provides the data set?	Centers for Disease Control and Prevention, National Center for Health Statistics.
How are the data gathered?	Sampled hospitals are short-stay general or children's general hospitals located in all states and Washington DC, with an average length of stay of fewer than 30 days and six or more beds staffed for patients use. Federal, military, and Veteran's Administration hospitals are excluded, as are hospital units of institutions. Data from sampled hospital discharges are obtained on the demographic characteristics and physician's diagnoses.
What documentation is available describing data collection procedures?	See http://www.cdc.gov/nchs/nhds/nhds_collection.htm for data collection documentation.
What types of data relevant for children's environmental health indicators are available from this database?	Relevant data include physician's diagnoses for discharges from hospitals, as well as demographic information.
What is the spatial representation of the database (national or other)?	NHDS sampling procedures provide nationally representative data, and may also be analyzed by four broad geographic regions: North, Midwest, South and West. Analysis of data for any other geographic area (state, patient zip code) is possible only by special arrangement with the NCHS Research Data Center.
Are raw data (individual measurements or survey responses) available?	Individual hospital discharge data are available. Some survey responses are not publicly released.
How are database files obtained?	http://www.cdc.gov/nchs/nhds/nhds_questionnaires.htm.

National Hospital Discharge Survey (NHDS)
(Used for Indicator H3)

Are there any known data quality or data analysis concerns?	The survey is designed to represent in-patient discharges to short-stay general or children's general hospitals, excluding federal and military hospitals. Data are obtained from a detailed complex survey sampling scheme including samplings of hospitals and discharges within hospitals. Survey responses must be appropriately weighted using the provided analysis weights to obtain national estimates. The public release version includes coefficients for variance estimation equations for approximate variance estimation. The available data are for discharges and not admissions. Some age and sex values were imputed.
What documentation is available describing quality assurance procedures?	http://www.cdc.gov/nchs/data/series/sr_01/sr01_039.pdf includes a description of the quality assurance procedures.
For what years are data available?	1965–present.
What is the frequency of data collection?	Continuously throughout the year.
What is the frequency of data release?	Annually.
Are the data comparable across time and space?	Some diagnosis codes are not comparable from year to year due to annual revisions to the International Classification of Diseases (ICD-9).
Can the data be stratified by race/ethnicity, income, and location (region, state, county or other geographic unit)?	Data can be stratified by race and region (four regions only). NHDS does not release information on Hispanic ethnicity or income of patients due to high non-response rates for these items. Although race is reported, there are also high non-response rates for race.

National Survey of Lead and Allergens in Housing (NSLAH)
(Used for Indicator E6)

Brief description of the data set	A nationally representative sample of homes was selected for this survey. NSLAH measured levels of lead, lead hazards, allergens, and endotoxins in homes nationwide. The lead data included the levels of lead in paint, dust and soil, and levels of paint deterioration.
Who provides the data set?	National Institute of Environmental Health Sciences (NIEHS) and U.S. Department of Housing and Urban Development (HUD).
How are the data gathered?	A nationally representative sample of 1,984 housing units in which children could reside was drawn from 75 primary sampling units (metropolitan statistical areas or counties), and 831 eligible housing units were recruited and completed a survey. Measurements of lead paint and dust were gathered from the surveyed homes in specific rooms; soil lead was gathered from the surveyed homes through core sampling.

National Survey of Lead and Allergens in Housing (NSLAH)
(Used for Indicator E6)

What documentation is available describing data collection procedures?	National Survey of Lead and Allergens in Housing. Final Report. Volume I. Analysis of Lead Hazards. April 2001. At http://www.nchh.org/Portals/0/Contents/Article0312.pdf. National Survey of Lead and Allergens in Housing. Draft Final Report. Volume II. Design and Methodology. March 2001.
What types of data relevant for children's environmental health indicators are available from this database?	Relevant information includes lead-based paint hazards in housing (prevalence, deteriorated, loadings), dust lead, soil lead (children's play areas, yard), indoor allergens (dust mite, cockroach, cat, dog, mouse, Alternaria), endotoxins, race, ethnicity, age, sex, income, asthma and allergies health history, housing characteristics (building age; heating, cooling, and cooking equipment), pets, and pests. Pesticide data were not collected.
What is the spatial representation of the database (national or other)?	National.
Are raw data (individual measurements or survey responses) available?	Not currently.
How are database files obtained?	Data have not been publicly released. HUD provided data files directly to EPA for purposes of developing an indicator for America's Children and the Environment. Summary tables are obtained from: National Survey of Lead and Allergens in Housing. Final Report. Volume I. Analysis of Lead Hazards. April 2001. At http://www.nchh.org/Portals/0/Contents/Article0312.pdf. NSLAH data summaries are also available in "American Healthy Homes Survey, Final Report, Lead and Arsenic Findings," June 2009, Draft Final Report (not yet publicly released).
Are there any known data quality or data analysis concerns?	http://www.nchh.org/Portals/0/Contents/Article0312.pdf. Chapter 7 of the study report outlines sources of error in data collection and analysis. Concerns include: response rate, non-response bias, and measurement errors.
What documentation is available describing quality assurance procedures?	http://www.nchh.org/Portals/0/Contents/Article0312.pdf. Chapter 7, sections 7.4 ("Quality of Field Data Collection") and section 7.5 ("Paint Testing Quality Assurance"), pages 7-32 through 7-36.
For what years are data available?	The main field study (survey and in-home lead) was conducted in 1998-1999, with an augmentation of the soil sampling in 2000.
What is the frequency of data collection?	Data were collected once, from 1998-1999, with an augmentation of the soil sampling in 2000.
What is the frequency of data release?	Raw data have not been publicly released. The report was published in April 2001.
Are the data comparable across time and space?	As a one-time survey, time comparisons within the NSLAH are not possible, but NSLAH results can be compared with the later AHHS survey (2005-2006). Geographic comparisons should be possible using the raw data, since the same data were collected at all homes.

National Survey of Lead and Allergens in Housing (NSLAH)
(Used for Indicator E6)

Can the data be stratified by race/ethnicity, income, and location (region, state, county or other geographic unit)?	Data can be stratified by residents' age, race, ethnicity, and household income, as well as by census region. Data can also be stratified by year of home construction, and by the housing type (rented or owned). However, estimates of lead hazards in the home for children ages 0 to 5 years broken out by race/ethnicity and income are not statistically reliable, due to the relatively small number of homes in each group.

National Vital Statistics System (NVSS)
(Used for Indicators H12 and H13)

Brief description of the data set	The National Vital Statistics System (NVSS) collects and disseminates data on births, deaths, marriages, divorces, and fetal deaths from vital event registration systems. The results of NVSS provide nearly complete data to track these vital statistics nationwide.
Who provides the data set?	Centers for Disease Control and Prevention, National Center for Health Statistics.
How are the data gathered?	Data are obtained from birth, death, marriage and divorce certificates collected by the various jurisdictions legally responsible for registration of these events.
What documentation is available describing data collection procedures?	See http://www.cdc.gov/nchs/data_access/Vitalstatsonline.htm for user's guides by calendar year.
What types of data relevant for children's environmental health indicators are available from this database?	Relevant data include births, deaths, marriages, divorces, demographic information, cause of mortality, and state and county (data prior to 2004 only). Birth data include birth order, period of gestation, method of delivery, birth weight, abnormal conditions of the newborn, and congenital abnormalities.
What is the spatial representation of the database (national or other)?	Nearly complete national registration data have been collected since 1985. State and county locations are recorded until 2004.
Are raw data (individual measurements or survey responses) available?	Data for each calendar year are available for download and analysis from (http://www.cdc.gov/nchs/data_access/Vitalstatsonline.htm) with records for each birth, death, marriage, or divorce certificate. Annual and monthly reports from the NVSS are available (http://www.cdc.gov/nchs/nvss/nvss_products.htm). Raw NVSS data are also available from the National Bureau of Economic Research at http://www.nber.org/data/#demographic Personal identification data (e.g., names) is not available.
How are database files obtained?	Raw data: http://www.cdc.gov/nchs/data_access/Vitalstatsonline.htm and http://www.nber.org/data/#demographic. Queriable, less detailed data set including births, deaths, and fetal deaths, with broad response categories: CDC WONDER at http://wonder.cdc.gov Prebuilt or user-built birth data tables are available at http://www.cdc.gov/nchs/VitalStats.htm.

National Vital Statistics System (NVSS)
(Used for Indicators H12 and H13)

Are there any known data quality or data analysis concerns?	For approximately 0.5% of the birth records, the mother's race was not stated and in those cases the mother's race was statistically imputed. From 2003, some states allowed reporting of multiple races, and in those cases the multiple race was bridged to a primary race using statistical methods.
What documentation is available describing quality assurance procedures?	See http://www.cdc.gov/nchs/data_access/Vitalstatsonline.htm for user's guides by calendar year.
For what years are data available?	Online data: Births 1968-2009. Mortality multiple cause 1968-2009. Fetal death 1982-2006.
What is the frequency of data collection?	Continuous.
What is the frequency of data release?	Annually.
Are the data comparable across time and space?	Some response variables have response categories that have changed over time. Cause of mortality International Classification of Diseases coding systems have changed over time. Birth certificate categories changed between the 1989 and 2003 versions of the birth certificates.
Can the data be stratified by race/ethnicity, income, and location (region, state, county or other geographic unit)?	Data can be stratified by race and ethnicity. State and county data are complete prior to 1989, contain county and city information only for counties with populations above 100,000 for 1989 to 2004, and contain no location information from 2005 forward. There are no income data.

Pesticide Data Program (PDP)
(Used for Indicator E9)

Brief description of the data set	The Pesticide Data Program (PDP), initiated in 1991, focuses on measuring pesticide residues in foods that are important parts of children's diets, including apples, apple juice, bananas, carrots, grapes, green beans, orange juice, peaches, pears, potatoes, and tomatoes. Samples are collected from food distribution centers in 10 states across the country. Different foods are sampled each year and then analyzed in various state and federal laboratories for the presence of residues of about 300 pesticides and similar chemicals.
Who provides the data set?	U.S. Department of Agriculture, Agricultural Marketing Service.
How are the data gathered?	Food and water samples are collected by the participating states. Food samples are prepared as if for consumption (washed, peeled, etc.). The pesticide residues are measured at state and federal laboratories, and compiled into a database managed by USDA.
What documentation is available describing data collection procedures?	Standard operating procedures, including data collection, are described here: http://www.ams.usda.gov/AMSv1.0/ams.fetchTemplateData.do?template=TemplateG&topNav=&leftNav=ScienceandLaboratories&page=PDPProgramSOPs&description=PDP+Standard+Operating+Procedures+(SOPs)&acct=pestcddataprg.

Pesticide Data Program (PDP)
(Used for Indicator E9)

What types of data relevant for children's environmental health indicators are available from this database?	Relevant data include pesticide residue concentrations measured in samples of fruits, vegetables, grains, and other food and drink products, particularly foods most likely consumed by infants and young children.
What is the spatial representation of the database (national or other)?	National. In 2009, sampling services for food samples were provided by 10 states (California, Colorado, Florida, Maryland, Michigan, New York, Ohio, Texas, Washington, and Wisconsin). Approximately half of the U.S. population resides in these 10 states.
Are raw data (individual measurements or survey responses) available?	Individual food and drink sample data are available.
How are database files obtained?	Data files are freely available from: http://www.ams.usda.gov/AMSv1.0/ams.fetchTemplateData.do?template=Template G&topNav=&leftNav=ScienceandLaboratories&page=PDPDownloadData/Reports&description=Download+PDP+Data/Reports&acct=pestcddataprg.
Are there any known data quality or data analysis concerns?	Detection limits vary by pesticide, laboratory, commodity and over time. The list of commodities sampled varies from year to year. The set of pesticides analyzed has generally expanded over time.
What documentation is available describing quality assurance procedures?	http://www.ams.usda.gov/AMSv1.0/ams.fetchTemplateData.do?template=TemplateG&topNav=&leftNav=ScienceandLaboratories&page=PDPProgramSOPs&description=PDP+Standard+Operating+Procedures+(SOPs)&acct=pestcddataprg includes documentation on quality assurance/quality control.
For what years are data available?	1992 – present.
What is the frequency of data collection?	Annually.
What is the frequency of data release?	Annually.
Are the data comparable across time and space?	Detection limits vary by pesticide, laboratory, commodity and over time. The list of commodities sampled varies considerably from year to year. The set of pesticides analyzed has also varied with time.
Can the data be stratified by race/ethnicity, income, and location (region, state, county or other geographic unit)?	Data can be stratified by state where sample is collected and state or country of origin.

Safe Drinking Water Information System Federal Version (SDWIS/FED)
(Used for Indicators E7 and E8)

Brief description of the dataset	SDWIS/FED is EPA's national database that manages and collects public water system information from states, including reports of drinking water standard violations, reporting and monitoring violations, and other basic information, such as water system location, type, and population served. (http://water.epa.gov/scitech/datait/databases/drink/sdwisfed/basicinformation.cfm)
Who provides the data set?	U.S. Environmental Protection Agency, Office of Ground Water and Drinking Water.
How are the data gathered?	Violation data for all public water systems are provided by states and EPA regions. Public water systems are required to follow treatment and reporting requirements, to measure contaminant levels, and to report violations of standards.
What documentation is available describing data collection procedures?	Information is available at http://water.epa.gov/scitech/datait/databases/drink/sdwisfed/basicinformation.cfm
What types of data relevant for children's environmental health indicators are available from this database?	Relevant data include violations of national standards for drinking water—due to contaminant levels exceeding allowable levels, violations of treatment requirements, or violations of monitoring and reporting requirements—and total population served by each public water system.
What is the spatial representation of the database (national or other)?	SDWIS/FED includes data for all public water systems in the United States.
Are raw data (individual measurements or survey responses) available?	Separate reports for each violation of drinking water standards or monitoring and reporting requirements for individual public water systems are available; measured contaminant levels are not available in SDWIS/FED.
How are database files obtained?	SDWIS/FED violation and inventory data were obtained from OGWDW staff who compiled the data into a dataset listing the water system, state, violation type and code, chemical contaminant code, violation dates, and the population served
Are there any known data quality or data analysis concerns?	The estimated number of people served by each public water system is approximate. Estimates are updated when there is a significant change in a water system's population. Some water systems serve more than one state (the primary state is reported) and water systems often serve more than one county. Many people obtain drinking water from more than one public water system. Although the data are largely accurate, EPA is aware of underreporting of some violation data in SDWIS/FED. Several states have recently found and corrected significant errors in their violation databases.
What documentation is available describing quality assurance procedures?	EPA routinely evaluates drinking water programs by conducting data verification audits, which evaluate state compliance decisions and reporting to SDWIS/FED. Every three years, the agency prepares summary evaluations based on the data verification. These evaluations are available at: http://www.epa.gov/safewater/databases/sdwis/datareliability.html.
For what years are data available?	1976 – present.
What is the frequency of data collection?	Quarterly.
What is the frequency of data release?	Quarterly.

Safe Drinking Water Information System Federal Version (SDWIS/FED)
(Used for Indicators E7 and E8)

Are the data comparable across time and space?	Violations across time are often not comparable because of changes in regulations and changes in drinking water standards (maximum contaminant levels), and variability over time in monitoring and reporting violations. Data may not be geographically comparable due to variations in state enforcement and database quality.
Can the data be stratified by race/ethnicity, income, and location (region, state, county or other geographic unit)?	Data can be stratified by state and county, with some uncertainty because boundaries of public water systems do not coincide with state and county boundaries. The state and county reported in SDWIS/FED are the primary state and county served by the water system. Data cannot be stratified by demographic characteristics because SDWIS/FED reports only the total population served by a public water system, without any demographic information.

Surveillance, Epidemiology, and End Results (SEER) Program
(Used for Indicators H4 and H5)

Brief description of the data set	The Surveillance, Epidemiology, and End Results (SEER) program is an authoritative source of information on cancer incidence and mortality in the United States. SEER collects and publishes cancer data from a set of 17 population-based regional cancer registries located throughout the country.
Who provides the data set?	National Cancer Institute.
How are the data gathered?	Data on all diagnosed cancer cases in the geographical area for a cancer registry are compiled each year and submitted to SEER. Mortality data for all causes of death in the entire US are collected by the National Center for Health Statistics. Population data are provided by the Census Bureau.
What documentation is available describing data collection procedures?	See http://seer.cancer.gov/index.html for detailed description of SEER organization and data collection practices.
What types of data relevant for children's environmental health indicators are available from this database?	Relevant data include cancer incidence and mortality (including cancer type, tumor site, tumor morphology, and stage at diagnosis, first course of treatment, and follow-up for vital status), demographic information, and state and county.
What is the spatial representation of the database (national or other)?	The most recent SEER database for cancer incidence has 18 population-based cancer registries in 14 states and covers 28% of the U.S. population. A subset of the current SEER includes 13 population-based cancer registries in 10 states and covers 14% of the U.S. population. The registries include: the Alaska Native, Atlanta, Connecticut, Detroit, Hawaii, Iowa, Los Angeles, New Mexico, Rural Georgia, San Francisco-Oakland, San Jose-Monterey, Seattle-Puget Sound, and Utah tumor registries. These data are taken to represent cancer incidence for the entire United States. See below for further discussion. The SEER database also includes national mortality data for all causes of death from the National Vital Statistics System.
Are raw data (individual measurements or survey responses) available?	Yes.

Surveillance, Epidemiology, and End Results (SEER) Program
(Used for Indicators H4 and H5)

How are database files obtained?	http://seer.cancer.gov/data/access.html includes various methods of accessing SEER data. Raw data for each person can be obtained. For ACE, annual summary cancer incidence and mortality rate data were obtained using SEER*Stat software available from the same website.
Are there any known data quality or data analysis concerns?	The population covered by SEER is comparable to the general U.S. population with regard to measures of poverty and education. The SEER population tends to be somewhat more urban and has a higher proportion of foreign-born persons than the general U.S. population. Cancer mortality data for North Dakota and South Carolina have significant percentages of persons with unknown ethnicity.
What documentation is available describing quality assurance procedures?	http://seer.cancer.gov/qi/index.html provides information on SEER quality improvement.
For what years are data available?	Data are available from the original 9 SEER registries from 1973–present, but over time the coverage of SEER has increased to cover more individuals and geographic regions. See below for further discussion.
What is the frequency of data collection?	Annually.
What is the frequency of data release?	Annually.
Are the data comparable across time and space?	The national coverage has increased over time from 9 to 18 cancer registries. Time comparisons should be between the same set of registries. Thus, long-term trend comparisons use SEER 9 (the original 9 registries) beginning with 1973 and cover the smallest percentage (9.5% in 2000) of the U.S. population. The full set of registries (SEER 18) has the broadest coverage (28%), but provides data only from the year 2000 forward. SEER 13 covers 14% of the population and provides data from 1992 forward. Population coverage varies by state.
	Over time the cancer classifications used by SEER have changed. As scientific knowledge has improved, some cancers that were once more generally classified are now given a more exact definition. However, with each annual update SEER updates the current and previous years' data to reflect the latest classification scheme. The one exception would be for conditions that are now classified as malignant cancers but were not previously and were therefore not registered by the SEER cancer registries for earlier years. This applies only to a limited number of rare tumor types, so it is not expected to contribute to changes in cancer incidence over time.
Can the data be stratified by race/ethnicity, income, and location (region, state, county or other geographic unit)?	The data can be stratified by race and ethnicity, as well as median county income. Incidence data within the given SEER registry can be geographically stratified by state and county Mortality data can be geographically stratified by state and county.

Texas Birth Defects Registry
(Used for Measure S1)

Brief description of the data set	Since 1994, the Texas Birth Defects Epidemiology and Surveillance Branch has maintained the Texas Birth Defects Registry, a population-based birth defects surveillance system. Through multiple sources of information, the Registry monitors all births in Texas (approximately 380,000 each year) and identifies cases of birth defects. The Texas Registry staff routinely visit all hospitals and birthing centers where affected babies are delivered or treated. There they review logs and discharge lists to find potential cases, and then review medical records of the potential cases to identify actual cases with birth defects.
Who provides the data set?	Texas Department of State Health Services.

How are the data gathered?	The Texas Birth Defects Registry uses active surveillance:
	▪ Does not require reporting by hospitals or medical professionals.
	▪ Trained program staff regularly visit medical facilities.
	  ▪ Have legislative authority to review all relevant records.
	  ▪ Review log books, hospital discharge lists, and other records to identify potential cases.
	  ▪ Review medical charts for potential cases to identify those with birth defects.
	▪ Program staff use medical charts for each potential birth defect identified. Records in the birth defect registry are matched to birth certificates and fetal death certificates filed with the Vital Statistics Unit of Texas DSHS to gather demographic data.

What documentation is available describing data collection procedures?	Methods report available at: http://www.dshs.state.tx.us/birthdefects/Data/99-04_Methods.pdf.
What types of data relevant for children's environmental health indicators are available from this database?	Relevant data include the following birth defects: central nervous system defects; ear and eye defects; cardiac and circulatory defects; respiratory defects; oral clefts; gastrointestinal defects; genitourinary defects, including hypospadias; musculoskeletal defects; and chromosomal defects.
What is the spatial representation of the database (national or other)?	Prior to 1999: selected health service regions of Texas. 1999 onward: entire state of Texas.
Are raw data (individual measurements or survey responses) available?	Raw data for 1996-2007 are available through special request.
How are database files obtained?	Routinely published tabulations of data for 1995-2007 (by birth defect, overall and broken down by selected demographic factors) can be accessed at: http://www.dshs.state.tx.us/birthdefects/Data/reports.shtm. A queriable database of data for 1999-2006, where users can design their own tabulations, can be found at: http://soupfin.tdh.state.tx.us/bdefdoc.htm. Other tabulations or raw data are also available through 2007, by written request. Go to http://www.dshs.state.tx.us/birthdefects/Data/reports.shtm and click on "Birth Defects Data Request and Access Policy."

Texas Birth Defects Registry
(Used for Measure S1)

Are there any known data quality or data analysis concerns?	Registry only includes birth defects diagnosed within one year of delivery (with the exception of fetal alcohol syndrome). Secondly, diagnoses made outside Texas or in Texas facilities that staff members do not have access to are excluded.
	Due to flooding during June 2001, several hospitals in Houston lost medical records. An estimated 50 fetuses and infants were born during this time with diagnosed birth defects at the affected hospitals.
	Data collected from medical records are subject to differences in clinical practice.
What documentation is available describing quality assurance procedures?	An article in *Birth Defects Research Part A: Clinical and Molecular Teratology* highlights quality issues:
	Miller, E. 2006. Evaluation of the Texas Birth Defects Registry: An active surveillance system. *Birth Defects Research Part A: Clinical and Molecular Teratology.* 76(11): 787-792.
	See: http://www3.interscience.wiley.com/journal/113455770/abstract.
For what years are data available?	1996-2007.
What is the frequency of data collection?	Ongoing.
What is the frequency of data release?	Annual.
Are the data comparable across time and space?	Yes, generally. However, data from different locations may not be comparable due to differences in clinical practice. Identification of some birth defects may change over time as more sensitive examinations and technologies lead to more accurate recording of birth defects and/or better diagnosis of subtle defects. Prior to 1999, only certain regions were included in the registry.
Can the data be stratified by race/ethnicity, income, and location (region, state, county or other geographic unit)?	Using the interactive data query system (http://soupfin.tdh.state.tx.us/bdefdoc.htm), data can be stratified by mother's race/ethnicity, mother's age group, infant's sex, and geographical unit (statewide, public health region, county, and border residence status.)

Appendix C: Alignment of ACE3 Indicators with Healthy People 2020 Objectives

Appendix C: Alignment of ACE3 Indicators with Healthy People 2020 Objectives

Healthy People 2020 (www.healthypeople.gov), an initiative of the U.S. Department of Health and Human Services, provides science-based, 10-year national objectives for improving the health of all Americans. This appendix provides examples of the alignment of the topics and indicators presented in *America's Children and the Environment, Third Edition (ACE3)* with Healthy People 2020 objectives.

Objectives in Healthy People 2020 are organized by topic area; the table below provides a key to the topic area acronyms used in the objective titles.

EH	Environmental Health
TU	Tobacco Use
RD	Respiratory Disease
MICH	Maternal Infant and Child Health
NWS	Nutrition and Weight Status
PA	Physical Activity

Environments and Contaminants

Criteria Air Pollutants

ACE3 Indicators

- E1: Percentage of children ages 0 to 17 years living in counties with pollutant concentrations above the level of the current air quality standards, 1999–2009
- E2: Percentage of children ages 0 to 17 years living in counties with 8-hour ozone and 24-hour PM2.5 concentrations above the levels of air quality standards, by frequency of occurrence, 2009
- E3: Percentage of days with good, moderate, or unhealthy air quality for children ages 0 to 17 years, 1999–2009

Healthy People 2020 Objective

- EH-1: Reduce the number of days the Air Quality Index (AQI) exceeds 100

Hazardous Air Pollutants

ACE3 Indicator

■ E4: Percentage of children ages 0 to 17 years living in census tracts where estimated hazardous air pollutant concentrations were greater than health benchmarks in 2005

Healthy People 2020 Objective

■ EH-3: Reduce air toxic emissions to decrease the risk of adverse health effects caused by airborne toxics
 ▪ EH-3.1: Mobile sources
 ▪ EH-3.2: Area sources
 ▪ EH-3.3: Major sources

Indoor Environments

ACE3 Indicators

■ E5: Percentage of children ages 0 to 6 years regularly exposed to environmental tobacco smoke in the home, by family income, 1994, 2005, and 2010
■ E6: Percentage of children ages 0 to 5 years living in homes with interior lead hazards, 1998–1999 and 2005–2006

Healthy People 2020 Objectives

■ TU–11: Reduce the proportion of nonsmokers exposed to secondhand smoke
 ▪ TU–11.1: Children aged 3 to 11 years
■ EH-8: Reduce blood lead levels in children
 ▪ EH-8.1: Eliminate elevated blood lead levels in children
 ▪ EH-8.2: Reduce the mean blood lead levels in children
■ EH-18: Reduce the number of U.S. homes that are found to have lead-based paint or related hazards
 ▪ EH-18.1: Reduce the number of U.S. homes that are found to have lead-based paint
 ▪ EH-18.2: Reduce the number of U.S. homes that have paint-lead hazards
 ▪ EH-18.3: Reduce the number of U.S. homes that have dust-lead hazards

Drinking Water Contaminants

ACE3 Indicators

■ E7: Estimated percentage of children ages 0 to 17 years served by community water systems that did not meet all applicable health-based drinking water standards, 1993–2009

- E8: Estimated percentage of children ages 0 to 17 years served by community water systems with violations of drinking water monitoring and reporting requirements, 1993–2009

Healthy People 2020 Objectives

- EH-4: Increase the proportion of persons served by community water systems who receive a supply of drinking water that meets the regulations of the Safe Drinking Water Act
- EH-5: Reduce waterborne disease outbreaks arising from water intended for drinking among persons served by community water systems

Chemicals in Food

ACE3 Indicator

- E9: Percentage of sampled apples, carrots, grapes, and tomatoes with detectable residues of organophosphate pesticides, 1998–2009

Healthy People 2020 Objective

- None

Contaminated Lands

ACE3 Indicators

- E10: Percentage of children ages 0-17 years living within one mile of Superfund and Corrective Action sites that may not have all human health protective measures in place, 2009
- E11: Distribution by race/ethnicity and family income of children living near selected contaminated lands in 2009, compared with the distribution by race/ethnicity and income of children in the general U.S. population

Healthy People 2020 Objective

- EH-9: Minimize the risks to human health and the environment posed by hazardous sites

Biomonitoring

Lead

ACE3 Indicators

- B1: Lead in children ages 1 to 5 years: Median and 95th percentile concentrations in blood, 1976–2010
- B2: Lead in children ages 1 to 5 years: Median concentrations in blood, by race/ethnicity and family income, 2007–2010

Healthy People 2020 Objectives

- EH-8: Reduce blood lead levels in children
  - EH-8.1: Eliminate elevated blood lead levels in children
  - EH-8.2: Reduce the mean blood lead levels in children
- EH-16.7: Inspect drinking water outlets for lead
- EH-17: (Developmental) Increase the proportion of persons living in pre-1978 housing that has been tested for the presence of lead-based paint or related hazards
  - EH-17.1: (Developmental) Increase the proportion of pre-1978 housing that has been tested for the presence of lead-based paint
  - EH17.2: (Developmental) Increase the proportion of pre-1978 housing that has been tested for the presence of paint-lead hazards
  - EH-17.3: (Developmental) Increase the proportion of pre-1978 housing that has been tested for the presence of lead in dust
  - EH-17.4: (Developmental) Increase the proportion of pre-1978 housing that has been tested for the presence of lead in soil
- EH-18: Reduce the number of U.S. homes that are found to have lead-based paint or related hazards
  - EH-18.1: Reduce the number of U.S. homes that are found to have lead-based paint
  - EH-18.2: Reduce the number of U.S. homes that have paint-lead hazards
  - EH-18.3: Reduce the number of U.S. homes that have dust-lead hazards
  - EH-18.4: Reduce the number of U.S. homes that have soil-lead hazards
- EH-20: Reduce exposure to selected environmental chemicals in the population, as measured by blood and urine concentrations of the substances or their metabolites
  - EH-20.3: Lead
- EH-22: Increase the number of States, Territories, Tribes, and the District of Columbia that monitor diseases or conditions that can be caused by exposure to environmental hazards
  - EH-22.1: Lead poisoning

Mercury

ACE3 Indicator

- B3: Mercury in women ages 16 to 49 years: Median and 95th percentile concentrations in blood, 1999–2010

Healthy People 2020 Objectives

- EH-20: Reduce exposure to selected environmental chemicals in the population, as measured by blood and urine concentrations of the substances or their metabolites
  - EH-20.4: Mercury, children aged 1 to 5 years

- EH-20.5: Mercury, females aged 16 to 49 years
- EH-22: Increase the number of States, Territories, Tribes, and the District of Columbia that monitor diseases or conditions that can be caused by exposure to environmental hazards
  - EH-22.3: Mercury poisoning

Cotinine

ACE3 Indicators

- B4: Cotinine in nonsmoking children ages 3 to 17 years: Median and 95th percentile concentrations in blood serum, 1988–2010
- B5: Cotinine in nonsmoking women ages 16 to 49 years: Median and 95th percentile concentrations in blood serum, 1988–2010

Healthy People 2020 Objectives

- TU–11: Reduce the proportion of nonsmokers exposed to secondhand smoke
  - TU–11.1 Children aged 3 to 11 years
- TU–13: Establish laws in States, District of Columbia, Territories, and Tribes on smoke-free indoor air that prohibit smoking in public places and worksites
  - TU–13.6 Commercial daycare centers
  - TU–13.11 Vehicles with children

Perfluorochemicals (PFCs)

ACE3 Indicator

- B6: Perfluorochemicals in women ages 16 to 49 years: Median concentrations in blood serum, 1999–2008

Healthy People 2020 Objective

- None

Polychlorinated biphenyls (PCBs)

ACE3 Indicator

- B7: PCBs in women ages 16 to 49 years: Median concentrations in blood serum, by race/ethnicity and family income, 2001–2004

Healthy People 2020 Objective

- EH-20: Reduce exposure to selected environmental chemicals in the population, as measured by blood and urine concentrations of the substances or their metabolites.

- EH-20.12 PCB 153, representative of nondioxin-like PCBs.
- EH-20.13 PCB 126, representative of dioxin-like PCBs.

Polybrominated diphenyl ethers (PBDEs)

ACE3 Indicator

- B8: PBDEs in women ages 16 to 49 years: Median concentrations in blood serum, by race/ethnicity and family income, 2003–2004

Healthy People 2020 Objective

- EH-20: Reduce exposure to selected environmental chemicals in the population, as measured by blood and urine concentrations of the substances or their metabolites
  - EH-20.18: BDE 47 (2,2',4,4'-tetrabromodiphenyl ether)

Phthalates

ACE3 Indicators

- B9: Phthalate metabolites in women ages 16 to 49 years: Median concentrations in urine, 1999–2008
- B10: Phthalate metabolites in children ages 6 to 17 years: Median concentrations in urine, 1999–2008

Healthy People 2020 Objective

- EH-20: Reduce exposure to selected environmental chemicals in the population, as measured by blood and urine concentrations of the substances or their metabolites
  - EH-20.17: Mono-n-butyl phthalate

Bisphenol A (BPA)

ACE3 Indicators

- B11: Bisphenol A in women ages 16 to 49 years: Median and 95th percentile concentrations in urine, 2003–2010
- B12: Bisphenol A in children ages 6 to 17 years: Median and 95th percentile concentrations in urine, 2003–2010

Healthy People 2020 Objective

- EH-20: Reduce exposure to selected environmental chemicals in the population, as measured by blood and urine concentrations of the substances or their metabolites
  - EH-20.15: Bisphenol A

Perchlorate

ACE3 Indicator

- B13: Perchlorate in women ages 16 to 49 years: Median and 95th percentile concentrations in urine, 2001–2008

Healthy People 2020 Objective

- EH-20: Reduce exposure to selected environmental chemicals in the population, as measured by blood and urine concentrations of the substances or their metabolites.
  - EH-20.16: Perchlorate

Health

Respiratory Diseases

ACE3 Indicators

- H1: Percentage of children ages 0 to 17 years with asthma, 1997–2010
- H2: Percentage of children ages 0 to 17 years reported to have current asthma, by race/ethnicity and family income, 2007–2010
- H3: Children's emergency room visits and hospitalizations for asthma and other respiratory causes, ages 0 to 17 years, 1996–2008

Healthy People 2020 Objectives

- RD–1: Reduce asthma deaths
  - RD–1.1: Children and adults under age 35 years
- RD–2: Reduce hospitalizations for asthma
  - RD–2.1: Children under age 5 years
- RD–3: Reduce hospital emergency department visits for asthma
  - RD–3.1: Children under age 5 years
- RD–4: Reduce activity limitations among persons with current asthma
- RD–5: Reduce the proportion of persons with asthma who miss school or work days
  - RD–5.1: Reduce the proportion of children aged 5 to 17 years with asthma who miss school days
- RD–6: Increase the proportion of persons with current asthma who receive formal patient education
- RD–7: Increase the proportion of persons with current asthma who receive appropriate asthma care according to National Asthma Education and Prevention Program (NAEPP) guidelines

- RD–7.1: Persons with current asthma who receive written asthma management plans from their health care provider

- RD–7.2: Persons with current asthma with prescribed inhalers who receive instruction on their use

- RD–7.3: Persons with current asthma who receive education about appropriate response to an asthma episode, including recognizing early signs and symptoms or monitoring peak flow results

- RD–7.4: Increase the proportion of persons with current asthma who do not use more than one canister of short-acting inhaled beta agonist per month

- RD–7.5: Persons with current asthma who have been advised by a health professional to change things in their home, school, and work environments to reduce exposure to irritants or allergens to which they are sensitive

- RD– 7.6: (Developmental) Persons with current asthma who have had at least one routine follow-up visit in the past 12 months

- RD– 7.7: (Developmental) Persons with current asthma whose doctor assessed their asthma control in the past 12 months

- RD–8: Increase the numbers of States, Territories, and the District of Columbia with a comprehensive asthma surveillance system for tracking asthma cases, illness, and disability at the State level

Childhood Cancer

ACE3 Indicators

- H4: Cancer incidence and mortality for children ages 0 to 19 years, 1992–2009
- H5: Cancer incidence for children ages 0 to 19 years by type, 1992–2006

Healthy People 2020 Objectives

- C–1: Reduce the overall cancer death rate
- C–12: Increase the number of central, population-based registries from the 50 States and the District of Columbia that capture case information on at least 95 percent of the expected number of reportable cancers
- C–20: Increase the proportion of persons who participate in behaviors that reduce their exposure to harmful ultraviolet (UV) irradiation and avoid sunburn
  - C–20.1: (Developmental) Reduce the proportion of adolescents in grades 9 through 12 who report sunburn
  - C–20.3: Reduce the proportion of adolescents in grades 9 through 12 who report using artificial sources of ultraviolet light for tanning
  - C–20.5: Increase the proportion of adolescents in grades 9 through 12 who follow protective measures that may reduce the risk of skin cancer

- ECBP–4: Increase the proportion of elementary, middle, and senior high schools that provide school health education to promote personal health and wellness in the following areas: hand washing or hand hygiene; oral health; growth and development; sun safety and skin cancer prevention; benefits of rest and sleep; ways to prevent vision and hearing loss; and the importance of health screenings and checkups
  - ECBP–4.4: Sun safety or skin cancer prevention

Neurodevelopmental Disorders

ACE3 Indicators

- H6: Percentage of children ages 5 to 17 years reported to have attention-deficit/hyperactivity disorder, by sex, 1997–2010
- H7: Percentage of children ages 5 to 17 years reported to have a learning disability, by sex, 1997–2010
- H8: Percentage of children ages 5 to 17 years reported to have autism, 1997–2010
- H9: Percentage of children ages 5 to 17 years reported to have intellectual disability (mental retardation), 1997–2010

Healthy People 2020 Objectives

- EMC–2.4: Increase the proportion of parents who receive information from their doctors or other health care professionals when they have a concern about their children's learning, development, or behavior
- MICH–29: Increase the proportion of young children with an Autism Spectrum Disorder (ASD) and other developmental delays who are screened, evaluated, and enrolled in early intervention services in a timely manner
  - MICH–29.1: Increase the proportion of young children who are screened for an Autism Spectrum Disorder (ASD) and other developmental delays by 24 months of age
  - MICH–29.2: Increase the proportion of children with an ASD with a first evaluation by 36 months of age
  - MICH–29.3: Increase the proportion of children with an ASD enrolled in special services by 48 months of age
  - MICH–29.4: (Developmental) Increase the proportion of children with a developmental delay with a first evaluation by 36 months of age
  - MICH–29.5: (Developmental) Increase the proportion of children with a developmental delay enrolled in special services by 48 months of age

Obesity

ACE3 Indicators

- H10: Percentage of children ages 2 to 17 years who were obese, 1976–2008

- H11: Percentage of children ages 2 to 17 years who were obese, by race/ethnicity and family income, 2005–2008

Healthy People 2020 Objectives

- NWS–5: Increase the proportion of primary care physicians who regularly measure the body mass index of their patients
  - NWS–5.2: Increase the proportion of primary care physicians who regularly assess body mass index (BMI) for age and sex in their child or adolescent patients
- NWS–10: Reduce the proportion of children and adolescents who are considered obese.
  - NWS–10.1: Children aged 2 to 5 years
  - NWS–10.2: Children aged 6 to 11 years
  - NWS–10.3: Adolescents aged 12 to 19 years
- PA–13: (Developmental) Increase the proportion of trips made by walking
  - PA–13.2: Children and adolescents aged 5 to 15 years, trips to school of 1 mile or less
- PA–14: (Developmental) Increase the proportion of trips made by bicycling
  - PA–14.2: Children and adolescents aged 5 to 15 years, trips to school of 2 miles or less
- PA–11: Increase the proportion of physician office visits that include counseling or education related to physical activity
  - PA–11.2: Increase the proportion of physician visits made by all child and adult patients that include counseling about exercise
- PA–15: (Developmental) Increase legislative policies for the built environment that enhance access to and availability of physical activity opportunities

Adverse Birth Outcomes

ACE3 Indicators

- H12: Percentage of babies born preterm, by race/ethnicity, 1993–2008
- H13: Percentage of babies born at term with low birth weight, by race/ethnicity, 1993–2008

Healthy People 2020 Objectives

- MICH–9: Reduce preterm births
  - MICH–9.1: Total preterm births
  - MICH–9.2: Late preterm or live births at 34 to 36 weeks of gestation
  - MICH–9.3: Live births at 32 to 33 weeks of gestation
  - MICH–9.4: Very preterm or live births at less than 32 weeks of gestation
- MICH–8: Reduce low birth weight (LBW) and very low birth weight (VLBW)
  - MICH–8.1: Low birth weight (LBW)
  - MICH–8.2: Very low birth weight (VLBW)

Supplementary Topics

Birth Defects

ACE3 Measure

- S1: Birth defects in Texas, 1999–2007

Healthy People 2020 Objective

- MICH–28: Reduce occurrence of neural tube defects
  - MICH–28.1: Reduce the occurrence of spina bifida
  - MICH–28.2: Reduce occurrence of anencephaly

Contaminants in Schools and Child Care Facilities

ACE3 Measures

- S2: Percentage of environmental and personal media samples with detectable pesticides in child care facilities, 2001
- S3: Percentage of environmental and personal media samples with detectable industrial chemicals in child care facilities, 2001
- S4: Pesticides used inside California schools by commercial applicators, 2002–2007

Healthy People 2020 Objective

- EH-16: Increase the proportion of the Nation's elementary, middle, and high schools that have official school policies and engage in practices that promote a healthy and safe physical school environment
  - EH-16.1: Have an indoor air quality management program
  - EH-16.4: Reduce exposure to pesticides by using spot treatments and baiting rather than widespread application of pesticide
  - EH-16.5: Reduce exposure to pesticides by marking areas to be treated with pesticides
  - EH-16.6: Reduce exposure to pesticides by informing students and staff prior to application of the pesticide